T0140596

Studies in Systems, Decision and Control

Volume 154

The series "Studies in Systems, Decision and Control" (SSDC) covers both new developments and advances, as well as the state of the art, in the various areas of broadly perceived systems, decision making and control—quickly, up to date and with a high quality. The intent is to cover the theory, applications, and perspectives on the state of the art and future developments relevant to systems, decision making, control, complex processes and related areas, as embedded in the fields of engineering, computer science, physics, economics, social and life sciences, as well as the paradigms and methodologies behind them. The series contains monographs, textbooks, lecture notes and edited volumes in systems, decision making and control spanning the areas of Cyber-Physical Systems, Autonomous Systems, Sensor Networks, Control Systems, Energy Systems, Automotive Systems, Biological Systems, Vehicular Networking and Connected Vehicles, Aerospace Systems, Automation, Manufacturing, Smart Grids, Nonlinear Systems, Power Systems, Robotics, Social Systems, Economic Systems and other. Of particular value to both the contributors and the readership are the short publication timeframe and the world-wide distribution and exposure which enable both a wide and rapid dissemination of research output.
** Indexing: The books of this series are submitted to ISI, SCOPUS, DBLP, Ulrichs, MathSciNet, Current Mathematical Publications, Mathematical Reviews, Zentralblatt Math: MetaPress and Springerlink.

More information about this series at http://www.springer.com/series/13304

Miguel Túñez-López
Valentín-Alejandro Martínez-Fernández
Xosé López-García · Xosé Rúas-Araújo
Francisco Campos-Freire
Editors

Communication: Innovation & Quality

Editors
Miguel Túñez-López
Faculty of Communication Sciences
Universidade de Santiago de Compostela
Santiago de Compostela
Spain

Valentín-Alejandro Martínez-Fernández
Faculty of Economics
University of A Coruña
A Coruña
Spain

Xosé López-García
Faculty of Communication Sciences
Universidade de Santiago de Compostela
Santiago de Compostela
Spain

Xosé Rúas-Araújo
Faculty of Social Sciences
and Communication
Universidad de Vigo
Pontevedra
Spain

Francisco Campos-Freire
Faculty of Communication Sciences
Universidade de Santiago de Compostela
Santiago de Compostela
Spain

ISSN 2198-4182 ISSN 2198-4190 (electronic)
Studies in Systems, Decision and Control
ISBN 978-3-030-06314-6 ISBN 978-3-319-91860-0 (eBook)
https://doi.org/10.1007/978-3-319-91860-0

Softcover re-print of the Hardcover 1st edition 2019

Printed on acid-free paper

This Springer imprint is published by the registered company Springer Nature Switzerland AG
The registered company address is: Gewerbestrasse 11, 6330 Cham, Switzerland

Editorial Project Developed By:

This editorial project was developed by:

- The Research Project on *Indicators related to broadcasters' governance, funding, accountability, funding, innovation, quality and public service applicable to Spain in the digital context* (Reference CSO2015-66543-P) a project for encouraging Scientific and Technical Research, a state subprogram of Knowledge Creation from the Spanish Ministry of Economy and Competitiveness, co-funded by the European Regional Development Fund (ERDF) from the European Union.
- The Research Project *Uses and informative preferences in the new media map in Spain: journalism models for mobile devices* (Reference CSO2015-64662-C4-4-R), from the Spanish Ministry of Economy and Competitiveness. The project is co-funded by the ERDF.
- International Research Network on Communication Management (R2014/026 XESCOM), supported by the Regional Ministry of Culture, Education and Organization from the Xunta de Galicia (Spain).

Editorial coordination: Miguel Túñez-López

Linguistic review and translation into English: Andrea Valencia-Bermúdez and Ana Gabriela Nogueira-Frazão

Book layout, formatting review and coordination with the authors and with the Springer: Sabela Direito-Rebollal

Editorial revision and final review: Marta Rodríguez-Castro and Sara Pérez-Seijo

Contents

About the Editors

Miguel Túñez-López, Ph.D. did his Journalism from the Universitat Autónoma de Barcelona, Spain. He is the Professor of undergraduate and postgraduate courses of the Universidade de Santiago de Compostela, Spain. He is expert in communication management in organisations and news making. He coordinates the Ph.D. programme in Information and Contemporary Information from the Universidade de Santiago de Compostela, Santiago de Compostela, Spain.

Valentín-Alejandro Martínez-Fernández is Tenured Professor at the Faculty of Economics of the Universidade de A Coruña. He has been Head of the Galician newspaper of *El Ideal Gallego*, and executive in many news companies. He has a degree in Information Science from the Universidad Complutense de Madrid. He also has an MBA in Business Administration from the Universidade de A Coruña and Ph.D. in Information Science from the Universidad Complutense de Madrid, A Coruña, Spain.

Xosé López-García is Professor of Journalism at the Universidade de Santiago de Compostela, Spain and has Ph.D. in history and journalism. He coordinates the research group Novos Medios. Among his research lines, there is the study of digital and printed media, analysis of impact of technology in mediated communication, analysis of the performance of cultural industries, and the combined strategy of printed and online products in the society of knowledge. Santiago de Compostela, Spain.

Xosé Rúas-Araújo is Tenured Lecturer of Techniques of Electoral and Institutional Communication at the Faculty of Social Sciences from the Universidade de Vigo. He is the principal investigator of the neuro-communication area of the research group CP2: Persuasive Communication from the Universidade de Vigo, Pontevedra, Spain.

Francisco Campos-Freire is Professor of Journalism at the Faculty of Communication Sciences of the Universidade de Santiago de Compostela, Spain, has an MBA. degree in Industrial and Service Business Management and a Ph.D. in Communication and Contemporary History, also at the Universidade de Santiago de Compostela. His research lines focus on the study of Cultural and Communication Industries, management of informative and audiovisual business and communication policies. Santiago de Compostela, Spain.

Innovation for Quality

The media ecosystem is currently suffering disruptive changes marked by how users use and consume contents. Digital media are entering their second decade of life involved in changing dynamics related to new technologies, as well as innovation processes in content production, distribution and commercialisation.

Public service broadcasters are not stranger to these changes. In fact, in the last few years of the second decade of the twenty-first century, studies, debates and news on public service revolve around five main pillars or challenges, which could be summed up as adapting this audiovisual service to the present social, economic, technological, consumption, access and competitiveness reality of mass media.

The European model of public broadcasting, which was born in the twentieth century and will turn a hundred in 2022 (BBC, London), has significantly changed and will have to even more to adapt to the present scenario and the new media ecology, dominated by a hyper-fragmentation of production and a mega-concentration of distribution platforms for audiovisual contents.

This book, *Communication: Innovation & Quality*, collects advances in different research areas: uses and information preferences in journalism, on one side, and public audiovisual policies in the European Union, related to governance, funding, accountability, innovation, quality and public service, on the other side. The impetus to research and disseminating initiatives in both areas thus converge, promoted within the debate activities of two projects. The first one is 'News uses and preferences in the new media map in Spain: media mobiles for mobile devices'. The second one is entitled, 'Indicators related to broadcasters' governance, funding, accountability, innovation, quality and public service applicable to Spain in the digital context'.

The study on news uses in the new media map is drawn from the observation of the performance of traditional (legacy) and emergent media outlets (native), especially the development of their web strategies and the study on the evolution of digital media in the framework of convergence (CSO 2009-13713-C05) and on the innovation processes launched to adapt content and business models to the context of crisis faced by media companies (CSO 2012-38467-C03). In order to advance in the study of digital media, the focus then moves to the analysis from the media and their offer (digital media) to users and their demands (customer media), so the research is opened to other platforms apart from the web.

The main objective of the project 'News uses and preferences in the new media map in Spain: media mobiles for mobile devices' consists in identifying and analysing contents produced by the media and distinguished because of users' acceptance in terms of diffusion, comments and evaluations, as well as the study of platforms.

Goals are clear: (a) Identifying the models behind contents and news narratives for different multi-screen devices that characterise the network society within a period dominated by access and mobility; (b) preparing a typology of innovative contents and narratives that achieve greater acceptance in the different digital platforms; and (c) analysing similarities and differences of contents and narratives typologies used for different devices and platforms.

The end result should be to prepare a report containing contents, narratives and tools adding value through the integration of languages, new narratives and the use of crossmedia and/or transmedia journalism. This report should also address the issue of the development of pieces with data journalism techniques, information displaying, as well as topics and resources, prescribers, sources and new trends that might arise in this unstoppable evolution characterising cyberjournalism in last years.

The project for assessing public broadcasting in Europe confirms that the access to distribution of audiovisual services is very competitive and diverse, penetrated by foreign brands and concentrated in 44 parent companies, 26 of which belong to the European Union, 15 to the EE.UU. and three to other countries. Only one in every 10 foreign channels broadcasted in the EU is a public service channel.

The challenges for European public service media may be presented as questions to future. Is the survival of public audiovisual services, within the context of a new digital media ecology, in risk? Is funding of public service media sustainable? Is there a need to reform the governance system to adapt it to the social requirements and to improve its function, representation, transparency, accountability and reputation? What innovation process should be faced by public service media in the next decade? And, do the value of public service media need integrated indicators to assess it?

This book, edited by professors Miguel Túñez-López (Universidade de Santiago de Compostela, Spain), Valentín-Alejandro Martínez-Fernández (Universidade da Coruña, Spain), Xosé López-García (Universidade de Santiago de Compostela, Spain), Xosé Rúas-Araújo (Universidade de Vigo, Spain) and Francisco Campos-Freire (Universidade de Santiago de Compostela, Spain), brings together experts from Europe and America working on four main areas developed in the two above-referenced research lines: broadcasting and the audiovisual sector, journalism and cyberjournalism, corporate and institutional communication, and education. Overall, works selected through an expert review process are an in-depth look from the academic rigour that, undoubtedly, contribute to provide a reliable media scene and its lines of future development.

Part I
Television and Audiovisual Sector

The Trend of Assessment Indicators for Public Service Media in Europe

Francisco Campos-Freire, Rosario de Mateo-Pérez and Marta Rodríguez-Castro

Abstract The creation of indicators to assess the provision of public audiovisual services is a strategy that has increased over the last years, both in Europe and other parts of the world, with the aim of reinforcing credibility and supporting the sustainability of their funding through the improvement of their governance, transparency, participation, innovation and accountability. The chapter, which starts from the evolution from public broadcasting to public media service, addresses the issue of public value tests and indicators as one of the current trends in the research and management agenda of these media of social communication. It is concluded that composed indicators, as quantitative and qualitative statistical tools, are demanded because of their utility for good practices in governance and new ways of accountability with the aim of reinforcing legitimacy and assessing both the effectiveness and efficiency of public service media, against the competition of many other media within the context of the new ecology of social communication.

Keywords Governance · Accountability · Corporate social responsibility
Value indicators *per* public media service

1 Introduction: Validity of PSM

The present ecosystem of social media puts into question and, at the same time, raises the need for revitalization of the validity of public service media. They emerged in Europe almost a century ago at the English studies 2LO of Marconi—Strand.

F. Campos-Freire (✉) · M. Rodríguez-Castro
Universidade de Santiago de Compostela, Santiago de Compostela, Spain
e-mail: francisco.campos@usc.es

M. Rodríguez-Castro
e-mail: m.rodriguez.castro@usc.es

R. de Mateo-Pérez
Universitat Autònoma de Barcelona, Barcelona, Spain
e-mail: rosario.demateo@uab.cat

M. Túñez-López et al. (eds.), *Communication: Innovation & Quality*, Studies in Systems, Decision and Control 154, https://doi.org/10.1007/978-3-319-91860-0_1

The British Broadcasting Company (BBC) was finally established, on October 18, 1922, and its Director General, the Scottish engineer John Charles Walsham Reith was appointed some months after. Few years later, the Nippon Hoso Kyokai (NHK, 1926), the Australian Broadcasting Corporation (ABC, 1929) and the Canadian Broadcasting Corporation (CBC, 1936) were created, becoming global benchmarks in public broadcasting. Throughout those years, broadcasting systems and social communication experimented great changes, especially since the arrival of the Internet and the introduction of digital technologies in the 21st century.

The road ahead for the BBC is clear, at least in the medium term, since it was approved in 2016 the Royal Charter, in force from January 1, 2017 to 2027, which establishes its strategy, governance and funding for that period. But this is not the case of other public service broadcasters, whose strategic and funding projections are much more uncertain. In any case, the discussion on the validity of public service media and their adaptation to the digital area remains on the table since the beginning of the present century. It is a matter of finding the new ethos of public service media and assess their social capital (Horowitz and Car 2015).

Theoretical contributions are abundant. Many books and articles in scientific journals published in the last years deal with the transition and adaptation of Public Service Broadcasting—PSB and its evolution towards Public Service Media—PSM. However, the uncertainty regarding whether there is a path and how and what is the future path for public service media has not been completely cleared (Gulyás and Hammer 2013), although there is a broad coincidence that its social function remains necessary in today's digital society (Consejo de Europa 2009; Iosifidis 2010; Jakubowicz 2010; Weeds 2013; Fuchs 2014; Bustamante 2015; Horowitz 2015; Tambini 2015; Tremblay 2016; Donders 2017).

There are major challenges that public service media organizations should face in their process of adaptation to the provision of audiovisual services imposed by the digital ecosystem. These challenges could be summarized as follows (Campos-Freire and Valencia-Bermúdez 2016):

1. Review of regulation, remit, mission, vision and strategy, at least, for the next ten years.
2. Readjustment of their governance models to the requirements of today's society.
3. Review, stabilization and sustainability of their funding models.
4. Ensuring and guaranteeing an institutional framework of independence for public service media, free from governmental, partisan, union and other lobbying pressures.
5. Clarifying, in the long term, the unknown about the availability of use of the necessary digital spectrum, both in terrestrial networks and different platforms, in order to meet the fundamental principle of universality of access.
6. Modernization of business management strategies and systems.
7. Developing systems of regulation, control, self-regulation and co-regulation committed to social responsibility, transparency, accountability and citizen participation as real owners or shareholders of the social capital, represented by these institutions.

8. Adaptation of contents and programming to the demands of the various forms of consumption and access (online and streaming) and different screens, combining the essential principles of public service media: universality, excellence, independence, diversity, innovation and accountability.
9. Promoting talent, innovation and quality as the engine of strategic change and opening up creativity to audience participation.
10. Finding ways and indicators to assess the social contribution of public service media to better communicate their public value.

These challenges to the existence and current validity of public service media respond to the weaknesses and shortcomings appreciated after the review of research published in last years and the analysis of corporate reports by these organizations (Campos-Freire 2017). The theoretical diagnosis is quite coincident about the need and urgency to address regulatory, governance and funding reforms, but practical and specific applications in corporations are much slower and diffuse.

Tambini (2015), following a comparative research of the *Mapping Digital Media* regarding public service media from 56 countries, notes the evidence that: (a) the audience for public service media is in decline; (b) their funding is also in decline; (c) their remit is contested, and their standards are under threat; (d) digitization undermines the traditional regulatory framework for PSM; (e) the intervention of governments and political parties enlarges their deterioration, generating a demand of independent public service media. Marzal and Zallo (2016) also add the mutations provoked in the traditional public audiovisual service by socio-technological mutations: digital convergence, audience fragmentation, diversity of access, creation of virtual communities and emergence of cheaper and more effective new online media.

Burri explores, in various works (2015a, b), the ability of traditional public broadcasters to adapt their remit and principles to the new multi-platform communication ecosystem, to new ways of production and distribution, and access to contents, as well as to changes in consumption patterns, prioritizing some type of editorial intelligence that links users to programming, in order to effectively meet the objectives diversity of public interest. Van Dijck and Poell (2015), and Martin (2016) consider mobility, connection and audience participation useful tools through digital social networks, even though Carpentier (2011) and Hasebrink (2011) doubt that public corporations want to use them properly for participation, because of the loss of ideological control that these might entail.

Other authors also raise the need for adaptation of public broadcasters to technological convergence (Humphreys 2010; Larrondo et al. 2016); the deinstitutionalization and adaptation to globalization (Horowitz 2015); the search for new forms of legitimacy (Pava and Krausz 1997; Powell 2016); the transformation, safeguarding the key values of public services (Trappel 2016); the preservation of the digital terrestrial spectrum and presence in multichannel platforms to maintain universality (AETHA 2014; Lamy 2016; Ala-Fossi and Lax 2016; Campos-Freire 2016); the search for new and better ways of communicating public value (Bruun 2016); and the assurance of funding sustainability (Campos-Freire 2012; Picard and Siciliani 2013; Burnley 2015; Bonini and Pais 2016; Richeri 2016).

Recognising the need and current validity of public service media organizations, because they address issues that other do not sufficiently deal with, and also generate relevant externalities for society not provided by intervention-free markets (Weeds 2013), state and regional corporations that manage PSM are facing, according to Donders (2017), five complex paradoxes.

The first of those paradoxes is the relationship between PSM and politics and politicians (should they be independent or spokespeople of government policy? To legitimate, control, or hold political power accountable?). The second one is related to competitors ("pax media" or polarization? Collaboration with private competitors or competence at risk of attacks by market distortions?). The third paradox has to do with institutions, looking for empathy through shared values and focusing internally. The fourth is about the dilemma of audience marketing and the balancing between quantitative and qualitative data. And the last one is on the risk of internal hypocrisy without doing anything in favour to put them into practice.

2 Public Good and Social Value

The consideration of public broadcasting as a pure public good, original matrix conception with which this service was born and developed during much of the 20th century, is currently under consideration (Horowitz and Car 2015; Picard and Siciliani 2013; Weeds 2013) and it is necessary to refocus it within the digital society by appealing to its value and social capital (Bordieu 1986, 1993; Coleman 1990; Putnam 1993; Lin 2001; Campos-Freire 2016). In the economic theory, we would say that the condition of Samuelson (1954) and the essence of the democratic civic right of access to independent information creates room for grounding (Stiglitz and Rosengard 2016) and re-establishing the public audiovisual services as a public good of social interest, necessary to cover the externalities that the market does not sufficiently address.

Public service media need a change in the approach to value creation, reorienting it towards its contribution to social capital. This form of capital is the expression of collaboration among the different groups of a community, whose sources of sociability are knowledge, recognition, mutual trust, effective rules and social networks. Public audiovisual services should demonstrate that, despite not being strictly efficient in terms of commercial profitability according to the logics of the market, it can be for cohesion and democratic, social, cultural and economic development of a country or community, establishing guarantees for their independence, accessibility, freedom of expression, contrast and range of views; the promotion of society, in general, and the promotion of the creative industry in the areas in which it operates, while contributing to the preservation of pluralism and diversity.

The founding principles of the PSM remit (universality, cohesion, independence, diversity, innovation, quality, transparency, participation and accountability) should be adapted and combined according to the social demands and the new ecology of the media not only provided by communication technologies but also by funding and competition conditions, which allow convergent regulations and markets. The responsibility and weight of that reshape rests on their governance and funding and, eventually, on the legitimacy of their public value.

Public service media legitimacy has to be based on their fundamental contribution to the social capital and on the recognition of their public value by society (Arriaza et al. 2015), preserving the equity of its primary social good (Rawles 1986). It is the challenge ahead for governance of public service media organizations: to respond to increasingly diverse and complex social demands while integrating cooperation and interaction of citizens (Mayntz 2000; Schmitter 2001). Legitimacy and governance are closely interconnected, as the present conception of the first one goes far beyond the legal existence and regulation of the second one. According to the modern theory, the legitimacy of governance and public value should be based on their social recognition to support the existence and funding of public service media.

That social recognition requires an open governance and an accountability where public service values may be perceived. This leads to the value axiology and theory, a field in which professors Picard (2012) and Lowe (2016) have worked, and to which many researchers have been incorporated, aiming at identifying models and indicators to assess the qualities and characteristics of communication services and institutions (Campos-Freire 2016).

3 Governance and Accountability

Governance is defined as a multilevel organizational system of distribution of powers and coordination of players, social groups and institutions for a better, more participative, democratic and transparent corporate management, in fragmented, complex, systemic and uncertain environments (Moore 1995; Mayntz 2000; Prats 2003; Le Gales 2007; Rhodes 2007; Olsen 2008; Klepper 2014) and is suitable for representing internal and external control of public service media (Coppens and Saeys 2006) through mechanisms for participation, relation and interaction with citizens (Puppis 2010; Puppis et al. 2014; Fengler et al. 2014).

The legislation of public broadcasting services regulates bureaucratic and uniformly their governance, without great changes over time nor differences between countries, even though their particularities and funding models are diverse. The structure of their governance is organized around internal bodies (boards of management, supervision and control, executive management, advisory councils and entities in connection with audiences and in dialogue with professionals), and externals (through sectorial and convergent regulatory bodies and commissions on parliamentary control and on funding and economic issues).

The profile and representation of governance is of a political-partisan and parliamentary nature—more or less measured depending on the country, but which is still contaminated by the *lottizzazione*, the "Italian anomaly" of political power-sharing that Betino Craxi and Berlusconi made evident in the 80's of the last century (Chimenti 2007; Padovani 2010; Zaccaria 2016). That way, the institutional representation of such governance becomes the channel of transmission between political and media powers, breaking independence of origin and action required by audiovisual media services to ensure their independence.

The European system of governance for public service media may be classified (Campos-Freire 2017) in three categories according to the characteristics of their functional assignment to the departments of the respective governments, election system of governing bodies, funding, control and accountability; in such a way that the affiliation may be really and organically independent of the current government, dependent of the corresponding Culture Department and the less independent model, linked to the Presidency and the Spokesperson of the government.

In the most independent model, audiovisual bodies of governments, steering and administration are chosen by an independent body outside the executive or parliamentary power (Audiovisual Media Councils). Others make this election through the Parliament, by an absolute or enhanced majority (2/3 or 3/5 majority); and by direct appointment of the President of the Government. The funding models are: (a) independent from government, multiannual and linked to a long charter (Royal Charter of the BBC, 2017–2027); (b) independent and adopted by the Parliament for not less than five years (Germany, for the ARD and ZDF); and (c) annual charters agreed by governments through subsidies (Spain, RTVE and regional broadcasters).

According to the same gradual map of three models, the control is exercised by: independent bodies, Parliament and government departments. And also accountability: (a) to a specialised independent body and society, with public participation and assessment indicators; (b) to a commission on parliamentary control; and (c) directly to the government. If control and accountability lack the necessary independence to counterbalance governmental and political-partisan interference, professional conflicts because of the manipulation of public service media deteriorate the reputation of these services, as reflected in the analysis of several European countries—including Spain—Monitoring Media Pluralism (Brogi et al. 2017).

What is evident in the analysis of public service media governance is the lack of open systems of participation, transparency and control opened to society beyond political and bureaucratic systems. Some corporations publish strategic declarations and documents of intentions, but the transition from theory to practice is still medium low. In order to regain or reinforce their reputation, many broadcasters have developed from 2007 public value tests and self-regulatory systems, as accountability on Corporate Social Responsibility, which are required by competitions rules imposed from Brussels. However, the weakest development link is in the integration of participation and interaction with society, despite the possibilities offered by tools of new social technologies.

Contrary to the bureaucratic and traditional conception that still prevails in the governance of public service media, new theoretical and practical approaches directly link this term both to policies and to satisfaction of social needs (Brower 2015), adapting behaviours to the characteristic of the media ecosystem (Peters 2016), enhancing innovation (Torfil and Ansell 2017) and placing public interest (Ginosar and Krispil 2016) and user participation in the centre of attention (Hutchinson 2015; Azurmendi 2015). Professors Muñoz Saldaña and Azurmendi (2016) call for inclusion of participation, openness, coverage, transparency, inclusiveness, citizen participation and active citizenship in the goals of public service media governance.

Contemporary research on this concept clarifies some aspects or characteristics that confuses governance with governability, decision-making, executive behaviour and strategy, which correspond more to the area of management (McGrath and Whitty 2015). Similarly, other authors (Barnett 2016) characterize the participatory dimension of governance by the dysfunction of the operability and effectiveness that may cause on the management of organizations. But governance requires not only participation, dialogue, consensus and interaction with stakeholders, but also accountability and corporate responsibility.

If governance is the multi-level system of distribution of powers and coordination of players that interact with an organization, accountability is the set of mechanisms that allow explaining and justifying objectives and achieved results to stakeholders with the aim of improving their performance (Schedler 2008). Public service media accountability is institutionally delimited in its own regulation although it is a horizontal accountability. Vertical accountability, on the other hand, is dome from citizenry and its sanctions are moral and reputable through public opinion (O' Donnell 2004; Rios Ramírez et al. 2014; Jacobs and Schillemans 2016).

The State organization itself endows with inter-institutional mechanisms, through the separation of powers (executive, parliamentary and judiciary) for self-control and horizontal accountability. The institutionalized control through the higher state bodies of division of powers (Parliament and Court of Justice) is known as accountability of balance and carries administrative and criminal penalties to the detected dysfunctions of deviations. But there is also other accountability assigned through certain institutions (Audit Commissions, Ombudsmen), whose opinions and sanctions are elevated to previous higher bodies. In this second category of horizontal accountability, assigned or delegated, organizations seek to preserve their independence through statutory institutions not dependent on either the government or the monitored body.

Public broadcasters carry out horizontal accountability through their own Board of Directors to government departments to which they are attached, the corresponding Control Committees of the Parliament, independent audiovisual and communication regulators and Audit Councils. Some of them in Europe (BBC, ORF, YLE, ZDF, VRT, SVT) develop vertical accountability systems, trying to find a more direct interaction with stakeholders and society to better communicate the public value of their social contribution (EBU 2015; ORF 2017). This last approach of communication on accountability of the social contribution of public value could be considered as social accountability, a third gradual type, apart from horizontal and vertical accountability.

Accountability focused on groups or stakeholders and openness to audience participation—increased through the possibilities of interaction offered by networks and new communication tools (Gacía González and Alende Castro 2015) are growing theoretical trends (Wurff and Schönbach 2014) and approaches to good practice increasingly recognised by corporations—with more or less tradition according to the countries' media models (Almirón et al. 2016)—as a tool to support and justify their funding systems (Benson et al. 2017). Two accountability systems developed by some broadcasters, both vertical and horizontal, are Corporate Social Responsibility reports and Public Value Tests.

4 CSR and Public Value Tests

Corporate Social Responsibility (RSC) is a self-regulatory system linked to horizontal accountability, transparency, communication of social values (Vázquez 2014), strengthening of credibility and recovery of trust by different stakeholders that are related to and organizations, with optimistic, hopeful, cynical and pessimistic expectations and perceptions (Olkkonen 2017).

Despite the fact that CSR is born in the private sector, public service media have seen in it a strategy capable of satisfying society's demand for information and of communicating the social value of the public service. It is, therefore, a voluntary commitment of corporations with three basic components: credibility, utility and counteracting negative impacts by expressing ethical attitudes (Lee et al. 2016).

The implementation panorama of Corporate Social Responsibility in European public service media is mainly composed by public corporations that regularly report on CSR, with the aim of analysing the result and performance of their transversal strategies in this matter (Fernández-Lombao and Campos-Freire 2013). These public corporations are the German ZDF, the Finnish YLE, the Irish RTÉ, the Italian RAI, the Portuguese RTP, the British BBC and the Spanish RTVE. Five other corporations (ORF, RTBF, VRT, DR and FT) and some Spanish regional broadcasters (Galicia, Catalonia and the Basque Country) have performed some specific action of CSR until 2016.

Within this trend of accountability system, we can also include the so-called public value test, previous evaluation or *ex ante tests*, in this case a form of horizontal accountability, as they regulate and condition the funding of new services. It is a regulation tool applied to decision-making processes on the approval of new services within the activities of a public broadcasting service. To do so, the test balances two different aspects: on the one hand, an evaluation of the public value of the proposed new service is carried out; on the other hand, the impact on the market of the proposal is assessed. The joint assessment will result in the approval or rejection of the proposed new service.

The need to implement this kind *of ex ante* mechanisms is born within a context in which advances in new media made by public service media are stopped by protest coming from the private sector. Having public resources and being, in a certain way, free from pressure to obtain economic benefits, public service media had more room to experiment with the offer of new audiovisual services, within the wide possibilities that the changing media ecosystem was beginning to show. From the perspective of competition, this created a hurdle for private initiatives.

Protests from the commercial about this competitive advantage soon surpassed the national level and reached the European Commission, alleging market failures resulting from double provision of services and denouncing the possible double funding of public media. That, along with complaints about the lack of transparency in development processes of communication policies, triggered a series of research works on the financing and governance of public service media (Bardoel and Vochteloo 2011; Donders 2011; Donders and Raats 2012).

These investigations were favourable to the demands of the private sector. The European Commission, then, initiated a series of negotiations with Member States, mainly on the need to define the mission of public service and to provide better supervision over their funding and investment of resources in new media. These negotiations between the Community and the national level influenced the drafting of the "Communication from the Commission on the application of State aid rules to public service broadcasting", published in 2009. Paragraph 84 recommends the incorporation of an evaluation prior to the management of public service media, with the aim of not violating the guidelines of the Amsterdam Protocol.

Following the guidelines set out in the Communication from 2009, many European countries have chosen to adapt this *ex ante* procedure to their respective national contexts. However, in 2007, the United Kingdom had already implemented the Public Value Test, the first public assessment in Europe that, also, served as a basic outline for the design of the proposal from the European Commission.

The basic elements of the British Public Value Test are, therefore, shared by most of the public media that have adapted this type of procedures. First, the Public Value Test consists of a public value assessment (with a three-months' duration), which should include a 28-day public consultation, open to participation of representatives of citizens and different stakeholders who wish to express their opinion.

In parallel, the second part of the test is also carried out: the analysis of the potential market impact that would come with the implementation of the proposed new service. To make the assessment, voices from different stakeholders—including the commercial sector—must also be taken into account. After completing both evaluations, it will be checked whether the public value of the proposal justifies its market impact and, in that case, the new service will be approved.

The British test, together with the recommendations of the European Commission, provided the basic guidelines for the development of other *ex ante* tests in Europe. At the moment, 9 Member States (Germany, the Flemish Belgium, Ireland, Austria, Netherlands, Denmark, Sweden, Finland and Croatia) together with two non-Community countries (Norway and Iceland) have opted for the adaptation of a public value test, although with certain differences concerning both the procedure and the actual use of the test.

The number of evaluations carried out in each country, in fact, is one of the most striking differences: whereas in Germany 45 tests have already been completed, other Member States such as Sweden and Denmark have not yet released the procedure. The duration and budget of each of the test adaptations also fluctuates. In the United Kingdom and Germany, the evaluation can last up to 6 months and involve an expenditure up to 1 million euro; however, in countries such as Norway, it can be completed in less than 4 months and expending more modest amounts.

The dynamics of the test depend on the context. In the Netherlands, for instance, the procedure does not include a market impact assessment. Germany, on the other hand, structures its *Drei-Stufen-Test* starting from three questions: the relation of the proposal with the needs of society, the contribution of the proposal to the editorial competence and the necessary financial expense to carry it out. Another aspect lies in the political interference during the process, since whereas in countries such as United Kingdom and Austria, decision-making falls on an independent body or regulators, this task competes with political agents in tests such as the Norwegian and the Irish.

Public value tests can be also understood as double-edged weapons and their effectiveness depends on how their development is approached: as a tool that contribute to increase transparency and participation in decision-making processes, legitimating the activity of public media, or as a hostile tool designed for its control and restriction, strongly in favour of commercial interests.

5 Composed Indicators

Indicators are useful to help manage, administrate and control governance, particularly through accountability. There are many types of simple and composed indicators, more than 400 according to the OECD (2008). When the topic to be addressed is complex and multidimensional, such as governance, it should be tackled using composed indicators. Multidimensional or composed indicators are statistical tools of information prepared to assess performance in different areas, through the simplification of data, using aggregation arithmetic formulae of their relevant variables (OECD 2008; Becker et al. 2015).

The 10 steps that OECD (2008) proposed in its basic guide for the elaboration of indicators include: study of the theoretical framework, selection of data, multivariate analysis, import of missing data, normalization, weighting, aggregation of indicators, sensitivity tests, links with other measures and visualization. In the field of communication, in general, and the public audiovisual service, in particular, different approaches and models have been made in order to achieve indicators that express different categories (UNESCO 2008; Bucci et al. 2012; Lowe and Martin 2013; EBU 2014; Brogi et al. 2017).

The first research works on the usefulness of indicators go back to the concern for media's social responsibility and the need to transfer to data the accountability of those ethical commitments. These are the *Moyens d'Assurer la Responsabilité Social* (M*A*R*S*) or Media Accountability Systems (English version), proposed by Claude Jean Bertrand (1999). These systems are followed by proposals from different authors and organizations, such as Media Ethics, Media Act, Media Accountability, Media Sustainability, the media standards by de Global Reporting Initiative, EBU PSM Values, Media Pluralism Monitor, Ranking Digital Rights, State of the News Media de Pew Research Center, DAMIAN method for assessing convergence (Nooren et al. 2014), ISO systems and EFQM.

The European Broadcasting Union prepared in 2012 and published two years later (EBU 2014) a model for the assessment of public service media, a peer review mechanism characteristic of the quality review of scientific production, from six core values: universality, independence, excellence, diversity, accountability and innovation. The EBU tool was implemented by some Northern European corporations (Finland and Sweden) and served as a basis for a further model (EBU 2015), which is an assessment of the social contribution of public service media from the indicators of performance, scope, perception and impact.

The BBC (2004) and the ORF (2017) developed systems of indicators. In the British case, these are democratic, cultural, creative, educational, social-community and global values. In the case of Austria, the corporation assesses the global value of the corporation consisting of individual, social, national, international and corporate values. Broadcasters from Germany, Belgium, Norway and France also use specific systems of indicators for the qualitative assessment of their services and the communication of their social contribution.

For its part, the Media Pluralism Monitor assesses media systems through five areas (basic protection, market pluralism, political independence and social inclusion) and 20 indicators. The Digital Rights Ranking, a non-profit initiative created in 2013 by Rebecca Mackinnon and New America's Open Technology Institute, analyses the compliance with the digital social responsibility of 22 mega-corporations through 35 indicators.

Research conducted by professor Manuel Chaparro (2016), from the University of Malaga, is pioneering in Spain in the development of an indicators of the social profitability of communication (IRSCOM). Also, professor López-López et al. (2017) proposes 29 indicators, grouped into three blocks (institutional and economic information, and information production), for the analysis of transparency in public broadcasters.

For the analysis of information quality of digital media, Luis M. Romero Rodríguez et al. (2016) propose a structured model in three macro-areas and 75 dimensions of indicators, drawn from the consensus of 40 experts. The authors who lead this work participate in other research in progress on indicators of governance, funding, accountability, innovation, quality and public service of European public service media applicable to Spain, in the digital context.

6 Defining Indicators for Public Service Media

Indicators are useful for assessing intrinsic (Moore 2009), exchange and use values of public service media, which help governance to be accountable and communicate its social value. It is a matter of trying to demonstrate, in short, that results correspond to the readapted objectives of the mission, the cost is efficient and the use accepted. In such a way that indicators are measuring tools to transform, according to Chaparro et al. (2016). Exchange and use values may be assessed with quantitative indicators, but the intrinsic ones are much broader and diverse, requiring qualitative and composed tools of greater complexity.

Based on theories that analyse the respective systems of governance, accountability and funding of public service broadcasters, we conceive the establishment of composed indicators grouped into four categories and ten indicators for each of them: regulation, application or implementation, integration or combination with regard to stakeholders and models and, eventually, impact through perceived effects. This is the criterion with which we work in the project of indicators of governance, funding, accountability, innovation, quality and public service of European public service broadcasters applicable to Spain, in the digital context. Not without forgetting the next steps needed to be completed, to weigh and contrast the indicators.

Acknowledgements Results of this work correspond to the project *Indicators related to broadcasters' governance, funding, accountability, innovation, quality and public service applicable to Spain in the digital context* (Reference CSO2015-66543-P), belonging to the National Spanish Programme for Encouraging Excellent Scientific and Technical Research of the Spanish Ministry of Economy and Competitiveness, and co-funded by the European Regional Development Fund (ERDF). It is also part of the International Research Network on Communication Management (REDES 2016 G-1641 XESCOM), supported by the Ministry of Culture and Education of the Xunta de Galicia (Reference ED341D R2016/019).

References

Azurmendi, A.: Un espacio público por conquistar. La participación institucionalizada de las audiencias en las televisiones de proximidad españolas y europeas como indicador de transparencia para la gobernanza de los medios de comunicación. Derecom **19**, 1–32 (2015)

AETHA: Future use of the 470–694 MHz band. Report for Abertis, Ariva, BBC. BNE, EBU and TDT. In: http://www.bipt.be/public/files/en/21172/02-aetha_consulting_-_report_on_the_economic_benefits.pdf (2014)

Ala-Fossi, M., Lax, S.: The short future of public broadcasting: replacing digital terrestrial television with internet protocol? Int. Commun. Gaz. **78**(4), 365–382 (2016)

Almirón, N., Narberhaus, M., Mauri, M.: Mapping media accountability in stateless nations: The case of Catalonia. Catalan J.Commun. Cult. Stud. **8**, 2 (2016)

Arriaza, K., Novak, E., Kuhn, R.: Public Service Media in Europe. A Comparative Approach, Routledge-ECREA (2015)

Bardoel, J., Vochteloo, M.: Dutch public service broadcasting between bureaucratic burden and political choice. Implementing the Amsterdam Test in the Netherlands. In: Donders, K., Moe, H. (eds.) Exporting the Public Value Test: The Regulation of Public Broadcaster's New Media Services Across Europe. Nordicom (2011)

Barnett, M.: Accountability and global governance: the view from paternalism. Regul. Gov. **10**, 134–148 (2016)

BBC: Building Public Value. Renewing the BBC for a digital world. In: https://downloads.bbc.co.uk/aboutthebbc/policies/pdf/bpv.pdf (2004)

Becker, W., Paolo, P., Saisana, M., Saltelli, A.: Measuring the importance of variables in composite indicators. European Commission. In: https://www.researchgate.net/publication/283014579 (2015)

Benson, R., Powers, M., Neff, T.: Public media autonomy and accountability: best and worst policy practices in 12 leading democracies. Int. J. Commun. **11**, 1–22 (2017)

Bertrand, C.J.: L'Arsenal de la démocratie: médias, déontologie et M*A*R*S. Économica, Paris (1999)

Bonini, T., Pais, I.: Hacking public service media funding: a scenario for rethinking the licence fee as a form of civic crowdfunding. Paper presented at the RIPE@2016 Conference (2016)

Bordieu, P.: Sociology in Question. Sage, London (1993)

Bourdieu, P.: The Forms of Capital. In: Richards, J.G. (ed.) Handbook o Theory and Research for the Sociology of Education, pp. 241–258. Greenwood Press, New York (1986)
Brogi, E., Bania, K., Nenadic, I., Ostling, A., Parcu, P. L.: Monitoring media pluralism in Europe: application of the media pluralism monitor 2016 in the European Union, Montenegro and Turkey. Policy Report Centre for Media Pluralism and Media Freedom. European University Institute. The Spanish report was prepared by Pere Masip, Carlos Ruiz, Jaime Suau & Ángel García Castillejo (2017)
Brower, J.: Aportes epistemológicos para la comprensión de los conceptos de gobernabilidad y gobernanza. Revista Venezolana de Gerencia **20**(72), 630–646 (2015)
Bruun, H.: The prism of change: 'continuity' in public service television in the digital era. Nordicom Rev. **37**(2), 33–49 (2016)
Bucci, E., Chiaretti, M., Fiorini, A.M.: Indicadores de Qualidade nas Emissoras Públicas, uma avaliaçao contemporánea. Unesco, Brazil. In: http://unesdoc.unesco.org/images/0021/002166/216616por.pdf (2012)
Burnley, R.: Public fundic principles for service media. In: http://www3.ebu.ch/home (2015)
Burri, M.: Contemplating a "Public Service Navigator". In search of new- (and better-) functioning public service media. Int. J. Commun. **9**, 1341–1359 (2015a)
Burri, M.: Public Service Broadcasting 3.0: Legal design for the digital present. Routledge, London (2015b)
Bustamante, E.: Europa: Un servicio público multimedia para una nueva era. Diagnóstico y propuestas. Síntesis de trabajos 2009–2015. Grupo Turín (2015)
Campos-Freire, F.: Modelos de financiación de las televisiones públicas autonómicas. In: Miguel de Bustos, J.C., Casado del Rio, M.A., Televisiones autonómicas. Evolución y crisis del modelo público de proximidad, pp. 143–170. Gedisa, Barcelona (2012)
Campos-Freire, F.: El valor social de la TV abierta. In: Gutiérrez Montes, E. (coord.), Televisión Abierta. Situación actual y tendencias de la TDT, pp. 147–166. Colegio Oficial de Ingenieros de Telecomunicación, Madrid (2016)
Campos-Freire, F.: Situación actual y tendencias de la radiotelevisión pública en Europa. Coordination of research by the Federación de Organismos de Radiotelevisión Autonómica (FORTA), Spain, Madrid (2017)
Campos-Freire, F., Valencia-Bermúdez, A.: Los retos de la gobernanza, financiación y valor de las radiotelevisiones públicas. In: Felici, J.M., Rabadán, Pl, Castillo, J.I. (eds.) Los medios de comunicación públicos de proximidad en Europa. RTVV y la crisis de las televisiones públicas. Tirant Humanidades, Valencia (2016)
Carpentier, N.: Media and Participation: A Site of Ideological-democratic Struggle. Intellect, Bristol (2011)
Chaparro, M., Olmedo, S., Gabilondo, V.: El indicador de la rentabilidad social en comunicación (IRSCOM): medir para transformar. Universidad Complutense de Madrid, Cuadernos de Información y Comunicación (CIC) (2016)
Chimenti, A.: L'ordinamento radiotelevisivo italiano. Giappichelli Editorial, Turín (2007)
Coleman, J.S.: Social capital in the creation of human capital. Am. J. Sociol. **94**, 95–120 (1990)
Coppens, T., Saeys, F.: Enforcing performance: new approaches to govern public service broadcasting. Media Cult. & Soc. **28**(2), 261–284 (2016)
Consejo de Europa: Reykjavik declaration on media and new communication services. A new notion of media? In: http://teledetodos.es/index.php/documentacion/1-textos-legales-normas-decisiones-ue/79-declaracion-reykjavik-sobre-nuevos-medios-audiovisuales-29052009/file (2009)
Donders, K., Raats, T.: Measuring public value with the public value test: best of worst practice? In Janssen, J., Comprovoets, J. (eds.), Geographic Data and the Law. Defining New Changes. Leuven University Press, Leuven (2012)
Donders, K.: Looking at Practice to Find a Theory of Public Service Media. The Center for Media, Data and Society. Central European University, Budapest (2017)

Donders, K.: The public value test. A reasoned response or panic reaction? In: Donders, K., Moe, H. (eds.), Exporting the Public Value Test: The Regulation of Public Broadcaster's New Media Services Across Europe. Nordicom (2011)
EBU: Funding of public service media. Media Intelligence Service. European Broadcasting Union. In: http://www3.ebu.ch/home (2016)
EBU: PSM values review: the tool. In: https://www.ebu.ch/member-services/psm-contributionsociety (2014)
EBU: Public service media contribution to society. Media Intelligence Service In: www.ebu.ch/mis (2015)
Fengler, S., Eberwein, T., Mazzoleni, G., Porlezza, C., Russ-Mohl, S.: Journalists and media accountability. In: Lang, P. (eds.) An International Study of News People in the Digital Age, vol. 12 (2014)
Fernández Lombao, T., Campos Freire, F.: La Responsabilidad Social Corporativa en las radiotelevisiones públicas en Europa. In Cuadernos Info, Chile **33**, 145–157 (2013)
Fuchs, C.: Social media and the public sphere. Triple C Commun. Capitalism Critique **12**(1), 57–101 (2014)
García González, A., Alende Castro, S.: El concepto de *accountability* en Facebook. Accidente ferroviario de Santiago de Compostela. Estudios sobre el Mensaje Periodístico **21**(1) (2015)
Ginosar, A., Krispil, O.: Broadcasting regulation and the public interest: independent versus governmental agencies. J. Mass Commun. Q. **93**(4), 946–966 (2016)
Gulyás, A., Hammer, F.: Public Service Media in the Digital Are: International Perspectives. Cambridge Scholars Publishing (2013)
Hasebrink, U.: Giving the audience a voice: the role of research in making media regulation more responsive to the needs of the audience. J. Inf. Policy **1**(1), 321–336 (2011)
Horowitz, M.A., Car, V.: The future of public service media. Medijske Studije/Media Stud. **6**, 2–9, Zagreb (2015)
Horowitz, M.A.: Public service media and challenge of crossing borders: assessing new models. Medijske Studije/Media Stud. **6**, 80–91, Zagreb (2015a)
Humphreys, P.: Convergencia digital y política: el futuro de la radiodifusión pública en el Reino Unido y Alemania. Infoamérica **3–4**, 57–72 (2010)
Hutchinson, J.: From Fringe to formalisation: an experiment in fostering interactive public service media. Media Int. Aust. **155**(1), 5–15 (2015)
Iosifidis, P.: Servicio público de televisión en Europa. Retos y estrategias. Infoamérica, 3–4, 7–21. In: www.infoamerica.org/icr/n03_04/iosifidis.pdf (2010)
Jacobs, S., Schillemans, T.: Media and public accountability: typology and exploration. Policy Polit. **44**(1), 23–40 (18) (2016)
Jakubowicz, K.: PSB 3.0: reinventing European PS. In: Iosifidis, P. (ed.) Reinventing Public Service Communication: European Broadcasters and Beyond, pp. 9–22. Palgrave, Basingstoke (2010)
Klepper, W. M.: Gobernanza y gestión del cambio del siglo XXI. En Reinventar la empresa en la era digital. In: https://www.bbvaopenmind.com/ (2014)
Lamy, P.: Results of the work of the high level group on the future use of the UHF band (470–790 MHz). Report to the European Commission. In: https://ec.europa.eu/digital-single-market/en/news/report-results-work-high-level-group-future-use-uhf-band (2016)
Larrondo, A., Domingo, D., Erdal, I.J., Masip, P., Van den Bulck, H.: Opportunities and limitations of newsroom convergence. A comparative study on European public service broadcasting organisations. Journalism Stud. **17**(3), 277–300 (2016)
Le Gales, P.: Gobernanza y cohesión social. In: P. Hall (dir.), Congreso Regiones Capitales, Comunidad de Madrid, 327–347 (2007)
Lee, C.G., Sung, J., Kim, J.K.: Corporate social responsibility of the media: instrument development and validation. Inf. Dev. **32**(3), 554–565 (2016)
Lin, N.: Social Capital: a theory of social structure and action. University Press, Cambridge (2001)

López-López, P.C., Puentes-Rivera, I., Rúas-Araújo, J.: Transparencia en televisiones públicas: desarrollo de indicadores y análisis de los casos de España y Chile. Revista Latina de Comunicación Social **72**, 253–272. In: http://www.revistalatinacs.org/072paper/1164/14es.html (2017)
Lowe, G., Martin, F.: The value and values of public service media. In: Martin, F., Lowe, G. (eds), The Value Public Service Media. Nordicom (2013)
Lowe, G.F.: What Value and Which Values? In Public Social Value. In: http://www.orf.at/ (2016)
Martin, F.R.: Mobile public service media in Australia: Ubiquity and its consequences. Int. Commun. Gaz **78**(4), 330–348 (2016)
Marzal F.J., Zallo E.Z.: Las televisiones públicas de proximidad ante los retos de la sociedad digital. Commun. Soc. **29**(4), 1–7, Pamplona (2016)
Mayntz, R.: Nuevos desafíos de la teoría de 'governance'. In Instituciones y Desarrollo **7**, 35–52. In: http://www.iigov/revista/revista7/docs/mayntz.htm (2000)
McGrath, S.K., Whitty, S.J.: Redefining governance: from confusion to certainty and clarity. Int. J. Managing Projects Bus. **8**(4), 755–787 (2015)
Moore, M.: Creating Public Value—Strategic Management in Government. Harvard University Press, Cambridge (1995)
Moore, G.E.: El valor intrínseco. Facultad de Filosofía de la Universidad Complutense de Madrid (2009)
Muñoz Saldaña, M., Azurmendi, A.: El papel de las televisiones públicas autonómicas en el desarrollo de la gobernanza multinivel en Europa. Commun. Soc. **29**(4), 45–58 (2016)
Nooren, P., Koers, W., Bangma, M., Berkers, Boertjes, E.: Regulation in the Converged Media-Internet-Telecom Value Web. Introducing the Damian Method for Systematic Analysis of the Interdependences Between Services, Organisations and Regulation. TNO Publications. In: www.publications.tno.nl/publication/.../NhocfJ/TNO-2014-R11482.pdf (2014)
O' Donnell, G.: Accountability horizontal: la institucionalización legal de la desconfianza política. Revista Española de Ciencia Política, 11–31 (2004)
OECD: Handbook on constructing composite indicators. Methodology and user guide. OECD & European Commission. In: http://composite-indicators.frc.ec.europa.eu (2008)
Olkkonen, L.: A conceptual foundation for expectations of corporate responsibility. Corp. Commun. Int. J. **22**(1), 19–35 (2017)
Olsen, J.P.: The ups and downs of bureaucratic organization. Annu. Rev. Polit. Sci. **11**, 13–37 (2008)
Padovani, C.: El pluralismo de la información televisiva en Italia. Infoamércia **3–4**, 173–188 (2010)
Pava, M.L., Krausz, J.: Criteria for evaluating the legitimacy of corporate social responsibility. J. Bus. Ethics **16**(3), 337–347 (1997)
Peters, G.B.: Governance and the media: exploring the linkages. Policy Polit **44**(1), 9–22 (14) (2016)
Picard, R.G., Siciliani, P.: Is There Still a Place for Public Service Broadcasting? Effects of the Changing Economics of Broadcasting. Reuters Institute for the Study of Journalism, Oxford (2013)
Picard, R.G.: La creación de valor y el futuro de las empresas informativas. Por qué y cómo el periodismo debe cambiar para seguir siendo relevante en el siglo XXI. Media XXI, Lisbon (2012)
Powell, A.: Hacking in the public interest: Authority, legitimacy, means, and ends. New Media Soc. **18**(4), 600–616 (2016)
Prats, J.: De la burocracia al management, del management a la gobernanza. INAP, Madrid (2003)
Puppis, M.: Media governance: a new concept for the analysis of media policy and regulation. Commun. Cult. Critique **3**(2), 134–149 (2010)
Puppis, M., Maggetti, M., Gilardi, F., Biela, J., Papadopoulos, Y.: The political communication of independent regulatory agencies. Swiss Polit. Sci. Rev. **20**(3), 388–400 (2014)
Putnam, R.: The prosperous community: social capital and public life. Am. Prospect **4**(13) (1993)
Rawles, J.: La justicia como equidad y otros ensayos. Tecnos, Madrid (1986)
ORF Public Value Report 2016/17. Media Intelligence Service. In: www.ebu.ch/mis (2017)
Rhodes, M.: Understanding governance: ten years on. Organ. Stud. **28**(8), 1243–1264 (2007)

Richeri, G.: Una prospettiva dei cambiamenti in atto. Presentación del número monográfico sobre "Futuro de la television". Revista de la Asociación Española de Investigación de la Comunicación **3**(6), 2–10. In: http://www.revistaeic.eu/index.php/raeic/article/view/60 (2016)

Rios Ramírez, A.R., Arbeláez, A.C., Suárez Valencia, Vélez, L.F.: Accountability: aproximación conceptual desde la filosofía política y la ciencia política. Colomb. Int. **82** (2014)

Romero Rodríguez, L.M., De Casas Moreno, P., Torres, A.: Dimensiones e indicadores de la calidad informativa en los medios digitales. Comunicar **49** (2016)

Samuelson, P.A.: The pure theory of public expenditure. Rev. Econ. Stat. **XXXVI**, 387–388 (1954)

Schedler, A.: ¿Qué es la rendición de cuentas? In: Instituto Federal de Acceso a la Información Pública, México (2008)

Schmitter C.: What is there to legitimize in the European Union…and how might this be accomplished?. Reihe Politikwissenschaft/ Polit. Sci. Ser. **75** (2001)

Stiglitz, J.E., Rosengard, J.K.: La economía del sector público, 4th edn. Bosch, Barcelona (2016)

Tambini, D.: Five theses on public media and digitization: from a 56-country study. Int. J. Commun. **9** (2015)

Torfing, J., Ansell, Ch.: Strengthening political leadership and policy innovation through the expansion of collaborative forms of governance. Public Management Review, 19(1) (2017)

Trappel, J.: Taking the public service remit forward across the digital boundary. Int. J. Digital Telev. **7**(3), 273–295 (2016)

Tremblay, G.: Public service media in the age of digital networks. Can. J. Commun. **41**(1) (2016)

UNESCO: Indicadores de desarrollo mediático: marco para evaluar el desarrollo de los medios de comunicación social. Paris. In: http://unesdoc.UNESCO.org/images/0016/001631/163102s.pdf (2008)

Van Dijck, J., Poell, T.: Making public television social? Public service broadcasting and the challenges of social media. Telev. New Media **16**, 148–164 (2015)

Vanhaeght, A.-S., Donders, K.: Do interaction, co-creation and participation find their way from PSM literature to PSM policy and strategy? A comparative case study analysis of Flanders, the Netherlands, France and the UK. Medijske Studije/Media Stud. **6**. In: https://www.researchgate.net/publication/311717921_Do_interaction_co-creation_and_participation_find_their_way_from_PSM_literature_to_PSM_policy_and_strategy_A_comparative_case_study_analysis_of_Flanders_the_Netherlands_France_and_the_UK (2015)

Vázquez, B.L.: Evolución e impacto de la comunicación de valores responsables. Caso de estudio en España. Historia y Comunicación Social **19**, 511–523 (2014)

Weeds, H.: Digitisation, Programme Quality and Public Service Broadcasting. In: Picard, R.G., Siciliani, P. (eds.) Is There Still a Place for Public Service Broadcasting? Effects of the Changing Economics of Broadcasting. Reuters Institute for the Study of Journalism, Oxford (2013)

Wurff, R., Schönbach, K.: Audience expectations of media accountability in the Netherlands. Journalism Stud. **15**(2) (2014)

Zaccaria, R.: Diritto dell'informazione e della comunicazione (9ª edición). CEDAM, Italy (2016)

Francisco Campos-Freire Professor of Journalism at the Faculty of Communication Sciences of the Universidade de Santiago de Compostela (Spain), has a MBA degree in Industrial and Service Business Management and a Ph.D. in Communication and Contemporary History, also at the Universidade de Santiago de Compostela. His research lines focus on the study of Cultural and Communication Industries, management of informative and audio-visual business, and communication policies. Santiago de Compostela, Spain.

Rosario de Mateo-Pérez Professor of Journalism at the Faculty of Communication Sciences of the Universitat Autónoma de Barcelona (Spain), has a degree in Economics, holds a Ph.D. in Communication Sciences, and is part of the EuroMedia Research Group. Mateo-Pérez main research line focuses on the Economics of Communications and has lectured on Media Management, on the Economics of Communication and on Business Communication. Barcelona, Spain.

Marta Rodríguez-Castro Degree on Audiovisual Communication from the Universidade de Santiago de Compostela (Spain) and Master's Degree on Research Applied to Media from the University Carlos III de Madrid, is currently a Ph.D. student on Contemporary Communication and Information at the Universidade de Santiago de Compostela and is in receipt of a FPU grant. Rodríguez-Castro research approaches European Public Service Media, focusing on public value tests and proximity media. Santiago de Compostela, Spain.

A System of Indicators for Evaluating Public Broadcasting Corporations

Olga Blasco-Blasco, Pedro J. Pérez and Luis E. Vila

Abstract We describe an operational procedure to build an integrated information system (IIS) for monitoring and evaluating the performance of public broadcasting corporations (PBC). The procedure grounds in a conceptual framework derived from evaluation models, and consists of two stages. The first stage is the selection, through expert judgment, of simple indicators that illustrate the most relevant aspects of the corporations. The second stage is the definition of a system of indicators related to each level of the hierarchy of objectives planned by the corporations. The system of indicators has two levels. The first level is a basic indicator system that allows description and monitoring of the reality of the activities of a corporation on a daily basis. The second level consists of a strategic indicator system, which expresses the consequences that the activities of a PBC generate in its environment, both as short-term effects and as medium and long-term impacts.

Keywords Information system · Indicators · Evaluation · Public broadcasting corporations

1 Introduction

Nowadays, the public media in European countries are striving not only to adapt to the ongoing technological changes motivated by the development of new digital environments, but also to improve the quality of their public service performance to cope with people's evolving demands. In this sense, the EU considers public broadcasting

O. Blasco-Blasco (✉) · P. J. Pérez · L. E. Vila
Universitat de València, Valencia, Spain
e-mail: Olga.Blasco@uv.es

P. J. Pérez
e-mail: Pedro.J.Perez@uv.es

L. E. Vila
e-mail: Luis.Vila@uv.es

M. Túñez-López et al. (eds.), *Communication: Innovation & Quality*, Studies in Systems, Decision and Control 154, https://doi.org/10.1007/978-3-319-91860-0_2

entities as "an essential element in the quality of the democratic process."[1] The question is reflected in strategic documents and memoranda of most European public radio and television entities. For example, the 'Specifications' document of France Télévisions states that "public service television is meant to be the benchmark for quality and innovation of programs, respect for human rights, pluralism and democratic debate, social inclusion and citizenship, and promotion of the French language".[2] In a similar line, RTÉ, in the new Five Year Strategy document, states that "Ireland needs strong, independent public service media more than ever. Stabilizing RTÉ will contribute in no small way to strengthening and enhancing Irish public life".[3]

In order to provide a high quality public service and to adapt to technological evolution and people's demands, public broadcasting corporations need to guarantee their financial sustainability, on the one hand, while striving to maintain or increase the audience shares, on the other hand. At the same time, and because of their public service nature, the corporations should remain accessible and accountable to society as a whole. For those reasons, among other, a public broadcasting corporation needs to be managed with due levels of transparency and accountability in its governance and decision-making processes. In this context, it is crucial to establish reliable information systems that provide all the relevant data to help the management staff in decision-making processes in diverse operational areas and at diverse levels of responsibility.

Undoubtedly, information systems permeate almost every aspect of people's lives today. However, there is a consensus among information specialists about the increasing complexity to draw a line around one specific application system to allow its management and control, considering that the relevant data are collected and distributed on a heterogeneous variety of information sources, most often autonomous, which may deter the exchange of information among them (Hasselbring 2000).

Particularly, we understand the idea of information system in a practical way, as a set of build-in components that work together to collect, store and process data, and to deliver the resulting pieces of information in an organized, structured, adequate and valuable way for a variety of potential stakeholders. According to the previous arguments, and with the objective to create an informational device to facilitate monitoring and evaluation of the performance of public media, the integrated information system we propose is designed and build up as to cover at least five key dimensions: value of public service, quality, innovation, financing, accountability and governance.

However, it is important to note that there is a conflict between the goals of the professionals building information systems and those of the professionals that will manage them, since the later focus mainly on the collection of information and their classification, as well as on the elaboration of search strategies for its recovery. The conflict has promoted a vision of the use of information from the perspective of the

[1] Amsterdam Protocol on Public Service Broadcasting, 17 June 1997.

[2] Specifications of France Télévisions, decree 2009-796, 23 June 2009.

[3] RTÉ, Strategy 2012–2017.

system itself that most often leaves aside the problems of the potential users, which should be never forgotten (Kuhlthau 1991).

After this introduction, the chapter is structured as follows. Section two describes the foundations and main characteristics of the Delphi technique proposed to obtain consistent input from a panel of experts about the potential of simple indicators to monitor and evaluate performance of public broadcasting corporations and we discuss the operational procedure for the selection of a set of simple indicators from those collected in one or more catalogs proposed by organizations such as M.A.R.S., Media Ethics, Media Act, Media Accountability, Media Sustainability (MSI), or UNESCO. The selection should illustrate accurately the most relevant aspects of the activities carried out by public broadcasting corporations and the characteristics of the contexts within which they operate. In section three we describe and discus how to elaborate an operational two-layer system of indicators tailored to the informational needs of diverse types of decision-makers in public broadcasting corporations. Finally, section four summarizes our proposal and concludes.

2 The Delphi Technique

The Delphi technique begun to be used in the middle of 20th century, as an instrument to induce consensus from divergent experts' opinions about a variety of forecasting issues linked to the operations of The Rand Corporation. The first investigations were implemented by Dalkey and Helmer (1963) in the form of expert consultations in several rounds, to facilitate a consensus from the views of a group of experts in particular areas of knowledge. In order to control the opinions, several questionnaires were passed to each expert in an individualized manner, so no meetings were necessary, avoiding face-to-face confrontations of the individuals involved in the process. With this technique, the researcher can interview, or pass a written questionnaire, to a number of experts in a chosen field without having to match them in time or space. In this way, the anonymity of the experts to each other is guaranteed, so it is not possible to condition others' opinions, eliminating the potential bias that could occur if there were personal contact among them (Landeta 1999).

Linstone and Turoff (1975) describe the Delphi technique as "a method for structuring a group communication process so that the process is effective in allowing a group of individuals, as a whole, to deal with a complex problem." The method is described as a systematic procedure that allows to obtain consensual information from the individual query; most often, the consensual information is expressed in qualitative terms. From the experts' assessments about one or more issues of interest, researchers may draw the corresponding conclusions after statistical treatment of the experts' responses. To carry out the analyses, the group of experts is consulted at least twice on each element or issue of interests. The opinions gathered in the first round are statistically summarized and offered as additional input to the experts for the second round, which may cause them to change their initial assessment, or to ratify their initial scores. The process of successive rounds ends when an acceptable degree of consensus about the issues object of consultation.

2.1 Characteristics and Phases

Among the most important features of Delphi method that have an influence on the quality of the results, we can mention (Linstone and Turoff 1975; Landeta 1999, 2006; Loo 2002; Coll-Serrano et al. 2012, 2013):

- Selection of experts. It is about choosing experts that are able, as a group, to contribute knowledge to represent all the aspects of the problem to be analyzed.
- Anonymity of participants. To eliminate the bias that may occur from personal contact, no expert should know who the other members of the panel are or, at least, the individual answers they provide.
- Iteration process. Researchers pass questionnaires more than once, or pass different questionnaires, to each expert in the panel. At the end of each phase, they offer the ratings added to the panel of experts.
- Controlled feedback. With the information of the previous phase, the experts can maintain or modify their opinions.
- Group statistical response. All responses are taken into account for the preparation of the final report. Questions are formulated in ways that facilitate statistical treatment of the data obtained from the answers.

Our proposal for implementation of the Delphi technique is structured in the following three phases (AECID 2009):

Initial phase. In this first phase, researchers focus the study on planning. The problem or process object of the research is clearly defined. In our case, the problem is which indicators to select, from the set of proposed public TV indicators, to provide data for our information system. First, the team that is responsible for coordination and supervision is formed; the context in which it is to be performed, is established and the panel of experts is selected. To be successful in this type of studies it is crucial to select adequately the experts or evaluators (Gordon 1994). Since the panel of experts evaluates qualitatively or quantitatively the indicators, the results obtained in the Delphi depend on the knowledge and the implication of the panel, that is to say, the success of the Delphi depends to a great extent on the quality of the experts.

Exploratory phase. This phase includes the preparation of the questionnaires, completion and compilation of the data of the different rounds of consultation, analysis of the results of each phase, feedback and adjustments until the responses of the experts stabilize. In general, the method allows several rounds to be performed. Loo (2002) argues that the researchers should launch new rounds until the consensus is reached or, alternatively, when the results are repetitive in two successive rounds. Although three to four rounds are typically used, often it is sufficient to perform only two rounds, since the most important changes occur between the first and second

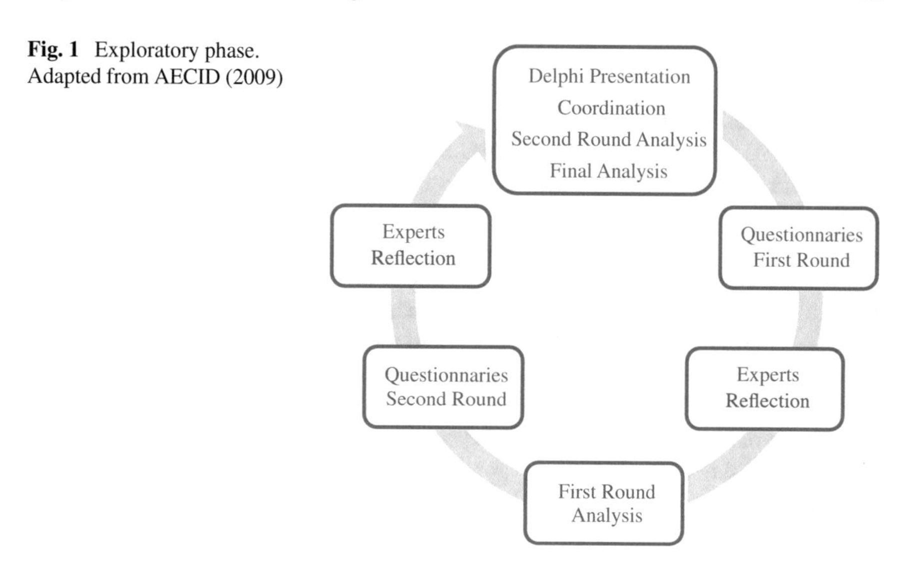

Fig. 1 Exploratory phase. Adapted from AECID (2009)

rounds (Zolingen and Klaasen 2003; Hanafin et al. 2007). The diagram in Fig. 1 shows the iterative nature of the exploratory phase.

Final phase. This phase is dedicated to the analysis of the data obtained and to the elaboration of a final report which includes the conclusions of the process.

In general terms, the steps in which the Delphi technique is developed, including the three phases are:

1. Design of the initial questionnaire by the researchers.
2. Selection of the group of experts covering all the areas included in the study.
3. Experts' response to the first round of questionnaires.
4. Analysis of the results of the first round by the researchers.
5. Sending the aggregate results that were obtained in the first round, and sending the second round of the questionnaire. The experts can maintain their scores or modify them based on the results of the first round.
6. Analysis of the results obtained in the second round. If the level of consensus is too low to be acceptable, new rounds can be launched to achieve a satisfactory result.
7. Once the adequate level consensus has been reached on the objectives of the research, the emerging selection of simple indicators will provide input data for the construction of synthetic indicators related to each one of the strategic areas.

2.2 *Evaluation of the Quality of the Results*

The Delphi method provides valuable results when researchers think that only expert judgment can shed light on a particular research topic. This is one of the reasons why it was widely accepted and extended its use. With the application of this method it is possible to reduce the effects that experts with dominant personalities may have on group opinions, relevant information provides feedback at every step, and statistical results can be obtained quickly. However, applying the Delphi method is a complex matter and presents, among other, some weaknesses such as the fact that results would always depend on the choice of experts, i.e. who is part of the study and what prejudices he or she may have; the impunity of possible irresponsible actions on the part of the experts due to anonymity; the manipulation that can be done by the people doing the study; the large amount of time needed to complete the different rounds; and the difficulty of checking the accuracy and reliability of the method in particular situations. In addition, applying Delphi can lead to problems if the experts are not rigorously selected, it may also happen that having to fill in the questionnaires in different rounds, as is a long time, the experts leave without completing the study; or that the researchers do not perform rigorous analysis of the results obtained. To avoid weaknesses, therefore, experts should be well selected to value correctly; researchers have to be cautious with the questions that are asked in the questionnaires avoiding the ambiguity and asking closed questions, and rigorous data analysis has to be done (Landeta 2006; Gordon 1994).

It is also appropriate to consider the number of experts who will participate in each strategic area, since there is no consensus on how many people have to be part of the study. The number of experts will depend, among other things, on the complexity of the problem and the heterogeneity of the indicators to be evaluated, but according to the literature the minimum is about seven experts per issue (Dalkey et al. 1969; Wilhelm 2001; Landeta 1999). What needs to be ensured is a diversity of experts that guarantees the quality of the process even in the case that the incidence of dropouts during the consultation is significant (Hallowell and Gambatese 2010).

2.3 Selection of Simple Indicators for Public Broadcasting Corporations Using the Delphi Method

We propose the application of the Delphi method to gather the opinions of a group of experts on the current situation of public broadcasting companies in terms of governance, financing, accountability, innovation, quality and public service value. More precisely, we propose to use those opinions to evaluate existing simple indicators compiled in different catalogs, and to select those that would fuel the construction of composite indicators tailored to evaluate the advances of public broadcasting corporations towards their planned objectives.

In this case, to apply the Delphi method the actions shown in Fig. 2 must be followed and they are developed as follows:

Presentation of the project and selection of initial indicators. In our case, the purpose is to generate knowledge base to illustrate the situation of European broadcasting companies regarding each one of the strategic areas under analysis from simple

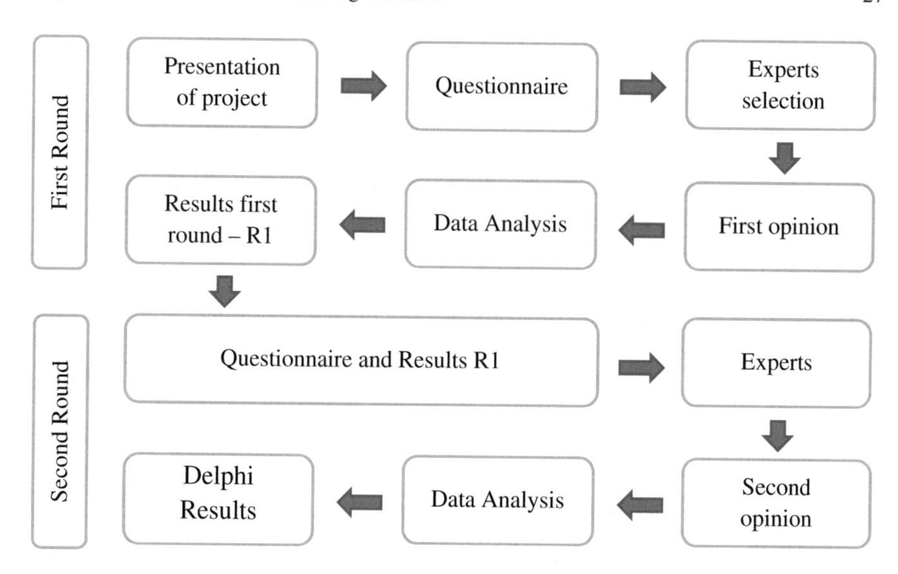

Fig. 2 Delphi Implementation

indicators available in catalogs. For the initial listing design, catalogs as M.A.R.S, Media Ethics, Media Act, Media Accountability, Media Sustainability (MSI), and UNESCO, can be consulted. The advantage of working with a list of indicators for each strategic area is that the selection of experts can achieve a greater specificity. In addition, shorter, area-specific questionnaires may prevent the experts from leaving before the process is finished.

Design of the questionnaires. The questionnaires should provide all the information necessary for the experts to know the objectives of the study and how many rounds are of consultation are initially planned. Questionnaires must be well structured and clearly worded. We recommend multiple choice questions, while open-answer questions must be avoided. The use of a single measurement scale for the answers would make much easier for the experts the task to fill in all the questions. Many studies use the Likert scales to rank the opinions of the experts about the characteristics of simple indicators regarding operational criteria such as relevance, specificity, and feasibility.

Selection of panel of experts. One of the keys to the success of the Delphi method is the adequate selection of the people who are going to participate in the assessment of the indicators (Landeta 2006). Individually, they must be able to assess the totality of the indicators related to at least one strategic area. As a panel, the group of experts must be able to provide the researchers with sufficient knowledge and experience to achieve the purpose of the consultation, which is to select the best set of simple indicators to provide data about the reality of public broadcasting corporations. Another practical issue is what can be done to prevent dropouts among experts before the consultation is completed.

Regarding the number of experts who are going to participate in each strategic area, our proposal is to consider a minimum of eight experts per area under the premise that the consultation uses structured questionnaires. Although in absolute terms the minimum number of experts in the panel is about 40, for each strategic area the panel size is within the recommendations made by Okoli and Pawlowski (2004).

First round of completion of the questionnaires. An online survey can be proposed in which the expert person can only access with a user name and a personal password. The implementation of online questionnaires has the advantage that the experts can answer the questions at their convenience from his home or office without having to move. An individualized follow-up may be useful to increase the pace in the collection of data, and to reduce the cost time of processing of the information; additionally, the use of kind reminders guarantees a high response rate.

Data Analysis. The use of statistical techniques allows researchers to describe the assessment of experts in the first phase of consultation. Dimension-reduction multivariate techniques generate the feedback information to be reported to the experts along with the next round of consultation. The feedback report can help the experts in the panel to change or to confirm their previous answers in the next round to achieve a certain degree of informed consensus.

Second round of completion of questionnaires. At the beginning of the second round an individualized report is sent to each expert who participated, containing the global results obtained in the first round. At the same time, the questionnaire is made available to fill in it again, taking into account that now the experts know the statistical results of the previous round.

Data Analysis. A new statistical analysis of the responses from every strategic area is performed again, using the same techniques as in the first round. Once the analyses have been carried out, the researchers re-examine the scores given by the experts to evaluate the degree of consensus among them. If the objective has been achieved, the final results will be presented, otherwise a new round would be carried out and the process repeated until an acceptable degree of consensus is achieved.

Delphi Results. After analyzing the data collected in the final round, the information obtained will serve to select the list of simple indicators for each area that will be used in the design and construction of composite indicators related to the hierarchy of objectives stated in the strategic plans of diverse public broadcasting corporations.

3 Building a Two-Layer System of Composite Indicators

This section contains an operational procedure to generate a system of indicators oriented to support monitoring and evaluation of the performance of public broadcasting corporations according to their own strategic plans. Strategic Planning can be understood as a systematic process for developing and implementing actions to achieve corporate goals or objectives. Ideally, every organization should have a strategic plan that guides their activities in function of their objectives. For companies or corporations, the strategic plan guides and provides a general direction for their projects and activities over time.

A system of indicators could be seen as a tool oriented both to support the processes of planning and decision making and to monitor and evaluate activities, projects and policies. The design of a complete system of indicators for any entity, in our case public broadcasting corporations, first needs a clear definition of the priority objectives and strategic areas. Obviously, the priority objectives will be, in general, specific for each organization because they respond to the reality and rationale of each particular corporation. On the contrary, the strategic areas would be considered common for all the corporations targeted; specifically, five strategic areas will be considered: governance, financing, accountability, innovation, quality and value.

The cross-intersection between the five strategic areas and the set of priority objectives, contingent for each corporation, defines a matrix of order 5 by N (where N is the number of specific objectives for a given corporation) that must be covered and monitored by the system of indicators proposed.

The system of indicators for monitoring and evaluation of corporate performance should be based on the nature of the information available to diverse types of decision-makers and involves the definition of suitable operational indexes for selecting a structured set containing the most valuable simple indicators.

3.1 Theoretical Framework

The system of indicators will be structured following a theoretical framework resulting from the integration of two well-known corporate evaluation models: (i) the Logical Framework Analysis (LFA) and (ii) the CIPP evaluation model.

The Logical Framework Analysis (Sartorius 1991; Bakewell and Garbutt 2005) is a tool used in the context of objective planning that allows presenting, in a logical and systematic way, the hierarchy of objectives of an entity and its causal relationships. The LFA allows the evaluation of the results according to the degree of fulfillment of the previously planned objectives.

The CIPP evaluation model (Worthen and Sanders 1987; Stufellbeam and Shinkfield 2002) understands evaluation as a process designed to generate information that allows decision-makers to evaluate the different alternatives available during the implementation of different processes and actions.

The combination of LFA and CIIP defines the indicator system into two layers or subsystems: The basic system and strategic system. The basic system is oriented towards short-term planning and daily management, and will be composed of indicators linked to the monitoring of the evolution of resources, processes and products, while the strategic system focuses specifically on the effects of the activities and on the permanent impacts generated through them; that is, it is oriented to strategic planning in the medium and long term.

The indicators that conform the basic system are usually indicators focused on a single aspect to be measured related to short-term operational issues. The indicators of the strategic system can also be partial indicators, but since their objective is to offer information on aspects of reality with multiple dimensions (effects and impacts), they are most often composite indicators that reflect in an aggregated way the different dimensions to be monitored and evaluated.

3.2 From the Delphi Selection of Indicators to the Information Set of the System

The second step towards the definition and construction of a system of indicators for public television corporations begins after Delphi is completed. The results provided by the experts about the suitability of simple indicators provide the guidelines that will feed and inform the analytical methods applied latter on; that is, the Delphi final results must be processed and analyzed with the purpose of generating a final list of simple indicators, or information set, that will eventually be used to compose the system of indicators.

This process of selection and prioritization of the partial indicators will be based on three principles or instruments: (i) the definition of synthetic indexes of relative operability for each partial indicator (ii) the typology of the indicators (resource, process, product, effect and impact), and (iii) the choice of feasibility thresholds for each priority action.

The scores provided by the experts who participated in Delphi in three key characteristics, relevance, efficacy and feasibility, will be processed using multivariate statistical techniques to construct for each individual indicator a compound operating index with the objective to select the final set of partial indicators. It seems obvious that indicators with higher values in the operability index should be more likely to be selected and be part of the indicator system, but it seems also reasonable to introduce additional criteria in the selection process to avoid potential deficiencies. In particular, it should be avoided that any relevant aspect of the corporation ceases to be monitored, it will be imposed that all crosses between strategic areas and priority objectives must have at least an adequate number of indicators, with a minimum number of two. For that reason, the selection of indicators is carried out by establishing a minimum value or threshold of feasibility, different by priority action, which delimits which indicators are the most feasible according to experts' judgement.

This selection methodology should provide the final set of partial indicators that cover in a balanced way each dimension of the cross-matrix of key strategic areas and priority actions. This final list of partial indicators is the starting point for the next stage of the proposal, where the information contained in the partial indicators will be aggregated in a reduced set of synthetic indexes that will serve to monitor governance, financing, accountability, innovation, quality and public service value of public audio-visual corporations in the new digital era.

3.3 From Simple Indicators to Composite Indexes

The ultimate objective of a complete system of indicators is to have reliable and truthful information about the entity or object to be monitored. Having a comprehensive system of indicators that enables and facilitates the evaluation and favors the adaptation of the activity of a public body to its assigned objectives, in theory, allows the managers to optimize effectiveness and the level of efficiency in the use of the public resources; but, potentially, due to the possible multiplicity of partial indicators to report on the same or related aspects about the evolution of the corporation, it would be possible that the information provided for two partial indicators would be contradictory, and therefore, it seems convenient to equip the System of Indicators with a set of composite or synthetic indicators that enable and favor a rapid and complete understanding of the evolution of the organization and the degree of fulfillment of its objectives.

According to Hammond et al. (1995), a composite indicator is a construct that provides a clue to a matter of larger significance or makes perceptible a trend or phenomenon that is not immediately detectable. We can think that the main objective of a composite indicator is to summarize in a single measure the information of a set of related simple indicators, facilitating in this way the understanding of the subject being analyzed. The principal advantages of synthetic index are that they can simplify the interpretation of complex topics, facilitate the evaluation of policies and programs, as well as accountability, and allow comparison between units of analysis and over time.

The use of composite indicators is nowadays generalized (Saisana and Tarantola 2002) but it also hides potential drawbacks, because reducing and synthesizing a set of indicators may simplify the interpretation of results, but can also give a wrong message if they have deficiencies in his definition and construction. Furthermore, the information and message provided by the composite indicators in isolation can lead to simplistic conclusions, therefore, they are not a substitute for the partial indicators, but must be used in combination with them to obtain useful, informative and appropriate conclusions.

Composite indicators are increasingly used in very diverse areas; therefore, there are different approaches for its definition and construction. In practice, it is difficult to integrate individual indicators in a way that accurately reflects the reality to be monitored. The starting point should always be a thorough knowledge of the problem or concept to be addressed. The OECD statistical office produced a Handbook with a listing of good practices that has become de facto the principal guide for designing, construction and use of composite indicators. The purpose of the Handbook is to provide builders and users of composite indicators with a set of recommendations on how to design, develop and disseminate composite indicators. In particular, the Handbook guides constructors and users by highlighting the technical problems and common pitfalls to be avoided (Joint Research Centre-European Commission 2008).

Following the methodology proposed in the Handbook to build composite indicators, we will follow a process consisting of several steps. The first and most important stage to build composite indicator is the definition of the conceptual framework. It is the most important because it provides the basis for the other stages: it provides the basis for selecting variables, weights and methods of aggregation so that the synthetic index adequately reflects the structure of the phenomenon or reality to be analyzed. At the end, obtaining meaningful composite indicators for the objectives in the planning is a task that depends on the conceptual framework considered. If a dimension of the phenomenon to be analyzed is omitted at this stage, the composite indicator could lead to simplistic and/or erroneous conclusions. In our case, as in the initial stage of the process, the expert judgment has been used intensively through the Delphi to select the most relevant partial indicators, it is expected that the initial set of indicators would be adequate. Once the conceptual framework and the initial set of information have been determined, the most common situation is that the data present certain problems that may need to be addressed, such as missing data, atypical observations, etc.

Before we proceed to the weighting and aggregation of the individual indicators to construct the composite indicator, it is convenient to analyse the statistical structure of the database and compare the results with the theoretical conceptual framework in order to detect possible differences or similarities. Multivariate analysis can be useful to evaluate the suitability of the data set, since it is possible that the statistical similarity of the partial indicators can lead to a situation rich in indicators but poor in information. These techniques allow the identification of subgroups of individual indicators that are statistically 'similar'. In some situations, the results of this stage may imply rethinking of the construction of the synthetic index, having to go back to previous stages of the process.

Individual indicators will usually be expressed on a variety of scales; for that reason, the normalization of the dataset is a prior step previous to any type of aggregation of simple indicators; as usual, the standardization procedure should be consistent with the theoretical conceptual framework and statistical properties of the data, and robustness tests are necessary to evaluate their impact on the results.

Another important step in the process of constructing composite indicators is the weighting and aggregation of the various partial indicators. There are a variety of methods and procedures for selecting weights (DEA (data envelopment analysis), regression analysis, principal components, unobservable component models etc.), but one of the most used is weighted linear aggregation (Joint Research Centre-European Commission 2008). Obviously, the theoretical framework and the properties of the data must be taken into account, and the final weights should reflect the relative importance of individual indicators in the construction of the particular composite indicator. We must be aware that regardless of the process and method used to choose the weights, in the end the weights represent value judgments that often respond and show the objectives that are intended to be achieved with the construction of the composite indicator.

Once the weights corresponding to the different partial indicators are obtained, the aggregation method to be used must be determined. Again, we have that there exist different methods of aggregation. These can be classified into additive aggregation methods, geometric methods and multi-criteria aggregation methods. Before proceeding to the next step, the compatibility between the weighting methods and the aggregation methods should be analysed.

Finally, it is really convenient to undertake a statistical treatment to verify the robustness of the composite indicator to changes in the elements of the previous stages before completing the process of constructing a composite indicator and making it available to the final users.

4 New Challenges

This chapter describes and discusses a methodological procedure aimed at building an integrated information system for monitoring and evaluating public broadcasting corporations. The resulting information system encompasses data that reflects what public broadcasting corporations are like, what types of activities they do, what results they obtain from them, and what consequences all that generates for society as a whole. The procedure for elaborating the information system follows a conceptual framework based on organizational evaluation models, and consequently takes into account from the context within which corporations operate to the long-term consequences of their activity. To build up the information system we propose a two-stage operational procedure. The first stage consists of the selection, through expert judgment, of simple indicators that illustrate the most relevant aspects of the activities carried out by public broadcasting corporations. One or more catalogs of simple indicators suggested by international organizations provide the basis for the selection process. The second stage is the composition of a system of indicators specifically related to each relevant dimension and to each level of the hierarchy of objectives planned by the corporations in their respective strategic planning processes. The indicators are, in general, informative constructs that serve the strategies of the corporation. The strategy is the whole set of planned actions that, starting from a concrete reality with certain potentialities, and using human and material resources, are oriented to the achievement of a structured set of pre-established corporate objectives. That is, the strategy is a set of action patterns that relate the present, observed reality with a desire for a future reality through a planning of hierarchical objectives in the short and long term.

The system of indicators itself is designed at two levels. The first level is a basic indicator system, which includes the indicators that allow the description and monitoring of the reality of the activities of the corporations on a daily basis. These indicators refer to the environment and the context within which the diverse corporations may operate, their internal structure, the resources they command, the processes they undertake, the services and products they offer, and the results they obtain with their activities. The basic indicator system provides the pieces of information needed to guide the ordinary management of corporate operations, and provides valuable knowledge base for future planning. The values of the basic indicators may be useful for defining typologies, or groupings of corporations with similarities in diverse aspects of interest for specific users; the resulting typologies would allow potential users to make reasonable comparisons of operational differences between corporations with similar characteristics. The second level of the system of indicators is the strategic indicator system, which should include indicators oriented to the planning, evaluation and benchmarking of the permanent operation of the broadcasting corporations. Strategic indicators express the consequences that the activities of a PBC generate in its environment, both in terms of its short-term effects and of its medium- and long-term impacts over the society that supports it. Strategic indicators are necessarily associated with planning and evaluation, and are oriented to support decision making at a level higher than that of ordinary operations management. An efficient way to monitor the progress towards the planned objectives is through appropriate sets of indicators. Therefore, the indicators mirror the objectives and these, in turn, can be of different types according to their hierarchical level in the planning of the PBC. In the planning process, the objectives follow an ascending hierarchy that classifies them into operational, specific, principal and strategic objectives. Thus, the indicators will have a different nature depending on the corresponding level of objectives whose degree of achievement we want to assess. The basic indicator system should contain indicators related to Context, Resources, Processes and Products, elements that correspond to the levels of operational and specific objectives. The Strategic System, in turn, should contain indicators related to Effects and Impacts, elements that correspond to the levels of main and strategic objectives, respectively. The basic system and the strategic system of indicators operate as two layers of the information system that guarantee that the whole system provides useful, reliable, consistent and appropriate information for decision-making about the reality of PBCs, at its different decision-making levels, and, consequently, oriented to the different stakeholders.

We understand the notion of an information system as the set of components that work together to collect, store and process data, and produce information in an organized, structured, adequate and valuable way for a variety of potential users. We believe that our proposal attempts to reduce the gap between traditional models of information-delivery systems and the natural process of the user when deciding to gather the data for a variety of potential uses. From this perspective, we propose to consider the information search process as a constructive activity of the final user to find meaningful information, and to give it a precise meaning within a personal or institutional frame of reference. In this sense, we highlight the fact that the efficiency of information-retrieval devices should take into account the integration of the results into the users' own work, as well as their evaluation of the usefulness of that information for problem solving and decision-making.

References

Bakewell, O., Garbutt, A.: The use and abuse of the logical framework approach INTRAC technical report, SIDA (2005)
Carrasco, S. (coord.): Cómo evaluar proyectos de cultura para el desarrollo: una aproximación metodológica a la construcción de indicadores. AECID, Madrid (2009)
Coll-Serrano, V., Carrasco-Arroyo, S., Blasco-Blasco, O., Vila-Lladosa, L.: Design of a basic system of indicators for monitoring and evaluating spanish cooperation's culture and development strategy. Eval. Rev. **36**, 272–302 (2012)
Coll-Serrano, V., Blasco-Blasco, O., Carrasco-Arroyo, S., Vila-Lladosa, L.: Un sistema de indicadores para el seguimiento y evaluación de la gestión sostenible del patrimonio cultural. Transinformaçao **25**, 55–63 (2013)
Dalkey, N., Brown B., Cochran S.: The Delphi method III: Use of self rating to improve group estimates. Research Memorandum, RM-6115-PR, CA: The Rand Corporation (1969)
Dalkey, N.C., Helmer, O.: An experimental application of the Delphi method to the use of experts. Manage. Sci. **9**, 458–467 (1963)
Gordon, T. J.: The Delphi method. AC/UNU Millennium Project, http://www.gerenciamento.ufba.br/downloads/delphi_method.pdf (1994)
Hallowell, M.R., Gambatese, J.A.: Qualitative research: application of the Delphi method to CEM research. J. Constr. Eng. Manage. **136**, 99–107 (2010)
Hammond, A. L., Adriaanse, A., Rodenburg, E., Bryant, D., Woodward, D.: Environmental indicators: a systematic approach to measuring and reporting on environmental policy performance in the context of sustainable development. World Resources Institute (1995)
Hanafin, S., Brooks, A.-M., Carroll, E., Fitzgerald, E., Gabhainn, S.N., Sixsmith, J.: Achieving consensus in developing a national set of child well-being indicators. Soc. Indic. Res. **80**, 79–104 (2007)
Hasselbring, W.: Information system integration. Commun. Assoc. Comput. Mach. **43**, 32–38 (2000)
Joint Research Centre-European Commission: Handbook on constructing composite indicators: Methodology and User guide. OECD (2008)
Kuhlthau, C.C.: Inside the search process: information seeking from the user's perspective. J. Am. Soc. Inf. Sci. **42**, 361–371 (1991)
Landeta, J.: El Método Delphi. Una Técnica de Previsión Para la Incertidumbre. Ariel, Barcelona (1999)
Landeta, J.: Current validity of the Delphi method in social sciences. Technol. Forecast. Soc. Chang. **73**, 467–482 (2006)
Linstone, H. A., Turoff, M.: The Delphi Method: Techniques and Applications. Addison-Wesley, Reading (1975)
Loo, R.: The Delphi method: a powerful tool for strategic management. Policing: Int. J. Police Strat. and Manag. **25**, 762–769 (2002)
Okoli, C., Pawlowski, S.D.: The Delphi method as a research tool: an example, design considerations and applications. Inf. Manag. **42**, 15–29 (2004)
Saisana, M., Tarantola, S.: State-of-the-art report on current methodologies and practices for composite indicator development. European Commission, EUR 20408 EN, Joint Research Centre (2002)
Sartorius, R.H.: The logical framework approach to project design and management. Am. J. Eval. **12**, 139–147 (1991)
Stufflebeam, D., Shinkfield, A.: Evaluación Sistemática: Guía Teórica y Práctica. Paidós Ibérica, Barcelona (2002)
Wilhelm, W.J.: Alchemy of the oracle: The Delphi technique. Delta Pi Epsilon J. **43**, 6–26 (2001)
Worthen, B.R., Sanders, J.R.: Educational Evaluation. Longman, New York (1987)
Zolingen, S.J., Klaasen, C.A.: Selection process in a Delphi study about key qualifications in senior secondary vocational education. Technol. Forecast. Soc. Chang. **70**, 317–340 (2003)

Olga Blasco-Blasco Ph.D. in Economics and Business Sciences. She is an associate professor of Quantitative Methods at the Universidad de Valencia. She began her research career in the field of Productive Efficiency. Currently, her research field is the application of Quantitative Methods applied to the Culture and Education, specifically, the design and implementation of cultural information systems. Cuenca, Spain.

Pedro J. Pérez Ph.D. in Economics, lecturer at the Universidad de Valencia since 1992. He was an associate research fellow at the University of Manchester through a Marie Curie grant. Currently researcher at MC2 (Quantitative methods for the measurement of Culture and Education). Research interests focused on business cycles, econometrics methods and the social and economic effects of educational and cultural investments both at individual and social levels. Zaragoza, Spain.

Luis E. Vila Doctor in Economics and Business Science. Research interest focused on the social and economic effects of educational and cultural investments at individual and social levels. Main publications include papers on the relationship between higher education investment, innovation and economic performance, the labour market effects of educational and cultural expansions, the consequences of education and skill mismatches and the development of professional competencies through university education. Valencia, Spain.

Transparency in Public Service Television Broadcasters

Paulo Carlos López-López, Mónica López-Golán and Nancy Graciela Ulloa-Erazo

Abstract One of the features of current society is linked to the ever-increasing demand that both public and private institutions account for their daily activities through the publication of information and data related to them. In this sense, the media are dually obliged, because they are companies and they also work with a public service body, as is the case of information. This chapter takes on the need for conceptually systematizing what we know as accountability, governance and transparency regarding public service television broadcasters, and we will do so by means of a comprehensive bibliographic review. In addition, the main transparency assessment models for these institutions are here highlighted and scored, describing their implementation already fulfilled in various geographical areas, together with their different results but with just a single reading: society—and law—goes far ahead of the institutions' answer.

Keywords Media · Public service television · Accountability · Transparency Information

1 Introduction

Transparency is directly linked to democracy. In contemporary societies, as the information and knowledge technologies develop, this dynamic becomes a requirement to all kinds of institutions and organizations, even television broadcasters. From this, we can safely state that free access to information must be understandable, appropriate, complete and up-to-date. Moreover, this access shall promote dialogue and

P. C. López-López (✉) · M. López-Golán · N. G. Ulloa-Erazo
Pontificia Universidad Católica Del Ecuador Sede Ibarra (PUCESI), Ibarra, Ecuador
e-mail: pclopez@pucesi.edu.ec

M. López-Golán
e-mail: molopez@pucesi.edu.ec

N. G. Ulloa-Erazo
e-mail: nulloa@pucesi.edu.ec

M. Túñez-López et al. (eds.), *Communication: Innovation & Quality*, Studies in Systems, Decision and Control 154, https://doi.org/10.1007/978-3-319-91860-0_3

feedback among these organizations, the media and the general public. The Special Study on the Right of Access to Information, published by the Inter-American Commission on Human Rights (2007) indicates that "access to information is an essential tool to combat corruption, (…) and improve the quality of our democracies" (p. 1). The media, and in particular public service television broadcasters, should somehow contribute with the good governance agenda. Consequently, it is necessary to establish their own internal codes of conduct in order to ensure their independence, the strength of their resources and their willingness to promote this type of information.

These two concepts of accountability and access to public information have a new meaning in current society, and the need that public radio and television broadcasting services assume transparency as the direction vector for their daily performance becomes mandatory. There are indicators and models to assess the compliance of these obligations, which are designed according to legally enforceable and ethically required criteria. Heikkilä et al. (2012) were the ones who distinguished three types of transparency in the media: before the publication (information about the company), during the publication (explanation of the news preparation), and after the publication (responsibility assumption). In this regard, the International Center for Media and the Public Agenda (ICMPA) classifies the transparency requirements in five areas (Redondo and Campos-Domínguez 2016): corrections; property; personnel policy; information policy, and production and participation processes.

2 Transparency and the Media

2.1 *Accountability and Access to Public Information*

Transparency is closely related to the measuring field and to the process management regarding the organizations, since it is an intangible asset (Túñez 2012), and hence also regarding the media. It involves starting a formal, clear and visible process about public information. Societies need to identify with the growth, innovation, culture and path trends regarding public and private organizations, NGO's and the media. Accountability is an argument that commonly guarantees the reality in order to detect growth and decrease rates. This process requires that organizations' activities become transparent in administrative, managerial and budgetary terms. Thus, within the framework of transparency, state and private entities establish close links with the citizens, who have full rights to access the information.

Accountability is currently part of a broad concept of social responsibility. In some cases, governments promote this area within their 'best practices', which leads both to governing entities and public and private servants to prove that decisions are in line with people's priority needs. Transparency is a very broad and decisive concept for the impact, stability and successful presence regarding the institutional image. This can be summed up as the access to information held by the authorities and the promotion of participatory mechanisms and practices that reduce the chances of corruption (Pérez Bravo 2004).

That is why three basic concepts appear in the reality of the processes required by the current society to be held accountable for actions (Suárez-Villegas et al. 2017). The first concept is transparency, the second self-regulation and the third participation. The willingness of the organizations to cope with the society that implicitly and explicitly must be linked to these concepts in order for the organizations to be effective will be measured at these three levels.

2.2 Civic Participation

Civic participation is established as an influence process when it comes to the decisions made by public authorities. Such behavior is also understood as an individual or collective right to have a say, and it combines democratic trends that join access to information structures and freedom of opinion. An example of this in the Latin American context is the *Vive Digital Plan*, a Colombian participatory experience promoting online forums on topics of national interest through its website, and in which citizens have real guidance on public policies. Although the *Vive Digital Plan* aimed at boosting the widespread growth of the use of the Internet, it emerged because this website became a strategy that records the civic participation, taking a leap towards democratic processes. In the case of the United States and through their domestic legislation, partnerships between citizens and several representatives are created, who analyze the law enforcement and their proposals for solutions. Brazil and many other countries develop participation mechanisms through the collaboration itself and the citizens' answers. Another example, also found in the American field, has as its point of reference the Organic Law of Transparency and Access to Public Information, issued in Ecuador in May 2000. As a necessary condition for participation, article 7 on Dissemination of Public Information mentions: "due to the transparency that all the state institutions are required to demonstrate (…), they shall disseminate through an information portal or website (…) the following up-to-date minimum information" (p. 4).

The fact of guaranteeing that citizens participate in these areas of socialization is the objective pursued from the legitimization of information. The companies and the media must necessarily generate areas for interactivity in civil society. Within the public television field, the BBC for example opened two consultations in 2015 as part of the production process of the *Report on the renewal of the BBC charter*, which expired in December 2016. The issues raised included the need to create new ways of audience participation. In another context, the Law on Access to Public Information (2011) from El Salvador, recognizes the process of active transparency from the information asymmetries existing between the state and its citizens, so it defines the term transparency as "the duty to act in accordance with the law, together with the duty of openness and advertising that the public servants have in the exercise of their competences (...) to give effect to the right of everyone to know and monitor the governmental management" (p. 11).

The societies shall observe the public and private administration entities. This means that besides the fact of enabling the interested parties to permanently assess the information and compare the management results, citizens strike a balance in the development of an institutional framework and in the transparency performance. This will enable people to undertake a citizen control over public management based on the information provided by media and institutions.

2.3 *Transparency in the Media Companies*

Amidst the multiple and profound changes that are taking place in journalism, the accountability systems have undergone a clear evolution in recent years, from reporting their activity to generating social interaction processes. The aim is to assess the public response and the impact that the media has in the construction of the social, cultural or environmental values regarding the specific tasks of civil society. The media, within the new models of consumption, must inevitably take advantage of the potential of the new information and communication technologies. Taking this into account, traditional accountability instruments must also be adapted to the digital environment so as to foster new forms of transparency and information quality control, enabling citizens to participate and give their opinion on the quality of information and communication. It is necessary to extend the analysis from the complexity and the crises of the administrations, together with their struggle to achieve the image of "good governance". Thus, this reality becomes a series of strategies that, if they are not supported by an effective decision making process and disciplinary contributions, an absence of democratic legitimacy is then introduced. This construction perspective of an image of social responsibility, regarding mainly the media and the information professionals, can be expressed through an effective process of accountability which provides information on the degree of compliance with values such as pluralism, equality and informative truth. In the case of Spain, this premise helps to detect the journalistic cultures existing in the different Autonomous Communities, obtaining results that allow the integration of values itself (Suárez-Villegas et al. 2017).

The development of democratic political systems has favored the expansion of transparency and an increasing number of the countries with specific legislation. López Cepeda et al. (2013), in their research on the corporate transparency of audiovisual media, point out that "the growth in the number of states with laws on the access to information and on information transparency highlights the growing need for a regulation in which the citizens are provided with this control mechanism for public management" (p. 46). In the case of public service television, Muñoz and Azurmendi (2016) refer to the multilevel governance of the Basque Country, Catalonia and Galicia, and include an analysis of concepts that lead to a series of principles that are especially present in the shaping of the European audiovisual public services.

Some of the objectives that these media shall meet are: accessibility, openness, transparency, dialogue with the institutions, pluralism, and the promotion of diversity and cultural richness".

Currently, the political and economic reality that is globally taking place causes the citizens' confidence in the institutions and in the media to go through some very critical times, prioritizing the fact of showing that the public resource management is rational and transparent. Citizens are required to be co-responsible in the decision making processes, so that they are able to defend the importance of public broadcasting services in social and democratic terms.

3 Public Service Television Broadcasters, Governance and Interest Groups

"Public broadcasting's only raison d'être is public service" (UNESCO 2001). This legal expression refers to various activities that the state supplies to the citizens in order to increase their quality of life. Therefore, in the case of public service broadcasters, the features would provide "democratic guarantees; encouragement of civic participation; social, cultural, identity, and social welfare function; territorial, economic and development balance function; driver for the audiovisual industry; creative innovation and experimentation; humanistic and moralizing purposes, and knowledge dissemination and socializing function" (De Moragas and Prado 2000). In short, we can consider public service television broadcasters a social, training and educational tool available to everyone.

However, the importance of public service television broadcasters has not always been obvious, nor is the same concept all around the world. While in Europe they have been awarded an important role in the protection of social pluralism, cultural enrichment of all individuals, or the dissemination of content in order to contribute to the democratic processes, in the United States, the television broadcasting system has never adopted any public service mission (Iosifidis 2010). The representation of public service television broadcasters in the United States has always been negligible, contrasting with a mainly commercial and market-oriented system. In the same way as in the American context, Latin American television broadcasters have had a "strong commercial influence" (Becerra et al. 2013) contrasting with the scarce development of public channels. Public service television in Latin America is distinguished by a "long, splintered and, lastly, unsuccessful history of public media" (Waisbord 2014), and governments have used them as institutions with propaganda purposes. A different public media system was articulated in each of the regions mentioned above, but there are also various models of public service television within these regions. In this way, in Europe, despite the fact that there are a set of common obligations to fulfill its objective of public service, there are also differences concerning the financial field, the functions to be performed, the internal organization or the entity's control (Salvador Martínez 2006), also depending on the regulatory system to which broadcasters are subject or the country's own customs.

The most recent history of public service television broadcasters all around the world is related to the changes brought about by the digital age. This has led to a reorganization of the corporations' structure, adapted to the media convergence ecosystem, as well as to an update in regulations with principles aimed at legislating this reality and a new way of management related to the most modern concept of governance. In this sense, there is a trend towards a "new way of managing, more collaborative and based on a more open relationship to society, on non-hierarchical coordination mechanisms, on new collective practices and on more participatory forms, which are going increasingly further away from the principle of authority and unilateral imposition" (Campos Freire and López Cepeda 2011). Besides, with a clear intention to give greater credibility and confidence to television as a public service, it is important to consider the transparency measures taken into consideration by these media as "one of the fundamental elements of governance" (Fernández Lombao 2015).

"The mother of all public service television broadcasters", the BBC, is in a process of governance rethinking, so as to improve the quality of public service (Campos Freire 2015a, b). In this approach, the communications regulator in the United Kingdom, OFCOM, has competence as regards the external monitoring of its functions as a public service. Internally, a board of directors is created with a majority of independent directors that will allow for supervision of the executive management. A complaints committee is established which forces the broadcaster to listen to the complaints from spectators and a financial monitoring is carried out by the National Audit Office (Campos Freire 2015a, b). However, the strategic direction undertaken by the BBC has nothing to do with the majority of governance systems from the rest of the public service television broadcasters within the European and Latin American framework. The traditional structure used in the current models of public service television broadcasters calls for a renewal that helps strengthen the strategic values of a public service television broadcaster: to raise the independence, "to increase commitment and diversity", "to accelerate innovation and development", "to maintain universal coverage" and "to comply with its responsibilities as a public service" (EBU 2015). To do this, the reformulation of the BBC's governance points out that it would be important to aim the board commitments at performing monitoring tasks, including a minority of elected representatives by the government and an independent majority with recognized competence in the audiovisual field in order to safeguard the independence of public service television broadcasting. In terms of financing, the models for public service television should be "transparent, stable, and sustainable" (Becerra et al. 2013), without a complete dependence by the state to avoid pressure from financers and with guarantees of viability to avoid high levels of debt. Technologically, public service television broadcasters should use all the tools available to increase the sector's levels of transparency. In this respect, websites have become one of the main drivers for the publication of the information demanded by the interest groups (Fernández Lombao 2015). For their part, chats and social media are the new channels of information exchange between television broadcasters and audience.

As noted, many of the measures which must be taken into account by public service television broadcasters for the promotion of a high-quality public service are linked to the technological development required by innovation. Public service television broadcasters shall also have an audience who take part in their activities as one of the major interest groups involved in the audiovisual industry from the public sector. Millions of citizens around the world have a public service television broadcaster. As the main owners and responsible for their financing, “they have also the right to get to know all about its management” (Campos Freire and López Cepeda 2011).

4 Financing Sources in Public Service Television

The public media landscape shows a very different reality in relation to financing models, which is one of the main requirements in the field of transparency. In this way, the Public Broadcasting Service (PBS), an American Organization for public service television which holds a network of operators with the objective of creating educational and informational content, has a varied financing system coming from different sources. The main means of financing comes from the affiliated stations (47%). Federal Funding is another source responsible for financing public channels, with an average allocation of 24%. The remaining percentage comes from the sale of educational products, royalties, broadcasting rights and investment income (Becerra et al. 2013).

The Latin American public service television context shows an evolution towards a mixed model of financing, and the income comes from the public and private sectors. It is the case of countries such as Brazil, Costa Rica and Uruguay. In this way, public channels as *TV Cultura* from Brazil, specialized in artistic and cultural content, “apart from the public resources, it uses advertising to ensure the maintenance of the channel” (López-Golán and Ulloa-Erazo 2016). When it comes to the public television service in other countries, as it is the case of Chile, which is represented by TVN (*Televisión Nacional de Chile*), the service responds to a self-financing scheme, mainly based on advertising sales and other services offered to private companies. This financing system contributes to an empowerment that turns TVN into a respected and diverse channel with quality content. However, this financing model is submitted for debate because of its lack of durability, “requiring a new state capitalization” (Sturm and Nalvarte 2016). Without leaving the Latin American region, we encounter the system implemented in Colombia in order to finance the public service television sector. In this country, by law, it is the sector itself that is responsible for providing the necessary economic resources to finance public service television. Thus, private television channels with audience leadership such as *Caracol* or RCN devote 1.5% of their income to the Television Development Fund (FONTV). Television broadcasters through cable or satellite also contribute a percentage per subscriber to the same fund (López-Golán and Ulloa-Erazo 2016). The alliance established between the Colombian State and the private television sector is beneficial to all those involved.

Citizens are ensured access to public television service without the need for a direct payment, not being oriented towards the market requirements by profitability needs, and the private sector is sure that public service television will not be marketed (Arenas 2015).

In the European public sector there is also a diversification of income from the point of view of the financing. Public funds represent the main source of income for 23 European countries. This is the case of the Netherlands, Spain or Luxembourg. The payment of a direct tax by the television owners, known as TV tax, is the financing method for countries such as France, Italy, Germany or the United Kingdom. The evolution of income through the tax payment, despite being abolished in some countries, has been positive in recent years. From 2010 onwards, and in comparison to 2014, when public media were funded with 66% through this financing source, the evolution has experienced a growth of 3.5% (EBU 2015). The remaining income for the European public service television broadcasters comes from advertising or, to a lesser extent, from selling the program rights in the international market.

In general, the public service television broadcasters with a solid system of financing reflect healthy democracies. Hence the importance given by corporations of moving towards the search for new financing sources in order not to depend only on state revenue or the payment of the direct tax.

5 Quality Content Without Giving up on Audience

Television broadcasters need to enhance and develop their content and services in terms of plurality, diversity and quality, in addition to disclosing their processes. Taking into account that one of their main functions is the "establishment of high quality standards for the entertainment, education and training sectors" (Iosifidis 2010), it is important that public service television broadcasters design varied programs which are addressed to all citizens and have high quality standards without giving up on the idea that public media can have obtain massive audience. It would therefore make little sense to finance a public service television broadcaster with state money if nobody is going to watch it. On the contrary, television broadcasters should neither give up on content quality by the fact that they get audience at any price. However, before making appraisals, it is important to know how quality is measured and assessed in the television context. Usually, public media associate quality with the descriptive value of the program, with diversity and with the capacity of representation of reality.

Bearing this in mind, together with the main mission of public service television broadcasters, it is important to broadcast events of general interest, programs that promote diversity, formats that encourage new languages consistent with the interests of the current audience, content that helps shape a citizenship, and spaces that enable the approach of public debate (Becerra et al. 2013), but all this without disregarding the entertainment in pursuit of producing boring content. Therefore, public media have the challenge of making educational, social, and cultural programs that are entertaining and obtaining a good response from the audience. Although the main objective of a public service broadcaster should never be achieving high levels of audience, there are cases in which the governments that foster, through calls for proposals that include economic prizes, the creation of new content for public media.

6 Enabling Transparency in Television Broadcasters

As information companies, television broadcasters have specific features that make it impossible to address the public information generated in the same way as public administrations or even other companies. In this sense, the fact of acting based not only on classical criteria of profitability, but also with public service content as is information (regulated with comprehensive legislation throughout the world), forces to establish a series of stricter indicators in order to be able to audit this type of entities in a holistic manner.

6.1 Governance and Interest Groups: Institutional and Economic Information

Governance in institutions, and in this case public service television broadcasters, is understood as the mechanism that promotes, manages and takes decisions: in other words, it establishes who is in charge. Governance has an integrated character and it "includes the processes and bodies that regulate the economic, social and democratic management of television broadcasters" (De Bustos 2015). The notion of governance is a broader and more open concept than the one of government (Campos Freire 2015a, b), that is, an integrated coordination system which is appropriate in the field of the public media management (Coppens and Saeys 2006). In this manner, governance in the field of public service television broadcasters is subject to regulation policies, especially at the European level (Puppis 2010), which place a heavy burden on this type of entities from an advertising point of view (because of their restriction).

In this regard, there are several studies, as the *Media-Act-Media Accountability Systems in Europe and Beyond* (2010–2014), involving 14 European and African universities that have analyzed the accountability and governance systems (Suárez-Villegas et al. 2017) based on the journalistic models established by Hallin and Mancini (2008). These authors have developed a very important contribution when it comes to categorizing the media systems and their relationship with the political system through a cross-variable. The authors established five categories that affect the accountability and transparency themselves: the Mediterranean model (Italy, Spain, France, Portugal and Greece); the North European model (Denmark, Sweden, Switzerland, Finland, the Netherlands and Norway); the liberal model (Great Britain, USA and Canada), and later expanded to the Eastern countries (Romania and Poland) and to the transition countries in the Arab world (Figueras Maz et al. 2013).

Table 1 Levels of institutional information

Information	Level
Institutional	First: legality
	Second: government
	Third: management
	Fourth: social responsibility

Table 2 Levels of economic Information

Information	Level
Economic	First: inputs
	Second: outputs
	Third: labor relations
	Fourth: accountability

The main challenge in the field of governance of public service television broadcasters is the resolution of the conflict between control and independence, searching a management guided by democratic and quality principles, on the basis of a transparent and free action. In order to put the field of governance into practice, a specification in two areas is necessary, such as the institutional and the economic one. In this way, the institutionalized part of the public service television itself shall be fixed at four major dimensions: firstly, legality, i.e. the need to systematize the regional, national or international laws affecting the body; secondly, the government, understood in a broad sense, and including its designation, incompatibilities, agendas or advertising of its meetings and events; then, everything that affects the management of the organization itself, including the clear visualization of the organization chart, roles, responsibility assumption and accountability; and, finally, social responsibility, also understood in a broad sense. It is important to know the actions carried out by the public service television and that cause an impact, evaluate it from the perspective of civil society and get to know the existence of a code of best practices in the corporate sector, which is strongly focused on prevention of corruption and on good governance (Table 1).

Another important area in the field of governance is the economic one, and it is probably the more regulated area within the different national laws and the international recommendations. The need to make the economic information coming from public budgets available for all citizens is considered one of the most precious values for the recovery of the legitimacy lost by broadcasters. Thus, four levels of analysis can be established when developing indicators for the execution of audits: firstly, the inputs given to the institution (national budgets, grants, advertising); secondly, the outputs, through the contractor profile, investment in technology, list of beneficiary companies or outside production; a third level is focused on labor relations and recruitment (lists of jobs, selection processes, trade unions or agreements); and, finally, accountability, either specifying the budgetary execution, indicating the business model or the rights operation process (Table 2).

6.2 *Information and Content Production*

The information production processes control as a guarantee for journalistic honesty is a major element of transparency and good governance in television broadcasters. This precept has a number of clear limitations, as for example the inability to audit the entire content broadcasted by the television services. Be that as it may, what we have to take into account is the framework that is established for the information production itself, and that this framework shall be carried out in an ethical manner and in accordance with legislation. In this regard, besides the problem regarding the news production, credibility is another intangible asset that is below minimum, and it is directly proportional to transparency (Meir and Reimer 2011): the greater the credibility, the greater the transparency; the less the transparency, the less the credibility. In recent years, and exacerbated by the economic crisis, a widening gap between citizens and the media has occurred when it comes to liaising between the political elite, the organizational actors and relevant events (Manfredi Sánchez 2010). In this sense, according to the Report on Journalism from 2016 (Informe de la Profesión Periodística 2016), the television is the medium with a greater lack of rigor and of information quality, where 79.3% of employees themselves shows sensationalism and lack of control in the processes. This is further supported by the credibility that citizens give to information in general, scoring the United States (32%), Spain (34%) and Italy (36%) the worst in a ranking led by Finland (68%), Germany (60%), and Denmark (57%), as established in the annual report on media from the Reuters Institute at the University of Oxford (2015).

When establishing the indicators it is necessary to assess the information production and the control of the journalistic creation processes (not of the content itself). We can distinguish here three levels of information: firstly, the knowledge on the existence, composition and selection criteria of a Drafting Committee or a similar body, as well as the editorial approach, a self-regulatory code of ethics; secondly, the incorporation of the legal and ethical precepts regarding the presence of minority and/or vulnerable groups; and, finally, the universal access to disabled people, especially with hearing or visual impairment (Table 3).

Table 3 Levels of production control

Information	Level
Production	First: institutionalization
	Second: inclusion
	Third: universalization

6.3 Audience Institutionalized Participation

Another important point is the institutionalized participation of the spectators in the television services. The extent of involvement would grant them the status of citizen and not simply audience, and it would be linked to the achievement of rights, the ability to have an impact on the model and the content, and not their function as mere recipients or a transmission belts. Thus, one of the most commonly used rankings (Azurmendi and Muñoz-Saldaña 2016) to standardize this type of participation is the one used by Carpentier (2011), which mainly distinguishes two models: firstly, participation in the media through their participation in its structure, as for example administration, financing or management ("structural-related participation"), and, secondly, participation related to content ("content-related participation"). Examples of structural-related participation occur in the British Audience Councils (Azurmendi 2015) and the content-related participation is much more often regulated through advisory boards, as it may be the case of Spanish public service television broadcasters (Fernández-Alonso and Fernández-Viso 2012).

According to this ranking, the citizens' institutionalized participation is materialized in several fields, where those related to the mechanisms are placed on a first level: effective channels for participation by means of a website adapted to mobile devices or the different social media. On a second level we can find a more formal kind of participation, as for example the right of access to public information and the authorized areas for enquiries and complaints. Finally, the third level is much more related to the claims presented individually or collectively in an institutional, economic or content production field, with quality control mechanisms or program assessments (Table 4).

7 Assessment Models

Regarding the transparency assessment in television broadcasters, the Internet seems to be the appropriate place to publish public information (Fernández Lombao 2015), and to provide the participation tools that can be used by the citizens due to their inclusive, universal and accessible nature. There are a large number of indicator developments that enable comprehensive audits in terms of organizations and institutions in general and media and television broadcasters in particular.

Table 4 Levels of participation

Information	Level
Participation	First: support
	Second: enquiry
	Third: input

A large proportion of the international research on media transparency has focused mainly on governance and economic issues, neglecting the control of the information production processes, that is: the information duality of this type of journalistic companies, as bodies that also work with a public service asset. From a typological point of view, two rankings regarding accountability and transparency in the media have been in use since the beginning of the 21st century: the ones which used to differentiate the traditional analogue mechanisms, such as letters to the editor or ethical codes, from the innovative mechanisms used by applied technologies (Redondo and Campos Domínguez 2016) (Table 5).

UNESCO (2012) has created indicator models (with a total of 19 basic criteria) of media development in order to assess quality in public service broadcasters, being the latter based on a series of criteria regarding "transparency, cultural diversity, geographical coverage and platforms supply, journalism public patterns, financing schemes, audience satisfaction, experimentation and innovation, and technical standards" (Campos Freire 2015a, b). Similarly, the European Broadcasting Union published (2014) a different indicator model (24) for managing its own public audiovisual service (*PSM Values Review The Tool*), based on the principles of diversity, innovation, independence, accountability, excellence and universality. On this basis, the authors Valencia-Bermúdez and Campos Freire (2016) developed value indicators for the regional public service television broadcasters (RTV) in Spain, applying to in EITB, CCMA and CRTVG and grouped in three criteria: "transparency", "audience" and "informing".

In this line of media transparency rates, Redondo and Campos Domínguez (2016) develop a complete, useful and integrated measurement method containing 62 items applied to all of the web media. This method includes five categories: economy, the editorial, production, interactivity and usage. In a more concrete way, López-López et al. (2017a) launched 29 indicators that specifically address the television sector, distinguishing three areas: institutional information, economic field and information production, and being updated to 58 in a recent publication (López-López et al. 2017a, b, c). The areas have the same structure as before, but include a broader field in each of them: institutional information, governance and interest groups; economic, labor and infrastructure field; information production and content access.

8 Conclusion: What Television Broadcasters Think

As it has already been said, plenty of quality, transparency and value models and indicators have been developed in recent years so as to assess the level of openness regarding the television broadcasters or their corporations. This applies to worldwide broadcasters, going from Latin America (Rodríguez and Villanueva 2010) through an integrated model which, generally speaking, audited television broadcasters from Australia (Australian Broadcasting Corporation, ABC); Azerbaijan (Public Television and Radio Broadcasting Company ITV); Canada (Canadian Broadcasting Corporation, CBC); Chile (*Televisión Nacional de Chile*, TVN); Spain (*Radio Tele-*

Table 5 Assessment model comparison * √ = fully met level, P = partially met level, X = non met level

Assessment model	Institution (levels)				Economy (levels)				Information (levels)			Participation (levels)		
	1st	2nd	3rd	4th	1st	2nd	3rd	4th	1st	2nd	3rd	1st	2nd	3rd
UNESCO	P	√	P	X	X	X	X	√	√	√	√	X	√	√
Valencia-Bermúdez and Campos Freire	X	P	P	X	√	P	X	√	X	√	√	X	√	√
López-López, Puentes-Rivera and Rúas-Araújo	√	√	P	X	√	P	√	X	√	√	X	P	P	X
Redondo and Campos-Dominguez	√	√	√	X	√	√	X	P	√	√	X	√	√	√
López-López, López-Golán and Puentes-Rivera	√	√	√	P	√	√	√	P	√	√	√	√	√	√

visión Española, RTVE); the United States (Public Broadcasting Service, PBS), France (*France Télévisions*, FT); the United Kingdom (British Broadcasting Corporation, BBC) or South Africa (South African Broadcasting Corporation, SABC), to other more recent broadcasters at a European and Anglo-Saxon level (Campos Freire 2015a, b). In this sense, and under the latest reports and models published, three of them have been selected to assess the compliance of television broadcasters with the model requirements: the model by Valencia-Bermúdez and Campos Freire; the one by López-López, Puentes-Rivera and Rúas-Araújo, and the one created by López-López, López-Golán and Puentes-Rivera. All of them have been applied on a pilot basis to several public service television broadcasters or bodies, showing different results depending on the tested area: Spain, the Andean Community, Chile and Portugal.

The second model analyzed, presented by Valencia-Bermúdez and Campos-Freire, is based on the concept of synthetic value indicator and it was designed to carry out a multidimensional audit of the broadcasting public services based on the UNESCO and EBU (2015) standard-setting instruments: even though it does not comprehensively address all the public information (as this is not its objective), it outlines a number of issues on accountability and the participation that encourages transformation of audience into citizens (Jakubowicz 2009), as well as the strengthening of the existing mechanisms in order to move towards a participatory democracy. When implementing this in the television corporations from the historical nationalities and regions of Spain (CRTVG, EiTB and CCMA), the conclusions shown by the authors are clear. On one hand, information in the fields of activity, management, corporate social responsibility, financial transparency and best practices fully meets the requirements, largely due to the legal obligations existing at national and regional levels. However, beyond this cooperation, it has not so far been possible to go one step further and promote dialogical accountability with the audience, with a certain feedback and assessing the confidence that is generated.

The third model, presented by López-López, Puentes-Rivera and Rúas-Araújo, tackles transparency in a triple way by means of 29 indicators, although manifesting obvious limitations when it comes to assessing the civic participation and being very general and lenient in some of the indicators. Thus, the first assessment (public service television broadcasters from Chile and Spain) drew the conclusion that those broadcasters had a high degree of compliance (nearly 70%) with the indicators proposed, being the Latin American television broadcaster the one showing greater efficiency and accessibility in the advertising of labor relations or in the audience participation. For its part, the Spanish RTVE's website provides little information in the institutional field and regarding the data related to the production and access to the content. The economic contributions are the most detailed ones on the website, especially on the budgetary implementation and procurement processes, being this information insufficient and opaque in the case of the Chilean television broadcaster. This model was also applied to the television broadcasters from Ecuador, Peru, Bolivia and Colombia (López-López et al. 2017a, b, c), where this model, besides being characterized by the political unity and its territorial proximity, it is also known in recent years by having promoted laws concerning transparency and access to public information. Nevertheless, its results are mixed: the Peruvian National Institute of Radio and Television yields better results, followed by Television and Radio of Ecuador, Bolivia TV and National Radio and Television of Colombia. In the field of economic transparency, the Bolivian and Peruvian entities must make efforts in this regard (the Bolivian case is very poor). Colombia shows a clear deficit in information production and content access, being the Ecuadorian broadcaster the one with better accountability from an economic perspective, and the one with worse institutional information.

The fifth model, recently created by López-López, López-Golán and Puentes-Rivera, corrects the deficiencies from the previous one through the incorporation of the governance model, the interest groups and the infrastructure to the indicators. This enables greater integration and a more holistic audit, considering transparency in the broadest sense. The pilot implementation of this model has been carried out once again at the Spanish RTVE, obtaining a significantly lower score than before, and at the Portuguese broadcaster. Thus, neither TVE nor RTP meet the requirements when applying much more defined and demanding criteria, obtaining the best results in information production process and content access regarding economic, employment and infrastructures indicators. On the negative side there is the institutionalized participation regarding the audience and interest groups. Furthermore, when it comes to accessing and reading, both formats lack regulated and easy procedures to be able to access information and code it in a timely manner.

In short, when assessing transparency in public service television broadcasters, integrated assessment models need to be considered in order to holistically audit the dual information generated by the broadcasters, being highly demanding in terms of the degree of compliance. A sole conclusion can be drawn taking into consideration the clear differences among geographical areas:

a. The evolution of information and knowledge technologies enable accountability in public service television broadcasters.
b. The requirements coming from society are far ahead of the legal requirements.
c. From a legal perspective, the requirements are aimed at making the economic and institutional management more transparent, neglecting the participation institutionalization processes and the production control of quality information.
d. Public service television broadcasters, through their entities or corporations, give a delayed response to the requirements by the society and the law.
e. From the model point of view, the self-financing provides independence and greater rigor in accountability processes.
f. The independence of television broadcasters from the public authorities is a necessary condition, but not enough, for the improvement of transparency processes.

References

Asociación de la Prensa de Madrid: Informe de la Profesión Periodística 2016. In: http://www.apmadrid.es/publicaciones/informe-anual-de-la-profesion/ (2016)

Arenas, P.: El futuro de la televisión pública educativa y cultural. El caso de Señal Colombia. Boletín Cultural y Bibliográfico **XLIX**(87), 41–52 (2015)

Azurmendi, A.: Un espacio público por conquistar. La participación institucionalizada de las audiencias en las televisiones de proximidad españolas y europeas como indicador de transparencia para la gobernanza de los medios de comunicación. Derecom **19**, 1–32 (2015)

Azurmendi, A., Muñoz-Saldaña, M.: Participación del público en televisiones públicas autonómicas: una propuesta a partir de la reforma 2016 de la BBC. El profesional de la información **5**, 803–881 (2016)

Becerra, M., García Castillejo, A., Santamaría, O., Arroyo, L.: Cajas Mágicas. El renacimiento de la televisión pública en América Latina. Tecnos, Madrid (2013)

Campos Freire, F.: Financiación e indicadores de gobernanza de la radiotelevisión pública en Europa. In: Marzal, J.J., Izquierdo, J., Casero-Ripollés, A. (eds.), La crisis de la televisión pública: el caso de RTVV y los retos de una nueva gobernanza, pp. 189–216. Universitat Autònoma de Barcelona, Servei de Publicacions, Barcelona (2015a)

Campos Freire, F.: Reformar la Gobernanza de la radiotelevisión pública. In Congreso CICOM Televisiones Autonómicas 2015 FCom. In: https://congresocicom2015tvautonomicas.files.wordpress.com/2016/10/francisco-campos.pdf (2015b)

Campos Freire, F., López Cepeda, A.M.: La nueva gobernanza y la televisión pública en España. Revista de Economía Política das Tecnologías da Informaçao e da Comunicaçao **XIII**(1), 66–85 (2011)

Carpentier, N.: Media and Participation: A Site of Ideological-Democratic Struggle. Intellect Books, Bristol (2011)

Comisión Interamericana de Derechos Humanos: Estudio especial sobre el derecho de acceso a la información. In: http://www.cidh.oas.org/relatoria/section/Estudio%20Especial%20sobre%20el%20derecho%20dev%20Acceso%20a%20la%20Informacion.pdf (2007)

Coppens, T., Saeys, F.: Enforcing performance: new approaches to govern public service broadcasting. Media Cult. Soc. **28**(2), 261–284 (2006)

De Bustos, J.C.: La gobernanza en las televisiones públicas autonómicas españolas. In Congreso CICOM Televisiones Autonómicas 2015 FCom. In: https://congresocicom2015tvautonomicas.files.wordpress.com/2016/07/miguel-de-bustos1.pdf (2015)

De Moragas, M., Prado, E.: La televisión pública en la era digital. Centre d'Investigació de la Comunicació, Barcelona (2010)

EBU: Connecting to a networked society. Continous improvement of trust and return-on-society. The Vision2020 Project Team: Ruurd Bierman, Francesca Cimino, Nicola Frank, Nicoletta Iacobacci, Andra Leurdijk, Mike Mullane, Sue Neilen, Michelle Roverelli, Roberto Suárez Candel, Bram Tullemans, Michael Wagner. In: https://www.ebu.ch/files/live/sites/ebu/files/Publications/EBU-Vision2020-Networked-Society_EN.pdf (2015)

European Broadcasting Union: PSM Values Review: The Tool. EBU, Ginebra (2014)

Fernández-Alonso, M., Fernández-Viso, A.: Internal pluralism in the governance of public service broadcasters in Spain and the role of social groups and professionals: the case of RTVE. Commun. Soc. **25**(2), 203–230 (2012)

Fernández Lombao, T.: La página web como espacio de transparencia de las radiotelevisiones públicas. In: Mateos, C., Herreros, J. (eds.) Cuadernos Artesanos de Comunicación. Ediciones Universidad de La Laguna: Tenerife. In: http://www.revistalatinacs.org/15SLCS/2015_libro/026_Fernandez.pdf (2015)

Figueras Maz, M., Narberhaus Martínez, M., Vegas, X.: Transparencia y rendimiento de cuentas en la información periodística. In: https://www.researchgate.net/profile/Marta_Narberhaus/publication/262583785_Transparencia_y_rendimiento_de_cuentas_en_la_informacion_periodistica/links/55bfe2ab08ae9289a09b6401.pdf?origin=publication_list (2013)

Hallin, D., Mancine, P.: Sistemas mediáticos comparados. Hacer Editorial, Barcelona (2008)
Heikkilä, H.H., Domingo, D., Pies, J., Glowacki, M., Kus, M., Baisnée, O.: Media Accountability Goes Online. A Transnational Study no Emerging Practices and Innovations (Media ACT Working paper). Journalism Research and Development Centre and Universidad de Tampere, Tampere (2012)
Iosifidis, P.: Retos y estrategias. Servicio público de televisión en Europa. Infoamérica. In: http://www.infoamerica.org/icr/n03_04/iosifidis.pdf (2010)
Jakubowicz, K.: Public service broadcasting in the 21st century: what chance for a new beginning? In: Lowe, G.F., Bardoel, J. (eds.) From Public Service Broadcasting to Public Service Media, pp. 29–49. Nordicom Göteborgs Universitet, Göteborg (2009)
Ley Orgánica de Transparencia y Acceso a la Información Pública: Ley 24 Registro Oficial Suplemento 337 de 18 de mayo del 2004 de la Asamblea Nacional del Ecuador. In: http://www.seguridad.gob.ec/wpcontent/uploads/downloads/2015/04/ley_organica_de_transparencia_y_acceso_a_la_informacion_publica.pdf (2004)
Ley de Acceso a la Información Pública: Tomo 391 de la Asamblea Legislativa de El Salvador, publicada el 30 de marzo del 2011. In: http://www.fiscalia.gob.sv/wp-content/uploads/portal-transparencia/Ley-de-Acceso-a-la-Informacion-Publica.pdf (2011)
López Cepeda, A., Manfredi, J., et al.: Análisis de la transparencia de las páginas web de los principales medios de comunicación audiovisuales en España. Trípodos **32**, 45–61 (2013)
López-Golán, M., Ulloa-Erazo, N.: Modelos de financiación de los canales de televisión pública en Latinoamérica. In: Mateos, C., Herreros J. (eds.), Cuadernos Artesanos de Comunicación. Ediciones Universidad de La Laguna, Tenerife. http://www.revistalatinacs.org/16SLCS/2017_libro/042_Lopez.pdf (2016)
López-López, P.C., Puentes-Rivera, I., Rúas-Araújo, J.: Transparencia en televisiones públicas: desarrollo de indicadores y análisis de los casos de España y Chile. Revista Latina de Comunicación Social **72**, 253–272 (2017a)
López-López, P.C., López-Golán, M., Puentes-Rivera, I.: Hipertransparencia y nuevas tecnologías: análisis de la información pública en TVE y RPT. Springer, Cham (2017b)
López-López, P.C., Ulloa-Erazo, N., Puentes-Rivera, I.: Transparency and new technologies: accountability of public television broadcasters in the Andean countries. In: Rocha Á., Correia A., Adeli H., Reis L., Costanzo, S. (eds.), Recent Advances in Information Systems and Technologies. WorldCIST 2017. Advances in Intelligent Systems and Computing, vol. 571, pp. 73–82. Springer, Cham (2017c)
Manfredi Sánchez, J.L.: Periodismo y Transparencia informativa. In: http://www.apmadrid.es/wp-content/uploads/2010/04/doc_vapm20100422175405.pdf (2010)
Meier, K., Reimer, J.: Transparanz im journalismus. Instrumente, konfliktpotentiale. Wirkung. Publizistik **56**, 133–155 (2011)
Muñoz, S.M., Azurmendi, A.: El papel de las televisiones públicas autonómicas en el desarrollo de la gobernanza multinivel en Europa. Commun. Soc. **29**(4), 45–58 (2016). https://doi.org/10.15581/003.29.4
Pérez Bravo, M.J.: Discusión parlamentaria en los temas de transparencia en el marco de las reformas del Estado: Análisis del acuerdo parlamentario 2003. Tesis para optar al grado de Magíster en Gestión y Políticas Públicas. Universidad de Chile, Santiago de Chile (2004)
Puppis, M.: Media governance: a new concept for the analysis of media policy and regulation. Commun. Cult. Critique **3**, 134–149 (2010)
Redondo García, M., Campos-Domínguez, E.: La transparencia mediática como mecanismo de autorregulación: análisis de su presencia en las webs de los principales medios españoles. Ámbitos. Revista Internacional de Comunicación **32**, 1–19 (2016)
Rodríguez, N., Villanueva, E.: Medios de Servicio Público y transparencia. Quipus, Quito (2010)
Salvador Martínez, M.: La Televisión Pública en la Unión Europea. Revista Española de Derecho Constitucional **48**, 315–319 (2006)
Suárez-Villegas, J.C., Rodríguez-Martínez, R., Mauri-Ríos, M., López-Meri, A.: Accountability y culturas periodísticas en España. Impacto y propuesta de buenas prácticas en los medios de

comunicación españoles (MediaACES). Revista Latina de Comunicación Social **72**, 32–330, https://doi.org/10.4185/rlcs-2017-1167 (2017)
Sturm, A., Nalvarte, P.: Medios Públicos de América Latina buscan modelos exitosos de financiación. Journalism in the Americas. The University of Texas at Austin. In: https://knightcenter.utexas.edu/es/blog/00-17441-medios-publicos-de-america-latina-buscan-modelos-exitosos-de-financiacion-segundo-arti?utm_source=feedburner&utm_medium=feed&utm_campaign=Feed%3A+kcbloges+%28Periodismo+en+las+Am%C3%A9ricas%29 (2016)
Túñez, M.: La gestión de la comunicación en las organizaciones. Comunicación Social Ediciones y Publicaciones, Zamora (2012)
UNESCO: La Radio y Televisión Pública ¿Por qué? ¿Cómo?. In: http://unesdoc.unesco.org/images/0012/001240/124058so.pdf (2001)
UNESCO: Quality Indicators for Public Broadcasters—Contemporary Evaluation. UNESCO, Brasilia (2012)
Valencia-Bermúdez, A., Campos Freire, F.: Value indicators for regional broadcasters: accountability in EITB, CCMA and CRTVG. Commun. Soc. **29**(4), 59–68 (2016)
Waisbord, S.: Who speaks for public media in Latin America? The World Bank. In: http://blogs.worldbank.org/latinamerica/who-speaks-public-media-latin-america (2014)

Paulo Carlos López-López Doctor in Communication and professor-researcher at Pontificia Universidad Católica del Ecuador Sede Ibarra. He has a Diploma of Advanced Studies from the Doctoral Program of Communication and Journalism. BA in Journalism and Political Sciences and Administration. Specialist in Political Communication and Transparency. METACOM leader. Ibarra, Ecuador.

Mónica López-Golán Ph.D. in Communication from the Universidade de Santiago de Compostela (2014). Current Director of the School of Social Communication from the Pontificia Universidad Católica (Ibarra Campus). Chair Professor at the audiovisual area from the same University. Former television producer at Gestmusic Endemol S.A.U. in programmes like *Tardes con Ana*. Ibarra, Ecuador.

Nancy Graciela Ulloa-Erazo Ph.D. in Communication from the Universidade de Santiago de Compostela, he works as a university professor. She is an institutional communicator at the Pontificia Universidad Católica del Ecuador Ibarra Headquarters. She has formed several scientific committees and organizing committees for national and international research events. Researcher in media, organizational communication and ICT. Ibarra, Ecuador.

Programming Strategies in European Public Service Broadcasters

Sabela Direito-Rebollal, Diana Lago-Vázquez and Ana-Isabel Rodríguez-Vázquez

Abstract In the present context, in which public service media (PSM) are facing constant criticism due to their lack of neutrality, European broadcasters should reformulate their mission, giving value to their remit through independence, diversity, excellence, transparency and innovation. In this regard, and taking into account that one of the particularities of PSM is based on offering a programming addressed to all audiences, the challenge focuses on finding an offer that meets the formation, education and entertainment needs of the entire society. The control over the genre info-show and the encouragement of national fiction define the main trend of a diverse European public service television, not only in funding systems and audiences, but also in contents with which these broadcasters try to maintain their competitiveness within the new convergent and digital scenario.

Keywords Programming · Television genres · Audiences · Public service media Public television

1 Introduction

The digital and networked society (Castells 2001) has provided a reconfiguration of the media ecosystem in general and public service media, in particular. The transition from a model based on unidirectional transmission of contents to a multimedia and polymedia environment (Bardoel and Lowe 2007), in which products flow through multiple media channels and where the participative culture of spectators opens up

S. Direito-Rebollal (✉) · D. Lago-Vázquez · A.-I. Rodríguez-Vázquez
Universidade de Santiago de Compostela, A Coruña, Spain
e-mail: sabela.direito@usc.es

D. Lago-Vázquez
e-mail: diana.lago@rai.usc.es

A.-I. Rodríguez-Vázquez
e-mail: anaisabel.rodriguez.vazquez@usc.es

M. Túñez-López et al. (eds.), *Communication: Innovation & Quality*, Studies in Systems, Decision and Control 154, https://doi.org/10.1007/978-3-319-91860-0_4

new possibilities of interaction between media and consumers (Jenkins 2006), defines a complex restructuring process to which public service media should adapt.

The change in the nature of the concept itself and the transition from public service broadcasting (PSB) to public service media (PSM) reflect not only the openness to the ensemble of media organizations, but also the transformations that public corporations are facing as regards audiences, programming offer and distribution strategies (Crusafón 2017).

In this regard, PSM organizations are facing a change in consumption habits of spectators, who abandon their passive role as couch potatoes to become active and multi-tasking users. The massive audience characterising the beginning live together with the fragmented, segmented, micro, individual and connected audience (Prado 2015), so PSM should reorient their programming with the aim of targeting both generalist audiences and niche markets.

Also, within a context of increasing competition, in which the number of digital channels and online services has multiplied and the consumption is increasingly made through mobile and connected devices (EBU 2012), public service media begin to be on complementing their programming offer with other services that enhance the TV experience of users, thereby encouraging their engagement. The design of inclusive strategies and multiplatform distribution as ways to increase access possibilities, and the connection between spectators and the public corporation of the media (Franquet 2017) are approached as one of the challenges that would condition the future development of PSM.

In this way, if each decade calls for a reorientation of PSM to ensure its survival (Enli 2008), in the current convergent and digital scenario, these changes are to adapt both offered services and contents and their management, production and distribution mechanisms to the possibilities enabled by new information and communication technologies (Suárez-Candel 2012).

2 Redefining Public Broadcasting in the Digital Age

The transition from the analogical television to the digital television and the convergence between networks, terminals, media and devices (Syvertsen 2006; Álvarez Monzoncillo 2011) have led to major changes in the television paradigm, resulting in new predictions about its extinction caused by the Internet. However, if the historical transformations occurred with the arrival of the cinema and the radio are analysed, it is observed how up to now any communication technology has not completely removed previous ones, but, instead, there was a redefinition of their characteristic functions (Prado 2010).

In this regard, the adaptation of television to technological innovations is defined in terms of hybridization, insofar as television establishes a synergy with the Internet, mobiles devices and social networks (González-Neira and Quintas-Froufe 2016). From a first lean period of available contents, when television was the only receiving terminal, we have gone to a phase in which supply is increased and access to contents

Fig. 1 Average TV daily consumption time for European citizens (hours and minutes). *Source* Prepared by authors with EBU data (2016)

Fig. 2 Average TV daily consumption time for young Europeans (hours and minutes). *Source* Prepared by authors with EBU data (2016)

is produced through various devices—mobile phones, tablets, laptops, among others—that allow to separate television consumption from time and place constraints (Prado 2015).

In this new period of "everything digital and networked" (Prado 2015), the TV viewing remains stable, with small annual variations that do not involve decreases over three minutes (Fig. 1). The trend is practically widespread in all EU countries, with the exception of the United Kingdom, Denmark and Ireland, which fell by 26, 23 and 11 min respectively, over a five-year period (2011–2015). The remaining countries present sustained increases and small declines. That decline is more pronounced among young people, who consumed an average of 2 h and 6 min per day, in 2015, 17 min less, compared with 2010 (Fig. 2). In fact, these new generations—to whom Howe and Strauss (2000) named millennials—are who embrace faster alterations, changing their consumption patterns towards those established by the OTT and the VoD consumption. In overall terms, 76% of young Europeans watch television and 43% connect with public channels, although their penetration is much lower than in other age groups (EBU 2016).

The search for connection with these new audiences arises as a fundamental issue for PSM, especially within a context where both their habits of consumption and their approach to television varied towards a more individual and personal use of the media (Suárez-Candel 2007). From linear and producer-controlled flow, in which broadcasters controlled broadcasting times in programming and viewing experiences of viewers, we have gone to a circular and user-generated flow, where users decide the devices and platforms in which they consume audiovisual products, as well as the link they establish with them (Quintas-Froufe and González-Neira 2016; Marinelli and Andó 2017). The active participation of spectators-Internet surfers (Izquierdo 2012) in the creation and distribution of contents through the web characterizes their media experience, which is modified as the audience become involved in production and flow process, which were, up to now, reserved for the media. The traditional unidirectional flow now becomes bidirectional and the roles of senders and receivers are constantly exchanged between the media and the prosumer audience (Toffler 1980; Islas 2008), who begins to interact and dialogue with the public media corporation.

In this regard, and stressing the words by Franquet (2017) when she quotes Harrison and Wessels (2005), "the new forms of online participation can be understood as a continuation of public service values in the digital platform" (p. 189). Therefore, PSM should adapt themselves to the needs of their audiences, redefining their functions and goals as well as the contents and services they offer (Suárez-Candel 2007).

Competing against private competence and maintaining the competitiveness and identity in the new digital context are two of the keys that will guarantee the continuity of PSM within a convulsive environment where their role as guarantor of diverse and plural systems is not only contested, but also their existence. To the debate on the lack of independence and credibility of PSM due to their proximity to the State (Steemers 2017), criticism on their lack of public service vocation, the absence of a sound funding model and the commitment to a commercial and homogenous programming—where audience data are prioritized over the quality of offered contents—is also on the table (Guerrero 2007; León 2007). The progressive budgetary cuts, accentuated by the economic crisis, and the consequent loss of quality of programmed products, and the deregulation of the commercial sector and the direct influence of governments, only aggravate a situation in which PSM fight for their survival (Barnett 2006).

Faced with the erosion of their public service remit, European televisions must be committed to independence, diversity, excellence, transparency and innovation, as basic elements (EBU 2013). However, in that redefinition of public ownership media, it is not enough to reinforce this ideology; rather, it is necessary for PSM to add to their historical functions—as guarantors of pluralism and democracy, identity and diversity, quality in programming and cultural, political and social needs of citizens (Moragas and Prado 2000; Iosifidis 2010)—those derived from the technological transformations, globalisation and the development of the information society (Moragas and Prado 2001).

In the era of digitization and media convergence, the mission of public broadcasters is materialised in a series of challenges that means being the driving force at the confluence with other sectors, counteracting the concentration process and the weight acquired by oligopolies (Prado and Fernández 2006; Manfredi 2006a). Fleeing from the strictly economist view in order to produce quality contents (Manfredi 2006a) and of interest for majorities and minorities (Prado and Fernández 2006), is essential within an online environment in which public service media have to guarantee universal access to audiovisual products and act as mediators and guide in the new media (Manfredi 2006b).

3 The Value of DTT Within the Multichannel Scenario

In the 1970s and 1980s, European public televisions were confronted with deregulation and liberalization caused by the introduction of private operators in the audiovisual market, but the current structural changes faced by PSM in the digital age are no less important.

The analogue switch-off and the migration towards digital transmissions marked the beginning of a complex process to which European public service media have been gradually adding to, since the end of the last century. The United Kingdom (1998), Sweden (1999) and Spain (2000) were the driver forces behind the launch of the Digital Terrestrial Television—although with failed attempts that caused a freeze on the digital transition and a delay of the analogue blackout until 2010 in the Spanish case and 2011 in the United Kingdom. This way, Europe was inserted in the process of digitization previously implemented in countries such as the United States and Japan.

However, in spite of the promises with which the arrival of the DTT was sponsored, the reality reflects that it was finally consolidated as “another modality of public service provision through a new digital technology” (Fuertes and Marenghi 2012, p. 98). In this regard, terrestrial television offered new possibilities for the media to design programming in different channels and with cultural and social interactive services (Miguel de Bustos and Garitaonandía 2007), that contributed to enrich the TV experience of the audience. But this enrichment was not materialised beyond a multi-channel and multi-track sound, a panoramic format and a higher quality in the signal emitted (García-Leiva 2011).

The commitment to the implementation of technology and the attention to new platforms rather than the content offered (Román 2012), reflect the trend of broadcasters in a new multichannel scenario where digital terrestrial television remains the main distribution and reception platform in European households (32%), followed by cable (30%), satellite (23%) and IPTV (15%) (OBS 2015).

4 Programming Strategies in European PSM: BBC, RTVE and RAI

The programming offer of European public broadcasters is strongly conditioned by digital transformations, an increase in private competition and the recent emergence of streaming platforms, audience segmentation and the socio-economic context of each country. The data presented in this chapter have been analysed in full knowledge that many of programming decisions of channels have and economic and contextual background outside the strictly programmatic scope.

In order to carry out a study as precise as possible, there was a five-month analysis period in which the content of the main generalist channels from the most relevant public broadcasters was observed: RTVE, BBC and RAI. The programming grids

Table 1 Analysed channels and time periods

Channel	Country	November 16 (Week)	December 16 (Week)	January 17 (Week)	February 17 (Week)	March 17 (Week)
La 1 (RTVE)	Spain	21–27	12–18	16–22	13–19	20–26
BBC One (BBC)	UK	21–27	12–18	16–22	13–19	20–26
RAI Uno (RAI)	Italy	21–27	12–18	16–22	13–19	20–26

Source Prepared by authors

for a full week have been extracted—from November 2016 to March 2017—and the contents have been classified according to the time slot, the number of emissions, the duration in minutes and the genre to which those contents belong (Table 1). The classification made by Prado and Delgado (2010) has been taken as a reference, including slight modifications, such as the removal of the young macro-genre and the incorporation of some genres—as lottery programs and talent shows—that were not specified in the original classification.

The breakdown of the programmatic offering of these three channels allows a sketch of the current strategies used by European PSM: what kind of contents are still at the top of the TV guide, what emerging genres are carving themselves a niche and what of them have become weaknesses. Three different strategies are identified in the three channels analysed, with some common factors that highlight current trends in programming, but also with many other differences.

4.1 Shared Trends: Information, Fiction and Info-Show

A first view at the results of the analysis confirms that macro-genres of information and fiction remain the most powerful bets of the PSM's offer, followed by a more recent genre called info-show (Figs. 3 and 4). In minutes, information account for 49% of the BBC's guide, followed by the info-show (24%) and fiction (11%). In the case of the Italian Rai Uno, information is also the dominant genre of its offer (35%), followed by the info-show (32%) and fiction (9%). In contrast, fiction dominates the programming in RTVE (52%), followed by information (40%) and info-show (5%).

The most forgotten genres are children, educational and religious contents, only present on the BBC and the RAI with the retransmission of the Sunday masses and a religious magazine before or after the sermon: *A sua immagine*—RAI—and *Songs of Praise* and *The Big Questions*—BBC.

The multiple transformations of the media have led to the hybridization of genres and the appearance of contents that mixes interviews, debates, reports, games,

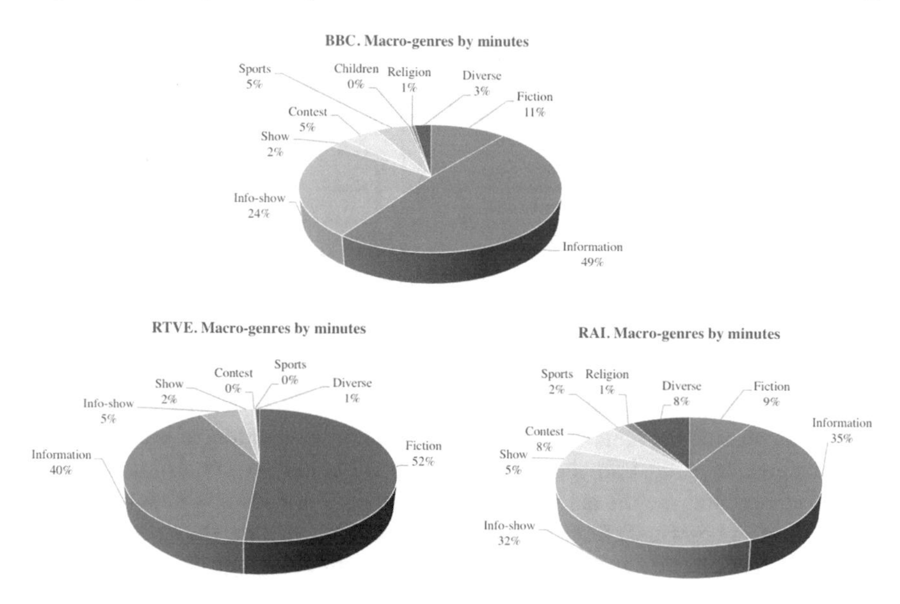

Fig. 3 Macro-genres offer by minutes. *Source* Prepared by authors

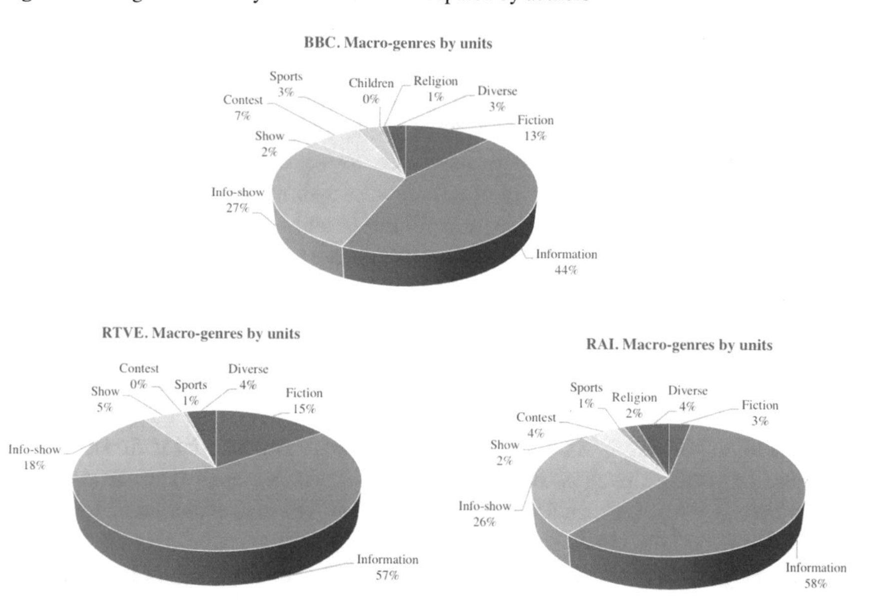

Fig. 4 Macro-genres offer by units. *Source* Prepared by authors

varieties and fiction. These programs within the macro-genre info-show are having a great presence in the programming of the public channels. Both in La 1 from RTVE, the BBC One and the RAI Uno, the info-show appears among the three positions of more offered genres. By units, i.e. the number of emissions, the BBC would be the

Table 2 Info-show offer in BBC One

BBC info-show	Offer (%)	Offer (Minutes)
Docuseries	11	1275
Reality game	9	1065
Reality show	69	8460
Talent show	1	175
Talk show	10	1210

Source Prepared by authors

channel that better includes the info-show, but by minutes of emission, the Italian channel would take the position. La 1 from RTVE and RAI Uno present a similar scheme as regards the number of emissions, being three-quarters of their programming dominated by the info-show, although they differ in the volume of fiction offer.

The start genres are the magazine in all its versions and the talk-show. The magazine, whether news and current affairs or thematic, has become the trend of present television. Its main advantage is to contain different formats within the same program, which offers multiple possibilities. The most widespread version is the morning magazine, a genre usually of long duration broadcast in the morning and with information and entertainment contents. *Las mañanas de La 1* from RTVE is a clear example of this kind of programs. *Tempo & Denaro* from the RAI and *The Andrew Marr Show* from BBC may be also categorized into this genre.

On the other hand, talk-shows are being done with the guides of the European public television. Here we again differentiate between the most informative programs, which prioritize the content of conversations about entertaining and the most relaxed talk-shows with a higher degree of humour and whose topics revolve around the world of entertainment. One of the flagship programs of the BBC One included in this category is *The Graham Norton Show*, where the colloquium is mainly led by actors and musicians. In RTVE, *Los desayunos de TVE* is a more informative talk-show broadcast every morning and mainly oriented to political issues. In the same vein, the morning slot of the Italian channel combines multiple informative sections like newscasts of short duration with the current affairs talk-show *Unomattina*.

The Italian television is the only one of the analysed ones that includes a genre in disuse in public channels: the court show (*Torto o ragione*? *Il verdetto finale*). This format is based on recreating trials with a charging part, a defendant part and a pseudo-judge and the aim is to resolve little lawsuits.

In the English case, the reality show dominates the info-show offer (Table 2). Contents such as *The Apprentice* (version of the American reality), *Escape to the Country* and *Flog it!* are frequent in its programming, encouraging a television oriented to TV reality as the main offer of entertainment. However, Spanish and Italian reality shows are broadcast in prime time, while the BBC positions this type of formats in other slots of lower concentration of audience—usually morning hours—establishing information and fiction as the cornerstones of its programming.

4.2 *Fiction: Strengths and Weaknesses of a Key Genre*

The relevance of fiction in terms of audience and brand image is indisputable, so it is not uncommon for it to play a fundamental role in the programming offer of RTVE, RAI and the BBC.

In this regard, the Spanish corporation stands out, as it has a wide offer of fiction series broadcast in the afternoon and in prime time. The series *Acacias 38* and *Seis Hermanas* have become an essential element in the programming of La 1, after several failed attempts to remain competitive in that slot with other genres such as contests. *Olmos y Robles* and *Víctor Ros* are some of the strong series in prime time, broadcast during November and cancelled at the beginning of the following year. Despite the great effort made by the public broadcaster to enhance its fiction, audience data are not favourable, resulting this to be one of its greatest weaknesses, especially in terms of unprofitable investment and loss of credibility and fidelity of an already damaged audience.

In the British case, the emphasis is on national versus foreign production (Fig. 5), prioritizing the latter the Australian production over any other market. *The Doctor Blake Mysteries* and *The Conde* are some examples of Australian series broadcast after-dinner.

In contrast to the Spanish case, the BBC has managed to crown its fiction as a greater representative of its personal brand. Divided into serials and afternoon soap operas (*EastEnders, Doctors*) and prime time series (*Casualty, My Mother and Other Strangers*), the British channel manages to offer fiction content that convinces the audience. The sitcom is also common (*Peter Kay's Car Share, Uncle, Not Going Out*), a short-time fiction genre mainly based on humour that is not so widespread in Spain and Italy.

On the other hand, the RAI Uno also prioritizes national production, but its fiction offer is limited. The series of continuity are few in comparison with the rest of the analysed public broadcasters and include more foreign production. Among these, three series stand out: *Che Dio ci aiuti*—comedy about a convent whose third season is broadcast in prime time, *Sorelle*—family drama also released in prime time—and

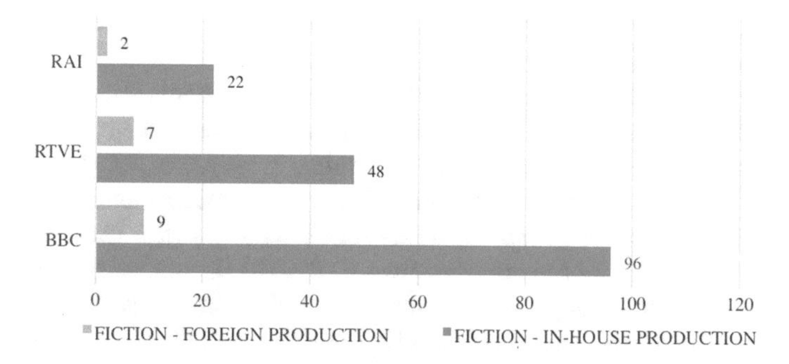

Fig. 5 Offer of in-house and foreign fiction. *Source* Prepared by authors

Un passo dal cielo—about a forest commander who collaborates with the Neapolitan police. This series of Italian production was dubbed by RTVE and broadcast in the second channel of the Spanish corporation, La 2. It is one of the most successful Italian series at the international level.

4.3 Adaptations of Spanish Fiction in the Italian Market

The data extracted from the guide of the RAI Uno show the relevance of the Spanish fiction market in the Italian offer, since all foreign series broadcast during the analysed periods were adaptations of Spanish series. In particular, we are talking about *Pulseras Rojas* (*Braccialetti Rossi*) and *Médico de Familia* (*Medici in Famiglia*), two series of continuity that were finalised in Spain, but that have counted on a long international trajectory—adaptations and dubbed emissions—in other countries. Although their emission in the sample months constituted repositions in late night slots, both are key elements of RAI Uno. In the case of *Medici in Famiglia*, it is one of the more long-life series of the channel still broadcast (11 seasons so far) and *Braccialetti Rossi* had one of the most successful premieres as regards audience—with three broadcast seasons.

4.4 Documentary, BBC's Personal Brand

One of the strengths of the BBC, which is also a weakness for RTVE, is the documentary genre. *America's Hate Preachers, Getting High God?, Six Wives with Lucy Worsley, The Scottish Bounty Hunter* are some of the documentary broadcast between November 2016 and March 2017 as part of a varied offering that gives a high strategic value to the processing of in-depth information. They are broadcasted regularly in different slots—although it prioritizes the late night—and in two types of formats: unique emission and documentary series. Most of them are national productions. The success of these programs has made them a key genre for the BBC, equating it with fiction and enhancing its public service status by offering a wide range of information programs (Table 3).

In the case of the Italian broadcaster, the documentary offer is also significant, but it is more set aside to slots of low audience figures. Although the case of the BBC shows the possibilities of documentary in the public broadcaster, RTVE hardly emits this type of contents. In the analysed period, only two documentaries were identified: *Belle Époque* and *Ochéntame otra vez*—documentary series. The programming strategy of the Spanish corporation involves turning its documentary offer into the second channel of the group (La 2), weakening the diversity of its offer in the main channel, compared to the other analysed European broadcasters.

Table 3 Information offer in BBC One

BBC information	Offer (number of broadcasts)	Offer (minutes)
Newscast	282	20809
Specialised newscast (Weather)	60	255
Current affairs	10	377
Debate	2	120
Interview	2	120
Report	2	120
Documentary	14	775
Magazine	12	690
Mini magazine	1	30
News magazine	22	972

Source Prepared by authors

4.5 *Sports and Kids' Genres: Great Absences*

Albeit to a different extent, the three channels under analysis have a scarce offer of sports and children contents. First, the continued decline in sports broadcasts in programming of the public broadcaster is directly related to the increase of competence in the acquisition of emission rights and economic difficulties to cope with the coverage of events such as football matches, Formula 1 events, etc. Only the Italian channel, RAI Uno, stands out for the live broadcast of the Formula 1 World Championship. The BBC focuses on football (Champions) and rugby and sports news (*Final Score, Match of the day*). RTVE has limited its emission of sports content to the Club World Cup in its main channel, whose emission rights acquired in November 2016 after paying about 1 million euros, according to the piece of news published in November 2016 by *El Mundo*, a Spanish newspaper.

On the other hand, children content has had less presence in generalist public broadcasters in recent years, until they become residual programs on the grids. In fact, only the BBC has this genre included in its programming offer and uses it sporadically, without having regular broadcast programs. In all three cases, this kind of content has been isolated to thematic channels and excluded of the generalist ones as they are not as competitive as the private offer.

5 Different Programming Strategies in Pursuit of a Quality Public Television

Technological innovations have revolutionized television, which is now facing greater competition as a result of deregulation and has also led to a major change in programming strategies of channels. Currently, programming of public televisions tends to the homogenization of contents and the increase of competence has boosted the worsening of the programmatic offer with greater percentage of mixed genres and entertainment to lower costs and attract audience. The strategies of La 1 from RTVE, BBC One and RAI Uno are similar, since all of them prioritize information genres, fiction and the info-show. However, there are particularities that make the difference in their image as public television.

The BBC is characterized by offering a greater diversity of content and a great effort to have a combined programming with in-depth information programs, but also with a high level of entertainment and a very powerful national production. Its numerous reality shows do not prevent the documentary and other informative genres have their place and get good audience figures.

The Italian RAI Uno bets on the information genre, although talk shows and magazines are the ones that have greater presence. Although its fictional offer is limited, its fiction is durable, powerful and indispensable to maintain the current status of the channel.

Finally, the Spanish broadcaster takes the worst part. Marked by the continuous controversies of last years and serious funding problems, RTVE also suffers the wear and tear of its programming. At the same time as the budget went down, so did the credibility in the eyes of the audience, which became evident, above all, in the fiction genre. The series that were audience and emblems of the channel, are now suffering failures and premature cancellations. Even to maintain the most successful formats, such as the series *El Ministerio del Tiempo*, has had to make use of coproduction—in this case with the famous streaming platform Netflix.

Although the television landscape presents more challenges than ever before, especially for public television, it is clear that there are formulas that continue to work and that guaranteeing quality, rich and diverse programming is a good point to compete with the private offer.

Acknowledgements This chapter is produced within the framework of the project *Indicators related to broadcasters' governance, funding, accountability, innovation, quality and public service applicable to Spain in the digital context* (Reference CSO2015-66543-P) and *Uses and informative preferences in the new media map in Spain: journalism models for mobile devices* (Reference CSO2015-64662-C4-4-R), supported by the Ministry of Economy and Finance from the Spanish Government and co-funded by the European Regional Development Fund (ERDF) from the European Union. Also, it belongs to the activities developed by the International Research Network on Communication Management (XESCOM) supported by the Ministry of Culture and Education of the Xunta de Galicia (reference ED341D R2016/019). The author Sabela Direito-Rebollal is beneficiary of the program for the training of university professors (FPU-15/02557) funded by the Ministry of Education, Culture and Sports from the Spanish Government.

References

Álvarez Monzoncillo, J.M.: Los grandes cambios: ocio audiovisual multiplataforma. In: Álvarez Monzoncillo, J.M. (ed.) La televisión etiquetada: nuevas audiencias, nuevos negocios, pp. 62–81. Ariel, Madrid (2011)

Bardoel, J., Lowe, G.F.: From public service broadcasting to public service media. In: Lowe, G.F., Bardoel, J. (eds.) From Public Service Broadcasting to Public Service Media, pp. 9–28. Nordicom, Göteborg (2007)

Barnett, S.: Public Service Broadcasting: a manifesto for survival in the multimedia age (a case study of the BBC's new charter). Paper delivered to the RIPE conference in Amsterdam. In: goo.gl/keBDhw (2006)

Castells, M.: La galaxia Internet. Plaza & Janés, Barcelona (2001)

Crusafón, C.: El espacio europeo de radiodifusión pública: definición, características y retos de futuro. In: Marzal, J., López, P., Izquierdo, J. (eds.) Los medios de comunicación públicos de proximidad en Europa: RTVV y la crisis de las televisiones públicas, pp. 57–69. Tirant Humanidades, Valencia (2017)

EBU: Vision 2020. Connecting to a networked society. In: goo.gl/4gpi3 K (2012)

EBU: Funding of Public Service Media 2013. In: goo.gl/XVsgZ6 (2013)

EBU: Audience Trends Television 2016. In: goo.gl/zKVRtd (2016)

Enli, G.S.: Redefining public service broadcasting: multi-platform participation. Convergence **14**(1), 105–120 (2008)

Franquet, R.: Estrategias de los servicios públicos de medios audiovisuales en el escenario crossmedia. In: Marzal, J., López, P., Izquierdo, J. (eds.) Los medios de comunicación públicos de proximidad en Europa: RTVV y la crisis de las televisiones públicas, pp. 179–193. Tirant Humanidades, Valencia (2017)

Fuertes, M., Marenghi, P.: Televisión digital terrestre: caracterización, antecedentes e importancia. In: Albornoz, L.A, García, M.T Leiva (eds.) La televisión digital terrestre. Experiencias nacionales y diversidad en Europa, América y Asia, pp. 24–42. La Crujía, Buenos Aires (2012)

García-Leiva, M.T.: Entre las promesas y los resultados: notas sobre los retos de futuro de la TDT. adComunica. Revista de Estrategias, Tendencias e Innovación en Comunicación **1**, 33–48 (2011)

González-Neira, A., Quintas-Froufe, N.: El comportamiento de la audiencia lineal, social y en diferido de las series de ficción españolas. Revista de la Asociación Española de Investigación de la Comunicación **3**(6), 27–33 (2016)

Guerrero, E.: Los géneros de entretenimiento en TVE: audiencias, programación y producción. In: Moreno, E., Giménez, E., Etayo, C., Gutiérrez, R., Sánchez, C., Guerrero, J.E. (eds.) Los desafíos de la televisión pública en Europa, pp. 385–408. EUNSA, Pamplona (2007)

Harrison, J., Wessels, B.: A new public service communication environment? Public service broadcasting values in the reconfiguring media. New Media Soc **7**(6), 834–885 (2005)

Howe, N., Strauss, W.: Millennials Rising. The Next Great Generation. Vintage Books, New York (2000)

Iosifidis, P.: Retos y estrategias: servicio público de televisión en Europa. Infoamérica **3–4**, 7–21 (2010)

Islas, J.O.: El prosumidor. El actor comunicativo de la sociedad de la ubicuidad. Palabra Clave. **11**(1), 29–39 (2008)

Izquierdo, J.: Distribución online de contenidos audiovisuales: análisis de 3 modelos de negocio. El profesional de la información **21**(4), 385–390 (2012)

Jenkins, H.: Convergence Culture: Where Old and New Media Collide. NYU Press, New York (2006)

León, B.: La programación de las televisiones públicas en Europa. La estrategia de la adaptación. In: Moreno, E., Giménez, E., Etayo, C., Gutiérrez, R., Sánchez, C., Guerrero J.E. (eds.) Los desafíos de la televisión pública en Europa, pp. 75–82. EUNSA, Pamplona (2007)

Manfredi, J.L.: Servicio público en el entorno digital: teoría y práctica. Quaderns del CAC **23–24**, 179–186 (2006)

Manfredi, J.L.: El reto digital de las televisiones públicas en Europa. Las estrategias de la BBC y de RTVE. Telos, 68. In: goo.gl/OqftsQ (2006b)
Marinelli, A., Andò, R.: From linearity to circulation: how tv flow is changing in networked media space. Tecnoscienza. Ital. J. Sci. Technol. Stud. **7**(2), 103–127 (2017)
Miguel de Bustos, J.C., Garitaonandía, C.: La televisión pública: el motor de la TDT. In: Moreno, E., Giménez, E., Etayo, C., Gutiérrez, R., Sánchez, C., Guerrero J.E. (eds.) Los desafíos de la televisión pública en Europa, pp. 729–741. EUNSA, Pamplona (2007)
Moragas, M., Prado, E.: La televisió pública a l'era digital. Pòrtic, Barcelona (2000)
Moragas, M., Prado, E.: Repensar la televisión pública en el contexto digital. Portal de la comunicación. In: goo.gl/6eV84 J (2001)
OBS: Yearbook 2015. Key trends. Television, Cinema, Video and On-demand Audiovisual Services-The Pan-European Picture. OBS, Strasbourg (2015)
Prado, E.: Introducción: contenidos y servicios para la televisión digital. Telos **84**, 47–51 (2010)
Prado, E.: Indispensabilidad de la televisión pública en el universo convergente. In: Marzal, J., Izquierdo, J., Casero, A. (eds.) La crisis de la televisión pública: el caso de RTVV y los retos de una nueva gobernanza, pp. 39–59. Universitat Autònoma de Barcelona, Bellatera, Barcelona (2015)
Prado, E., Delgado, M.: La televisión generalista en la era digital: tendencias internacionales de programación. Telos **84**, 52–64 (2010)
Prado, E., Fernández, D.: The role of public service broadcasters in the era of convergence: a case study of Televisió de Catalunya. Commun. Strat. **62**, 49–69 (2006)
Quintas-Froufe, N., González-Neira, A.: Consumo televisivo y su medición en España: camino hacia las audiencias híbridas. El profesional de la información **25**(3), 376–383 (2016)
Román, M.: TDT en España y el dividendo digital. Estudios sobre el mensaje periodístico **18**, 801–809 (2012)
Steemers, J.: Servicio público de radiodifusión en el siglo XXI. La recuperación de la iniciativa. In: Marzal, J., López, P., Izquierdo, J. (eds.) Los medios de comunicación públicos de proximidad en Europa: RTVV y la crisis de las televisiones públicas, pp. 27–41. Tirant Humanidades, Valencia (2017)
Suárez-Candel, R.: La televisión pública como precursora de la implantación de la TDT: la situación de Catalunya. In: Moreno, E., Giménez, E., Etayo, C., Gutiérrez, R., Sánchez, C., Guerrero, J.E. (eds.) Los desafíos de la televisión pública en Europa, pp. 755–771. EUNSA, Pamplona (2007)
Suárez-Candel, R.: Adapting Public Service to the Multiplatform Scenario: Challenges, Opportunities and Risks. Hans-Bredow-Institut, Hamburg (2012)
Syvertsen, T.: Television and multi-platform media hybrids: corporate strategies and regulatory dilemmas. In: Marcinkowski, F., Meier, W.A., Trappel, J. (eds.) Media and Democracy: Experience from Europe, pp. 253–273. Haupt Verlag, Bern (2006)
Toffler, A.: The Third Wave. Bantam Books, New York (1980)

Sabela Direito-Rebollal Degree in Audiovisual Communication from the Universidade de Santiago de Compostela (USC). Master in Communication and Creative Industries (USC) and Degree in Movie and TV Script from the Madrid Film Institute. She is a Ph.D. candidate at the USC and was visiting scholar at the University of Hull (United Kingdom). Her field of research is the area of generalist and scientific social networks. Santiago de Compostela, Spain.

Diana Lago-Vázquez Degree in Journalism at the Universidade de Santiago de Compostela (USC) and Master in Communication and Creative Industries (USC). She is a Ph.D. candidate at the USC and was visiting scholar at the University of Hull (United Kingdom). Her fields of research focus on social network analysis within the field of communication and information, paying special attention to television. Santiago de Compostela, Spain.

Ana-Isabel Rodríguez-Vázquez Ph.D. in Communication at the Universidade de Santiago de Compostela. She has a BA in Information Science from the Universidad Complutense de Madrid (UCM). After more than a decade working in the media, she started to work at the USC in 2003. She belongs to the research group Audiovisual Studies (GEA) at the USC, and her main fields of research have to do with information, programming and audiences. Santiago de Compostela, Spain.

Constructing Reality Through Audiences

Valentín-Alejandro Martínez-Fernández and María-Dolores Mahauad-Burneo

Abstract The media played an essential role in moving linearly the message to a set of receivers unified in audience. They represented the "mirror" to reflect the mediated reality. Technologies, those that have given rise to the digital arena, have transformed the relation transmitter-receiver, through a process of networked simultaneous communication, where media-mediation is contingent. At the same time, technologies have provided a space where the interaction and proactivity of the receiver prevail, making themselves up as issuers through the spreading of a message. In the digital universe, the interaction of receivers, constituted in "liquid audiences", define the new reality, where the truth is transmuted and propagated without the "before" deviating the "now" and, without this, is an amendment to the former. In the digital society, subjected to immediacy, reality is not grounded, projected and reflected; it is constructed "at the moment" by audiences, without "truth" being a necessary element, since it can be dispensable. Hence the "collective imaginary", catalysed in audiences, transcends unreality to be "true" and become a new construct of reality.

Keywords Social media · Audiences · Digital network · Reality · Truth

1 The Mediated Reality

Reality is a perception, subjective, where the linkage of the Fact articulated by reason leads to individual or collective certainty of what happened. It is, therefore, an illusion endowed with the attribute of truth to make it irrefutable, because, otherwise, it would lose the status of real by adopting the role of imaginary. Thus, it is something true and therefore undeniable, susceptible of being shared and generate a common halo

V.-A. Martínez-Fernández (✉)
Universidad de A Coruña, A Coruña, Spain
e-mail: valentin.martinez@udc.es

M.-D. Mahauad-Burneo
Universidad Técnica Particular de Loja, Loja, Ecuador
e-mail: mdmahauad@utpl.edu.ec

M. Túñez-López et al. (eds.), *Communication: Innovation & Quality*, Studies in Systems, Decision and Control 154, https://doi.org/10.1007/978-3-319-91860-0_5

to those who catalyse in it the vision and introspection of something incorporated to its life and able to be projected. In this regard, Zambrano (1978) points out that, "when reality chooses the awakening, the truth assists it with its simple presence. And if this were not so, reality could not be supported and presented to men with its character of reality".

It is true that there may be different approaches to reality. That is to say, the way in which the Fact is presented and acts on the individual, as well as the way in which it rationalizes it and agree with others. And from that consensus arises an interpretative key of something that obeys the existence of a phenomenon, since without the essential requirement of that there is no reality.

From the social vision of men, Ortega (1999) talks about "historical reality", which reflects everything that happens in the world of Being, from their human conception. In short, it is not a collection of loose or disjointed facts, since it has a "strict anatomy" and a defined structure. Hence, I postulate the need of this type of reality to be studied and am against the pretension of social media to establish themselves as mediation platforms in the social visualization of reality, as these "do not give us news of this reality".

It is true that Ortega referred mainly to newspapers and more specifically to information reflected on them and distorted by biases of their respective "editorial guidelines" and ideological or simply business orientations. The medium is not the object of criticism, but the instrumentation of it in its mediating role. An instrumentation that becomes important in Gabriel Tarde's vision on the effect of mediation in newspapers, as Bauman (2017) points out in his posthumous work *Retrotopía*, it was late when he realized that the crowd was being replaced by a new social formation, self-organized virtually, around the daily newspaper.

The problem, therefore, is not what the medium says, but the assumption of the certainty of what it says and how that certainty is transformed into the base of the real and is assumed to internalize it transfigured into an unquestionable idea in its existence. A question to which Lippmann (1997) already tried to give answer in 1922 with his reference work *Public Opinion.*

Together with the advent of radio and television, where the word and the image strengthen the perception of reality, through the "listened" and "visualized" truth, there is a certain connection between the individual and the Fact, so that the former becomes a spectator, a virtually ubiquitous witness of what happened. Without moving from his/her own physical space, the individual is able to "witness" what is happening in the world and, thanks to the means, he/she becomes accessible to objectify his/her vision of what happened and make it real. In this regard, Arendt (2000) considers that "the presence of others who see what we see and hear what we hear assures us of the reality of the world and ourselves" and, undoubtedly, that seeing and hearing is done through social media and becomes common. In such a way, the historical reality is opposed to fiction, although fiction is itself a certain type of reality.

Thus it emerges the media truth and the mediated reality. It is not anymore what it IS but what the social media say it IS, and it is accepted as such by the person, first individually and then, often, almost simultaneously, as a collective certainty that generates a common opinion which transcends the individual, in which the truth given by the majority of those who show the reality plays a relevant role. And it is precisely reality that leads individuals to adopt a given behaviour, individually or collectively.

The mediated reality is not addressed to individual subjects; it is intended for these grouped into a new social entity: the audience. These are the ones that not only "validate" that reality with their attitudes and behaviours, but also come to build the reality itself with their perception and interpretation of the mediated Fact. And that reality brings together people in masses oriented in uniform directions (Le Bon 1995).

The important thing is not anymore what happened, not even *how* and *why*, the essences of the Fact. What is significant is the way it is translated to the audiences and the way the latter or a part of them accept, refer and visualize reality in their environments; the way in which they give rise to a public opinion with a great ability to influence own and collective behaviours. Although sometimes that opinion presents a misleading appearance. This is the focus of the "spiral of silence" proposed by the political scientist Elisabeth Noelle-Neumann (2003), which affirms that when the subject considers that the public opinion reflect their attitude in a majoritarian sense, even if they differ from that attitude, it consists in adopting silence as a way to not feel integrated into the dissident minority, while avoiding the risk of exclusion, and, thereby, making reality what simply and objectively is unrealistic.

In any case, according to Debord (2009), the media reality has been a key element in the construction of the so-called *entertainment society* and where social media adopt their "most overwhelming superficial manifestation" due to the control they exercise in the projection of the Fact. A projection mainly made unilaterally. Entertainment thus become a basic nutritional element for audiences and in its individualization conforms the experiential future of the subject.

We are therefore facing a solid reality, where social media play an essential role in moving the message (the Fact) from the sender to the receiver. They mediate between one and the other. Both need them. They are essential, by no means contingent. Without them, you can Be, but not Be ubiquitously, even though through third parties, where the Fact is produced.

One can be part of reality, but not becoming aware of Reality.

Therefore, what the media do not collect, transmit and reflect, does not exist for audiences and what does not exist is clearly not real. Hence the power, term coined by Burke, of the media as influencers of social behaviours or advisers through mediated reality.

One of the characteristics of the media reality, in a world of solid audiences, consists of the "filter effect" that social media, almost preponderantly, exercise over what happened and the way it, understood as Fact, reach audiences and these interpret the Fact as an indissoluble element of the construct of that reality to which they accede and integrate in their vital senses. And, in turn, that filtered action of what Is and How it Is, takes place outside the focus of transparency.

The filter appears to be a "black box" by which the Fact is "collected" and reflected in a mirror where audiences see themselves and they take them for granted, without indicating how the mirror is and if this, as would be the case of the concave or convex, gives rise to a certain distortion. Thus, "a deforming mirror reflects a deforming mirror, which, in turn, reflects reality" (Pariser 2017).

What you see is what it is; and what exists and what is existent is true. And the collective assent of the real triggers its transmutation into Reality.

2 The Transparent Reality

The development of communication technologies has led to the generation of the Network, of the digital universe, in which audiences have become liquid, as they do not maintain the same shape for a long time, as they are immersed in that "liquid life" enunciated by Bauman (2016), where everything moves without a precise direction, the immediacy is constant and replaces the strong horizon by uncertainty and the systemic fear of change. A "liquid life" which is made transparent in the digital arena to which we are irretrievably driven to live technologies and where, precisely because of transparency, the media filter has been eliminated to reconvert media reality in "interacted reality", by creating new forms of action and interaction and, from them, new types of social relations (Thompson 1998).

Technologies have had a profound effect in the mediation of the media by generating a media convergence consisting of something far more significant than mere technological change.

For Jenkins (2008), convergence implies a clear alteration in the relationship between existing technologies, as well as industries, markets, contents and, of course, audiences. Perhaps in the case of the latter, the relationship with audiences would be the most significant aspect of this alteration, since the audience itself is getting away from the dependence on the media to visualize, perceive and embrace reality.

The Network is the digital environment in which social media emerge and develops. There, the individual enjoys the potentiality of being in itself a media with multiple performance of the functions of sender, propagator and receiver of the Fact, without more effort than the action of a click, which allows the individual to break free from the measurement between what happened and what is perceived, as well as the validation of the truth by traditional social media that made the Fact unquestionable for audiences to which they were addressed, once accepted and agreed.

As noted by Ferraris (2017), the Network is "an empire that ver sunsets and the fact of having a smartphone in the pocket means to have a world at our fingertips, but also, automatically, being in the hands of the world". In this regard, two decades ago, Thompson (1998) saw in the newcomer "digital universe" a favourable space to give rise to what he called "media ownership", a kind of "public property" of individuals, action and events; in short, the Fact, whose main characteristic no longer consists in sharing a common place.

Now that "validation of the truth" is made in the direct relation between those who together make up the new digital audiences; a relation based on trust in "the other", almost abstract, since, from that "other", is not usually known anything other than that reflected on the Network, but whose vision of the Fact, when it matches with the others' vision, integrated in a common audience, becomes what happened and, therefore, the Fact, the real thing that, linked to other Facts, represents the Reality.

Here it is the axiom of truth, the need for audiences to rely on the certainty of the Fact. A trust that before, in the period of solid audiences, was placed in the social role of the media and that now, in the present arena of liquid audiences, emerges in the attitude of interaction with the "other", even when that "other" may be a machine or robot, and whose behaviour is indistinguishable from a person, since what is significant is not the "personality" but the interaction, feeling "interacting" and "integrated" in the idea which Eco advanced decades ago (1995). In short, being for being.

On the Net, the Reality becomes flat, according to Friedman (2006), since physical distances are transmuted into virtual ones and even time volatilizes the past to create a permanent present dominated by immediacy (Martínez-Fernández 2017).

Everything is on the Net. Everything is visible. Ubiquity is its essence. Anything is overlooked. The Fact is omnipresent and what happened is permanently moved to a continuous present. Everything is transparent. Although this does not guarantee the truth either, but rather the opposite according to Han (2013). The author notes that the hyperinformation and hypercommunication provide evidence of the lack of truth, and even the lack of being. More information, more communication does not erase the fundamental imprecision of the whole. Rather it aggravates it. Although this German philosopher of Korean origin does not fail to recognize that "transparency removes things from all charm and forbids fantasy to weave their possibilities, so any reality cannot reward us its loss".

The Net is like that "Hawk-Eye" to which Cruz (2017) refers, to which nothing escapes and everything is seen accurately, plainly. However, it cannot be considered a space, a place where privacy can fit, as Cruz (2017) also points out, it is a place of places, without the delimitation of physical places, a digital panopticon, where the individual is an observer in turn observed in a single movement, but that observation is what gives rise to an "assumed reality".

A panoptic composed of what Han (2014) calls "the digital swarm", where in the opposite sense of the mediated reality, audiences are being watered down in the individuality of those who integrate them, because unlike audiences of the mass society, they lack "soul", "spirit", without giving rise to "us". Thus, while social mass media brought people together in their audiences, the Net isolates them.

However, this isolation does not imply a collective misunderstanding of what happened, rather the opposite. The Fact is exhibited, as in a big square, once constructed interactively in a constant and circular action of constructing-deconstructing-constructing. That action becomes reality and its nature is charged in the immediacy of the moment. There is no need, therefore, for the role of traditional social media.

3 The Media *De-mediatisation*

Together with the Net, the media "de-mediatisation" or "disintermediation" arrived, conceived in a vision of broader scope as it was described by Jon Pareles, a New York Times critic, who estimates that this occurred throughout the decade of 2000 (Pariser 2017). A disintermediation that has also touched businesses, professions, etcetera.

Linearity, simplified in the Fact-Medium-Audiences, is replaced by the neuronal process in which the Fact is propagated, according to the terminology used by Jenkins (2015), by the individual action of the subjects that, synaptically, come together in extensive and complex audiences, connected in the temporal space of instantaneity and with the immediate ability to mutate.

The Fact is constructed being subject to the action of the constant moment, since the "before" has lost the ability to distort the "now" and it is exactly the "now" that enjoys the attribute of certainty; the "now" excludes and replaces by itself the "orthodox truths" of the "before", which, in turn, is transformed into canons where no dispute is admissible (Bauman 2015).

As a result, the reality is not grounded, projected, reflected. It is simply conformed by digital audiences at the same time of the interaction between those who make it, without the truth being understood as the objectification of the Fact, which constitutes a necessary a relevant and necessary element. It is simply something contingent and, thus, dispensable.

Therefore, the "collective thinking" in the Net, catalysed in the audiences, transforms the unreal, as a virtual illusion, to endow it with certainty, to make it true and as such to transform it into a new construct of reality.

Argemi (2017) provides an interesting perspective on the way audiences create reality. According to the author, "in the digital conversation and in the media, a social construction of knowledge can be generated, which is not the same as the construction of reality". It refers to the different frames that audiences can give to the fact, as well as the subjects that conform audiences, for which it is based on the idea formulated by Goffman (2007). Although the fact that there are several and different frames of a Fact does not prevent a dominant convergence in the Network where the realities conform the Reality accepted through the implicit consensus of the assumption of the same.

For this reason, it is possible to distinguish between the reality submitted to the objectivised reality and the reality accepted through the truth based on the belief. For the former, reflection is required and this needs time. The latter reality, through the action of immediacy which governs the network, where the substantial cannot escape the reduction of the ephemeral, does not give rise to reflection, since digital audiences are thoughtless.

In the behaviour of digital audiences, Lippmann's view of the way audiences acted (2011) has gained renewed importance, despite the fact that almost a century has passed. The author considered that audiences, to whom he called the public, were extremely malleable and easy to manipulate by accepting false information as certain, but in that certainty the reality was reflected. Well, that malleability has been accentuated in liquid audiences, in which the moment, the instantaneous accelerates the disposition of emotion against reason (Damasio 2009), and imposes it in its vision of the real and therefore of its reality, perhaps as a result of what has come to be called "naïve realism" (Pariser 2017).

On the Net, unlike our relationship with traditional social media, we are not observers to test our individual vision of what happened and then integrated it, more or less quickly, in the common opinion of the audiences or audiences to which we belong and give form. In the Net we are simply observed in order to be induced not only to adopt an attitude, after accepting as true what is meant to be translated as real, but to shape our own identity.

Nothing that we do on the Net goes unnoticed. We will always have the "Hawk's eye" upon us, to which we have referred. Every action is going to be scrutinised to identify our beliefs and, according to them, to look for the algorithmic connection with other subjects similar to us, with whom we form an audience, in order to present before our eyes, the Fact not according to what happened but according to how we wanted that would have happened. And so, in a silent process, our identity mutates towards what we want it to be on the part of those who control the traces of the Net and, with it, towards the creation of the Fact ad hoc that, by itself, decisively contributes to build an ad hoc reality.

References

Arendt, H.: The Portable Hannah Arendt. Penguin, New York (2000)
Argemí, M.: El sentido del rumor. Cuando las redes sociales ganan a las encuestas. Península, Barcelona (2017)
Bauman, Z.: La globalización. Consecuencias humanas. Fondo de Cultura Económica, México (2015)
Bauman, Z.: Vida líquida. Paidós, Barcelona (2016)
Bauman, Z.: Retropía. Paidós, Barcelona (2017)
Cruz, M.: El ojo del halcón. Arpa, Barcelona (2017)
Damasio, A.: El error de Descartes. Crítica, Barcelona (2009)
Debord, G.: La sociedad del espectáculo. Pre-Textos, Valencia (2009)
Eco, H.: Apocalípticos e integrados. Tusquets, Barcelona (1995)
Ferraris, M.: Movilización total. Herder, Bercelona (2017)
Friedman, T.: La tierra es plana. Martínez Roca, Barcelona (2006)
Goffman, E.: Frame Analysis: los marcos de la experiencia. Centro de Investigaciones Sociológicas, Madrid (2007)
Han, B.-C.: La sociedad de la transparencia. Herder, Barcelona (2013)
Han, B.-C.: En el enjambre. Herder, Barcelona (2014)
Jenkins, H.: Convergence Culture: la cultura de la convergecia de los medios de comunicación. Paidós, Barcelona (2008)
Jenkins, H., Ford. S., Grenn, J.: Cultura Transmedia. La creación de contenido y valor en una cultura en red. Paidós, Barcelona (2015)
Le Bon, G.: Psicología de las masas. Morata, Madrid (1995)
Lippmann, W.: Public Opinion. Free Press, New York (1997)
Lippmann, W.: Público fantasma. Geneuve Ediciones, Palma de Mallorca (2011)

Martínez-Fernández, V.A.: Immediacy and Metamedia. Time Dimension on Networks. Springer, Switzerland (2017)
Noelle-Neumann, E.: La espiral del silencio. Paidós, Barcelona (2003)
Ortega y Gasset, J.: La rebelión de las masas. Espasa, Madrid (1999)
Pariser, E.: El filtro burbuja. Cómo la Red decide lo que leemos y lo que pensamos. Tauris, Barcelona (2017)
Thompson, J.B.: Los media y la modernidad: una teoría de los medios de comunicación. Paidós, Barcelona (1998)
Zambrano, M.: Claros del bosque. Seix Barral, Barcelona (1978)

Valentín-Alejandro Martínez-Fernández Tenured Professor at the Faculty of Economics and Communication Sciences, both from the Universidade de A Coruña. He has been Head of the Galician newspaper of *El Ideal Gallego*, and executive in many news companies. He has a degree in Information Science from the Universidad Complutense de Madrid. He also has a MBA in Business Administration from the Universidade de A Coruña. Ph.D. in Information Science from the Universidad Complutense de Madrid. A Coruña, Spain.

María-Dolores Mahauad-Burneo Director of the Department of Entrepreneurial Sciences from the Universidad Técnica Particular de Loja, MBA in Economics for Agrifood and Environment from the Universidad Politécnica de Valencia. BA in Engineering and Business Administration from the Universidad Técnica Particular de Loja—Ecuador. Member of the Research Group Innovation and Entrepreneurship. Loja, Ecuador.

Audiovisual Regulatory Authorities in the European Union

Tania Fernández-Lombao and Paulo Faustino

Abstract All EU member states have independent regulatory authorities for the audiovisual sectors which are, in general, responsible for the concession of licenses, the supervision of public and private media, and the enforcement of punishments for non-compliance with the rules. Despite sharing a good number of capabilities, differences exist in the ways these bodies proceed, the number of members that make them up, and their processes of appointment. Upon these foundations, lies the key that distinguishes the degree of independence and the democratic well-being of their mission.

Keywords Regulatory authorities · Audiovisual · European union

1 An Introduction to European Public Media

The Declaration of the Rights of Man and of the Citizen (France, 1789) shows in article 11 that the freedom of communication of thoughts and of opinions is "one of the most precious freedoms of Man: every citizen will be able to speak, write, and print freely, without prejudice to respond to the abuse of this freedom in cases determined by the law."

The European Broadcasting Union in 1993 identified public broadcasting with programs for all as, the general base service with thematic ranges, a focal point for democratic debate, a point of free access for the public to major happenings, a reference in matters of quality, an abundance of original productions and an innovative spirit, a cultural shop window, a contribution to the re-strengthening of European

T. Fernández-Lombao (✉)
Universidade de Santiago de Compostela, A Coruña, Spain
e-mail: tania.fernandez.lombao@usc.es

P. Faustino
Universidade do Porto, Porto, Portugal
e-mail: faustino.paulo@gmail.com

M. Túñez-López et al. (eds.), *Communication: Innovation & Quality*, Studies in Systems, Decision and Control 154, https://doi.org/10.1007/978-3-319-91860-0_6

identity, and of its social and cultural values, a machine for investigation and for economic development.

Public broadcasting services survived under the argument that the market cannot cover all the media needs that the public media deal with, that state control is anti-democratic and that an alternative in the centre ground would be necessary. Nowadays, the context of technological convergence and social and economic globalization add new ingredients to the debate (Centre for Social Media 2009; Copé 2008; Downie and Schudson 2009). The process of transition began at the start of the 21st century. Whilst the literature of the eighties and nineties of the previous century tinted the future of public broadcasting services with pessimism, seeing the proliferation of channels, financial cuts and the loss of a sense of their mission, as forces for dissolution, the start of the year 2000 however brought with it opportunities for reinvention based on convergence and multiple platforms (Brugger and Burns 2011).

Despite the differences between the audiovisual policies unveiled by the states which today make up the European Union, we speak of a European model of television as opposed to the American one, with the European institution acting as the cradle of a philosophy for breaking down barriers. Such is the case that television from the old continent was born and developed under one same sign: the vocation of public service faced with the American model of industry in the hands of private interests and subject to the laws of commerce (Martínez 1999).

Mary Debret, in an article from 2014 in the scientific review Television & New Media, shows that innovation is germinating in public broadcasting bodies all over the world under the umbrella term 'public service media', a concept supported by Karen Donders—which is to say, those that expand through new platforms, and experiment with interactive content. However, this new phenomenon brings with it certain risks for the media which is funded using public funds, as at the same time it provokes hostility from their commercial rivals, new and old, and it invokes new obstacles and normative points of reference in order to demonstrate its public value.

2 Governance as a Guarantee for the Future

The concept of governance is a new way of managing which is more cooperative, open, transparent and participative, different to the old hierarchical model, in which public powers exerted their sovereignty over the public (López Cepeda 2010). "In modern governance, state and non-state institutions, public and private actors, participate and often cooperate in the planning and application of public policies. The structure of modern governance is not characterised by hierarchy, but rather by cooperative autonomous actors and by networks between organizations" (Mayntz 2000: 1).

Governance is a synonym of "governability (administration and management) but, currently, this concept must also be viewed in the convergence of the triangle of State, market and civil society. In the context of this information society, governance must not only contemplate the dimension of representation, but also of participation, which electronic networks permit and promote" (Campos 2009: 61)

The lack of independence that the organisms for management and control of the public service media have, turns governance into a desire for the future that would facilitate the recovery of the reputation of these broadcasting bodies in Europe after passing through a crisis of credibility, which was more or less intense depending on the country.

In fact, the Commissions for Parliamentary Control have become bodies of little support that set political positions instead of being bodies for debate about the public media, and without also taking on the problem of private companies. In the Administrative Councils for public services there are distinct cases, but many of them are full of communications managers from various parties, ending up as mere echo chambers of the Commissions for Parliamentary Control or vice versa, with functions which are often limited or generic (the criteria for programming remains reserved for the general directors of public broadcasting bodies) most of the time ending up simply being deliberative or political games (Zallo 2011).

Together with these bodies, there are the Advisory Council, Defenders of the Audience and Spectators, and the Tribunals or Chambers of Accounts, which have the aim of policing the functioning of public broadcasting in concrete aspects: programming, rights of the spectators or management of the budgets for public entities, respectively.

However, this control remains weak without the creation of independent authorities that police the activity of publicly owned media, as well as private media (López Cepeda 2010).

According to Carmen Fernández-Miranda Campoamor and Alfonso Fernández-Miranda Campoamor (2003), independence is sustained by two axes, origin and exercise. In the first case, a body or complex of bodies has independence when its members are elected without interference from any other state power and taking into account its training and experience in similar positions. In the next case, it refers to the fact that the existence of a body and its functions cannot be altered or interrupted by any other State power.

The European Council established criteria for member states to take into account in relation to the naming, composition and functioning of the Audiovisual Councils: economic independence, power and capabilities, and a system of responsibility.

García Castrillejo (2006) demonstrated that audiovisual authorities must be independent of both public powers and of the agents present in the audiovisual buyers market. This independence will be sustained through the election of the members of the body in charge of taking decisions, in the financing and the decision making by a reinforced majority that obliges parliamentary groups with representation on the cameras to reach agreements. As for powers and capabilities, the audiovisual authority will play a regulatory and disciplinary role of sanctioning.

3 Audiovisual Councils: The Reason Why They Are Important

Audiovisual councils are the result of a system of public service media self-regulation which must perform four basic functions, according to Aznar and Villanueva (2000): (1) to publicly formulate the rules and ethical values that media activity must follow; (2) to contribute in terms of professional and social working conditions—which make it possible to follow the rules, ethical demands and deontology of communications; (3) to make the public aware of those cases which do not follow said criteria; (4) the study, discussion and judgement of the conflicting actions of the media will help whoever controls them and the public as a whole to understand and learn about their moral dimension.

These organs were created originally to arbitrate a just distribution of licences for radio and television services and for the administration of the radio spectrum, which is limited. The objective was to avoid political interference from the governments of the time who guaranteed the observance of laws for the public and private media. These functions were widened and nowadays they sustain the ruling of audiovisual content, advertising and the protection of minors.

Zallo (2006) identifies three phases in the regulation of broadcasting in Europe. In the first stage, the agent was the State, which was faced with a double tendency. On the one hand, to extend culture and information as mechanisms for socialization and democratization and, on the other hand, the perception of successive governments that they were a powerful tool of influence for social control and its perpetuation. With all the interventionism and politicization of the system, they seriously damaged the reputation of the public media, whilst the private media was able to act with freedom.

In the second phase, the rising institution was that of the market. In the decade of the eighties the central public monopoly was broken with liberalization and the emergence of private, autonomous operators, new management criteria were introduced in the systems, which tended to focus on advertising and entertainment. This model found its way over to the public media, which also resorted to commercial advertising as a means of funding.

Lastly, in the third phase the most characteristic subject is that of civil society. Against this backdrop, high audiovisual authority rises, independent of the Administration, political parties and economic powers, with a legal personality that guarantees pluralism. The Superior Audiovisual Council of France, constituted in 1989, is the paradigm.

The European Commission in 1999 established some guidelines for creating a regulatory framework which would cover all networks of communication and services associated with the access to radio electric space. The four statements were as follows: (1) regulatory bodies must not depend on the Government or on operators; (2) aspects in relation to content are essentially to do with the State, and as such, must conform with the principle of subsidiary, the reglamentation of content is the responsibility of member states; (3) technological convergence requires greater cooperation between the implicated regulators; (4) the regulatory bodies can contribute to the development of self-regulatory methods (Carniel 2009).

The Committee of European ministers in 2008 adopted a Declaration on the independence and the functioning of regulatory audiovisual authorities, in which it reiterated the need for States to create regulatory bodies with independence in respect to governments and the market, and to bring about their concrete application. This recommendation is not a duty, but a good number of the member states supported the creation of bodies with functional autonomy.

In May 2016 the European Commission proposed an update of the Audiovisual Communication Services Directive of 2010. In addition to actions related to European cinema, the protection of minors and the fight against inciting hatred, it projects a new focus for audiovisual regulatory bodies. Concretely, it urges member states to have regulatory authorities which are truly independent of governments and the sector, and which can perform their role in an optimum way.

Furthermore, it proposes that the European Union legislate over the functions of the Group of European Regulators for Audiovisual Media Services (YERGA), composed of twenty-eight audiovisual state regulators, which will consist of evaluating the co regulatory codes of conduct and advising the European Commission. Carniel (2008) presented a list of the elemental criteria which must govern audiovisual regulatory bodies with their respective variables based on the characteristics of each case (Table 1).

Their mission is that society has a radio and television system which is free, plural, responsible and suitable for the need for information, communication, cultural and linguistic promotion and entertainment. Furthermore, it avoids politicized interventionism, instead following collective values and it marks the guidelines for the development of the audiovisual media in a series of deontological minimums related to freedom, pluralism and service to society, without lessening freedom of expression (Zallo 2006).

As for the model of election, the appropriate selection of members is made among people of recognised prestige, of differentiated training and with guarantees of full independence in exercising their functions. Avoiding ideological social control and the political party system requires a wide consensus to reach a balanced election. The norm is to propose people without any link to other parties, balanced in the composition and rules of incompatibilities. With the will to protect independence, it is desirable to fulfil the requisite of wide majorities for the election of members.

When referring to the authority of an audiovisual regulatory body, it is measured by its scope for intervention and its integration; its capabilities if it regulates or only controls, policies and advises; and by what it regulates—the content, the market, agents and citizens' rights; acceptance and influence of decisions, reports and recommendations.

Table 1 Characteristics of audiovisual regulatory bodies (Carniel 2008)

Independent Audiovisual Regulatory Body	
Elemental Criteria	Variables
A public body created with rules, expressed legally	Law, decree or other judicial mechanism in force in the particular country
Own legal personality	Legal doctrine o the State in question
Collegial Structure Members have recognised link with the material, fixed with their term of office, a system of incompatibility with other activities	Number of members, time of mandate, form of election, form of exclusion and everything relating to the internal working of the body, such as for example the majority necessary for a vote
Economic and financial autonomy Political provision of accounts and activities	Sources of income, elaboration of budgets, output of accounts, administration and heritage
Proportional, territorial scope	By state of divided, according to the structure of the state
Competition of the market and of content Competition between public and private media	Distribution of capabilities with other entities of the same configuration
Availability of all legal means of intervention, normative legal authority, inspecting and arbitral legal authority, fiscal and sanctioning legal authority	Proposition of laws and development of rules starting from a legal framework Indirect or direct use of sanctions
Regulatory acts put an end to the administrative route, resources being passible before the legal power	Court or legal petition in accordance with the corresponding legal framework

4 Paradigmatic Cases in the European Union

All member states of the EU have bodies of control in the audiovisual sector, except Spain, where the National Commission of Markets and Competition performs these functions. The most salient cases are those of the ALM in Germany, CSA in France, AGCOM in Italy, ERC in Portugal and Ofcom in Great Britain.

4.1 Arbeitsgemeinschaft Diere Landesmedienanstalten (ALM), Germany

In Germany radio and television are the responsibility of the Federal States, there exists a total of fourteen regulatory authorities, while Berlin and Branenburg share. All these authorities are associated in Arbeitsgemeinschaft diere Landesmedienanstalten (ALM) or the Consortium of Media Authorities, and their members are chosen by qualified majorities in the Parliaments. Their aim of control and supervision is reduced to private media.

The ALM operates through four organs: The Commission of Regulatory Matters, The Conference of Media Directors, The Conference of Presidents and the General Conference. The first includes the representatives of the fourteen authorities of the Länder, which have as capabilities the concession of licenses, the supervision of national broadcasting bodies, and the development of digital broadcasting. The Directors Conference protects the interests of the authorities, is responsible for the exchange of information with broadcasting bodies and solicits the opinions of experts. The Conference of Presidents deliberates over the ethics and politics of the media and the General Conference tends to matters of programming.

With respect to the public service broadcasters ZDF and ARD, supervision is undertaken by the Council of Diffusion, an internal body, which reinforces its actions through an external body, the Rechtsaufsicht, which is limited to checking that the law has been followed. Its members are chosen by Parliament who are generally qualified, and it obliges a consensus to be formed between political parties for the appointment of independent professionals.

4.2 Comisión Nacional de los Mercados y la Competencia (CNMC), Spain

The Comisión Nacional de los Mercados y la Competencia acts as the audiovisual independent authority in Spain. It supervises the smooth running of the audiovisual communications market. Amongst its other functions are controlling all content, ensuring these fulfil the public service assigned to public media, resolving conflicts and enforcing punishments for non-compliance with legislation (Table 2).

This is a public body with its own legal entity, independent from the Government and under parliamentary surveillance, which began to operate in October 2013. It performs its functions through two main bodies: the Council (*Consello*) and the Presidency (*Presidencia*).

The first is a collegiate decision-making authority that is composed by ten members who are named by the Government at the Economy and Competition Department's proposal. These are people who are widely renowned and professionally competent in the Commission's field. Its term runs for six non extendable years and it has to comply with a strict incompatibilities' policy.

4.3 Conseil Supérieur de l'Audiovisuel (CSA), France

Law 89–25 from 1989, which reformed the existing Law 86–1067 from 1986, documents the creation of the Conseil Supérieur de l'Audiovisuel. This is an independent authority whose purpose is to guarantee freedom of audiovisual communication in France. It replaced both the Haute Autorité de la Communication Audiovisuelle

Table 2 Audiovisual regulatory authorities in European Union

Audiovisual regulatory authorities in European Union				
Country	Authority	Functions	No. of members	Election
Germany	Arbeitsgemeinschaft der Landesmedien-anstalte	Concession of licenses, supervision of public and private media and development of digital broadcasting	14 authorities of the Länder	Qualified majority in the Parliament
Spain	Comisión Nacional de los Mercados y la Competencia	Control content, ensure this fulfills the public service, resolve conflicts and enforce punishments for non-compliance with legislation	10	Government
France	Conseil Supérieur de l'Audiovisuel	Concession of licenses, ensure respect for pluralism and honesty in the supply of information, provide frequencies to operators, control the radio spectrum and protect children's rights	9	3 President of the Republic, 3 President of the National Assembly and 3 President of the Senate
Italy	Autoritá per le Garanzie nelle Comunicazioni	Concession of licenses, ensure freedom of competition, grant key rights and copyright	10	First Minister
Portugal	Entidade Reguladora para a Comunicação Social	Concession of licenses, supervision of public and private media, draw up norms, perform consultation and monitoring	5	The National Assembly appoints 4 and these pick the fifth member
UK	Office of Communications	Concession of licenses, enforce punishments for non-compliance with legislation, produce codes of practice and monitoring	9	The Government appoints 6, who suggest other tree from the own Ofcom's staff
Austria	Austrian Communications Authority	Management of the radio spectrum, supervision of private media and monitoring	5	Federal Presidency
Belgium	Conseil Supérieur de l'Audiovisuel (Comunidad francesa)	Concession of licenses, enforce legislation and punish non-compliance	44	Government

(continued)

Table 2 (continued)

Audiovisual regulatory authorities in European Union				
Country	Authority	Functions	No. of members	Election
	Vlaamse Regulator voor de Media (Comunidad flamenca)	Concession of licenses, regulation of public and private media, mediation in conflicts of interest and deal with breaches of legislation	3	Government
	Medienrat (Comunidad alemana)	Concession of licenses, supervision of programmes and complaints' management	ALM	Qualified majority in the Parliament
Bulgaria	Council for Electronic Media	Concession of licenses and supervision of public and private media	5	3 are appointed by the National Assembly and 2 by the President of the Republic
Cyprus	Cyprus Radio television Authority	Concession of licenses, monitoring audiovisual media broadcasting services, on-demand services, authors' rights and methodologies to determine audience share	7	Government
Croatia	Agency for Electronic Media	Concession of licenses, support launching new radio and tv channels both traditional and on-demand	6	Parliament, at the Government's proposal
Denmark	Agency for Culture and Palaces	Concession of licenses, supervision of public and private media, complaints' management and enforce punishment for non-compliance with legislation	8	7 are appointed by the Ministry of Culture and 1 by the audience
Slovakia	Council for Broadcasting and Retransmission	Management of frequencies, concession of licenses, supervision of public and private media, complaints' management and enforce punishment for non-compliance with legislation	9	Parliament

(continued)

Table 2 (continued)

Audiovisual regulatory authorities in European Union				
Country	Authority	Functions	No. of members	Election
Slovenia	Agency for Communication Networks and Services	Control the compliance to ethical and professional standards, ensure social pluralism, issue licenses and supervise public and private media	5	Government
Estonia	Estonian Technical Surveillance Authority	Concession of licenses and authorisations for frequencies, supervise the media and control networks of digital communications	5	–
Finland	Finnish Communications Regulatory Authority	Concession of licenses and electronic authorisations, survey the running of public and private media, penalise infringements in law and perform consultation	7	–
Greece	National Council of Radio and Television	Concession of licenses, supervision of public and private media, penalisation and issue consultation reports	7	Presidency of the Parliament
Hungary	National Media and Infocommunications Authority	Concession of licenses, control of public and private media and complaints' management	7	2/3 of the National Assembly
Ireland	Broadcasting Authority of Ireland	Concession of licenses, exert control over public and private media, penalise breaches in legislation and issue consultation reports	9	The Government appoints 5 and other 4 are proposed by a parliamentary comission
Latvia	National Electronic Media Council	Concession of licenses, supervision of public and private media, penalise breaches in legislation and perform consultation	5	Parliament

(continued)

Table 2 (continued)

Audiovisual regulatory authorities in European Union				
Country	Authority	Functions	No. of members	Election
Lithuania	Radio and Television Commision of Lithuania	Concession of licenses and authorisation of frequencies, supervision of public and private media, complaints' management, publication of norms and codes and perform consultation	11	2 are appointed by the Presidency of the Republic, 3 by the Parliament under the recommendation of a Culture Comission, 1 by the Epicospal Conference, 3 by the Lithuanian Artists Association, 1by the Lithuanian Association of Journalists and 1by the Lithuanian Union of Journalists
Malta	Broadcasting Authority of Malta	Concession of licenses, supervision of public and private media, penalise breaches in legislation, publication of codes and norms and perform consultation	5	Presidency of the Government at the First Minister proposal
The Netherlands	Commissariaat voor de Media	Concession of licenses, supervision of public and private media, publication of guidelines for action and advise	3	Department of Education
Poland	National Broadcasting Council	Concession of licenses, supervision of legislation fulfilment and control media providing a public service	–	President of the Government

(continued)

Table 2 (continued)

Audiovisual regulatory authorities in European Union				
Country	Authority	Functions	No. of members	Election
Czech Republic	Council for Radio and TV Broadcasting	Concession of licenses and extension of authorisations, supervision of public and private media, complaints' management, penalise breaches in legislation and perform consultation	11	Presidency of the Republic from an initial proposal by the political parties represented in the Parliament
Romania	National Audiovisual Council	Concession of licenses, exert control over public and private media, publication of codes and norms, punishment, monitoring and consultation	11	Congress and Senate
Sweden	Swedish Broadcasting Authority	Concession of licenses, supervision of public and private media, complaints' management, penalise breaches in legislation and monitoring	11	Government

Source Prepared by the authors, 2017

and the Communications de la Comission Nationale, made up of nine and thirteen members, respectively.

Following the same structure as the Haute Autorité de la Communication Audiovisuelle, the Conseil Supérieur de l'Audiovisuel is made up of nine members, three of which are designated by the President of the Republic, three others by the President of the National Assembly and three by the Senate's President. Its members cannot, either directly nor indirectly, carry out duties, receive professional fees or have stakes in companies that manufacture audiovisual or cinematographic content, or in those related to editing, press, advertising and telecommunications. Moreover, during the period of its term and until one year after ceasing in their functions, they must refrain from taking a side publicly on questions that relate to the Council. For the next three years, members must not accept employment in any of the audiovisual communication companies subject to authorisation, to agreement or those that could potentially be penalised by the CSA. Furthermore, the Council independence is achieved "without bonds on the length of their term with the person responsible for the appointment (the CSA members' term is 6 years long), through the establishment of a partial renewal of the body, in thirds every two years, and limited by a rigid incompatibilities' policy" (López Cepeda 2010).

The body's functions range from ensuring respect to pluralism and honesty in the supply of information, to providing frequencies to operators. It also controls the broadcasting spectrum, and grants respect to the dignity of human beings and protects children. In addition, it deals with radio and television accessibility, namely for people with different hearing and visual capabilities. The CSA is also responsible for assessing programming and appointments, and it draws up and delivers opinions.

4.4 Autoritá per le Garanzie nelle Comunicazioni (AGCOM), Italy

The Autoritá per le Garanzie nelle Comunicazioni is an independent body of communication in Italy, which is made up of four different organisations: the Presidency, the Infrastructures and Networks Commission, the Commission for Services and Products and the Council. Its competences include telecommunications, radio and television.

It supervises the freedom of competition within the communications market and grants key rights on top of copyright and authors' rights since the year 2000. As in previous cases, this body delivers penalties in those instances when there is a violation of those children's rights that are protected by the broadcasting sector's Self-regulatory Code.

This body operates through two main dimensions: (1) to guarantee the rights of the operators providing solutions to any controversy, supervising the application of legislation and protecting authors' rights; (2) to guarantee the rights of users by ensuring the quality of the services, products and advertising, resolution of conflicts, regulation of access' rights, protection of disadvantaged groups and protection of the social, political and economical pluralism within the broadcasting sector.

The independence of this organisation is being disputed due to the model used to appoint its members. Each one of the two Commissions is made up of a Presidency and four Commissioners that are designated by the First Minister following a correlative allocation with the groups in the Parliament.

The AGCOM funding source is mixed. Around seventy-nine per cent of their budget comes from the State, another seventeen per cent comes from television taxes, and another ten per cent from taxes paid by other operators that own a license and the remaining from different sources (Azurmendi et al. 2009).

4.5 *Entidade Reguladora para a Comunicação Social (ERC), Portugal*

The Entidade Reguladora para a Comunicação Social (ERC) was created in 2005 in order to regulate and supervise the operations of social communication bodies in Portugal. It benefits from administrative and financial autonomy and is independent. This body is formed by a Regulatory Council, an Executive Direction, an Advisory Council and a Single Auditor. It is funded by the State Budget, taxes from members that carry out activities in media outlets, professional fees from radio and television operators, money coming from fines, sales of their own actives and financial investments. Its purpose is to grant licenses, to supervise all private and public media outlets, to draw up norms and to perform consultation and monitoring.

The Regulatory Council is formed by the presidency, the vice-presidency and three more members. The Republic's Assembly is in charge of designating four members that shall choose a fifth for a period of five non-extendable years.

In terms of incompatibilities, it must be outlined that the members are nominated amongst highly suitable, independent and both professionally and technically professional candidates who are not subject to bias. Such members are not allowed to be part of executive boards (in private companies, unions, confederations or business associations) within the social communication sector or the Government within the last two years of their being appointed. Their responsibilities range from appointing a president and a vice-president, defining the activities of the ERC and overseeing its compliance with the law.

The Executive Direction, made up by the presidency, the vice-presidency and the executive direction, handles administration and finances of the ERC, while the Advisory Council is a consultative body formed by sixteen members of social and audiovisual associations. These are appointed by the competent branches of the representative entities for a three year period and can be replaced at any time. Opinions issued by this body are non-binding. Lastly, the Single Auditor is the body responsible for legal surveillance and also ensures the ERC's efficient financial and patrimonial management. It is appointed by the Assembly of the Republic for a period of five non-extendable years.

4.6 *Office of Communications (Ofcom), UK*

The UK was the first European country to create a regulatory body for the audiovisual sector through the Independent Television Authority, which is a body that came from the 'Television Law' of 1954. This was followed by the Radio Authority, the Broadcasting Complaints Commission and the Broadcasting Standards Council. As a consequence, there were five regulatory bodies, an authority for radio communications and another for broadcast at the end of the last century (López Cepeda 2010).

The Office of Communication (Ofcom) was created in 2002, although its competences were not fully transferred until the Communications Act came into force in 2003. Its members are elected by the State Secretaries of Industry, Commerce, Culture, Media and Sports, on top of an independent audit. The number of people ranges from three to six and its executive members are appointed at the Ofcom management's proposal.

In terms of responsibilities, the Ofcom manages the radio spectrum in the UK, providing licenses on the latter and for all kinds of audiovisual services. Furthermore, it can punish operators for non-compliance with legislation and even suspend or remove licences. It also handles the extension of existing license periods, deals with criminal proceedings, produces codes of practice, monitors and delivers consultation.

This body has nine members, six of which are appointed by the Government. The Government also proposes three more members from the own Ofcom's staff.

When it comes to the public service broadcaster the BBC, and in accordance with the White Paper from 2006, the Ofcom is responsible for the surveillance of the activities undertaken by the public media corporation within the digital environment. A report underlining aspects such as its financial situation, digitalisation, programming, production, audience or investments is issued every five years.

5 Regulatory Authorities in Others State Members

There are two main regulatory authorities in the audiovisual market in Austria. The Austrian Communications Authority (KommAustria) is the body that handles public media audiovisual services. Moreover, it handles the management of the radio spectrum and the digital broadcasting and also supervises private channels.

This is an independent authority made up of five members appointed by the Federal Presidency at the Federal Government's proposal, in accordance with the main Commission of the National Council. Members are nominated for a period of six years and must be jurists and have at least five years experience in the legal field. The incompatibility regulation is broad in order to ensure the independence of its members and their functions.

In addition, the Telekom-Control-Comission (TKK) was founded in 1997 and is a collegiate authority in charge of regulating the telecommunications market, the competition within it, the procedure to allocate frequencies and the rates that are offered by telecommunications companies. This body is integrated by three main members and three substitutes who are appointed by the federal Government for a five-year period.

In Belgium, the regulatory authority for the French community is the Conseil Supérieur de l'Audiovisuel (CSA), which is responsible for regulating both public and private audiovisual media. It is made of the College of Opinion, a body delivers opinions on all audiovisual related matters, and the College of Authorisation and Control, a decision-making body that is responsible for allocating broadcasting licenses to radio stations and private televisions established within the French community, supervising the compliance with the law and penalising breaches of legislation.

The College of Opinion is made up of the presidency, three vice-presidencies and thirty members, each with a substitute, appointed by the Government for a period of five extendable years. Members must be professionals from different segments within the audiovisual sector and must represent miscellaneous ideologies and philosophical approaches.

The College of Authorisation and Control consists of a presidency, three vice-presidencies and six more members, three of which are appointed by the Council of the French Community and three by the Government. Their term lasts for four extendable years. Such. Members must be highly competent in the fields of Law, radio broadcasting and communication and shall represent different ideologies and philosophical approaches.

The regulatory authority for the Flemish community is the Vlaamse Regulator voor de Media (VRM). It has been taking care of media regulation, mediation, complaints management and breaches of legislation since 2006. Moreover, it deals with child protection issues, bias and audiovisual broadcasting licensing management. The VRM has three members in its administration board and an executive management appointed by the Flemish government.

The regulatory authority for the German community is the Medienrat, which is responsible to the German ALM. Its functions are approving licences for broadcasting media, supervising programmes and resolving complaints.

The Bulgarian audiovisual authority is the Council for Electronic Media (CEM) which is in charge of supervising the running of both public and private media. The digital communications networks are regulated by the Communications Regulation Commission. This is made up of five members, three of which are chosen by the National Assembly and two by the Presidency of the Republic. Their term lasts six years and they can be re-elected.

The Cyprus Radio-television Authority is an independent regulatory body that allocates licenses, monitors audiovisual media broadcasts, on-demand services, author's rights and devises a methodology for measuring audience share.

This body defines different categories that the allocated licences would fall into and the responsibilities that are associated with these in terms of service, evaluation criteria, fees for each licence and the payment of certain rights. At the same time it, deals with property restrictions, penalties for breaches in legislation and boundaries for each type of licence.

It is made up of seven members proposed by the Council of Ministers. Candidates come from various fields: from art, science and technology to people with a remarkable knowledge and experience in audiovisual media and high standards of professionalism. Their term lasts six extendable years.

The Agency for Electronic Media, in Croatia, is an external and independent body that regulates private and public media. It also issues licences, supports launching new television and radio channels both traditional and on-demand, and grants permits for satellite, cable and other kinds of broadcasting services. This body is integrated by six members, all recommended by the Government of the Republic and ultimately appointed by the Parliament. Their term lasts five years and they may be re-elected.

The Agency for Culture and Palaces from Denmark is the regulatory body for the audiovisual media and is responsible for the country's cultural heritage. Its main duties are to issue licenses, to supervise the public and private audiovisual media, deal with and solve complaints and also penalise any breach in legislation. This body is formed by eight members, seven of which are appointed by the Ministry of Culture amongst experts in the legal, financial, administrative, media-related and cultural fields, whilst the eighth is suggested by the organisation representing the audience.

The Council for Broadcasting and Retransmission is the audiovisual authority in Slovakia. It takes responsibility for the management of frequencies and issue authorisations to audiovisual media. It also supervises public and private media, deals with complaints and penalises infringements of law. This body is composed of nine members designated by the Parliament from professionals, who operate in civil associations and institutions related to the audiovisual sector. A term is six years long and they may be re-elected.

In Slovenia, the Agency for Communication Networks and Services (AKOS) has operated like an independent regulatory authority since 2011. Some of its responsibilities are to control the compliance to ethical and professional standards, to ensure social pluralism, to issue licenses and to supervise the efficient running of public and private media. This body is formed by five members that are named by the Government for a five-year extendable period.

In Estonia, the independent regulatory authority is the Estonian Technical Surveillance Authority (ETSA). Together with the Estonian Competition Authority, it replaces five government authorities that operated previously. It issues licences and authorisations for frequencies, supervises the media and controls networks of digital communications. It is formed of five members.

The Finnish Communications Regulatory Authority (FICORA) is an independent body whose responsibilities are to issue licences and electronic authorisations, to survey the running of public and private media and to penalise any infringements in law and perform consultation. This body is made up of seven members.

The independent monitoring body in Greece is the National Council of Radio and Televisi. There are seven members in this body, whose term lasts four renewable years. These are suggested by the Presidency of the Parliament in relation to the political representation within the Camera. Their role is to grant licenses, monitor public and private media, and to penalise and issue consultation reports.

The responsibilities assigned to Hungary's National Media and Infocommunications Authority (NMHH) are also those of issuing licenses, controlling public and private media and dealing with complaints. It is made up of a presidency and four members, all elected by the National Assembly with two thirds of the members physically present during the voting. Their term lasts nine years.

Ireland's Broadcasting Authority (BAI) is made up of nine members, five of which are elected by the Government and four suggested by a parliamentary commission. Their appointment is based around their knowledge and experience. Their term is five years long, and may be extended to five more. This body can allocate licences, exert control over public and private media, penalise breaches in legislation and issue consultation reports.

The independent regulation authority in Latvia is the National Electronic Media Council, which is made up of five members elected by the parliament, and based on a consultation with associations and foundations related to media, education, culture, science and human rights. Each member must have a University degree and at least five years experience, either professional or academic. Their term is five years long and there is no possibility of re-election. Their responsibilities consist of issuing licenses, supervising public and private media and enforce punishments for non-compliance as well as performing consultation.

The Radio and Television Commision of Lithuania (RCTL) has amongst its competences the concession of licenses and the authorisation of frequencies, the supervision of public and private media, dealing with complaints and producing codes as well as assuming consultation and monitoring responsibilities. This body is made up of eleven members, two of which are elected by the Presidency of the Republic, three by the Parliament under the recommendation of the Commission of Culture, another by the Epicospal Conference, three by the Lithuanian Artists Association, and the last two by the Lithuanian Journalists Association. Their term is four years long, and can be extended to further four.

The Broadcasting Authority in Malta is an independent audiovisual management body that includes five members, who operate for a period of three extendable years. They are recommended by the First Minister, and then appointed by the Presidency of the country (after checking with the leader of the opposition). The main responsibilities of this body are the concession of licenses, the supervision of public and private media, the enforcement of punishments for non-compliance with the rules, and the publication of codes and norms as well as performing consultation.

The Commissariaat voor de Media from the Netherlands focuses on the concession of licenses, the supervision of public and private media and the publication of guidelines for action. It also provides advice within the consultation area, on top of performing monitoring duties. This body is made up of three members suggested by the Ministry of Education, Culture and Science, and they carry out their term for a period of five years that are extendable to five more.

The National Broadcasting Council is the independent authority in the Netherlands that takes responsibility for the issuance of radio and tv licenses, supervises the legislation fulfilment and controls media that provide a public service. Its members are elected by the Presidency of the Government and the parliament for six years' period.

The Council for Radio and TV Broadcasting (RRTV) of the Czech Republic is the independent audiovisual authority in charge of issuing licenses, extending authorisations, supervising private and public media, dealing with complaints, punishing non-compliance and performing consultation tasks. This body includes eleven members all appointed by the Presidency of the Republic from an initial proposal made by those political parties that have representation in the parliament. Their term lasts six years, and they may be re-elected a second time.

The National Audiovisual Council in Romania is an independent body that regulates the audiovisual system and whose responsibilities include issuing licenses, controlling public and private media, publishing codes and norms, penalising, monitoring and performing consultation. It is a body that has eleven members appointed by the Congress and the Senate for a six years period, although they may be re-elected.

The Swedish Broadcasting Authority supervises the radio and television markets, including services on-demand and teletext services. It is an independent body that deals with the concession of licenses, the supervision of both private and public media, the management of complaints, the punishments related to non-compliance and the monitoring. This authority is made up of eleven members, chosen by the Government, for a period of three extendable years.

6 Regulatory Authorities in Others State Members

All EU member states develop, what they refer to in their respective legislations, as the creation of independent audiovisual authorities that look after the compliance of the competition rules and safeguard the provision of pluralism and independence.

For European countries as a whole, such organisations share a number of responsibilities that outline their mission: the management of the radio and audiovisual spectrum through the concession of licenses, the extension of authorisations, the supervision of public and private media, the monitoring of contents and programming and the enforcement of penalties for non-compliance with the rules.

On top of these obligations, some of them also issue regulations or guidelines for action in the audiovisual sector—i.e. the Ofcom from the UK and the National Council of Radio and Television from Greece. Others take on consultation tasks, such as the Finnish Communications Regulatory Authority from Finland, or the National Electronic Media Council from Latvia or the Council for Radio and TV Broadcasting from the Czech Republic.

In terms of number of members, the range is from three to eleven, except for the Conseil Supérieur de l'Audiovisuel (CSA) from Belgium, related to the French community, which has forty-four members distributed throughout the various commissions that make up the body. The authorities with the lowest number of members are the Vlaamse Regulator voor de Media from Belgium –Flemish community- and the Commissariaat voor de Media from the Netherlands. On the opposite side with eleven members are the Radio and Television Commission from Latvia, the Council for Radio and TV Broadcasting from the Czech Republic, the National Audiovisual Council from Romania and the Swedish Broadcasting Authority from Sweden.

When it comes to the selection process, authorities from Germany, Portugal, Bulgaria, Croatia, Slovakia, Hungary, Latvia and Romania turn to their parliaments, but only Germany and Hungary require the need to achieve a qualified majority.

The Government is responsible for appointing members of independent audio-visual bodies in Spain, the UK, Belgium, Cyprus, Denmark, Slovenia, Ireland, the Netherlands and Sweden. Moreover, in the Czech Republic, Poland, Malta, Austria and Italy, the person who embodies the presidency of the government is also in charge of appointing people for those high positions.

As these are bodies that execute direct control over the concession of licenses and supervise the enforcement of legislation, they must base their internal operations on principles of governance, independence and transparency. Therefore, the election of their members is key to guarantee the smooth running of such institutions and ensure they conduct their work with objectivity.

Qualified parliamentary majorities shall become the most reliable mechanism to secure the highest degree of independence in order to avoid from appointments that originate from the political party that is in power at that time. This is the reason why this formula, which currently represents a minority within the comparative table, is the one that must cross borders and establish itself in all regulatory authorities of all EU member states.

References

Aznar, H., Villanueva, E., Zeid, F.A.: Deontología y autorregulación informativa: ensayos desde una perspectiva comparada. Universidad Iberoamericana (2000)

Azurmendi, A., De la Iglesia, A., Salvador, M.A., Muñoz, M.: El Consejo Audiovisual de Navarra. Sus funciones y competencias en el contexto europeo de regulación del sector audiovisual. Pamplona: Consejo Audiovisual de Navarra (2009)

Brugger, N., Burns, M.: Public Service Broadcasters on the Web: A Comprehensive History. Peter Lang, New York (2011)

Campos, F.: La RSC es una oportunidad para mejorar la credibilidad y confianza de la prensa. In XI Congreso de la Sociedad Española de periodística (SEP). El drama del Periodismo. Narracción e información en la cultura del espectáculo. Murcia, 24–25 April (2009)

Carniel, R.: Organismos independientes de regulación del audiovisual: Los primeros cinco años de la OFCOM y el desafío de la convergencia. Ruta **2**, 8–34 (2009)

Carniel, R.: Las actuaciones de los organismos reguladores del audiovisual. Los casos de España y del Reino Unido (2003–2007). Universitat Autònoma de Barcelona, Bellaterra (2008)

Center for Social Media: Public media 2.0: Dynamic, engaged publics. American University, Washington (2009)

Copé, J.F.: Commision rapport pour la nouvelle télévision publique. La Documentation Française, Paris (2008)

Debrett, M.: Tools for citizenship? Public service media as a site for civic engagement. An Australian case study. Telev. New Media **6**(16), 557–575 (2014)

Downie, L.J., Schudson, M.: The reconstruction of American journalism. Columbia J. Rev. **48**(4), 28–55 (2009)

García Castillejo, Á.: Una laguna fundamental del sistema democrático. El Consejo Estatal de Medios Audiovisuales de España. Telos, 68 (2006)

López Cepeda, A.: Órganos de control e xestión da radiotelevisión pública estatal e autonómica en España. Ph.D. dissertation, Universidade de Santiago de Compostela, Santiago de Compostela (2010)
Martínez, L.: La problemática del modelo europeo de televisión. Formats: Revista de Comunicació Audiovisual, 2 (1999)
Mayntz, R.: Nuevos Desafíos de la Teoría de Governance. In Instituciones y Desarrollo, **7**, 35–51 (2000)
Zallo, R.: Políticas de comunicación audiovisual (y para después) de la crisis económica en España. In: Campos, F. (ed.) El nuevo escenario mediático. Comunicación Social, Zamora-Sevilla (2011)
Zallo, R.: Dos modelos opuestos. Consejos del audiovisual en las Comunidades Autónomas. Telos, 68 (2006)

Tania Fernández-Lombao Journalist and Ph.D. in Communication from the Universidade de Santiago de Compostela. Head of the operational management area of programs at Radio Galega and professor in the Degree of Journalism at the USC. In the field of research, she works as a member of the group Novos Medios from the same university, and she focuses on public service in the European Union. Lugo, Spain.

Paulo Faustino Ph.D. in Media Studies. President of Media XXI/Formalpress (consulting, research & publishing company) and director of the Media XXI Magazine. He published 6 books (2 as co-author) and several scientific articles in academic journals. He is an editorial board member of the International Journal on Media Management and a board member of the European Media Management Association (EMMA). Porto, Portugal.

Innovation, Transmedia and Neuroscience in Television

Beatriz Legerén Lago and Verónica Crespo-Pereira

Abstract This chapter explains how the physiology of the human brain affects the way fiction products are processed and how emotions have become the most efficient vehicle for transmitting information on transmedia platforms. Understanding this can improve the development and profitability of audio visual products. The way in which humans process information is analysed, before reflecting on the value of emotions as a way of transmitting information in stories. The way in which emotions are applied in televisual and transmedia productions has been reviewed. Finally, the role of digitalisation as a source of innovation is explored, as it enables the creation of non-linear, organic transmedia works that enrich the television experience and provide emotional experiences to the audience. Organic transmedia projects, such as Norwegian production *Skam*, will be used to illustrate the findings in this chapter.

Keywords Neuroscience · Storytelling · Narrative · Organic transmedia Emotions

1 How We Think

Ever since the apparition of the human race, the brain has evolved to develop mechanisms to enable the transmission of thought in a way that can be understood by others. First came concepts, then language allowed concepts to be expressed through stories. Grammar and syntax, the rules and principles that are the building blocks of language and the relationship between words, appeared with the introduction of written language. Since then, new ways of expression and storytelling have been developed, but stories have always been at their centre.

B. Legerén Lago (✉) · V. Crespo-Pereira (✉)
Universidade de Vigo, Pontevedra, Spain
e-mail: blegeren@uvigo.es

V. Crespo-Pereira
e-mail: veronicacrespopereira@gmail.com

M. Túñez-López et al. (eds.), *Communication: Innovation & Quality*, Studies in Systems, Decision and Control 154, https://doi.org/10.1007/978-3-319-91860-0_7

Human beings are predisposed to process information better in story form. Evolutionist biologists like Pinker (1997) confirm that human being prefer stories to other forms of communications because they stimulate the brain. Renvoisé and Morin (2006) explain in detail the reasons behind this preference. The brain is structured in three interconnected parts, each of which has a specific function. The new brain processes data and shares deductions with the other two parts. The middle brain processes feelings and emotions and also shares information with the other two parts. Finally, the primitive brain, also known as the reptilian brain, makes its decisions based on information received from the new brain and the middle brain. The primitive brain is the result of millions of years of evolution and is identified as the "survival brain". It does not understand complex messages; it acts by instinct and prefers images to words (Morin 2011). The primitive brain is vital because it is the oldest part of our brain and it is the part that connects with the stories we are told (Renvoisé and Morin 2006).

Stories are the backbone of entertainment industries such as television, film or video games. We are all attracted to well-told stories; as their content and the way there are told determine their impact on the audience (Renvoisé and Morin 2006). The structure that underpins the concepts in stories is the map to understand, give meaning, remember and plan our lives (Haven 2007). Mandler (1984) states that "experiences not framed in story form suffer loss in memory".

What exactly are stories and what are its components? In Fisher's narrative paradigm (1984), all meaningful communication is defined as forms of storytelling or reporting of events. Information presented in a story form is better remembered than if it is presented in other narrative forms.

A story is not what is being narrated, but the way it is structured. In his Poetics, Aristotle stated that a story has to have a beginning, a middle and an end. The elements that form a story need also to be taken into account. After analysing various works considered as good examples of storytelling, Johnson (1999, cited in Haven 2007: 78) identified seven elements that all successful stories must have:

1. Goals: Actions are directed toward goals.
2. Motives: Actions and goals are explained by giving reasons why they were identified and performed.
3. Agents: Some character must adopt the goal and perform the action.
4. Contextual Circumstances: Actions are embedded within a morally significant context that determines their character.
5. Interactions with Others: Events unfold as protagonist interacts with other characters in the story.
6. Meaningful Existence: All goals and actions are part of an intentional hierarchy of thoughts and actions that leads to an attempt to lead a meaningful, fulfilled life.
7. Responsibility: Agents must be responsible for their actions.

To summarize, a story is the narration of related events that are experienced by an agent (character) in a fictional environment (a representation of the real world) to achieve an objective; they must follow a logical sequence consistent with the principle of causality (each event has a "before" it comes from and an "after" it leads to). So, if stories are composed of characters, events, a sequence, an objective, a meaning and a motive, what makes them interesting?

2 The Value of Emotions to Transmit Knowledge

Human beings are prepared to understand each other through narrative structures that are capable of creating emotional responses; Bondebjerg (2014) states that "the ancient Greeks understood the connection between form, person and emotions when they defined good communication as creating a relation between ethos (personal credibility), logos (the power of arguments) and pathos (the power of emotions)".

The different elements of a production (editing, direction, narrative) are designed and orchestrated to offer specific sensory experiences and responses in the audience. Shimamura (2013a, b) uses the I-SKE framework to explain the components and process of the filmic experience whereby the author's intentions with the work will give rise to emotions in the audience/spectators/receivers: the author's intentions (I) with the work will provoke the sensory process (S) that, through knowledge (K) will cause the emotions (E). Emotions have been shown to be the main trigger to enhance attention, memory and attitudes. According to Pinker (1997), "humans seek meaning through feelings".

Cognitive psychology studies structures, processes and representations and analyses how people perceive, learn and remember information. This discipline assumes that the human mind works in a similar way to a computer: input–coding–processing–storage–decoding–output. This metaphor is only functional, as computers operate with symbols following rules and use simple operations.

The similarities of this process with the human mind cause psychologists to use computers as the model to formulate psychological hypotheses and develop theoretical interpretations. Depending on the program it uses, a computer will display one result or another. Similarly, our brain also processes information; but it's the emotions created by the stories that are the best method to transmit knowledge, stimulate learning and to even promote attitudes. Renvoisé and Morin (2006: 92) explain the capacity that stories have to move us in the following fashion:

> The impact a story produces is due to the fact that the primitive brain thinks that it is really you who is experiencing the story. Stories lead the audience to a world of sensory impressions that make it impossible to the primitive brain to distinguish reality from the story. The primitive brain perceives that it has had an experience in reality, even though it has heard that it is a story.

According to Renvoisé and Morin (2006) there are six key factors that ensure our message is received and understood by our primitive brain: we only remember the beginning and ending of things; we need tangible, non-abstract information; we are sensitive to contrasts (before/after, fast/slow); we are concerned with ourselves; we accept visual information better; and emotions trigger our primitive brain, impacting the processing and retention of information. Some of these elements that open the door of the primitive mind are the cornerstone of the entertainment industries: visual and emotional elements are the drivers of the film and television industries.

These authors have also analysed the way commercial messages need to be crafted so that they have an impact on this part of our brain; they suggest that to create good stories we should:

- Create a world of sensory stimuli using audio, visual and kinaesthetic signals to make the primitive brain that the story is real.
- Create a strong link between the story that is being told and the audience's environment: why should they be worried?
- Ensure that the stories that are told have a message.

We all like a good story, even if it is not true, because through its characters, its plot and the use of language they can create an emotional response within us. What role do emotions play in our engagement with stories? Do our beliefs influence how we enjoy them, even if we are conscious that they are not real?

One of the tools that cognitive psychologists use is the information processing paradigm, which has identified as top-down processing the capacity that previous knowledge has to awaken the senses and emotions. In other words, events are anticipated and this previous knowledge shapes our response to them. Several theories have been promulgated based on this idea that emotions are based on previous knowledge and beliefs. As Frome (2006) mentions, the first of these is called the illusion theory, that states that we are attracted to a story because we want to believe that it is true. Another of these theories is the pretend theory, which argues that our we pretend a story is true, even when we know it isn't real.

But human emotions are too complex a field to be analysed in a simplistic, isolated manner. Frome (2006: 18) proposed a framework to study emotions and how they affect us when interacting with a work and when acquiring knowledge, not only as spectators but in some cases as authors and participants in the same work. This is the hybrid illusion theory.

> The hybrid illusion theory helps us answer a question that the other theories could not: why films, literature, and videogames excel at producing different types of emotions. Each medium provides certain features of prototypical emotion-eliciting situations and fails to provide others. Film provides high visual resolution but no interactivity. Literature can expertly convey abstract concepts but cannot present visual information rapidly. Videogames provide interactivity but usually do not provide the depth of storytelling or characterization regularly achieved by other media. Consequently, each medium will excel at generating the emotions that most strongly rely on the features it can simulate or provide.

3 Use of Emotions to Create an Audiovisual Product

In traditional storytelling, it is the audience who must imagine the actions of the characters. In entertainment media, it is actors who represent these actions so that we can recognise ourselves in them. Bartsch and Hartmann (2015) identify some of the reasons why humans use entertainment media:

> Recent theory and research have drawn attention to non-hedonistic viewing motivations, suggesting that entertainment media are used by individuals not only to relax and to improve their mood but also to experience a sense of challenge, meaningfulness, and personal growth.

It is this sense of challenge, meaningfulness and personal growth that must be targeted when creating audiovisual projects, and the use of emotion is key for this. Stories are an effective instrument to impact our primitive brain by stimulating our capacity for empathy and learning. Thus, being able to generate empathy is key in the success of entertainment products. Scientists from the University of Parma, Italy, led by Giacomo Rizzolatti, discovered that our capacity to feel empathy and emotions is caused by mirror neurons. These neurons, which have been observed in primates, fire both when an animal acts and when the animal observes the same action performed by another; they "mirror" the behaviour of the other as if the observer were acting (Lieberman 2012). These neurons are very useful in the survival of the species, as

> From an evolutionist perspective, it seems that what is important is not empathy that is felt for the pain of others but rather the understanding that what is happening to them is important for survival. The capacity to simulate what has been observed is especially relevant for learning and social interaction, creating a shared action space, necessary for prosocial behaviour and interindividual relationships. (García García et al. 2011: 270)

Empathy is very relevant in the consumption of entertainment media. The audience is able to feel what others are experiencing; through the use of the moving image, actors' performances and camera techniques, they can identify with the characters in the story.

4 Organic Transmedia

In the words of Liz Rosenthal, CEO of Power to the Pixel (cited in Bernardo 2011), "technology is dramatically transforming the way that audiences experience stories". Three important forces (stories, social media and gamification) are coming together in almost all aspects of our lives, transforming entertainment into something more social than never before. Producers can no longer think about telling stories on only one format, they must embrace the multiple platforms that are now available: linear, multimedia, cross media and transmedia.

Crespo-Pereira et al. (2017) argue that neuroscience can provide a theoretical underpinning for the use of information and communication technology (ICT) to create a relevant emotional and cognitive connection with the audience to optimise the results of television products.

> The qualitative leap that cognitive neuroscience has experienced in recent decades, thanks to economic stimuli, technological advances, and multidisciplinary research [...] has significantly improved the understanding of the brain mechanisms that affect learning, memory, attention or emotion. The results of this research show the positive effects of including neuroscientific methodologies in the design and testing of audiovisual content and the potential of ICT in the learning process as it drives tactile, kinesic, interactive and personalised experiences.

These technologies promote the conversion of spectators into users, giving them the option of participating in the action. To paraphrase Bavelier et al. (2010), television shows that promote the spectator participation are best for cognitive development and audience engagement. In this sense, television products that make use of new platforms and formulas such as cross media or transmedia could become magnificent vehicles to thrill and engage audiences and to achieve great results regarding learning and persuasion.

Cross media uses multimedia narrative and different channels to ensure that the same story reaches a wider audience. In other words, one story many channels. Transmedia storytelling refers to "stories that unfold across multiple media platforms, with each medium making distinctive contributions to our understanding of the world, a more integrated approach to franchise development than models based on urtext and ancillary products" (Jenkins 2006: 293).

Transmedia storytelling creates a world by combining many stories, in different forms and across multiple channels. In his blog about transmedia journalism, Moloney (2014) highlights the power of training and collaboration of multidisciplinary teams to maximize the impact of stories and media products across several platforms:

> Producing transmedia requires partnerships and collaboration. Few journalists have all the skills to produce many stories in many forms all by themselves. This is a team effort, and few legacy news companies control more than one or two media channels. To truly target their audiences they will need to collaborate with the owners of other channels in a mutually beneficial manner. Journalism is no stranger to that collaboration either. For example, newspapers and television stations have partnered on stories for decades.

The main characteristics of a transmedia product are that they offer more depth that a linear narrative, so the complexity and degree of sophistication of the story will also be greater (Jenkins 2006). Specific content needs to be created for each platform; and each piece of content must provide a satisfactory user experience in and of itself. According to the Producers Guild of America, a transmedia project must be composed of three storylines belonging to the same fictional universe, to be developed across different platforms.

There are different phases in the creation of a Transmedia product. The first one is to use different channels to expand a brand or an idea that comes from the analogue world. An example of this could be Harry Potter: first came the books, then the films, the video games, the websites, and new books of the Harry Potter universe. The second one is to take a digital content and expand it across other platforms; Final Fantasy can be considered an example of this.

Organic transmedia is a term created by Bernardo (2011). It refers to the creation of a universe (storyworld) with well-defined characters and with multiple stories that are adapted to the channel on which they will be delivered. To be successful, an appropriate timeline must be defined for distributing the stories across each platform. The objective is to make the audience become the greatest asset of the project, as they identify with the stories (target group), they obtain rewards (to learn about and become part of the characters' lives), they become motivated (as they are involved in the different events experienced by the characters) and they are driven to action (posts on social media). This universe will expand with the creation of a television series, social media profiles, websites, etc.

One of the characteristics of an organic transmedia product is the interaction between the user/spectator and the author. Research on interactive stories show that these generate interest, excitement, and high gratification levels. A link is created between affective-emotional and cognitive matters, as users identify with the characters, share their experiences and, at the same time, create affective relationships with them. On a cognitive level, users are stimulated intellectually by the interactive stories and by entertaining mental exercises; as their attitude and role evolves from passive to active, they are able to experience the feelings of the characters in the stories (Soto-Sanfiel and Igartua 2015).

5 Emotions in the Creation of an Organic Transmedia Product: The Case of Skam

The creation of an organic transmedia product implies the use of many platforms, including social networks. These are currently considered to be the greatest expression of interactivity (bidirectionality), whereby the audience becomes an active agent in the story. It is precisely this sense of agency, as well as the application of emotions, that enables de creation of a deeper relationship with the project; the audience feels it is a part of it.

As an example we will reference successful Norwegian series *Skam* (in English, Shame) produced by public broadcaster NRK. It has four seasons composed of between ten (seasons 3 and 4) and twelve (season 2) episodes, with a duration that ranges from 20 to 50 min per episode. The reason for choosing a television series is that

> First, unlike other narratives found in films or novels, TV series are constituted as a narrative-in-progress. Second, the TV series audience is much more important in the production of the series and in artistic decisions: throughout the development of a series, the audience becomes an essential element of production and is much more powerful and active than in other formats or genres. [...] the feedback that the audience provides through various channels (including new media), the expectations that are generated or the impact on and debate about certain characters and storylines can be decisive. (Martínez and González 2016)

Skam follows the daily life of a group of teenagers at the Hartvig Nissen School in Oslo as they grapple with existential crises, sexual identity issues, questions about religion or eating disorders. Its creator, Julie Andem, wanted to design a series for teenagers that dealt with their problems using their own language and with their own rules. Her intention was to reengage with an audience group that had abandoned the public broadcaster in favour of new forms of entertainment.

Her first choice was to avoid conventional broadcasting and to use the possibilities offered by NRK3, an online channel, to release daily short content capsules in real time. There was no fixed release schedule for these capsules, which could be at any time of day or night; alerts were sent out on social media informing of each publication. In addition, a weekly compilation programme was broadcast every Friday.

Her production approach was based on the NABC (Needs/Approach/Benefit/Competition) model. Instead of relying on information from different sources, she conducted in-depth interviews with Norwegian teenagers before writing the scripts. The language used by the characters, which constantly include English words and expressions, reflects the real way Norwegian teenagers communicate.

The characters in the series face problems, questions and needs that reflect real-life issues of the intended target audience. To achieve this, each series focuses on one of the main characters and deals with specific topics. In order to be as realistic and credible as possible, storylines are updated to reflect the feedback from the actors and from the audience, sent via the series' website and social media profiles on Facebook and Tumblr.

In the first season, the main character is Eva; it focused on relationship difficulties, loneliness, identity, and belonging. Noora is the protagonist of the second season, which deals with a troubled relationship, feminism, eating disorders, and sexual assault. Isak takes the lead in the third season, which focuses on coming out difficulties, homosexuality, mental health illness and religion. The fourth focuses on the Islamic religion, forbidden love, and the Norwegian Russ celebratory period. Driven by Sana, this is the final season of the series; it premiered in April 2017.

Skam can be considered an organic transmedia production because the use of different platforms has been contemplated since the creation of the storyworld in which the action will take place. The different characters have individual social media profiles with which the audience can interact, adding realism and closeness to the plot. On-screen, text messages between characters can replace dialogues. Off-screen, the characters may exchange messages on Facebook or Instagram.

As has been seen, the new platforms are an essential part of the format: each episode is divided in daily capsules that are published daily, in "real time", on the NRK3 website. These are complemented with text messages or social media posts that are not included in the episode, using the characters' profiles on social media. In this way, the line between reality and fiction becomes blurred.

6 The Value of Stories

In the beginning, there were concepts. These evolved into stories to stimulate our brains. For stories to make sense, their content and the way they are told are key. The emotions that are transmitted through stories have become the best way to share knowledge. When creating a story, several key factors need to be taken into account to generate an emotional response in the audience; using tangible and visual information, establishing links with the audience and stimulating empathy are amongst the most important.

Information technology has enabled the development of stories that meet most, if not all, of these key factors mentioned above. Using multiple platforms, stories can come to life in different screens so that our emotions are more deeply engaged and we become part of the story; in other words, we become prosumers. In a product like *Skam*, we can become part of the story creation process, as we interact with the different items of content in which it has been structured. We become identified with the different characters and our experiences are intensified.

Human beings are predisposed to understand each other through narrative structures that can provoke emotional responses. The interactive and personalised content that is now possible to create using information and communication technologies offers new opportunities to attract and engage audiences. The creation of organic transmedia products that exploit and develop new stories and platforms will maximise the televisual experience and will optimise the audiences' cognitive and emotional processes.

References

Bartsch, A., Hartmann, T.: The role of cognitive and affective challenge in entertainment experience. Commun. Res. **44**, 29–53 (2015)

Bavelier, D., Green, C.S., Dye, M.W.G.: Children, wired: for better and for worse. Neuron **67**, 692–701 (2010)

Bernardo, N.: The Producer's Guide to Transmedia: How to Develop, Fund, Produce and Distribute Compelling Stories Across Multiple Platforms. BeActive Books, Lisbon (2011)

Bondebjerg, I.: Documentary and cognitive theory: narrative emotion and memory. Media Commun. **1**, 12–21 (2014)

Crespo Pereira, V., Martínez Fernández, V.A., Campos Freire, F.: Neuroscience for content innovation on European public service broadcasters. Comunicar **25**, 09–18 (2017)

Fisher, W.R.: Narration as a human communication paradigm: the case of public moral argument. Commun Monogr **51**, 1–22 (1984)

Frome, J.: Representation, reality, and emotions across media. Film Stud. **8**, 12–25 (2006)

García García, E., González Marqués, J., Maestú Unturbe, F.: Mirror neurons and theory of the mind in explaining empathy. Ansiedad y Estrés **17**, 265–279 (2011)

Haven, K.: Story Proof: The Science Behind the Startling Power of Story. Greenwood Publishing Group, Westport (2007)

Jenkins, H.: Convergence Culture: Where Old and New Media Collide. NYU Press, New York (2006)

Johnson, C.: Metaphor vs. conflation in the acquisition of polysemy: the case of see. In: Hiraga, M., et al. (eds.) Cultural, Psychological and Typological Issues in Cognitive Linguistics. Current Issues in Linguistic Theory, vol. 153, pp. 155–169. John Benjamin, Amsterdam (1999)
Lieberman, M.D.: A geographical history of social cognitive neuroscience. NeuroImage **61**, 432–436 (2012)
Mandler, J.M.: Stories, Scripts, and Scenes: Aspects of Schema Theory. Psychology Press, New York (1984)
Martínez, A.G., González, A.M.: Emotional Culture and TV Narratives. In: García, A. (ed.) Emotions in Contemporary TV Series, pp. 13–25. Palgrave Macmillan, London (2016)
Moloney, K.: https://transmediajournalism.org/
Morin, C.: Neuromarketing: the new science of consumer behavior. Society **48**, 131–135 (2011)
Pinker, S.: How the Mind Works. Penguin Books, New York (1997)
Renvoisé, P., Morin, C.: Neuromárketing: el nervio de la venta. Editorial UOC, Barcelona (2006)
Shimamura, A.P.: Psychocinematics. Exploring cognition at the movies. Oxford University Press, Oxford (2013a)
Shimamura, A.P.: Experiencing Art: In the Brain of the Beholder. Oxford University Press, Oxford (2013b)
Soto-Sanfiel, M.T., Igartua, J.J.: Cultural proximity and interactivity in the processes of narrative reception. Int. J. Arts Technol. **X**, 1–21 (2015)

Beatriz Legerén Lago Professor and researcher at the School of Communication and Social Studies of the Universidade de Vigo (Spain). Her research is focused on the design of interactive entertainment products and their use in non-entertainment environments. Before starting her academic career, she enjoyed a successful career as co-founder and CEO of an award-winning digital agency and videogame developer. She is actively involved in promoting the digital industry. Pontevedra, Spain.

Verónica Crespo-Pereira Degree in advertising and public relations from the Universidade de Vigo and Master's degree in audiovisual production and management from the Universidade de A Coruña. Her professional activity has been developed in the audiovisual TV sector. Nowadays, she works as a teacher and researcher at the Universidade de Vigo where she carries out her doctoral thesis about the implementation of neuroscience methodologies in the study of TV audience. Pontevedra, Spain.

Interactive and Transmedia Documentary: Production, Interface, Content and Representation

Jorge Vázquez-Herrero and Arnau Gifreu-Castells

Abstract In recent years the way we produce new media has changed dramatically. Interactive nonfiction narratives have transformed the processes of producing, distributing and showing documentaries, and especially the processes involved in how the viewer relates to the text. One of these new media forms of narrative expression is "interactive documentary". In this work, first, we explore the evolution of the documentary genre as an expression form, then we analyse the current state of development of the interactive documentary. Second, we analyse five representative projects from different critical perspectives: production, interface, content and modalities of representation. Finally, we outline a set of issues that emerged from the analysis performed that need to be resolved.

Keywords Interactive documentary · Transmedia documentary · Webdoc
Comparative analysis · Nonfiction · Digital storytelling

1 Introduction

As a new media object, interactive documentary challenges traditional methods of study and dissemination. Textual analysis, which has been a key method of documentary research since the 1970s, becomes difficult when there are several potential ways of relating to the documentary text. This new phenomenon needs to be studied from global, multidimensional and complementary perspectives, taking into account, for example, aesthetics, economics, sociology, technology, etc.

This work analyses a series of interactive documentary projects developed during the representative period of 2010 to 2014 from different perspectives. Projects

J. Vázquez-Herrero (✉)
Faculty of Communication Sciences, Universidade de Santiago de Compostela, A Coruña, Spain
e-mail: jorge.vazquez@usc.es

A. Gifreu-Castells
ERAM - University of Girona, Salt, Spain
e-mail: arnau.gifreu@eram.cat

M. Túñez-López et al. (eds.), *Communication: Innovation & Quality*, Studies in Systems, Decision and Control 154, https://doi.org/10.1007/978-3-319-91860-0_8

selected for the study are *Out My Window* (Cizek 2010), *Insitu* (Viviani 2011), Alma, A Tale of Violence (Dewever-Plana and Fougère 2012), A Short History of the Highrise (Cizek 2013) and *The And* (Adizes 2014). First, we explore the evolution of the documentary genre as an expression form, then we analyse the current state of development of the interactive documentary. Second, we will approach the five selected projects from different critical perspectives: production, interface, content and modalities of representation. Finally, we outline a set of issues that emerged from the analysis performed that need to be resolved.

2 Theoretical Framework

2.1 Changing Forms of Expression of the Documentary Genre

Throughout its history, the documentary genre has had different forms of expression. The initial predominant form was linear and audiovisual, which emerged with the invention of the cinematograph and film in 1895. Film was originally nonfiction—the so called *actualités* by the Lumière brothers—however, several authors like Georges Méliès, David W. Griffith and Dziga Vertov, among others, changed the course of narrative and fiction in film appeared in the early 20th Century.

Interactive narrative appeared during the last decades of the 20th Century thanks to the evolution of digital technology, expanding the possibilities of nonfiction audiovisual forms of expression and sparkling a period of experimentation with new formats. In this new context, the classical sequential composition—introduction, middle and end—broke down and user interaction in the narrative became important.

Today, transmedia storytelling, which coexists with the previous two forms, is a type of narrative which expands its content through different media and platforms, always seeking the complicity of prosumers (Scolari 2013). Pioneering transmedia documentaries like *Guernica, pintura de guerra* (CCRTVi and Haiku Media 2007) and *Herod's Lost Tomb* (Barrat and Grant 2008) should be mentioned, although the work with most impact produced in recent years is possibly *Highrise* (Cizek 2009–2015) by the National Film Board of Canada.

We are living in a period of innovation and reinvention of the documentary form. But this situation is not new, it has also arisen at different times in history, such as when printing was invented in the 15th Century, photography and optical toys were developed during the 19th Century, and particularly during the first decades of the 20th Century, when the artistic avant-garde proposed using audiovisual art in a disruptive way, breaking the rules and conventions established in a provocative and transgressive manner.

2.2 Approach to the Interactive Documentary

John Grierson, key British documentary school figure, was the first one to use the term "documentary" to describe the film *Moana* (Flaherty 1926) by describing it as "documentary value", and engendering it with "the creative treatment of reality" (Grierson 1926). The visual documentation Grierson regarded as "documentary evidence" was the recreation of the daily life of a Polynesian boy (Rabinowitz 1994).

In the late '80s of the last century, as a result of this hybridization between documentary film and interactive digital media, a new format, which is still in a phase of growth and adaptation to the environment, appeared in the communication ecosystem. It is called "interactive documentary", a genre in gestation that uses multiple media and platforms for production, distribution and exhibition, including the web (webdocumentary), hardware (installation), visual media (film and television), optical supports (CD-ROM, DVD-ROM, video disc) and several platforms (transmedia documentary), among others.

If the traditional definition of documentary is blurred and undergoes a continuous process of construction and deconstruction, the definition of interactive documentary is still in an earlier conceptual stage. Although there have been pioneering authors, like Glorianna Davenport (1995), who began experimenting with cinematic multimedia narratives in the mid-80s at MIT and who coined the concept of "evolving documentary" (1995: 1), research into this new hypertextual form of expression has taken off approximately since 2010 (Whitelaw 2002; Dinmore 2008; Renó 2008; Choi 2009; Almeida and Alvelos 2010; Renó and Renó 2011; Dovey and Rose 2012; Gaudenzi 2013; Nash et al. 2014; Uricchio 2015; Aston et al. 2017).

2.3 Early Experiments

The first simulation that is a clear precedent of this format was produced at the end of 1979 (*Aspen Movie Map*, MIT Media Lab), and various games were developed during the 1980s. At the end of the eighties, the first application called as an interactive documentary was created (*Moss Landing*, Florin 1989), and the following decade saw experimentation with formats for more than one support, mainly offline (CD-ROM and DVD-ROM), online and for interactive installation (*Voyager*, *RMN*, etc.). All this innovation was consolidated with the productions at the beginning, and mainly at the end, of the first decade of the 21st Century, and has continued up to the present where new tendencies are diversifying and examples of each category are increasing in number.

2.4 State of Development

In recent years, interactive documentary has experienced remarkable growth due to several favourable factors. Recently there has been a considerable number of interactive documentaries produced, enough to develop a systematic, differentiated and consistent conceptual framework. Besides initial experiments using optical technologies (Laserdisc, CD-ROM, DVD-ROM) in the early 90s in France and the United States, the most awarded and recognised works in this field are those by ARTE, with remarkable co-productions such as *Gaza/Sderot* (Szalat et al. 2008), *Prison Valley* (Dufresne and Brault 2010) and *Alma, A Tale of Violence* (Dewever-Plana and Fougère 2012), and the National Film Board of Canada, with productions such as *Highrise* (Cizek 2009–2015), *Welcome to Pine Point* (Simons and Shoebridge 2010) or *Bear 71* (Mendes and Allison 2012), or collaborations between these two organizations (*Fort McMoney*, Dufresne 2013; *In Limbo*, Viviani 2015; *Do not track*, Gaylor 2015).

As well as the works of the Canadian and French producers, there is currently much movement and development around this kind of documentary: production companies and agencies (Upian, Honkytonk Films, Submarine Channel, Helios Design Labs, Barret Films, etc.), digital newspapers (Le Monde, The New York Times, etc.), festivals and conferences (IDFA Doclab, i-Docs conference, interDocsBarcelona), research labs (MIT Open Documentary Lab, i-Docs Lab), platforms and initiatives (interDOC), interactive and transmedia markets (Cross Video Days, Sunny Side of the Doc, Sheffield DocFest, etc.), private funding (Tribeca Film Institute, Sundance Institute, etc.), indexes of projects (MIT Docubase, Docshift Summit, interDOC_indice), etc. This is just a partial state of the art because this institutional landscape is changing rapidly and the range of possibilities is constantly expanding, and now the changes are even larger because newspapers, NGOs, academia, multimedia companies, festivals, labs and museums have joined in.

3 Interactive and Transmedia Documentary in Recent Years

The work carried out by Vázquez-Herrero (2015) is taken as a precedent which provides a broad database of the interactive documentary's production. In this research, an exploration on the object of study between the years 2010 and 2014 is conducted, considering it the first five-year period of the current stage of consolidation of interactive documentary. Such research contributes with macroscopic-level data, useful to trace the contemporary tendencies.

From the list of 125 interactive documentaries analysed, the productions from France, the United States and Canada should be highlighted, as well as significant international coproduction, in which the contributions of France and Canada stand

Table 1 Projects selected in the period 2010–2014

Title	Year	Country	Authorship
Out My Window	2010	Canada	Katerina Cizek
Insitu	2011	France	Antoine Viviani
Alma, A Tale of Violence	2012	France	Miquel Dewever-Plana & Isabelle Fougère
A Short History of the Highrise	2013	Canada	Katerina Cizek
The And	2014	United States	Topaz Adizes

Source: Own elaboration

out. The most prolific company is the National Film Board of Canada (NFB), followed by the Franco-German channel Arte.

Another interesting approach in the search of a profile of the production is the yearly distribution, which reflects a continuous growth of the corpus. This evolution results in three phases: 13 products in 2010, an annual average of 22.5 products between 2011 and 2012, and of 33.5 between 2013 and 2014. Although this increase suffered an important boost in these five years, regarding the previous stage, the progression of the narrative non-fiction forms makes it possible for the growth and, thus, the development to be maintained. From 2009, the year in which significant projects like *The Big Issue* (Bollendorff and Colo 2009) or *PIB* (Choquette 2009) are released, or *Highrise* (Cizek 2009–2015) is initiated, the production with a greater corpus of productions is consolidated. Among these productions, five representative cases (Table 1), which come from the principal countries already detailed, are selected from the 2010–2014 period.

Out My Window (Canada, 2010) is a production by the National Film Board of Canada, directed by Katerina Cizek within the *Highrise* project (2009–2015). The documentary describes what is like to live in high-rise buildings in different places around the world, its cultural manifestations and most frightening realities. *Out My Window* received the First Prize at IDFA (2010), the Digital Emmy Award for Non-Fiction (2011) as well as other prizes (Fig. 1).

Insitu (France, 2011) is a coproduction of the French-German TV network Arte and Providences, directed by Antoine Viviani. This documentary shows urban life through different experiences such as poetry, magic and imagination, involving artists and citizens. *Insitu* was awarded First Prize at IDFA Doclab (2011) (Fig. 2).

Alma, A Tale of Violence (France, 2012) is a coproduction of Arte, the multimedia studio Upian and the photographic Agence Vu. *Alma, A Tale of Violence* was directed by Miquel Dewever-Plana and Isabelle Fougère and tells the story of a young girl who leaves her family to live in the *maras*, a violent gang reality in Guatemala. This documentary won the Innovation Award at Sheffield Doc/Fest (2013) and the First Prize in Multimedia at World Press Photo (2013), among others (Fig. 3).

Fig. 1 Navigation in *Out My Window*. *Source* Screenshot

Fig. 2 Access to the film or the collaborative map in *Insitu*. *Source* Screenshot

A Short History of the Highrise (Canada, 2013) is a coproduction of the National Film Board of Canada and The New York Times, directed by Katerina Cizek within the *Highrise* project (2009–2015). It uses public participation to focus on the current social consequences of living in large buildings. *A Short History of the Highrise* received the Innovation Award at Sheffield Doc/Fest (2014), the First Prize in Multimedia at World Press Photo (2014) and the Emmy New Approaches (2014) (Fig. 4).

Fig. 3 *Alma, A Tale of Violence* app. *Source* Screenshot

The And (United States, 2014) is a production of The Skin Deep, directed by Topaz Adizes. The documentary shows the dynamics of a relationship, and proves how important it is to speak sincerely. It was awarded First Prize in Multimedia at World Press Photo (2015) (Fig. 5).

3.1 Production, Distribution, Exhibition and Reception

The creation of a project of interactive or transmedia documentary presents some challenges; firstly, from the point of view of production. They are productions which require multidisciplinary equipment combining both audiovisual and technological competences. Among the professionals who are incorporated, there are profiles such as that of the web developer and designer, the creative technologist or the transmedia producer. There are also different models of production and new alliances among organizations.

The comparative analysis of these five selected projects gives us some insight into interactive documentary. Production in this kind of format has different origins and coproduction is often also relevant, for instance, with media (NFB, Arte) or French multimedia studios (Upian, Providences). Furthermore, we should highlight the participation of governmental organizations, such as the National Film Board

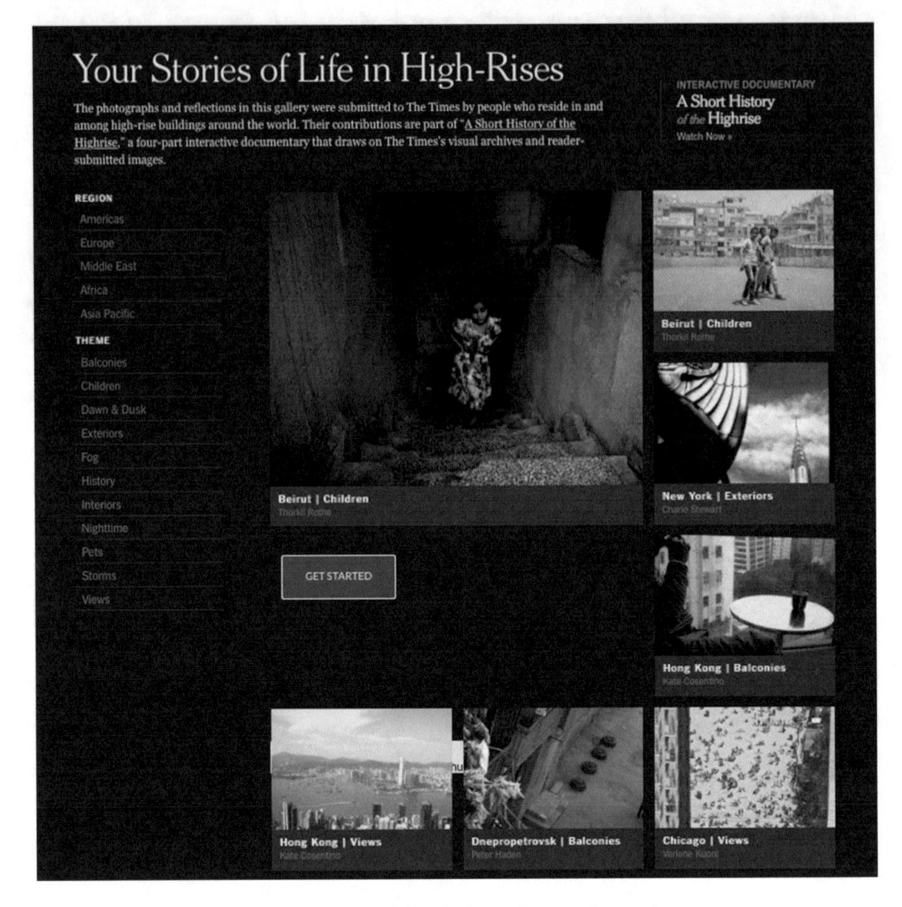

Fig. 4 Participation in *A Short History of the Highrise*. *Source* Screenshot

of Canada (*Out My Window, A Short History of the Highrise*), compared to the independent production of other projects, such as *The And*.

From the point of view of distribution, the fact that interactive documentary is closely related to the Web represents a new circulation, until now away from the commercial and conventional circuits for the film genre. Nonetheless, the conception of transmedia documentary also represents a rupture with the unique platform towards a combination of spaces where contents are diversified, independent and autonomous but maintaining relation.

Firstly, the main way to distribute these projects is through the Internet and festivals, or awards related to digital storytelling, such as International Documentary Film Festival Amsterdam, World Press Photo and Sheffield Doc/Fest. Secondly, the works are mainly exhibited through a web documentary (94% of the projects in 2010–2014 use this format as we observed in the macro analysis of 125 products); nevertheless, *Out my Window*, *Insitu* and *Alma. A Tale of Violence* use other media

Fig. 5 User interaction for personalization in *The And*. *Source* Screenshot

and platforms (feature documentary, apps, participative forums, etc.) to expand the story or get users to participate. *Alma. A Tale of Violence* used a transmedia strategy with a webdoc, TV, book and exhibitions.

It has drifted apart from the traditional consumption forms linked to the audiovisual products, as it can be concluded from webdoc. The demand to take part in the advancement of the story moves away from habitual passivity in front of the big screen or the television. On the other side, the progressive adaption to mobile devices opens new spaces and consumption forms.

Currently, the reception of interactive documentary tends to be individual, normally in front of a computer, with low effort for the user and a strong level of interaction. The user generally takes control of the progress of the story and participates in its expansion at different levels depending on the documentary.

3.2 *Interface*

The interface is constituted as a space where the user establishes contact with the content, generally by means of a screen: "content and interface merge into one entity, and no longer can be taken apart" (Manovich 2001). As a base of human-computer interaction (HCI), it works as "the environment of translation among subjects, their experiences, objectives and desires, and the technical devices" (Scolari 2015). It is a gateway which favours the continuity reality-representation between the user's position and the development of the story. It coordinates interaction and gives access to the navigation of the content.

That interface is a significant part of the interactive documentary because it is the intermediary between the author and the user, and it is also the connection between different media. For this reason, the interface design generally takes advantage of the entire screen to be more visual and immersive. Menus or timelines are also often included to show the position of the user and even the relationship between different levels of content.

We found that, generally, the author, in the beginning, suggests a linear story that can be changed by the user. The interactivity of interactive documentaries is usually limited due to the relationship between the author and the text. Sometimes strategies such as 360° spatial navigation, branchings or free exploration are developed. According to the analysis, we can say that projects, generally, have medium-high control established by the author, so user actions are usually limited by the creator.

3.3 Content

Interactive documentary gathers different media and languages which enrich the story and offer alternatives for the user. Opposed to a classic conception of multimedia, such as "the capacity, granted by digital support, of combining in a single message at least two of the three following elements: text, image and sound" (Salaverría 2005), the composition of the media has spread, nowadays, to virtual reality, 360° video, maps and geolocation, videomapping, etc.

The content is usually structured in several chapters organized by chronology, subjects, places or personal stories and in most of the cases follows the linear axial scheme. However, users have the option to access what they want. Within these interactive documentaries, we found that multimedia and mainly video assets are used extensively.

In the case of transmedia documentary, the proposal of different pieces, independent but related, represents a step beyond multimedia. This contemporary language includes the active user into a narrative universe open to expansion and maintains its defining feature of interactivity, both in navigation or with participation.

3.4 Modes of Representation of Reality

Lastly, the representation of reality in the documentary is achieved through the combination of different modalities. Nichols (1991, 2001) proposed for film documentary his modes of representation of reality, linking styles and practices along the history of the genre. In the poetic mode, associated to avant-garde, he aims at achieving a tone or affection above the diffusion of knowledge. In general terms, the expository mode refers to a discourse which is more characterized by rhetoric than by its aesthetics, with the figure of the omniscient narrator. In the observational mode he takes as a referent Cinéma Vérité and Direct Cinema, a direct observation on reality. The participatory mode adds the producer on the scene, which in the case of interactive documentary represents co-authorship shared with the user. The reflexive mode is based on raising awareness and on critical view. Lastly, the performative mode seeks expressivity with a more subjective meaning, closer to the artistic avant-gardes.

The combination gives rise to proposals of a greater complexity, as it can be observed in the selection of cases, where between three and five modes are combined. The reflexive mode appears in all of the products, and the observational and participatory modes were present in four of the five documentaries. The documentaries that use most modalities of representation of reality are *Insitu* and *A Short History of the Highrise*, both with five.

The first of them applies the poetic mode through a fragmented look of reality, with subjective impressions and experimental load. It also presents characteristics of the observational mode, by paying attention to what happens and looking at life as it is. In another level, it gets closer to the reflexive mode, since it presents a construction which seeks the spectator's reflection. Also performative, when it offers connections with the public's personal experience and when it gets interested in the subjective and affective dimensions, trying to express the producer and spectator's compromise with the world. The participatory mode appears when the users' contribution is taken into account. It is precisely by the users that the story is enlarged through their own vision of it.

A Short History of the Highrise is also complex regarding the forms of representing reality. From the expository mode, with the intention of documenting a historic reality through direct locution and experts; to the poetic, which incorporates subjective impressions provided by the readers of The New York Times. The participatory mode has an important role, that of inviting the spectator to get involved in the game of the documentary. In a superior level, also performative, which establishes a link with the user's subjective dimension. Due to the subject matter, the reflexive mode is present as well in an implicit message which, broadly speaking, calls the attention on the new way of life in the city.

4 Issues to Be Resolved

In the current state of development of interactive and transmedia documentary, a set of problems and shortcomings that can be used to stimulate further research in this area are identified:

(a) In accordance with the production findings, the industry must find new funding strategies and business models in order to finance these projects. NFB and Arte as public broadcasters are exceptional because they are interested in documentary film and have traditionally invested in innovation and culture. However, funding is usually independent, with a low budget coming from academia or non-profit institutions.

(b) Regarding reception of these works, it is imperative and urgent to study the new audiences and their relationships with interactive nonfiction narrative. Despite the development of digital metrics, like Google Analytics, the systems to quantify and identifies audiences need to be improved. This need is revealed by the lack of information on the audience response in terms of viewing, permanence and user participation in the cases studied.

(c) The author-text-interactor relationships show that different interaction systems are used in each interactive documentary, and in some cases they are so complex they are difficult to understand. This situation can cause confusion and even that users abandon the documentary experience. Usability and intuition are two fundamental characteristics of interactive systems to keep the user interested and stimulated. In this sense, usability—understood as the evaluation of how easy it is for the user to use the application (Nielsen 2003)—is a factor to take into account from the design of these works. Gamification could also help. This concept, applied to all those contexts that are not characteristic of the game, consists of "the process of game-thinking and game mechanics to engage users and solve problems" (Zichermann and Cunningham 2011). As has been seen in the project *The And* and others like *Réfugiés* (Huet and Siegel 2014) or *Pirate Fishing* (Ruhfus 2014), it has been used in interactive documentary.

(d) It is important to continue analysing projects to detect the transformations and innovations in interactive language and identify new modes of navigation and interaction. Language development in digital media must offer, in its current evolutionary maturity, deeper narratives that involve the user and become deep transformative experiences.

5 Consolidation and Diversification of the Interactive Documentary

Interactive documentary, as an expression form in the innovative field of interactive nonfiction, is now in a highly complex phase of consolidation and diversification. Interactivity is one of the most significant features of the format, which breaks down the linear narrative and allows participation, so that control and expansion of the story are handed over to the user.

Production is concentrated in France, Canada and the United States, together with a high degree of international coproduction and other countries with lower contributions. The business model is one of the challenges and the current production strategy is based on a public body (NFB) or associations between a broadcaster (Arte, France TV), traditional producers of documentary, multimedia studies and other stakeholders (newspapers, private institutions, academia, NGOs, etc.).

The Internet is the quintessential distribution medium and the webdoc is the most widespread form of exhibition. Although not as common in the field of nonfiction, we also detected very well designed and executed transmedia proposals, such as *Alma, A Tale of Violence* and *Insitu*. This multiplatform nature considers new audiences that flow across multiple devices and look for enriched experiences.

Interactivity, which facilitates the construction of more complex and personalized narratives, is made possible in the interactive documentary through multiple modes of navigation and interaction.

Substantial production of interactive documentaries in the period 2010–2014 and growing development since then suggests that in the coming years this expression form will become consolidated as an established and recognized interactive nonfiction format. Moreover, interactive documentary will find its place in the online media, like the multimedia report which is supported by The New York Times, The Washington Post and The Guardian, among others. It will also be interesting to explore its evolution within the gamification and didactic logics and dynamics, as well as its adaptation to mobile devices and virtual reality applications.

Acknowledgements The author Jorge Vázquez-Herrero is a beneficiary of the Faculty Training Program funded by the Ministry of Education, Culture and Sport (Government of Spain).
This article has been developed within the research project *Uses and informative preferences in the new media map in Spain: journalism models for mobile devices* (Reference: CSO2015-64662-C4-4-R) funded by Ministry of Economy and Competitiveness (Government of Spain) and co-funded by the ERDF structural fund, as well as it is part of the activities promoted through the International Research Network of Communication Management—XESCOM (Reference: ED341D R2016/019), supported by Consellería de Cultura, Educación e Ordenación Universitaria of Xunta de Galicia.

References

Adizes, T. (director): The And. The Skin Deep, USA. In: http://www.theand.us (2014)

Almeida, A., Alvelos, H.: An interactive documentary manifesto. In: Aylett, R., Lim, M. Y., Louchart, S., Petta, P., Riedl, M. (eds.), Interactive Storytelling: Third Joint Conference on Interactive Digital Storytelling, ICIDS 2010, Proceedings, pp. 123–128. Springer-Verlag, Berlin, Heidelberg, Germany (2010). ISBN: 9783642166389. https://doi.org/10.1007/978-3-642-16638-9_16

Aston, J., Gaudenzi, S., Rose, M. (eds.): i-Docs: The Evolving Practices of Interactive Documentary. Columbia University Press, New York, USA (2017). ISBN: 023118123X

Barrat, J., Grant, E. (director): Herod's Lost Tomb. National Geographic. In: http://channel.nationalgeographic.com/channel/content/herods-lost-tomb-3571/game/ (2008)

Choi, I.: Interactive Documentary: A Production Model for Nonfiction Multimedia Narratives. In: Intelligent Technologies for Interactive Entertainment, Third International Conference, INTETAIN 2009, Proceedings, 44–55. Springer-Verlag, Berlin, Germany. https://doi.org/10.1007/978-3-642-02315-6_5 (2009)

CCRTVi, Haiku Media (producers): Guernica, pintura de guerra (2007)

Cizek, K. (director): Out My Window. National Film Board of Canada, Canada. In: http://highrise.nfb.ca/onemillionthtower/ (2010)

Cizek, K. (director): A Short History of the Highrise. National Film Board of Canada, Canada. In: http://highrise.nfb.ca (2013)

Cizek, K. (director): Highrise. National Film Board of Canada (2009–2015)
Davenport, G., Murtaugh, M.: ConText: Towards the Evolving Documentary. In: ACM Multimedia 95. Electronic Proceedings, pp. 5–9. ACM, San Francisco, USA (1995). https://doi.org/10.1145/217279.215302
Dewever-Plana, M., Fougère, I. (directors): Alma, A Tale of Violence. Arte, Upian, Agence Vu, France. In: http://alma.arte.tv (2012)
Dinmore, S.: The Real Online: Imagining the Future of Documentary. Ph.D. Thesis, University of South Australia, School of Communication. Division of Education, Arts and Social Sciences, Australia (2008)
Dovey, J., Rose, M.: We're Happy and We Know it: Documentary:Data:Montage. Stud. Documentary Film **6**(2), 260–272. Taylor & Francis, London, UK, ISSN: 17503280, https://doi.org/10.1386/sdf.6.2.159_1 (2012)
Dufresne, D., Brault, P. (directors): Prison Valley. Arte, Upian. In: http://prisonvalley.arte.tv/ (2010)
Dufresne, D. (directors): Fort McMoney. Arte, National Film Board of Canada, Toxa. In: http://www.fortmcmoney.com (2013)
Flaherty, R.J. (director): Moana. Robert J. Flaherty (1926)
Florin, F. (director): Moss Landing. Apple Multimedia Lab (1989)
Gaudenzi, S.: The Living Documentary: from Representing Reality to Co-Creating Reality in Digital Interactive Documentary. Ph.D. Thesis, University of Goldsmiths, Centre for Cultural Studies, London, UK (2013)
Gaylor, B. (director): Do not track. Arte, BR, National Film Board of Canada, Upian. In: https://donottrack-doc.com (2015)
Grierson, J. (1926): Flaherty's Poetic Moana. The New York Sun (1926)
Huet, D., Siegel, L. (directors): Réfugiés. Arte. In: https://info.arte.tv/en/refugees (2014)
Manovich, L.: The Language of New Media. MIT Press, Cambridge, USA (2001). ISBN: 9780262133746
Mendes, J., Allison, L. (directors): Bear 71. National Film Board of Canada. In: http://bear71.nfb.ca (2012)
Nash, K., Hight, C., Summerhayes, C.: New Documentary Ecologies, Practices and Discourses. Palgrave Macmillan, London, United Kingdom, Emerging Platforms (2014)
Nichols, B.: Representing Reality. Indiana University Press, Bloomington, USA (1991). ISBN: 0253206812
Nichols, B.: Introduction to Documentary. Indiana University Press, Bloomington, USA (2001). ISBN: 0253339545
Nielsen, J.: Usability 101: Introduction to Usability. In: http://tfa.stanford.edu/download/IntroToUsability.pdf (2003)
Rabinowitz, P.: They Must Be Represented: The Politics of Documentary. Verso, London/New York, UK/USA (1994). ISBN: 9781859840252
Renó, D.P.: A montagem audiovisual como base narrativa para o cinema documentário interativo: novos estudos. Revista Latina de Comunicación Social **63**, 83–90 (2008). https://doi.org/10.4185/rlcs-63-2008-755-083-090
Renó, D.P., Renó, L.: Bogotá Atômica: o documentário interativo com estrutura algorítmica. Razón y Palabra, 76, 1–13, ISSN: 16054806 (2011)
Ruhfus, J. (director): Pirate Fishing. Al Jazeera. In: https://www.aljazeera.com/piratefishing (2014)
Salaverría, R.: Redacción periodística en Internet. EUNSA, Navarra, Spain (2005)
Scolari, C.: Narrativas transmedia: cuando todos los medios cuentan. Deusto Ediciones, Madrid, Spain (2013). ISBN: 9788423413362
Scolari, C.: Los ecos de McLuhan: ecología de los medios, semiótica e interfaces. Palabra Clave **18**(4), 1025–1056 (2015). https://doi.org/10.5294/pacla.2015.18.4.4
Simons, M., Shoebridge, P. (directors): Welcome to Pine Point. National Film Board of Canada. In: http://interactive.nfb.ca/#/pinepoint (2010)
Szalat, A., Ronez, J., Lotz, S. (directors): Gaza/Sderot. Alma Films, Trabelsi Productions, Arte, Bo Travail!, Ramattan Studios, Upian. In: http://gaza-sderot.arte.tv (2008)

Uricchio, W.: Mapping the intersection of two cultures: interactive documentary and digital journalism. MIT, Massachusetts, USA (2015)
Vázquez-Herrero, J.: Documental interactivo: un género multimedia en expansión. Estudio de desarrollo del género 2009–2014, aproximación a su definición y caracterización. Master's thesis, Universidad de Santiago de Compostela, A Coruña, Spain (2015)
Viviani, A. (director) (2011): Insitu. Arte, Providences, France. In: http://insitu.arte.tv/
Viviani, A. (director): In Limbo. Arte, National Film Board of Canada, Providences. In: https://inlimbo.tv/en/ (2015)
Whitelaw, M.: Playing Games with Reality: Only Fish Shall Visit and interactive documentary. In: Brunt, B. (comp.) Halfeti: Only Fish Shall Visit. Artspace, Sidney, Australia (2002)
Zichermann, G., Cunningham, C.: Gamification by Design: Implementing Game Mechanics in Web and Mobile Apps. O'Reilly, Sebastopol, Canada (2011)

Jorge Vázquez-Herrero Ph.D. Student in Communication and Contemporary Information from the Universidade de Santiago de Compostela, member of the Novos Medios research group (USC, Spain) and the Cátedra Latinoamericana de Narrativas Transmedia (Universidad Nacional de Rosario, Argentina). He does research on interactive nonfiction digital storytelling, focusing on the knowledge of interactive documentary and interactive narratives in online media. A Coruña, Spain.

Arnau Gifreu-Castells Research affiliate at the Open Documentary Lab (MIT) and part of the i-Docs group (University of the West of England). He has published various books and articles in his research area, interactive and transmedia non-fiction. He is a lecturer at ERAM (Universitat de Girona). He coordinates interDocsBarcelona (DocsBarcelona Documentary Film Festival) and collaborates with the Interactive Department at RTVE.ES. Girona, Spain.

Radiography of Public TV in Latin America

Catalina Mier-Sanmartín, Gabriela Coronel-Salas, Kruzkaya Ordóñez, Abel Suing and Carlos Ortiz

Abstract Public media have been a reality instituted in a great number of countries of the world. Europe has important references that have managed to build a solid public television system. In Latin America, channels fonded by the Estate generally had bad experiences because they have been more propagandistic television stations. Such public service system must consolidate an administrative, financial structure independent from its governments in order to permit and propel programming of public interest in which citizens are reflected. To date all television networks run a website as well as social networks. Television sees Facebook as a key emitter of informational events. From this, the 360° images, and especially the streaming, hatch the Network. Public TV in, terms of democracy, must be reinvented and converge for users are eager for new experiences.

Keywords Public television · Programming · Web · Social networks

C. Mier-Sanmartín (✉) · G. Coronel-Salas · K. Ordóñez · A. Suing · C. Ortiz
Universidad Técnica Particular de Loja, Loja, Ecuador
e-mail: cmier@utpl.edu.ec

G. Coronel-Salas
e-mail: glcoronel@utpl.edu.ec

K. Ordóñez
e-mail: kordonez@utpl.edu.ec

A. Suing
e-mail: arsuing@utpl.edu.ec

C. Ortiz
e-mail: ccortiz@utpl.edu.ec

M. Túñez-López et al. (eds.), *Communication: Innovation & Quality*, Studies in Systems, Decision and Control 154, https://doi.org/10.1007/978-3-319-91860-0_9

1 Introduction

To speak of public television is synonymous with autonomy of the political and financial power of a State. In Latin America government channels that have been established do not fulfill public service functions.

International organizations, such as UNESCO considers public media indispensable to foster media pluralism and in this spirit, they have ratified the imperatives of its complying with universality, independence and diversity principles. On the other hand, the responsibility to strengthen culture, education and reinforcing of freedom of expression, must be assumed by the Estate.

Some Latin American countries were deprived of having public service means with delivery logic. The media systems were configured by the private financial sector as providers of radio and television services, right from the time TV's first appeared.

The history of public TV, in the region, has been accompanied by low budgets and political pressures. In most countries, public channels have been more governmental and the editorial line responds to the current government, thereby changing with every new term. In the last decade, the public means were strengthened with the so-called "socialist" governments (Brazil, Venezuela, Chile, Argentina and Ecuador). Although it is attributed to neoliberalism and conservative sectors to urgently re-articulate and stop jeopardizing public media's existance. Take Ecuador's media for instance, as it could be threatened by its lack of a strong administrative, financial and citizen structure. Conversely, in the current media scenario, television breaks down in the face of political situations such as concentration of media, open marketing competition, cable TV content and the Internet. Television markets in Latin American and worldwide are currently submitted to such context. Also, the interest of the audience in relation to audiovisual products, social appreciation is done now through various media screens.

Recovering space for public service communication means fostering pluralism through legal structures that allows and guarantees access to information and that is, in accord to the spirit of The Bill of Magnes, so that everyone can freely express their opinions and constitute means of communication.

In different countries of Latin America, one can see a concentration of State media (radio, television, press) constituted by companies and public corporations with clear dependence on State budgets that, in turn, suffocate the media and end up becoming propagandistic, contrasting discourse with the content generated by private media.

In some countries, the channels have an advocate for audiences such as Ecuador Colombia, Argentina and Brazil, the latter, additionally, has advisory or central advice to the viewer. In other countries such as Bolivia, Uruguay or Venezuela, the contact is through the telephone, e-mail or attention centers to the listener. In Chile, Paraguay, Perú does not have any figure that allows the participation of the public.

2 Television Programming

Programing is done in close relationship with the audience as the content is conceived and selected according to the public. The available financial resources are a determining factor. The role of the programmer is to creatively and strategically materialize a grid that enables the positioning of the market by its televised offer. The actions taken around programming are linked to the marketing strategies with short or long term planning, according to the profitability and sudden changes in schedules, due to, perhaps, cancelled shows, new content, flexible possibilities at the time of programming.

There are several factors that affect programming and that determine the nature of the public television station. In this sense, we speak, in the first instance, of financing, that, according to Zallo "is committed to mixed financing, with recourse to advertising and collective funds, therefore avoiding contractions, such as public spending that paralyzes public services" (2013: 281). In the case of public educational television, its whole existence is marked by educational content whose purpose should not be media or economic power.

General public television, although framed in the concept of participatory practices with cordial philosophies (Hartley 2000) and share similar deontological codes, are not, however, far removed from the political power of the government of time, there are exceptions that mark a difference.

We must take into account that programing is a reflection of the business objectives and the communicative policy of each public channel; Taking into account mainly that televised products are at the service of the audience. In this context, available programming for public television as well as the type of shows delivered to the audiences, the origin of these products, determine how Public television channels in Latin America fulfill their public service role and impel independent national production and identity.

Considering that today channels must function on a fractional screen fee, as can be seen in Table 1, in financial terms, public television obtains most its allocation from State budgets and in other cases, additionally the laws have allowed them advertising revenue. On the other hand, public policies must be redesigned as technological advances and demands, such as Digital Terrestrial Television (DTT), require expansion and restructuring of the electrical spectrum.

2.1 *Profiles of Public Television Programming*

The programming public channels in Argentina, Bolivia Brazil, Chile, Colombia, Ecuador, Panama, Paraguay, Perú, Dominican Republic, Uruguay and Venezuela have established standardization schedules due to the diversity of shows, as demanded by the Law in each of these country.

Table 1 Profile of Latin American public media

Country	Company type public means	Administrative and management structures	Financing
[a]AR	Empresa Pública Radio y Televisión Argentina Sociedad del Estado (RTA SE): TV pública (Channel 7), Radio Nacional, RAE (Argentine Broadcasting Abroad)	Depends on the Federal Executive Branch	20% of taxes established by law. National budget appropriations. Sale of advertising Marketing of the production of its contents Sponsorships or donations
[a]BO	Radio and television do not belong to any company or media system: Radio Illimani; TV Bolivia; TVU (Canal 13)	Central Government Administration. The Directory of "BOLIVIA TV" is conformed by the Ministers of: Presidency, Planification of Economy, Development and Public Finance, Public Works Services and Housing. Education and Culture The minister of presidency presides the board and appoints the General Manager	The channel raises financing through: Capital contribution from the General Treasury only once; specific resources generated by the company; resources and assets from the Liquidation of the former National Bolivian Television Company—ENTB.d. Transfers of capital. The channel can coordinate the management of financing from other internal sources
[a]BR	Company Brasil de Comunicação (EBC): TV Brasil, TV Brasil Internacional, MEC AM and FM Radios In addition to Rio de Janeiro: National Radio AM and FM of Brasilia	EBC consists of three councils: administrative, financial and curatorial and executive management. The councils must regulate the management of the company in all its areas	The public system its ran by the State. Public broadcasters can raise external financing through program sponsorships, cultural support, institutional and legal publicity, services, licensing and sale of products and services
[a]CH	Radio and television do not belong to any company or media system: Televisión Nacional de Chile	It is an autonomous body, with legal personality and It is composed of eleven members, the President of the Republic is in charge of appointing the incumbent and the remaining members are named through the Senate	Chile does not receive resources from the State and his main source of financing is publicity

(continued)

Table 1 (continued)

Country	Company type public means	Administrative and management structures	Financing
[a]CO	Sistema de Medios Públicos de Colombia: Señal Colombia, Canal Institucional, Radio Nacional de Colombia, Radiónica, Señal Clásica, Señal Rock Colombia	The public media system is a part of the Ministry of Information Technologies and Communications. Formed by a board of directors composed of five members: Two appointed by the federal government. One elected among representatives of the regional television channels. One member gets chosen among professional communication associations (directors, producers, technicians, journalists, etc.). Lastly, one is picked among the Spectators Association	It is supported by government resources provided to the National Radio Television of Colombia (RTVC) Sponsorships and cultural supports The Fund for Television Development (FDTV) comes mainly from taxes on the services of broadcasting of the private sector
[a]EC	Empresa de Medios Públicos: TV Ecuador, Radio Pública, Diario El Telégrafo	The board is presided by: Secretary of Communication, as president of the board of the public company Radio and Television of Ecuador (RTVEcuador) Delegate from the executive board of directors. These appointments are based on numeral 1 Of article 7 of the Organic Law of Public Enterprises	Income is based on publicity sales Marketing of its communication products Funds obtained from donations, sponsorships, national and international cooperation
[a]PA	Radio and television do not belong to any company or media system: Radio Nacional, Radio Carlos Antonio López de Pilar, National Publicity	Television is subordinate to the General Directorate of State Media (DGME)	Publicity
[a]PE	Instituto Nacional de Radio y Televisión de Perú (IRTP): Radio Nacional, Radio La Crónica, TV Perú	It relies on the Federal Executive Power	Public Media Honorary Consultative Council, financed by the RTA SE itself, and formed by members of outstanding trajectory in the culture, education and communications of the country

(continued)

Table 1 (continued)

Country	Company type public means	Administrative and management structures	Financing
[a]UR	Radio and television do not belong to any company or media system Babel (97.1 FM); Classical (650 AM); Broadcaster of the South (94.7 FM); Radio Uruguay (1050 AM) National Television (TNU)	Functioning part of The Ministry of Education and Culture. Radios are managed by the radio television and Shows Officer (SODRE). The channel works under the Ministry of Education and Culture	
[a]VE	Radio and television do not belong to any company or media system Venezolana de Televisión	Subjected to the National Ministry of Communication and Information. The director gets appointed by a public executive	

Sources Brazilian Broadcasting Observatory, Mier (2015)
[a]Argentina, Bolivia, Brazil, Chile, Colombia, Ecuador, Panamá, Paraguay, Perú, República Dominicana, Uruguay, Venezuela

As depicted in Table 2, schedules are established in each of the Latin American countries according to a study conducted by a research group of Audiovisual Narratives of the UTPL, in 2016.

The profile of public TV programming in Latin America is defined mainly by informational spaces through the news, debates and informative specials. The news information is inscribed as one of the vital tools for delivering media to the audience and public market, as an informative showcase for the world.

Public TV that spend more time in minutes to informative content are public TV of Argentina, Bolivia TV, Public TV of Paraguay, TV Perú, TVN Chile, Ser TV and Ecuador TV. It is also noticed that there is a significant boost to the information programming provided by the multi-state public television channel TELESUR, which maintains an important (open) coverage in Latin America and which is strengthened by the active participation of its shareholders in the governments of Bolivia, Cuba, Ecuador, Nicaragua, Uruguay and Venezuela. There are cases in which information is transmitted through other genres such as the public channel "Señal Colombia" [5] the same that through its programing promotes the creation and innovation of audiovisuals via educational and cultural products. In Venezuela, Venezuelan Social Television prioritizes entertainment content, these spaces propel national independence and Latin American production (Table 3).

As a second revealing element is the presence of informative programs and actuality. More incidence is found in Ser TV of Panama, National Television of Uruguay, Bolivia TV and TV Brazil. In terms of a communicative ecosystem, these contents "on their own, demand the attention of journalists and, above all, the audience" (Cebrián 2014). They are available in magazines, cultural science and become a dif-

Table 2 Time schedules for countries

Country	Description	Law Article
[a]**AR** 06:00 a 22:00 22:00 a 06:00	To all public Adults	Ley de Servicios de Comunicación Audiovisual 26.522 de 2010
[a]**BR** 08:00 a 20:00 21:00 a 23:00	To all public Viewers ages 14 and 16	Ley General de Telecomunicaciones No. 9.4972 de 1997
[a]**CO** 15:55 a 16:55 08:00 a 10:00 06:00 a 15:55 16:55 a 22:30 22:30 a 00:00	Children (Mondays-Fridays) Viewers ages 14 and 16 Children (Saturdays & Sundays) Family Adults	Ley de Tecnologías de la Información y Telecomunicaciones No. 1341 de 2009
[a]**CR** 06:00 a 18:00 18:00 a 22:00 22:00 a 06:00	To all public Viewers ages 14–17 Adults	Ley de Espectáculos Públicos N° 7440 de 1994
[a]**EC** 06:00 a 18:00 18:00 a 22:00 22:00 a 05:00	To all public Viewers ages 14–17 Adults	Ley Orgánica de Comunicación 2013
[a]**PAR** 06: 21:00	Child protection programing	Ley de Telecomunicaciones. 1986
[a]**PE** 06:00 a 22:00 22:01 a 23:59 24:00 a 05:59	Child protection programing	Ley que modifica los artículos 40 y 70 de la ley 28,278. Ley de Radio y Televisión. 2015
[a]**UR** Señal hasta 21:30	Child protection programing	Decreto N° 169/005 de 31 de mayo de 2005. Artículo 1. 2005
[a]**VE** 07:00 a 19:00 19:00 a 23:00 23:00 a 05:00	To all public	Ley de Responsabilidad Social en Radio y Televisión. 2004

Source Observacom

[a] Argentina, Brazil, Colombia, Costa Rica, Ecuador, Paraguay, Perú, Uruguay, Venezuela

ferentiating point before private channels that base their programing on economic profitability instead of best servicing the public. With respect to the contents, their classification, variety and children's public channels, CERTV assigns more space to these issues. Also, Televisora Venezolana Social and Ecuador TV.

Educational contents are highlighted in channels such as TV Perú and Ecuador TV, strengthened by the introduction of DTT, which promotes, among its policies, proposals for cultural, educational, health and technological contents, according to laws and master plans to implement DTT in these territories. There are particularities to highlight in the programming of public channels such as the case of Public TV of Argentina that has given an impulse to the gastronomic contents due enactment of

Table 3 Programming genres

Information		Genres						
Country	Channel	News	Education	Fiction	Kids	Show	Gastron	Sport
[a]AR	Tv Pública	1950	1950	3900	750	600	600	0
[a]BO	Bolivia TV	1800	1800	0	275	0	575	450
[a]BR	TV Brasil	500	500	1000	1380	0	0	150
[a]CH	TVN	1575	1575	3150	0	1500	0	0
[a]CO	Señal Colombia	0	0	0	750	0	150	450
[a]EC	Ecuador TV	1200	1200	2400	1275	675	0	300
[a]PA	Ser TV	1500	1500	3000	525	0	0	0
[a]PAR	TV Pública	1800	1800	150	300	750	150	300
[a]PE	TV Perú	1590	1590	0	150	1020	0	0
[a]RD	CERTV	525	525	0	0	2250	300	150
[a]UR	TNU	900	900	0	600	1350	150	0
[a]VE	TVS	0	0	0	0	2250	300	300
	Telesur	2970	2970	0	0	0	0	570

Source Latin American Research Program
[a]Argentina, Bolivia, Brazil, Chile, Colombia, Ecuador, Panamá, Paraguay, Perú, República Dominicana, Uruguay, Venezuela

the Law on Audiovisual Communication Services "which generated a new map of the field of content production" with the inclusion of "social actors who were absent from the model of economic concentration of media" who assumed the challenge to "Produce content that is not only organized according to information needs, formative and recreational regions, but also have the imprint of constructing an image of their own from the identity and iconographic perspective" (González 2012:17).

The general programming of public channels are aligned with the communication policies established in each territory, this involves the participation of various actors that promote independent production with authentic audiovisual proposals, with independent brands. This is the principle under which public media acts, however, it should be noted that the public apparatus considers exchange of programming as one of the alternatives to strengthen public programming in Latin America. In Ecuador, public channels, before being required by the "Ley Orgánica de Comunicación" to strengthen audiovisual production systems, already had calls like "Mírame Ecuador" that sought new audiovisual proposals by independent producers.

The project was institutionalized four years after the beginning of the public television in the country and had the participation of 268 projects that came from 17 provinces of the country. It is clear the intention that the Latin American countries have in solidifing their national production, as all approved laws are directed towards this objective, in this respect Mastrini [9], one of the precursors for the approval of the Ley de Servicio de Comunicación Audiovisual of Argentina, maintains that "laws that have to do with the strengthening of a country's audiovisual sector should

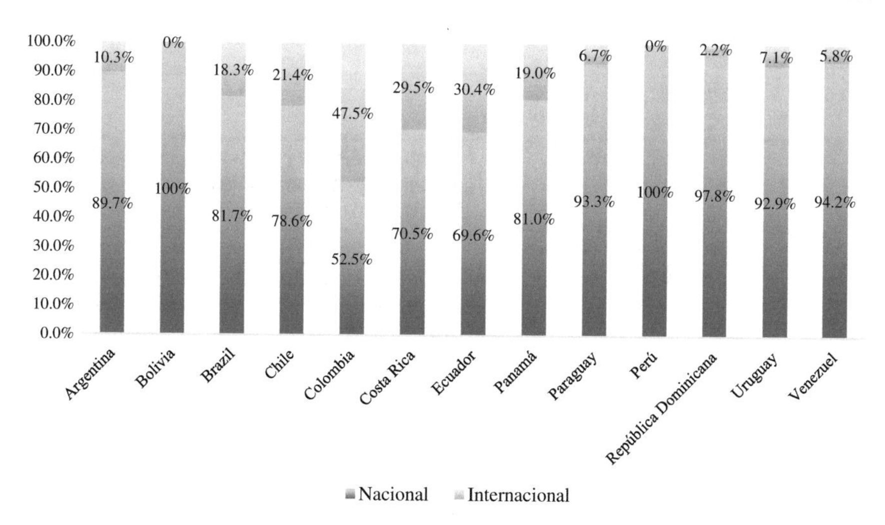

Fig. 1 Procedence of televised public production. *Source* Latin American Research Program (2017)

be considered as points of departure, not as points of arrival. Law is a condition of existence, it is something that allows certain issues to develop, however, law itself will not generate audiovisual production; this should create an environment that promotes the possibility of generating national audiovisual content." Public channels are those that bet most on national production, most of them build the programming grid with their own production. The most relevant cases are Bolivia and Perú, which have 100% of national production. However, Colombia, which in recent years has been able to consolidate its national television production, maintains similar percentages in terms of the origin of programming (Fig. 1).

The programs types imported the most are those of current affairs and soap operas, which are produced in other public channels from neighboring countries. All the informative programs are produced locally. Public channels have also become a strategic partner for independent audiovisual producers seeking new proposals for educational, cultural and informational programs. However, programmatic offers from public channels do not reach the desired levels of audience. Television Nacional de Chile has changed considerably since the dictatorship, in 1989 it remained the second most tuned channel in front of private TV; For 2013, it continued with that position (Fuenzalida and Pohlhammer 2013: 176), by 2015, it descended to the fourth place (Table 4).

3 The Digitization of Public TV

When viewers have the power to choose, television schedules do not remain the same as before. Netflix demands public broadcasters to reflect on their digital context. The transition to the information age is understood as a fulminating progress since its

Table 4 Audience on prime time (PT)

Country	Channel	Audience share (PT)	Description
[a]AR	Tv Pública	5.4	Fourth place, three private channels concentrate the largest percentage (Telefé, América 2 and El Trece)
[a]BO	Bolivia TV	No data	
[a]BR	TV Brasil	0.5	Ocupies the fifth place. The private channels Globo, SBT, Record, Rede TV exceed the rating of 10 to 20 points
[a]CH	TVN	8.2	In 2015 won fourth place, before Megavision, Universidad Católica Chile and Chilevisión.ColombiaSeñal Colombia
[a]CO	Señal Colombia	0.7	It falls under four private channels.
[a]EC	Medios Públicos EC	0.6	On 2010 the channel begins to be monitored by IBOPE and occupies the last place in rating
[a]PAR	Tv Pública	0.6	It occupies the seventh place in front of the private channels. SNT channel 9 and Telefuturo have more than 20 rating points
[a]PE	TV Perú	1.7	Private channels take the lead by 15 to 20 points
[a]UR	Televisión Nacional Uruguaya	2.9	Fourth and last place among the national television stations are occupied by public channel
[a]VE	Televisora Venezolana Social	1.8	According to Eurodata, the fifth and sixth place occupy Venezuelan television and Venezuelan Television respectively in the media panorama
	Televisora Venezolana	2.1	
[a]PA	Ser TV	No data	According to Eurodata, private channels Telemetro and TVN occupy approximately 24 rating points
[a]CR	Sinart Canal 13	1.3	Five channels of cut Private companies have higher ratios, 16 different points compared to Teletica channel 7
[a]RD	CERTV	No data	Private channels like Telesistema 11, registers in Eurodata (2015) rating of 18.5 points

Source Eurodata (2015) e IBOPE Ecuador, 2010

[a]Argentina, Bolivia, Brazil, Chile, Colombia, Ecuador, Paraguay, Perú, Uruguay, Venezuela, Panamá, Costa Rica, República Dominicana

inception, stands out for technical devices that transformed and brought men closer, shortening time and space in their communication; Such as telephone (1880), radio (1900), television (1930), computers (1940) and networks (1980), evolve day by day, furthermore, with the communicative needs of its users. Wolton clearly differentiates information from for it might be confused with communication. "Information is the message, while communication is the human relationship, much more complex," since the channels that are use to reach the public sometimes comply with short reporting without attracting spectators's attention entirety; it is complex since, at times, deciphering and understanding the message becomes a tedious and long process (2010, pp. 25–27).

Internet revolutionized communications since its inception. In the seventies, the idea of the US Army to keep in touch with its troops would not go unnoticed. From a Network with purely military purposes, ideas were generated that would change the contemporary world, that is, to keep two computers (and humans) connected. For this, in 1989, researchers Tim Bernes-Lee and Robert Cailliau, at the Conseil Européen pour la Recherche Nucléaire (CERN), succeeded in developing the World Wide Web, better known as the triple WWW. Table 5 shows Internet access in Central and Latin America. Different aspects can be mentioned, most importantly, the emergence of public access policies by governments, which, to a certain extent, have a low cost of connectivity.

Table 5 Population and internet users

Country	Population	%	Internet users	%	Facebook	%
[a]AR	43,833,328	10.12	34,785,206	12.02	29,000,000	12.20
[a]BO	10,969,649	2.53	4,600,000	1.59	4,600,000	1.94
[a]BR	206,050,242	47.59	139,111,185	48.06	111,000,000	46.70
[a]CH	17,650,114	4.08	14,108,392	4.87	12,000,000	5.05
[a]CO	48,593,405	11.22	28,475,560	9.84	26,000,000	10.94
[a]EC	16,080,778	3.71	13,471,736	4.65	9,700,000	4.08
[a]PAR	6,862,812	1.58	3,149,519	1.09	2,900,000	1.22
[a]PE	30,741,062	7.10	18,000,000	6.22	18,000,000	7.57
[a]UR	3,351,016	0.77	2,400,000	0.83	2,400,000	1.01
[a]VE	29,680,303	6.85	18,254,349	6.3	13,000,000	5.47
[a]PA	3,705,246	0.86	2,799,892	0.97	1,700,000	0.72
[a]CR	4,872,543	1.13	4,236,443	1.46	2,900,000	1.22
[a]RD	10,606,865	2.45	6,054,013	2.09	4,500,000	1.89
Total	432,997,363	100	289,446,295	100	237,700,000	100

Source Internet World Stats 2016

[a]Argentina, Bolivia, Brazil, Chile, Colombia, Ecuador, Paraguay, Perú, Uruguay, Venezuela, Panamá, Costa Rica, República Dominicana

3.1 Usability as Central Axis Based

Based on Eyetrack's research, Jakob Nielsen determined three components or movements: horizontal, at the top of the content, the second horizontal a little lower, and the third, a vertical movement on the left side of the screen. These are called reading patterns (E-F-L) (Franco 2008: 37–38). The main public television channels analyzed have an informative website, fulfilling the basic characteristics of: interactivity, multimedia and hypertextuality.

TV Pública Argentina, is the only one that has the "accessible version" (http://accessible.tvpublica.com.ar/), which blocks each of the categories that can be heard or seen by people with different capacities, improving the accessibility and usability of web services. It meets reading pattern F because it places the most important information at the top, allowing an orderly tour. In the case of TV Ciudad (Uruguay), it complies with the reading standards proposed by Nielsen, the image carousel stands out, which is prominent in relation to the logo and sections; This is a constant in much of the analyzed websites. As a structure, the background image is black, as a usability rule, this should not happen, since it prevents users from staying long periods of time on the site; The content is related to all the programming. TV Brasil, shows great images and headlines, fulfills the pattern of reading in "inverted L", when placing and giving more importance to the content at the top. The content is informative and not by last minute, showing its information grid through images, which allows to capture more attention of the user.

One of the main characteristics of TVN (Chile) web, is the continuous "flicker" of the "live streaming". At the same time, to complement with the latest news. It has an exclusive news channel 24 h a day, 7 days a week. It mixes correctly the hypertext, multimedia and immediacy.

There are certain cases that do not fully comply with the laws proposed by the experts, for instance, Bolivia TV, an informative page with a tweet as news headlines. It links in the home page to the embedded accounts of Twitter and Facebook, which denotes the little effort from a television web page in its structure and content architecture. It mentions at least 5 sections but none are redirected to any content. It does not meet reading standards. Like Televisora Venezolana Social, at first sight, shows old informative notes that lack hypertextuality and immediacy. It is more of a corporate website than a media. It links you to their social network account and broadcasts live programming (one month delayed updates). Something similar occurs with Public TV Paraguay and Ser TV, Panama, which despite having a photo gallery or "carousel" of images, which show the most important programming, is nuanced with general news that lack sense of immediacy.

Corporate sites like Señal Colombia and Medios Públicos Ecuador offer no updates or immediacy which are characterized by appearing on the grid's initial page, home or site start. For restrictions of IP Signal Colombia cannot be viewed from outside the country.

There are no networks that handle the generality or pull of Media Public, the networks are created separately: Radio, TV, Agency and Journal, each has its particularity. In 2016, the following medias were bound by decree, as part of one Public Company: El Telégrafo, El Tiempo, EcuadorTV, Public Radio, Agencia Andes, PP Digital, EditoGran, Citizen Council. In relation to usability, it is a corporate site, according to the source of its means. Its section in Kichwa stands out, showing information of general character and without losing periodicity. In this case, there are no networks that handle the generalities of public media. To fulfill the role of "being on the web", the State Corporation of Radio and Television, CERTV (Dominican Republic), which, like Medios Públicos Ecuador, owns several media, Canal 4RD, radio Dominican, radio Quisqueya and radio Santo Domingo. It has the icons of social networks on the web page, however, they do not automatically link you to them. In the case of Twitter, this account is tweet-proteced which goes to show the lack of interest for this platform, the same that focuses on corporate presence.

Live broadcasting is present in most public channels, especially TV Perú, reproducing programming from news reels, series and special programming. Images that cover much of the screen are shown, dividing the site into three parts; The first with prominent image, the second with "news" and the third, a reproduction of their programs. In addition, TV Perú app stands out, which, according to its description, "seeks to ensure an agile, timely and efficient experience for users, seeking interaction and constant participation, generating content of value." It can be accessed without any IP restrictions. It highlights an image that completely covers the site, making it like a TV screen; Its video library is a plus, which is organized by dates, categories and programs.

In terms of Vive, belonging to the Bolivarian Communication System of Information of Venezuela, although the website maintains a basic structure, with a head, image and last minute news, it shows its programs, grill and signals through Youtube live, viewable in all countries. TELESUR, has become one of the largest exponents of public television in the Region. Its site maintains a structure in accord with web media, focusing on the constant updating of live content, with left vision and ideology, making room for news on the Venezuelan government, without leaving aside matters of global character. TELESUR has great presence and followers in social networks. The Costa Rican media Sinart Channel 13, exposes its programming with breaking news; Its signal is transmitted live, and even immediatly accessed through social networks, its place at the top does not go unnoticed.

3.2 *Multiple Screens, New Bets*

Computers, cellphones, tables enclose the circuit of what is known as "multiple screens", where the movies and television turn into a window for content difusion. In this way, new platforms of consumption are born out of the Internet's (r)evolutions. The influence of services such as *Amazon Primer o Netflix*, this last mentioned, with over 100 millon subscribers works in 190 countries; proclaims itself "the world's leader on Internet **television network**. In which one can enjoy over 125 millons of hours in series and recent movies, including documentaries, movies and original series" (Netflix 2017). Users own the right to chose what they want to watch, when to watch it, pause it and continue it. Besides, the platform give users contents free from advertising.

The "boom" in phone apps and social networks has set foot to a major transformation in communication. From the time WWW, the evolution and explansion of the digital world has been unstappable. The initial objective being that through these tools, the Internet could be really accessible and useful to all people. These investigators knew that, in order for the Web to reach its max potencial, subyacent tecnologías should become global standards and applied, in the same way, everywhere in the world.

As you could see in Table 6, the Internet's growth its exponencial; all these television networks have a web page and social networks. *Facebook* in the last 5 years has

Table 6 Usability and social networks

Country	Name	Social Networks				Usability
		Facebook	Twitter	YouTube	Instagram	
[a]AR	Tv Pública	1,309,358	839 K	384,754	8844	Yes
[a]BO	Bolivia TV	109,570	18.3 K	8092	1300	No
[a]BR	TV Brasil	461,484	208 K	177,395	5524	Yes
[a]CH	TVN	1,646,061	2.8 M	366,919	122 K	Yes
[a]CO	Señal Colombia	351,006	641 K	52,762	31 K	Yes
[a]EC	Medios Públicos EC	No posee				
[a]PAR	Tv Pública	33,536	68.7 K	3101	2308	Yes
[a]PE	TV Perú	154,579	23.3 K	77,200	No	Yes
	Televisión Nacional Uruguaya	120,848	66.7 K	7220	2324	Yes
[a]UR	TV Ciudad	*35,954*	15.2 K	5439	3421	Yes
[a]VE	Vive Tv	23,144	367 K	6943	296	No
	Televisora Venezolana Social	34,674	480 K	23,134	83.1 K	Yes
	TELESUR	1,432,709	1.42 M	217,976	105 K	Yes
[a]PA	SER TV	17,763	34.4 K	3870	5802	No
[a]CR	Sinart Canal 13	141,861	41.5 K	0	2735	Yes
[a]RD	CERTV	No posee				

Source Websites and social networks (May 2017)
[a]Argentina, Bolivia, Brazil, Chile, Colombia, Ecuador, Paraguay, Perú, Uruguay, Venezuela, Panamá, Costa Rica, República Dominicana

evolved and expanded to all parts of the world, channels are perceived as key source of informative events. Since then, 360° images and mostly *streaming*, eclosed the web, whether through a *website* (key on a channel), *Periscope (Twitter)* and *"live" from Youtube, Instagram* and *Facebook*; are getting more attention for the emision of clips, previews trivias and even spoilers.

Measured-content is decisive at the time of selecting and public TV must reinvent itself and comply with the needs of the users (maintaining and captivating them) eager for new experiences; even more so, since having the Internet as its main competitor and allie.

In order to catch new audiences, public television must encourage the industrial audiovisual development in each of their countries. Such is the work done by networks and producers from American Latin television, exchanging and broadcasting audiovisual documentaries, series and short films with 20 participating Latin American countries. It is essential that producers take responsibility for the space that is used to present their productions, reflecting on the everyday challenges and specialised quality content work that TV viewers and web browsers require.

References

Brazilian Broadcasting Observatory. (2017) http://es.observatorioradiodifusao.net.br

Cebrián, M.: Divulgación audiovisual, multimedia y en red de la ciencia y la tecnología. Ciespal, Quito (2014)

Coronel-Salas. G.: Information Obtained Through the Review of Websites and Social Networks per Public Media Mayo (2017)

Eurodata (2015) & IBOPE Ecuador rating reports (2010)

Fuenzalida, V., Pohlhammer, P.: Chile: Cambios en la industria. In: Orozco, G., Vassallo, M. (eds.) Memoria Social y Ficción Televisiva en Países Iberoamericanos, (541). Editora Meridional Ltda, Porto Alegre (2013)

Franco, G.: Cómo escribir para la Web. Knight Center for Journalism, Austin (2008)

González, N.: Contenidos en la TV digital argentina. Estrategias y actores. Primeras Jornadas Iberoamericanas de Difusión y Capacitación sobre Televisión Digital Interactiva (2012). https://goo.gl/MoxHVw

Hartley, J.: Los usos de la televisión. Paidós, Barcelona (2000)

Internet World Stats (2016). http://www.internetworldstats.com

Latin American Research Program (2017)

Mastrini, G: Interview about Ley de Servicios Audiovisuales de Argentina. (C. Ortiz, interviewer) (2015)

Mier, C.: Creación y desarrollo de Ecuador TV. Universidad Santiago de Compostela. Santiago de Compostela, Thesis (2015)

Netflix: Media Netflix (2017). https://media.netflix.com

Observacom: Laws and regulations extracted (2017). http://www.observacom.org/

Señal Colombia: Programming schedule (2015). https://goo.gl/McxMs4

Wolton, D.: Informar no es Comunicar. Gedisa, Barcelona (2010)

Zallo, R.: Book Reviews Cajas mágicas. El renacimiento de la televisión pública en América Latina. Revista Zer, **18**, 277–309 (2013). https://goo.gl/4JCHMA

Catalina Mier-Sanmartín Ph.D. in Communication and Journalism from the Universidade de Santiago de Compostela. Degree in Social Communication at the Universidad Estatal de Cuenca, Research Professor at the Department of Communication Sciences from the Universidad Técnica Particular de Loja (UTPL), Academic Counsellor in the Iberian-American Program in Communication and Educational Technologies of the School of High Studies in Educational Communication ILCE-México, Member of the Iberian-American Network of Audiovisual Narratives, Ecuador chapter. Former Secretaries-General of CIESPAL. Author of some scientific articles in magazines and books. Lines of research: Public Television, edu-communication and social networks. Loja, Ecuador.

Gabriela Coronel-Salas Ph.D. in Communication and Journalism from the Universidade de Santiago de Compostela (Spain). MBA in Social Studies of Science and Technology from the Universidad de Salamanca (Spain). Research Professor at the Department of Communication Sciences from the Universidad Técnica Particular de Loja (UTPL). Her line of research is communication and technologies. Manager of technologies focused on the public dissemination of science online and social media. Member of Communication and Digital Culture Research Group. Loja, Ecuador.

Kruzkaya Ordóñez MBA in Communication and Creative Industries. Candidate Ph.D. of the Universidade de Santiago de Compostela. Tenured professor at the Department of Communication from the Universidad Técnica Particular de Loja, departmental section Audiovisual Narratives. Lines of research: local television, audiovisual media, digital communication, scientific divulgation. Postgraduate in Digital Documentation. Participation in the Network of Audiovisual Narratives and Thematic Network of Applications and Usability of Interactive Television in Latin America REDAUTI. Loja, Ecuador.

Abel Suing BA in economics from the Universidad Técnica Particular de Loja (UTPL), Ph.D. in Communication from the University of Santiago de Compostela. Professor and researcher at the Department of Communication Sciences from the UTPL, where he currently coordinates the MBA in Communication and Digital Culture. He is an integral part of the Research Group on Digital Culture and Communication (C+CD). Lines of research: Television, Communication Policies, Higher Education. Loja, Ecuador.

Carlos Ortiz Degree in Social Communication from the Technical University of Loja. Doctor in communication by the University Santiago de Compostela- Spain. Lecturer Researcher of the Chair of Television, in the Degree of Social Communication from the Universidad Técnica Particular de Loja. Audiovisual director on television. Director of MediaLab UTPL. Member of the Research Group in Communication and Digital Culture (C+CD). Research lines: Television, Audiovisual Narratives, Documentary. Loja, Ecuador.

Part II
Journalism and CyberJournalism

Laboratory Journalism

Xosé López-García, Alba Silva-Rodríguez and María-Cruz Negreira-Rey

Abstract Journalism, as a technique for modern communication which has been both a product and a part of society for more than a century and a half, has needed to pass through the laboratories of innovation in order to better respond to the challenges that the Internet age poses. This chapter (This research has been developed within the project *Uses and informative preferences in the new media map in Spain: journalism models for mobile devices* (Reference: CSO2015-64662-C4-4-R) funded by Ministry of Economy and Competitiveness (Government of Spain) and co-funded by the ERDF structural fund, as well as it is part of the activities promoted through the International Research Network of Communication Management—XESCOM (Reference: ED341D R2016/019), supported by Consellería de Cultura, Educación e Ordenación Universitaria of Xunta de Galicia.) reviews the innovative strategies developed in the principal international modes of reference through these laboratories of innovation.

Keywords Labs · Journalism · Innovation · Smartphones · Big data Ubiquitous journalism

1 Introduction

Journalism is sailing though the tempestuous seas of communication that flow in these times of the Internet age (Castells 1996) in search of vitamins that might help it confront old and new challenges. The transformations that have taken place in modern societies through the Internet and the configuration of a rejuvenated media

X. López-García (✉) · A. Silva-Rodríguez · M.-C. Negreira-Rey
Universidade de Santiago de Compostela, Santiago de Compostela, Spain
e-mail: xose.lopez.garcia@usc.es

A. Silva-Rodríguez
e-mail: alba.silva@usc.es

M.-C. Negreira-Rey
e-mail: mcruz.negreirarey@gmail.com

M. Túñez-López et al. (eds.), *Communication: Innovation & Quality*, Studies in Systems, Decision and Control 154, https://doi.org/10.1007/978-3-319-91860-0_10

ecosystem, with the multiplication of actors and new mediums, have led journalism to new unknown territories. Now trying to explore with tools from the past, through the difficulty of getting rid of old forms and approaches, which for many years have been fruitful, and with modern technologies which open doors to the scenes which will paint the picture of the future. Many businesses as well as the majority of professionals are trying not to remain indifferent so that this reformist and ground-breaking era does not drag them and leave them adrift.

The profound state of reconversion in which journalism finds itself, provoked by the processes of digitalization which have profoundly altered our processes of communication, and sharpened by the financial crisis around the year 2008, has altered the advertising market (Casero Ripollés 2012) and is eroding away the cement of its industrial bases (Curran 2010). In such an adverse context, with many dark clouds on the horizon, the challenges faced by journalism and by the businesses of the media ecosystem of this third millennium are related to the creation of value for consumers as well as for citizens and society (Picard 2010). It seems to be therefore a challenge that demands a change of objectives, strategies and links to society in order to achieve a change in direction that will secure its future and that of journalism in the short and long term.

The process of a shift in strategies in any case must be considered in the complete context of communications, where technologies mark the new rhythms not just of messages but also of processes. The dynamism that the new technological framework has brought and the nodes of connection in the Internet age, with users acting as receivers and producers are accelerating the transformation of our environment. In the middle of such reconfiguration the questions multiply reflecting on whether we find ourselves before an impending death foretold, or the start of a new revitalized age of journalism. What very few people will doubt is that we are in a distinct period of time, with new challenges and renewed threats.

2 The Question and Its State

The appearance and development of the Internet has had a great impact on journalism and on its evolution. Journalism has not just changed in its move away from traditional mediums and with the appearance of digital natives, but since the end of the 20th century it has also forced us to re-think different aspects of journalism (King and College 1997). The technological bundle that permitted the print press, once new, to thrive and develop, also took the first largely adopted non-printed publication, the online journal, and opened the path to the digitalization of the news (Bockzkowski 2004).

From the onset, technology has had an influence in at least four large areas: in the work of journalists and how they do it, in the actual content of the news, in the structuring and organization of editing offices, and in the relationships between the media, journalists, and Internet users, (Pavlik 2000). The appearance of active audiences generating content (Philo 2008; Meso et al. 2015), the new tools related to data journalism and virtual reality (Doyle et al. 2016), have led to changes in all sectors.

The appearance of cyber media, with its hypertextual, multimedia and interactive characteristics made the experimentation and elaboration of pieces of journalism which developed these dimensions possible. Thus intermediality (Deuze2004), interactivity (Schultz 1999; Chung 2008; Scolari 2008), and participation (Carpentier 2011), most captured the attention of journalists of the digital online media who were ready for innovation.

In the same way, there is academic research related to this field of study, which has grown stronger in Spain this millennium (Masip and Micó-Sanz 2010) and which at the international level counts on a wide range of scientific production (Masip et al. 2010). Starting from here, from numerous investigations in the first decade of the century, they began to form new questions about the future of journalism, and in particular about the effects of online journalism and the new communications ecosystem (Van Der Haak et al. 2012), and especially about the future in different sectors (Franklin 2011, 2016). The debate remains open and awaits new evidence.

3 Experimentation

The search for alternatives has brought laboratory tests back to the field of journalism. After a long period in which experimentation with products reined, thanks to the growth of the sector and big businesses' capacity to risk trying proposals before putting them on the market, the changes in the ecosystem have encouraged some companies, under the protection of programmes of investigation, development and innovation, to create their very own laboratories. The trail of great technologies continues in this way, many of which benefit from their own laboratories of invention, in which they have introduced their experimental work in the fields of content, linked with the media or on their own.

3.1 Innovation in the Media

Since the digital eruption in editing, and the media's leap towards the Internet, they have seen themselves obliged to redesign their strategies for the production and distribution of content, modifying their routines and creating new products and types of relationship to the public. Confronted with such a changing environment like that of the Internet and ICTs (information and communication technologies), the media have faced up to the need to innovate and to the dilemma of introducing new models even when the traditional ones remain more stable and profitable (Christensen 1997).

The lack of resources for the adoption of innovations in editing tend to be the principal reason for this tension, just as Paulussen (2016) notes by reviewing the main ethnographic studies about these processes. In this way, the author remembers that

Table 1 Characteristics of media innovation

Media innovations as objects of research	Media innovations as processes
1. Continuous need for newness	5. Interconnection of the innovation and diffusion/appropriate phase
2. Media innovations as high-risk products and processes	6. Media innovation processes require a long period of time
3. Close interaction between intangible (creative) and technological/organizational aspects	7. Close interactions between media innovations and established media
4. Overlap of media innovations as both product as well as process innovations	8. Media innovations continue to economic and social change processes, and meet the attributes of both economics as well as social innovations

Source Dogruel (2014: 58)

the media must face up to an investment in technological infrastructures, contracting new professional profiles orientated towards the digital talent or multidisciplinary training of journalist staff, as well as taking on greater workloads in the same time.

Even with all that, Cabrera González (2016: 24) acknowledges the need to understand that the media is, in addition to a business which seeks profit, also a service whose benefits and uses must satisfy the need of citizens to be informed. In this way, the author proposes a definition of innovation adapted to the field of media and cybermedia.

> Innovation as a process of creation or modification of the product or service that the media offers, through the integration of new technologies, routines and business models in its structure, organization, processes of production and diffusion of information, with the aim to widen and diversify its market with some kind of competitive advantage or creation of value. (Cabrera González 2016: 26)

For Storsul and Krumsvik (2013), media innovation could be classified in terms of the products, understood as new technologies and services offered by the media; the processes that incorporate changes in organizational and routine forms; brand and company placement in the market and paradigmatic innovation which completely renews the organization or the model of a particular newspaper.

On the other hand, Dogruel (2014) states that media innovation can be assumed and studied as an object and as a process, and summarises the main characteristics in eight points (Table 1).

As we see, innovation in the media is a complicated question in which, according to Westlund and Lewis (2014), two types of actors interrelate: humans, which includes journalists, agents, advertisers, computer technicians or collaborators, as well as the audiences who participates actively; and the technological actors such as algorithms, applications or CMS (content management systems).

Indeed, in recent years the media has had to assimilate multiple technological advances and adapt itself to them in order to create new products and informative formats. Gershon (2017) reviews the principle technological milestones and their consequences on the digital media, from the possibilities of interactivity or convergence, to mobility, artificial intelligence or virtual reality. All of this situates us in the scene of experimentation and current change, in which is it possible to highlight some innovative trends which are renovating journalistic forms.

Ubiquitous journalism. The journalism industry finds itself in constant transformation. The technological advances of recent years have made it possible for mobile devices to become a platform through which all kinds of content can be channelled. Factors such as ubiquitousness, personalization, localization, security, connectivity, and accessibility are the principal contributions of mobility to the journalism sector.

This disruptive paradigm reconfigures new dimensions, making it possible to be "present" in different spaces at the same time, to have access to and produce content from anywhere, with various devices, at any hour of the day. The development of mobile Internet has given a boost to the Internet age (Castells 1996) toward becoming an increasingly technological society where mobile devices have become indispensable tools for functioning.

Whilst the fixed telephone has played an important role in the concentration of financial and commercial activity in urban nuclei (Lasén 2002), the mobile phone has broken down spatial barriers in order to enter into mobility and increase communication capabilities (Geser 2004). Its quick implementation, facilitated by being a commercial technology made available to the masses (Aguado and Martínez 2006) has placed the mobile phone in all sectors. This popularization of mobile communication has modified the ways in which we connect and communicate (Campbell 2008) establishing a new framework not just for technologically-mediated communication, but also for journalism in our mobile society.

The challenges that mobile communication poses oblige us to rethink new informative products to the point of adapting ourselves to a new framework in which immediacy, personalization, permanent availability, simplicity and ubiquitousness take priority above all else. Mobile industry is currently one of the few sectors that are maintaining an unstoppable growth among cultural industries. Its expansive success has converted it into a point of interest for many sectors in general and for journalism in particular.

With the second decade of the third millennium, a new stage in mobile communications began. The protagonists were intelligent watches, new devices which integrate many functions in the universe of connectivity, with the Internet as its principal referent. The main actors in the media ecosystem didn't want to miss out on the possibilities of this new platform, and prepared content to make a reality the dream of being present on all platforms and devices.

Faced with this context, many forms of media are integrating these new devices in their editing (Paulussen 2016), considering as a priority the elaboration and management of informative pieces in the category of "mobile journalism" (or mojo journalism), and participation (Barnes 2016), which is reached at a high level thanks to mobile devices. The integration of mobile phones in the professional routine is prefigured "as a natural extension of the very technological skills of the online medium" (Aguado and Martínez 2006). This mobile journalism represents not just a great save in costs, but it also allows people to have information and images in any situation and at any moment.

Big data and data journalism. In recent years, information both poured into and produced from the Internet has not stopped increasing at an exponential rate. The volume, variety, and velocity of production of this enormous amount of data is what is given the name 'big data', which is related to the mining of data, and which also employs artificial intelligence and statistics for its analysis (Tascón 2013). According to Casero-Ripollés and López-Meri (2015), behind this great volume of data there are movements such as open government and open data which make available to citizens any type of information about the functioning of institutions and public administration, to which is added all the data generated by users and their own behaviour on the net. The access to and treatment of this big data is favouring the production of new types of reports and news, at the same time as it widens the functions of journalism as an anti-establishment movement (Casero-Ripollés and López Meri 2015).

The use of big data with journalistic aims has given rise to what today is known as data journalism. Flores Vivar (2012: 12) in fact situates its antecedents in the 1960s with the consolidation of a new journalism in the United States and the figure of the muckraker. However the more immediate precedents can be found in computer assisted journalism (CAJ) and precision journalism (Meyer 1993), introduced in the seventies by Philip Meyer and which came to be implanted by the end of the eighties in newspapers such as USA Today, The Washington Post, Los Angeles Times or The New York Times (Ferreras Rodríguez 2012).

Data journalism has been consolidated since 2007 (Ferreras Rodríguez 2012) and its development is based on journalistic investigation, technological knowledge, and a good law of access to information (Flores Vivar and Salinas Aguilar 2013: 19). Peiró and Guallar (2013: 26) define it as "the journalistic practice or speciality that uses data as a fundamental raw material (and not other sources of information), in order to create diverse informative products such as: articles with data, computer graphics or interactive visualizations with data and (open groups of data).

This discipline, which is also related to documentation journalism, computer graphics or fact checking (Peiró and Guallar 2013) introduces new possibilities for journalistic production, widening the range and the time taken for working on the subjects, facilitating multidisciplinary collaborations between professionals, innovation in the visualization and narrative of products etc., (Gray et al. 2012).

The production process of data journalism may follow the steps of the inverted pyramid proposed by Bradshaw (2011): data compliance, filtering, contextualization, combination and communication (visualization).

The introduction of this speciality to the media gave shape to the data journalist figure, defined by his or her knowledge not only about communication, but also about documentaries, IT, statistics, sociology or visual information (Ferreras Rodríguez 2012; Flores and Cebrián 2012; López-García et al. 2016). Zanchelli and Crucianelli (2012: 2) suggest four keys to the integration of data journalism teams in the newsrooms that follow the laboratories' general philosophy: locating the data journalism team close to the newsroom; encourage teamwork between journalists and software developers that can fill in the skills gap; and producing stories that show what data mean and its importance to the audience.

Immersive journalism. The features of the change in journalism in the last few years, driven by the current technologies and the new communications scenario, have been analysed by numerous authors, both from a general point of view linked to the format (Meyer 1991; Boczkowski 2004) and from the specific aspects that stem from the evolution of a network society where information circulates at great speed, almost in real time (Guallar 2010). In this environment, the production of information and its adaptation process acts as a prevailing logic of constant news' updating, normalising an uninterrupted cycle (Rodríguez Martínez et al. 2010; Casero-Ripollés 2012). In such context, wearable devices, especially smart watches, can offer an interesting option for journalistic products.

This is the reason why some authors (De la Peña et al. 2010; Pavlik 2001; Pryor 2000), consider that immersive journalism expresses itself in all its potential through technologies and virtual and immersive reality systems which allow not only visual but also sensory experimentation in a synthetic three-dimensional environment. The development of technologies that remove in a sensorial way this physical barrier is the most prevalent exploration in newsrooms when it comes to testing the idea of immersive journalism.

Immersive experiences in journalism, due to them being expensive and experimental, have been relegated to certain investigation projects, primarily from the USA (Domínguez 2013: 7). The Universities of Columbia and South California have been pioneers in practising such ground-breaking narrative.

Bots and journalism. Up to the moment, most innovations created by the media have been related to the distribution, not to the content creation. However, progress in areas related to the identification of patterns or the generation of a natural language have changed this situation.

The report *Journalism, media and technology predictions 2016* from the Reuters Institute indicates that there is software from companies such as Automated Insights or Narrative Science that is being used to create automated reports monitoring profits for Associated Press or sports reports for companies such as Yahoo (Newman 2016). The similarities between both were so significant that most readers could not distinguish between those that had been created for journalists and the ones that were made through algorithms.

Bots are not a new phenomenon and have been studied in a variety of human communication platforms. Researchers have focused their efforts in analysing the usage of bots in social media (Hwang et al. 2012). However, there is few investigations that address the usage of bots in the identification and expansion of newsworthy events. The media arena in which social media emerge is not only the result of human communication, but a confluence of human interaction with other humans in an automatic way (Larsson and Hallvard 2015).

There are Twitter bots that are already producing thousands of updates of 140 characters about stories of public interest and are becoming an opportunity for development for many journalistic companies (Lokot and Diakopoulos 2016; Diakopoulos and Koliska 2016). Media corporations such as The Washington Post have been experimenting with bots in order to attract younger users through contests and games in chats' applications such as kik47 (Newman 2016: 36).

Quartz, in collaboration with the Knight foundation, has recently launched Quartz Bot Studio. This involves experiencing with bots' applications, artificial intelligence and technologies that are related to journalism in new platforms. At the same time, media corporations such as The New York times or the Washington Post are also making incursions in the world of bots so that they get a more significant interaction with their audience.

3.2 *The Rise of Labs as Social Spaces for Innovation*

The development of such new models and journalistic narratives is the result of the integration of innovation processes in the media that, over the last few years, have resulted in the establishment of laboratories within the newsrooms. These inherit the philosophy of the so-called labs and medialabs, which have evolved towards the user-driven innovation model. This comes from "the Web 2.0, the generalisation of 'open innovation', the universalisation of mobile technology and its connection to the internet, as well as different ways of social innovation expressed as coworking spaces, crowdfunding or P2P economy" (Serra 2013: 284).

The reality of labs as we know it, has its origin in the predigital form of the scientific lab, the industrial lab and the design lab, whose models lay the foundations for the creation of digital technological labs (Sangüesa 2013: 264). Ruiz-Martín and Alcalá-Mellado (2016) distinguish between three main stages in the development of medialabs: the seventies represent the elementary steps, in which 'pioneer labs' started to emerge as centres of experimentation; between 1970 and 2000 the 'modern medialabs' arose massively. The likes of the MIT Medialab became a historical-cultural reference. Lastly, the 'current medialabs' came into being in the third millennium, featuring new models linked to the digital manufacturing of information and an evolution towards social issues.

According to Klimczuk (2013), scientists Nicholas Negroponte and Jerome Wiesner are considered the authors of the medialab concept, given that it was them who promoted the foundation in 1985 of the MIT Medialab in the Massachusetts Institute of Technology (Brand 1987). This centre established a new model for technological innovation and became an international reference for bringing together IT and design, through the cooperation of engineers, designers, developers, educators and artists.

In a broad sense, Klimczuk (2013: 130) defines medialabs as "interdisciplinary institutions combining tasks of traditional scientific, educational, cultural and artistic institutions". For Villar-Alé (2015: 279) the term lab refers to the "usage of Information and Communication Technologies (TIC) in a particular workspace, with a group and multidisciplinary work methodology, with activities and productions that can be placed anywhere between the social activism and the production of objects".

The labs' scenario is a complex one and within it coexist the digital technological labs and their democratized versions, such as the living labs, the citizen labs, the hacklabs or the world wide labs (Sangüesa 2013). Authors Romero-Frías and Robinson-García (2017: 31) turn to the suggestion of Tanaka (2011) for the classification of medialabs, according to which there is a distinction between industry labs, funded by companies for the purpose of development and research; media art labs, in which technology is used for artistic purposes; university labs, focused in innovation and entrepreneurship in the academic environment; and citizen labs, that encourage citizen involvement, do it yourself philosophy and social implication.

What seems common to their general evolution is the transition from technologies' labs to spaces where citizens intercede (Ruiz-Martín and Alcalá-Mellado 2016), partly due to the democratization of the access to technology (Romero-Frías and Robinson-García 2017). In this sense, labs have a social function that brings many benefits for those communities they operate in: they stimulate the production of network content and take advantage of scarce resources such as creativity, time or attention; they also have positive effects on the impact of social stock; and contribute to the design of different types of social, technological and economic innovations (Klimczuk 2013). This change in the culture of medilabs is translated into generalised transdisciplinarisation of processes, the use of open resources and licenses and citizen participation (Ruiz-Martín and Alcalá-Mellado 2016).

Authors Romero-Frías and Robinson-García (2017) recover the common agenda that Romero-Frías (2013) outlined for laboratory work programs and that could easily be applied to the activity of labs in the media:

> The analysis and participation in multiple digital cultures: the culture of the screen, oral, remix, visual, transmedia, prototype and design; free culture derived from free software; hacker ethics; interdisciplinary/transdisciplinary/multidisciplinary; the combination of transversality and specialization; co-creation and rethinking of forms of authorship and academic recognition; and entrepreneurship and innovation experiencing new ways of transferring knowledge and connections with society. (Romero-Frías and Robinson-García 2017: 30)

In the context of the media, Salaverría (2015: 398) defines medialabs as "units or departments dedicated to the research, experimentation, development and implementation of technological innovations and editorials". These are meant to be the headquarters and source of creativity, "this figure is making it easier for media companies to face corporately the study and implementation of novelties that arise from the impact of technology on their work" (Sábada Chalezquer and Salaverría Aliaga 2016: 42).

The labs from the perspective of the newsrooms. Newsrooms are investing in laboratories in order to experiment with new forms of stories, but many continue to rely on external partners that develop the most ambitious immersive projects.

Laboratories in the media kicked off at the end of the 20th century (Salaverría 2015) and in the second decade of the third millennium several media have innovation labs they use to try to respond to the many challenges that dissemination of information presents for different platforms whilst optimising the same for the mobile scenario.

International media such as The New York Times, The Washington Post, BBC, AFP, The Huffington Post or The Guardian have opted for these units to experiment quietly (López Hidalgo and Ufarte Ruiz 2016). The aforementioned means constitute a small part of the census of innovation laboratories located by Salaverría in 2015, who detected a total of 31 units.

Some of the media at the forefront of innovation are carrying out some thought provoking experiments from their labs. Such is the case of The Wall Street Journal with the Dow Jones Innovation Lab whose team launched the Daydream app in 2016. This application has been one of the pioneers in offering news in 3D. The newspaper is part of the wave of media that have been encouraged to experiment with new innovative narratives.

It is also important to highlight the work of new media companies such as Vox or Buzzfeed that convey stories in an innovative way using the latest technology to attract (and retain) new audiences. In June 2016 Buzzfeed sent a 360 camera to a protest against Donald Trump in San José (California) during the celebration of the presidential elections. The platform used to film the protest was designed by Ben Kreimer, a partner at the Open Lab laboratory. Kreimer helped develop 360° cameras and also experimented with cameras and drones, and converted the footage into actual virtual worlds (3D). In recent times the Buzzfeed team have been trying new ways of using automatic learning for the creation of narratives and are also looking for new ways to use technology through discrete wearable cameras.

Scherer, senior product designer at Vox Media Storytelling Studio, is part of a team of eight engineers and designers who develop new ways to tell stories through Vox Media publishing brands. Another interesting fact related to Vox Media is the launch of the Storytelling Studio. This is associated with the games laboratory of the American University and has ventured into the world of video games connected to journalism.

A trend that is starting to be monitored in newsrooms is the association with external companies in order to initiate collaborations for project development. Game developers actually do not rule out collaborating with newsrooms. Al Jazeera, for example, partnered with the National Film and Television School to develop #Hacked and with Altera Studio to create Pirate Fishing.

Even for news organizations with R&D labs, collaboration with game designers or VR production studies remains a viable option for certain specific stories and platforms. Working with external partners can pay for costs and streamline development while delivering impressive results.

The Washington Post has also partnered with the University of Texas to publish a virtual reality story about Mars. Moreover, The Guardian has collaborated with The Mill production studio to create several projects. Among them is "Underworld", a virtual exploration of the London sewage system created by Daydream.

The New York Times, under the NYTLabs, its R&D lab, has opted for the use of Google Cardboards to produce 360° video stories and stereo sound. Through the glasses and the VR application of the newspaper (NYT app), the American paper has taken its audience, among others, to the top of the World Trade Centre. The New

York Times' Daily 360 offers readers a new way to experience places they would not normally visit.

Data indicate that journalistic laboratories are an emerging trend and that the phenomenon has been expanding since 2010 (Salaverría 2015). Therefore, we are in front of a trend that some media will use to step into the network society and face the challenges.

4 An Opportunity

The launching of journalistic laboratories by both matrix (or traditional) media and digital native media is an opportunity to obtain alternative narratives, formats and technologies that innovate and operate in the network society without compromising the main principles of journalism. Experimentation in closed circuits, although expensive, offers greater chances of success for new languages and multimedia formats to be widely accepted and supported by users.

Perhaps what journalism needs is for these projects to guide their efforts towards the stories displaying a better quality, as the impression is that the same is becoming lost (Gómez Mompart et al. 2015) due to the current crisis situation.

Users of the third millennium live in a network society with a communicative environment that provides them with news and messages through very different channels that are open and elaborated in some cases by information professionals and in others by citizens who, in a context of active audiences, turned into amateur journalists. The citizens, who have always been interested in the news, access these messages not having, in many cases, media literacy or digital skills. News and journalism are constantly evolving (Hamilton and Tworek 2017), and its transformation is imposed through formulas that ensure quality, regardless of the format and the formula.

Journalism has a future and the future is digital and networked. There are many approaches and initiatives to carry out innovation, as well as experts who contribute with numerous proposals (García Avilés et al. 2016). Today we have digital journalism experiences that show that there are ways to secure the future of journalism (Deuze 2017) and laboratories are an ideal place to try out formulas that could potentially add value and improve existing models.

5 A Step Forward

Laboratories will not save journalism, but they can infuse it with vitamins and oxygen for its transformation. In the current digital environment, where all citizens are digitally literate and can take part and show their talent, it becomes clear that there are mechanisms that are more suitable to achieve such aims. Labs, in their different typologies and levels of collaboration, seem a good formula for the application of research, development and innovation in the network society. Rigorous laboratory work together with properly designed and tested products and technological innovation, will always be useful in a network society increasingly marked by technological developments that need to constantly renew models and formats to add value.

The current issues that affect journalism do not refer only to a good use of hardware and software, but mainly affect its ability to add value and present users with the quality information that is natural to journalism.

Current journalism has to come out stronger from the laboratories, where experimentation is beneficial, yet not enough. The main objective for innovation laboratories should not only be to experiment with new narratives and formats, but also to equip the media with a package of resources that guarantee the application of mechanisms that finally lead to a higher quality of journalistic work. That is the journalism that professionals and citizens would like to get out of the laboratories. At the moment, based on some contributions that emerge to the surface and all the information gathered through academic explorations, we are aware that journalism is in the laboratory and that there is also a sort of laboratory journalism.

Acknowledgments This research has been developed within the project *Uses and informative preferences in the new media map in Spain: journalism models for mobile devices* (Reference: CSO2015-64662-C4-4-R) funded by Ministry of Economy and Competitiveness (Government of Spain) and co-funded by the ERDF structural fund, as well as it is part of the activities promoted through the International Research Network of Communication Management- XESCOM (Reference: ED341D R2016/019), supported by Consellería de Cultura, Educación e Ordenación Universitaria of Xunta de Galicia.

References

Aguado, J. M., Martínez, I.: La mediatización de la telefonía móvil: de la intersección al consumo cultural. ZER, Revista de Estudios de Comunicación, 11, 319–343. In: http://www.ehu.eus/zer/hemeroteca/pdfs/zer20-15-aguado.pdf. Accessed 15 May 2017 (2006)

Barnes, R.: The ecology of participation. In: Witschge, T., Anderson, C.W., Domingo, D., Hermida, A. (eds.) The Sage Handbook Digital Journalism. Sage, New York (2016)

Boczkowski, P.J.: Digitizing the News: Innovation in Online Newspapers. MIT Press, Cambridge (2004)

Bradshaw, P.: The inverted pyramid of data journalism. Online Journalism Blog. In: https://onlinejournalismblog.com/2011/07/07/the-inverted-pyramid-of-data-journalism/. Accessed 15 May 2017 (2011)

Brand, S.: The Media Lab: Inventing the Future at MIT. Viking Penguin Inc, New York, USA (1987)

Cabrera González, M. A.: La innovación: Concepto y taxomización. In: Sábada Chalezquer, C. García Avilés, J. A., Martínez-Costa, M. P. (eds.), Innovación y desarrollo de los cibermedios en España, pp. 23–30. Ediciones Universidad de Navarra, Navarra (2016)

Campbell, W.: Mobile technology and the body: apparatgeist, fashion and function. In: Katz, J. (ed.) Handbook of Mobile Communication Studies. MIT Press, Cambridge, MA (2008)

Carpentier, N.: Media and Participation. A site of ideological-democratic struggle. Intellect, Bristol (2011)

Casero Ripollés, A.: Contenidos periodísticos y nuevos modelos de negocio: evaluación de servicios digitales. El Profesional de la Información, **21**(4), 341–346, (2012). https://doi.org/10.3145/epi.2012.jul.02

Casero-Ripollés, A., & López-Meri, A.: Redes sociales, periodismo de datos y democracia monitorizada. In: Campos Freire, F., Rúas Araújo, J. (eds.), Las redes sociales digitales en el ecosistema mediático, pp. 96–113. Sociedad Latina de Comunicación Social, La Laguna, Tenerife, (2015). http://dx.doi.org/10.4185/cac92

Castells, M.: The Information Age: Economy, Society and Culture, Vol. I: The Rise of the Network Society. Blackwell, Oxford (1996)

Christensen, C.: The Innovator's Dilema: When New Technologies Cause Great Firms to Fail. Harvard Business School Press, Boston, MA (1997)

Chung, D. S.: Interactive Features of Online Newspapers: Identifying Patterns and Predicting Use of Engaged Readers. J. Comput.-Media Commun. **123**(3), 658–679 (2008). http://onlinelibrary.wiley.com/doi/10.1111/j.1083-6101.2008.00414.x/full. Accessed 15 May 2017.

Curran, J.: The future of journalism. Journalism Stud **11**(2), 464–476. (2010). http://dx.doi.org/10.1080/14616701003722444

De la Peña, N., Weil, P., Llobera, J., Giannopoulos, E., Pomés, A., Spaniang, B., et al.: Immersive journalism: Immersive virtual reality for the first-person experience of news. Presence: Teleoperators and Virtual Environ. **19**(4), 291–301, (2010). https://doi.org/10.1162/pres_a_00005

Deuze, M.: What is multimedia journalism? Journalism Stud. **5** (2), 139–152, doi:http://dx.doi.org/10.1080/1461670042000211131 (2004)

Deuze, M.: Considering a possible future for Digital Journalism. Revista Mediterránea de Comunicación **8**(1), 9–18 (2017). http://dx.doi.org/10.14198/MEDCOM2017.8.1.1

Diakopoulus, N., & Koliska, M.: Algorithmic transparency in the News Media. Digital Journalism **3**(6), 1–20 (2016). http://dx.doi.org/10.1080/21670811.2016.1208053 (2016)

Dogruel, L.: What is so special about media innovations? A characterization of the field. The journal of media innovations, 1(1), 52–69 (2014). http://dx.doi.org/10.5617/jmi.v1i1.665

Domínguez, E.: Periodismo inmersivo. La influencia de la realidad virtual y del videojuego en los contenidos informativos. Editorial UOC, Barcelona (2013)

Doyle, P., Gelman, M., Gill, S.: Viewing the future? Virtual reality in Journalism. Knight Foundation (2016). https://knightfoundation.org/reports/vrjournalism. Accessed 10 May 2017.

Ferreras Rodríguez, E. M.: Nuevos perfiles profesionales: el periodista de datos. In C. Mateos Martín; C. Hernández; J. Herrero; S. Toledano Buendía; & A. Ardèvol Abreu (Eds.), Actas IV Congreso Internacional Latina de Comunicación Social. Sociedad Latina de Comunicación Social, La Laguna, Tenerife (2012). http://www.revistalatinacs.org/12SLCS/2012_actas/062_Ferreras.pdf. Accessed 15 May 2017

Flores Vivar, J. M.: Ecosistema del Periodismo de datos. Comunicação & Sociedade, **34**(1), 7–35 (2012). http://dx.doi.org/10.15603/2175-7755/cs.v34n1p7-35

Flores Vivar, J.M., Cebrián Herreros, M.: El Data Journalism en la construcción de mashups para medios digitales. In: Sabés, F., Verón, J.J. (eds.), El periodismo digital analizado desde la investigación procedente del ámbito académico, pp. 215–230. Asociación de Periodistas de Aragón, Zaragoza (2012). http://decimotercero.congresoperiodismo.com/pdf/Libroelectronico2012.pdf. Accessed 15 May 2017

Flores Vivar, J.M., Salinas Aguilar, S.: El periodismo de datos como especialización de las organizaciones de noticias en Internet. Correspondencias & Análisis **3**, 15–34 (2013). https://dialnet.unirioja.es/servlet/articulo?codigo=4739290. Accessed 15 May 2017

Franklin, B.: The Future of Journalism. Risk, threats and opportunities. Journalism Stud. **17**(7), 798–800 (2016). http://dx.doi.org/10.1080/1461670X.2016.1197641

Franklin, B.: The Future of Journalism. Routledge, London (2011)

García Avilés, J. A., Carvajal, M., Comín, M.: Cómo innovar en periodismo. Entrevistas a 27 profesionales. Universidad Miguel Hernández de Elche, Elche (2016)

Gershon, R.: Digital Media and Innovation. Management and Design Strategies in Communication. SAGE Publications, Thousand Oaks, California (2017)

Geser, H.: Towards a Sociological Theory of the Mobile Phone. Sociology in Switzerland: Sociology of the Mobile Phone. Resource document (2004). http://socio.ch/mobile/t_geser1.pdf. Accessed 10 May 2017

Gómez Mompart, J.L., Gutiérrez Lozano, J.F., Palau Sampaio, D.: Los periodistas españoles y la pérdida de calidad de la información: el juicio profesional. Comunicar **45**(23), 143–150 (2015)
Gray, J., Chambers, L., Bounegru, L.: The Data Journalism Handbook. O'Reilly Media, United States (2012)
Guallar, J.: Prensa digital en 2010. Anuario ThinkEPI, 2011. EPI SCP, Barcelona (2010)
Hamilton, J.M., Tworek, H.: The natural history of the news: An epigenetic study. Journalism **18**(4), 391–407 (2017)
Hwang, T., Pearce, I., Nanis, M.: Socialbots: voices from the fronts. Interactions **19**(2), 38–45 (2012)
King, E., College, L.: The Impact of the Internet on Thinking about Journalism. Electron. J. Commun. **7**(4) (1997)
Klimczuk, A.: The role of medialabs in regional cultural and innovative policy. In: Hittmár, Š. (ed.) Management trends in theory and practice, pp. 130–132. Faculty of Management Science and Informatics, University of Žilina, Žilina, EDIS (2013)
Larsson, A., Hallvard, M.: Bots or Journalists? News Sharing on Twitter. Commun.—Eur. J. Commun. Res. **40**(3), 361–370 (2015). http://dx.doi.org/10.1515/commun-2015-0014
Lasén, A.: The Social Shaping of Fixed and Mobile Networks: A Historical Comparison. Digital World Research Center (DWRC), University of Surrey. In: http://www.kiwanja.net/database/document/report_mobile_history.pdf. Accessed 15 May 2017 (2002)
Lokot, T., Diakopoulos, N.: News bots: automating news and information dissemination on Twitter. Digital Journalism **4**(6), 682–699 (2016). http://dx.doi.org/10.1080/21670811.2015.1081822
López Hidalgo, A., Ufarte Ruiz, M. J.: Laboratorios de periodismo en España. Nuevas narrativas y retos de futuro. Ámbitos, **34**, 1–12 (2016). http://www.redalyc.org/pdf/168/16848204007.pdf Accessed 10 May 2017
López-García, X., Toural-Bran, C., Rodríguez-Vázquez, A.I.: Software, estadística y gestión de bases de datos en el perfil del periodista de datos. El profesional de la información **25**(2), 286–294 (2016). http://dx.doi.org/10.3145/epi.2016.mar.16
Masip, P., Díaz Noci, J., Domingo, D., Micó Sanz, J. L., Salaverría, R.: Investigación internacional sobre ciberperiodismo: hipertexto, interactividad, multimedia y convergencia. El Profesional de la Información **19**(6), 568–576 (2010). https://doi.org/10.3145/epi.2010.nov.03
Masip, P., Micó-Sanz, J.L.: La investigación sobre el ciberperiodismo en España. Trayecto hacia la consolidación. El Profesional de la Información **19**(6), 577–584 (2010)
Meso, K., Agirreazkuenaga, I., Larrondo, A.: Active Audiences and Journalism. Analysis of the Quality and Regulation of the User Generated Contents. Universidad del País Vasco, Bilbao (2015)
Meyer, P.: The New Precision Journalism. Indiana University Press, Indiana (1991)
Meyer, P.: Periodismo de precisión: nuevas fronteras para la investigación periodística. Bosch, Barcelona (1993)
Newman, N.: Journalism, media and technology predictions 2016. Reuters, Oxford (2016)
Paulussen, S.: Innovation in the newsroom. In: Witschge, T., Anderso, C.W., Domingo, D., Hermida, A. (eds.) The Sage Handbook Digital Journalism. Sage, New York (2016)
Pavlik, J.: Journalism and new media. Columbia University Press, New York (2001)
Pavlik, J.: The impact of Technology on Journalism. Journalism Studies, 1(2), 229–237 (2000). http://www.tandfonline.com/doi/abs/10.1080/14616700050028226. Accessed 10 May 2017
Peiró, K., Guallar, J.: Introducció al periodisme de dades. Característiques, panoràmica i exemples. Ítem, 57, 23–37 (2013). http://eprints.rclis.org/28670. Accessed 10 May 2017
Philo, G.: Active audiences and the construction of public knowledge. Journalism Stud. **9**(4), 535–544 (2008). http://www.tandfonline.com/doi/abs/10.1080/14616700802114217. Accessed 11 May 2017
Picard, R.G.: Value Creation and the Future of News Organizations: Why and How Journalism Must Change to Remain Relevant in the Twenty-First Century. Media XXI, Lisbon (2010)
Pryor, L.: Immersive news technology: beyond convergence. USC Annenberg. Online journalism review (2000). http://www.ojr.org/ojr/technology/1017962893.php. Accessed 11 May 2017

Rodríguez Martínez, R., Codina, L., Pedraza Jiménez, R.: Cibermedios y web 2.0: modelo de análisis y resultados de aplicación. El Profesional de la Información **19**(1), 35–44 (2010). https://repositori.upf.edu/handle/10230/13140. Accessed 11 May 2017

Romero Frías, E., Robinson García, N.: Laboratorios sociales en universidades: Innovación e impacto en Medialab UGR. Comunicar **25**(51), 29–38. https://doi.org/10.3916/C51-2017-03. Accessed 2 May 2017

Romero-Frías, E.: Ciencias Sociales y Humanidades Digitales: una visión introductoria. In: Romero-Frías, E., Sánchez-González, M. (eds.), Ciencias Sociales y Humanidades Digitales. Técnicas, herramientas y experiencias de e-Research e investigación en colaboración. Cuadernos Artesanos de Comunicación, vol 61, pp 19–50. Sociedad Latina de Comunicación Social, La Laguna, Tenerife (2013)

Ruiz Martín, J.M., Alcalá Mellado, J.R.: Los cuatro ejes de la cultura participativa actual. De las plataformas virtuales al medialab. Icono 14, **14**(1), 95–122 (2016). http://dx.doi.org/10.7195/ri14.v14i1.904. Accessed 2 May 2017

Sábada Chalezquer, C., Salaverría Aliaga, R.: La innovación y los cibermedios: los labs. In: Sábada Chalezquer, C., García Avilés, J. A., Martínez-Costa, M. P. (eds.), Innovación y desarrollo de los cibermedios en España. Ediciones Universidad de Navarra, Navarra (2016)

Salaverría, R.: Los labs como fórmula de innovación en los medios. El profesional de la información **24**(4), 397–404 (2015). https://doi.org/10.3145/epi.2015.jul.06. Accessed 11 May 2017

Sangüesa, R.: La tecnocultura y su democratización: ruido, límites y oportunidades de los Labs. Revista CTS **23**(8), 259–282 (2013). http://www.revistacts.net/volumen-8-numero-23/131-dossier/546-la-tecnocultura-y-su-democratizacion-ruido-limites-y-oportunidades-de-los-labs. Accessed 18 May 2017

Schultz, T.: Interactive Options in Online Journalism: A Content Analysis of 100 U.S. Newspapers. J. Comput.-Mediated Commun. **5**(1) (1999). http://onlinelibrary.wiley.com/doi/10.1111/j.1083–6101.1999.tb00331.x/full. Accessed 18 May 2017

Scolari, C.: Hipermediaciones. Elementos para una teoría de la Comunicación Digital Interactiva. Gedisa, Barcelona (2008)

Serra, A.: Tres problemas sobre los laboratorios ciudadanos. Una mirada desde Europa. Revista CTS **23**(8), 283–298 (2013). http://www.revistacts.net/volumen-8-numero-23/131-dossier/541-tres-problemas-sobre-los-laboratorios-ciudadanos-una-mirada-desde-europa. Accessed 18 May 2017

Storsul, T., Krumsvik, A.H.: What is Media Innovation? In: Storsul, S., Krumsvik, A.H. (eds.) Media Innovation. A Multidisciplinary Study of Change. Nordicom, Göteborg (2013)

Tanaka, A.: Situating within Society: Blueprints and Strategies for Media Labs. In: Tanaka, A., et al. (eds.) A Blueprint for a Lab of the Future. Baltan Laboratories, Eindhoven (2011)

Tascón, M.: Big Data. Pasado, presente y futuro. Telos. Cuadernos de comunicación e innovación **95**, 47–50 (2013). https://dialnet.unirioja.es/servlet/articulo?codigo=4423775. Accessed 18 May 2017

Van Der Haak, B., Parks, M., Castells, M.: The Future of Journalism: Networked Journalism. Int. J. Commun. **6**(95), 2.923–2.938 (2012). http://ijoc.org/index.php/ijoc/article/view/1750. Accessed 18 May 2017

Villar Alé, R.: Procesos artísticos en laboratorios. Génesis y perspectivas. Universum **30**(1), 277–292 (2015). http://dx.doi.org/10.4067/S0718-23762015000100016. Accessed 18 May 2017

Westlund, O., Lewis, S.: Agents of Media Innovations: Actors, Actants and Audiences. J. Media Innov. **1**(2), 10–35 (2014). http://www.journals.uio.no/index.php/TJMI/article/view/856. Accessed 18 May 2017

Zanchelli, M., Crucianelli, S.: Integrando el Periodismo de Datos en las Salas de Redacción. Knight Journalism Fellow, Argentina (2012)

Xosé López-García Professor of Journalism at the Universidade de Santiago de Compostela (Spain), Ph.D. in history and journalism. He coordinates the research group Novos Medios. Among his research lines there is the study of digital and printed media, analysis of impact of technology in mediated communication, analysis of the performance of cultural industries, and the combined strategy of printed and online products in the society of knowledge. Santiago de Compostela, Spain.

Alba Silva-Rodríguez Professor of Journalism at the Universidade de Santiago de Compostela (Spain) and member of the research group Novos Medios. Her research lines focus on the analysis of strategies, rhetoric and technological formats for emergent markets in communication. She is secretary of the RAEIC journal (Spanish Journal of Research in Communication). Santiago de Compostela, Spain.

María-Cruz Negreira-Rey Ph.D. Student in Communication and Contemporary Information from the Universidade de Santiago de Compostela (Spain) and researcher in training at Novos Medios group (USC). She develops her research in the areas of online journalism and media of proximity, focusing on the hyperlocal online media in Spain and Portugal. Santiago de Compostela, Spain.

Proposal for a Common Framework to Assess Media Quality

Maria González-Gorosarri and Andoni Iturbe Tolosa

Abstract The assessment of journalistic quality has been conducted in different ways, according to the language employed. As a consequence, there is a vast number of overlapping concepts that impede both cross-national and crossmedia evaluations of Media Quality. This chapter aims to gather the European (German, English, Italian and Spanish) and American (English and Spanish) academic research tradition on journalistic quality. Consequently, a common frame to assess Media Quality will be designed, from a crossmedia perspective, following the most employed criteria in those regional research traditions.

Keywords Quality · Media quality · Journalistic quality · Television quality
Newspaper quality

1 Defining Media Quality

The question of journalistic quality comes to attention when meeting all essential elements to offer a standard product may not accomplish the requirements of such a product. Despite it might be a consequence of erroneous Media Marketing strategies or journalistic routines, it is taken for granted that not all journalism practice is by itself quality journalism (Ruß-Mohl 1994a, b: 60).

A broad set of several works, published in different languages, refer to Media Quality. As a result, some concepts are overlapped. Prof. McQuail identifies three levels of media operation. The present chapter aims to remark the importance of Professor McQuail's contributions as the basis for assessment of quality journalism. Firstly, *structure* deals with "all matters in relation to the media systems, including its form of organization and finance, ownership, form of regulation, infrastructure,

M. González-Gorosarri (✉) · A. Iturbe Tolosa
Universidad Del País Vasco (Spain)—UPV-EHU, Leioa, Spain
e-mail: maria.gonzalezgorosarri@ehu.eus

A. Iturbe Tolosa
e-mail: andoni.iturbe@ehu.eus

M. Túñez-López et al. (eds.), *Communication: Innovation & Quality*, Studies in Systems, Decision and Control 154, https://doi.org/10.1007/978-3-319-91860-0_11

distribution facilities and so on". It fits in the denomination System Quality. Secondly, *conduct* refers to "the manner of operation at the organizational level, including the methods of selecting and producing content, editorial decision-making, market policy, relations established with other agencies, procedures for accountability, and so on". Finally, *performance* alludes to "content to what is actually transmitted to an audience". The most researched field of media performance is News Quality (González-Gorosarri 2011; McQuail 1992: 87–98, 2010: 192–199).

As a result, Media Quality must be defined in the function of indicators, stemming from excellence principles, according to a given value system. Scholars argue that journalistic quality can effectively be evaluated against excellence indicators, since standards materialize the essential abstraction of the quality conception, as evidenced by diversity assessment. Accordingly, Media Quality refers to any conception of excellence applied to media. However, two different research goals are to be distinguished. On the one hand, quality is to be achieved following organizational goals (business or commercial success, fulfillment and audience satisfaction…), focusing on the resources management every firm deals with (Albers 1996; Leggatt 1996a, b; Hillve and Rosengren 1996). This perspective has been assessed in the economical field of the communication research. Market driven media company policies have been research from the perspective of journalistic excellence (Picard 2004). Other scholars have tried to define the correlation between newspaper circulation and news quality (Meyer 2004a, b; De Pablos and Mateos 2004) or company management and journalistic quality (Chen et al. 2005). However, online media turned not to validate such a correlation (Chyi and Yang 2009). Similarly, the 'impact of journalists' working conditions on the journalistic quality has been pointed out, driving attention to the "practices that diminish the social value of newspaper content" (Picard 2004; Ruß-Mohl 2005; Sjurts 2005; Weischenberg 2006b).

On the other hand, Media Quality is understood in terms of means or channel of publication, as traditionally researched by Media Studies. Thus, not only News Quality is to be assessed as media performance, but any group of news or other media content belongs to the category of performance. This chapter will focus on the legacy of Media Studies assessing Media Quality as a product. For such a goal, German scholars have distinguished several levels of evaluating media performance. First, every type of media (that is to say, every media product), occupies the broadest research level. Secondly, as media genres (quality newspapers, boulevard newspapers), various media outlets can be together assessed under such a wide-ranging category. Thirdly, every media outlet becomes a research goal itself. Fourthly, the quality of certain sections of newspapers can also be measured. Finally, the output of every author would be assessed. As a matter of quality of performance, all those evaluation levels are assessed according to the same principles. On the one hand, they are considered as a product. Thus, three main indicators assess them: currency, objectivity, and diversity. On the other hand, from the audience perspective, use and acceptance are measured (Beck et al. 2010; Beck et al. 2010:16-17; Weisechemberg (2006a, b) (Table 1).

Therefore, we will refer to the conception of product quality as considered in the field of Media Studies for the assessment of Media Quality in the present chapter.

Table 1 Matching of McQuail's and German scholars' assessment levels of journalistic quality

McQuail	Onion model or circle model	Hierarchical levels of media evaluation
Structure	Media system	Media system
Conduct	Media institutions (quality assurance)	
Performance	Media output	Type of Media (TV, newspaper)
		Media genres or subsystem of media types (Quality Newspapers, Boulevard press)
		Media product (BBC, FAZ...)
		Sections
	Media authors	Authors, actors, presenter...

Source Authors

2 Tv and Radio Quality

Programming—next to the television and radio system—is the second big area of reference for the discourses on the quality of radio and television. The term *quality*, on the one side, has to be considered as being a "relational concept". Pujadas Capdevila (2011) distinguishes four operative concepts: the economical and political system (in charge of regulating will also derive its principles from the media art values to accomplish, in terms of social goals); national belonging feelings (broadcasting becomes an instrument to gather a community); localism (to raise local production, focused on self-agenda), and democracy (the latter as a synonym for public service).

On the one hand, as it is derived from the Broadcasting System Quality, programming depends on the regulating financial and political system, as well as on the accompanying values in the pursuit of social accomplishments (Pujadas Capdevila 2008: 197–208; Rosengren 1996: 4–8). The television and radio system, on the other hand, are strategic and instrumental for stimulating a community and designing an emotional space, covered by a structure that is aligned with the idea of community building. In that sense, they are essential for those territories that don't count with a hegemonic language. In comparison with the Anglo-Saxon liberal tradition, and besides the protectionist tradition of the European system, Canada has been a paradigmatic case for its active role concerning downgraded languages.

Media System Quality have been specially preserved in Sweden, Wales, the United Kingdom, France and the European Union have had an active role in the preservation of the television and radio system. Broadcasting plays a crucial part when it comes to preserving identity (Medina Laverón 2006: 19–21; Mulgan 1990: 19–21; Pujadas Capdevila 2008: 197–208; Raboy 1996: 60; Tur Viñes, 2006: 175–177). Two more variables to take into account are, firstly, the increase in local productions striving to maintain their own agenda; and secondly, the fulfilment of democratic requirements and its financial and economic commitments (Bardoel 2003: 83; Gutiérrez Gea

2000b: 154; Medina Laverón 2006: 15–19, Mulgan 1990: 21–24; Pujadas Capdevila 2008: 197–208).

The concept of *Television Quality* refers to the diversity as well as to the formats that are being used. In the current context—each time more multi-platformed and multi-channelled—it's important to refine the indicators of quality of programming: a discourse that considers diversity as an identifier for the quality of the programming, and a discourse that conceives quality independently from the notion of diversity (Ishikawa 1996b: 95–96; Lasagni and Richeri 1996: 19; Lasagni and Richeri 2006: 21; Ojer Goñi 2009: 135–138; Pujadas Capdevila 2008: 231–253).

Quality of Programming alludes to elements such as "diversity as more channels and choice for the audience"; "diversity as reflection"; and "diversity as access", (McQuail 1992: 144–145). Regarding the quality of programming a distinction can be made between horizontal diversity and vertical diversity. The first refers to the degree of real diversity enjoyed by a spectator with access to different channels of a territory, as a result of the set of programming policies designed by each channel in a concrete time slot. Vertical diversity, on the other hand, emphasizes the diversity a particular channel is offering throughout its programming schedule.

Channel Quality, therefore, should not be confused with the concepts of "channel diversity" or "vertical diversity". On the one hand, we stick to the idea of "channel" (as programme producers and content programmers), for which Channel Quality derives from Quality of Programming.

The quality concerning programming, in any case, could be understood from Presentism, regarded for its capacity for innovation and anticipation, and from the recognition of a past that deserves to be preserved. In order to preserve and recognize the history of forgotten television of the United Kingdom, a project[1] (History of Television Drama, June 2017) investigates the history of forgotten television drama in the UK, by looking at productions that are "largely unknown, either because they were produced live and not recorded, or because they were recorded but subsequently wiped, junked, mislaid, or lost". In 1946 all drama was broadcast live and no recordings were made - in fact it was another seven years before the primitive system of recording live television resulted in the earliest surviving television dramas. By 1982 nearly all drama was pre-recorded and the practice of wiping and junking recordings, which occurred on a regular basis during the 1960s–'70s, had ceased. To the mission traditions of the BBC[2] "to enrich people's lives with programmes and services that inform, educate and entertain", we have to add the one of preserving and conserving its heritage—although the latter being an academic project—since its outcome also affects the idea of quality, as it:

> explores the production of television drama in the regions and nations of the UK (the English regions plus Northern Ireland, Scotland and Wales), from both regional BBC production centres and regional ITV companies, and considers dramas that may have just been transmitted in their region of production, as well as dramas that were networked.

[1]Retrived on May 10th, 2017. https://www.royalholloway.ac.uk/mediaarts/research/thehistoryofforgottentelevisiondrama/historyofforgottentvdrama.aspx.

[2]http://www.bbc.co.uk/aboutthebbc/insidethebbc/whoweare/mission_and_values.

A report of the BBC (1992) advocates for quality TV that "is intended to ensure an offer of programmes and services able of creating opportunities for education".

Therefore, one of the parameters that Television Quality measures is the diversity of formats. Pioneer studies such as the one of P. Hilve, quoted by Pujadas Capdevila (2011: 138), assess the quality of television from the relation that each channel establishes between "cultural diversity" and "diversity of genres". This relationship identifies, according to the author, the extent to which the programmation elaborated by each channel is reflecting social structure, as well as near societies and cultures.

In order to make Quality of Programming measurable, diversity has been the key word. There is, furthermore, a general agreement for composing an objective standard for quality. Quality of Programming, in this sense, would be the result of the balance in channel offer, audience access and programme variety. Although these aspects can be interrelated they are also believed to act independently (Fabbro 2006:25; Hillve and Rosengren 1996: 237–238; McQuail 1992: 144–145; Mulgan 1990: 26–28; Ojer Goñi 2009: 135–138; Pujadas Capdevila 2008: 239–253).

Currently, from the popular point of view, the so-called "Golden Age" of television receives more attention than the concept of quality, especially in The United States. American television has known three periods that have been categorized as "stages of the Golden Age". In the ongoing one, thanks to the quality of the fiction (denominated as *Drama Quality*) of HBO and/or Netflix series, the quality of American drama—although not only American—has become a commercial seam and an international reference. One of the paradigmatic cases is the series "Orange is the New Black". From the public statements made by Ted Sarandos—content director of Netflix—we can deduce that the three main principles of the company lie in creating "original, exclusive and global contents", this in over 190 countries, while aiming for its audience to recognize the quality and the degree of "addiction" to its audiovisual products.

The audience has been the object of study since the seventies, making use of audience data as a strategy for their concerns. Small communities such as Canada were worried about the risks for identity. By the way, the broadcasting purpose was pointed out as both professional accomplishment (Uj) and market driven (USA). Japan also took a leading role in studies on broadcasting, especially about influence on voter's intentions and children television. The following decade, through the sponsorship of NHK (Japanese Broadcasting Corporation), several scholars began a research on the assessment of Broadcasting Quality. It was carried out from 1990 to 1993, and involved the cooperation of scholars from the United States, Canada, the United Kingdom, Sweden and Japan. Their work is still considered as the baseline for every quality research on the subject (Eguchi and Ichinohe 1971: 7).

Studies of the nineties, as stated by Pujadas Capdevila (2011: 138), argue that the distribution of different genres also demonstrates that, compared with private channels, public service channels reflect better the society and culture. In that sense, this type of conclusions contain—Pujadas continues—a significant amount of suppositions that often confuse the mission granted to informative genres. Television studies of the 21st century, nevertheless, insist on the hybrid character of audiovisual genres and the irruption of formats such as reality TV that settled internationally on unequal manners—Spain being one of the few countries, alongside Brazil, where the format of "Big Brother" perseveres, for example–.

Public media continue to play a crucial part when aiming for the consolidation and promotion of minority languages. Basque Public Radio and Television (Eitb) was created in 1982 with the intention to potentiate and extend the Basque language as well as Basque culture. In this vein, the doctoral thesis of González-Gorosarri (2011: 705) arrives to the conclusion that the News Quality broadcasted by two Channels of ETB is similar to that of French and Spanish Media. Basque media, in other words, has decreased the abilities to produce news, although it would enhance journalistic activity and satisfy the informational needs of nearer social groups.

The concept of hypertelevision hinders both the definition of today's audiovisual genres and the concept of diversity.[3] A lot of contemporary audiovisual products, furthermore, have been created by global production powerhouses—in the case of Big Brother, by the Dutch production house Endemol, and the concept of quality, as it is outlined by Hilve and Rosengren (1996), depends on the balance between own and external production.

The leading publication on television studies (Ellis 2000) argues that there have been three stages in the history of television: the first era of scarcity; the second era of availability; and the third one, of plenty. Mentioning this point of view is useful in order to understand the evolution of the concept of quality and the diversity in the history of television.

Ellis (2000) considers that the first era of scarcity coincided with and promoted a period of standardized mass market consumerism. This was the stage of the development of public service broadcasting. The second era, namely of availability, corresponded with a much more diverse consumer market that, in the name of choice, accentuated and commodified every available difference between citizens. In the era of scarcity, on the other hand, television tended to present a diffuse and extensive process of working through. "Televisual working through seems to be a process of reconciliation based upon familiarity and repetition-in-difference" (2000: 81). Ellis believes that today's television finds itself in an era of uncertainty. He concludes that "Within its stable formats and developing narrative structures, television also shows a marked degree of discontinuity" (2000: 82).

Scholars have discussed the workability of the term broadcasting, because the "inevitable convergence of telecommunications, mass media, and computing technology amplifies the strategic importance of the Internet. Vilches (2013), in the same vein, talks about the combination of convergence and transmediality of contemporary television. Academic literature coincides in the coexistence of three stages: broadcasting (its strategy would be based on accumulating the audience in order to negotiate its value to the advertisers), narrowcasting (producing specialized content), and netcasting (convergent model).

[3]In this sense it is worth mentioning, amongst other trends, the rise of reality TV programmes and programmes of light entertainment, as well as the decrease in the amount of time dedicated to news broadcast and informative programmes.

Table 2 Hierarchical concepts deriving from broadcasting quality

Broadcasting system quality	Quality of programming		Channel quality	Programme quality
It matches the overall broadcasting quality of a political and legislative system	*Horizontal Quality* It refers to the goals all television channels as a whole must accomplish and the values they must transmit	*Vertical Quality* It reflects the outcome of programmes' quality in every media outlet	It combines accomplishment of the objectives in the articles of association and management goals fulfilment	Every programme is considered a complete assessment object

Source Authors

From this chapter we appeal to the line of Broadcasting Quality, while insisting that Broadcasting Quality investigation has never become a homogenous research object. Evidence for this is the great amount of confusion that exists concerning the concepts and denominations. First of all, from the traditional meaning of Broadcasting, diversity has been understood as more channels and choice for the audience, hence referring to the market offer. In this sense, the term Broadcasting Quality refers to the ability of programming in a different way. Programming a channel in the end is, as Guglielmo (2003) puts it, is an act of assemblage. It's true that the meaning of programming has been undergoing quite some changes, as in the way that the audience selects and creates its "own programming". Television went from being "appointment TV" to "TV everywhere" (multi-screens; elimination of prime time, more user control and a way of understanding zapping).

Therefore, the assessment of Broadcasting Quality has been conducted in different levels of evaluation (Table 2).

In fact, some concepts overlap each other. That would be the case of Television Quality and Channel Quality (Table 3).

The quality of the channel refers to the professional and ethical commitment of its professionals, as well as to its management and deontological practice. The German School has developed, in that sense, objective standards of deontological practice. Likewise, as González-Gorosarri (2011: 97) sets out, "German tradition (especially, that of sources) pleads for diversity, identity and conflict publication as derived standards to assess media performance from a deontological perspective".

3 Quality Newspapers

The first studies on excellence in journalism are closely tied to the moment where the reading consumption of the press began to decrease; namely from 1970 on. At that juncture, the main objective was to activate the concept of excellence in

Table 3 Matching of M. Wober's television quality and channel quality indicators

Television quality indicators	Channel quality indicators
• Enjoyable, pleasing	• Enjoyable programmes
• Informative, educational	• Information programmes • Educational programmes
• Good casting, professionalism	
• Interesting, stimulating	
• Value-presenting	
• For all the family, for every taste	• Programmes for children under five • *Confidence* that programmes non-suitable for all audiences will be broadcast after 9 pm
• Varied	• Wide programme range to choose from
• No quiz, neither prize	
• *Balanced in diversity*	
• Doesn't insult viewers' intelligence	
• Others	
	• Some religious programmes
	• *High cost programmes*

Source Authors

order to recover the position in the market. Therefore, research set out to consolidate the relationships between Newspaper Excellence and press distribution. One of the pioneering works here is Leo Bogart's study of 1977, in which he systematizes the following standards of quality: accuracy, impartiality of reporting, investigative enterprise, specialized staff skills, individuality of character, civic mindedness and literacy style (Bogart 2004: 40–46; Gladney 1990: 65-66; González-Gorosarri 2011; Meyer 2004a: 127). The concept of Newspaper Excellence nevertheless evaluates the media context as well as other objectives and values (i.e. influence, integrity, editorial independence, editorial courage,etc.).

The crisis of journalism—sales have dropped over the last decades—alongside the surge of the Internet and shifts in consumption, has changed the media landscape at the international level. In that sense we can detect a turning point, namely after 1990, where European academics started to show interest in the process of "tabloidization".

The habits of current journalism have absorbed a debate about the so-called "fast journalism". Slow journalism has emerged in response to the increasing importance of "fast and instantaneous journalism and the concerns about the deleterious effects of speed" (Le Masurier 2015: 138).

The quality of newspapers has been the object of various studies, and as it shows, newspapers have opted either to reinvent themselves or to offer an added value. The two following investigations on major media serve as examples reflecting these strategies. On the one hand, Vehkoo (2010: 68) considers that "many quality newsmagazines—*The Economist*, *The Spectator*, *The Atlantic Monthly*—are doing well, despite the recession. Perhaps the style of magazines is the way forward for newspapers. Many of them already publish glossy magazine supplements on weekends". Harbers' (2016: 507) research on the quality and innovative spirit of the Dutch daily newspaper *De Correspondent*, on the other hand, states: "What makes *De Correspondent* innovative is its explicit rejection of the objectivity regime as a suitable professional standard for our current digital and postmodern era".

As reported by Wijnberg, this professional framework has become outmoded as it cannot adapt to a post-industrial production logic and as such fails to engage a new generation of news consumers (Wijnberg 2013a, b; Broersma and Peters 2013).

Lublinksi (2010: 313) orients his research on the specialized correspondents' office in Science. He argues that new qualifications are needed: "a high level of journalistic proficiency is required of science journalists if they are to cooperate and compete with their colleagues from other beats". Vehkoo (2010: 68) declares that a democracy can do without newspapers, but that it cannot exist without quality journalism: "The problem at the moment is, as Clay Shirky puts it, that the old stuff gets broken faster than the new stuff is put in its place. We need to make sure that during the revolution of journalism we don't lose its most important standards and values".

According to Abramson (2010: 41) at *The New York Times* there is a fierce determination to protect the core of our news-gathering, including the most robust international and investigative coverage: "Quality journalism is produced on many platforms. I applaud the announcement that *The Huffington Post* will be underwriting original investigative reporting, perhaps giving work to journalists who have lost their jobs". Lacy and Fico (1990) established a base to systematize the criteria for Newspaper Excellence, with the intention to reduce the possible impact of individual bias on ranking the newspapers and the groups (Table 4).

4 Internet Quality

After the recent incorporation of the Internet in the habits of millions of users around the globe, Internet Quality has been studied with the same rigor as investigations conducted in the fields of Media Quality and Newspaper Excellence. The dimension and heterogeneity of the new medium, nevertheless, have supposed a genuine challenge for many researchers, especially after the expansion of many websites as a result of the open access of the Internet network. The audience, in that sense, had to address "old problems in a new media" and identify different "non-quality features": information overload, availability of vast quantities of useless information, the potential for inaccurate materials, the ephemeral nature of materials disseminated, and so on (Cooke 1999: 6; Ornelas Ley and López Ornelas 2007: 1–3).

The first criteria for evaluation of the Internet were determined largely by technological issues. At a later stage different criteria such as hypertextuality, interactivity, multimediality and crossmediality were assigned (Alexander and Tate 1999: 2;

Table 4 Newspaper excellence criteria, in relation to Gladney's classification

Quality dimensions		Bogart (1977)	Burgoon et al. (1982)	Gladney (1990)
Content Standards	*Diversity*	Variety of content	–	–
	News treatment	Accuracy	Accuracy	Accuracy
		Authority	–	Lack of sensationalism/Strong editorial page
		Breadth of coverage	Depth Sophistication of treatment	News interpretation
		Reflection of the entire home community	–	Strong local coverage/Common community values
		Comprehensiveness		Comprehensive coverage
		Vivid writing	Literary style	Good writing
		Fairness	–	–
	Form	Attractive makeup, packaging or appearance		Visual appeal
		Easy navigability		–
Organisational standards		Integrity		Integrity
		Balance	Impartiality	Editorial independence/Impartiality
		Diligence in discovery	Investigative enterprise	Editorial courage
		–	–	Decency
				Influence
				Staff enterprise
				Staff professionalism
				Community leadership

Source Authors

Barth 2004: 205–211; Fritch and Cromwell 2001: 499–502; Neuberger et al. 2009b: 252–253; Tena Parera 1999: 199–200).

After the promotion of the Internet as a window to the world, therefore, the parameter of Medium Quality took on as a means for the measuring of web pages. In like manner, it coincides with the content elements of the classical aspects of television programmes' assessments (form, content and the interrelation between them); for the evaluation of websites they are called respectively technical quality, content quality and communicative quality. Later on another characteristic joined these three, namely: interactivity (Table 5).

According to Neuberger (2014: 420), the heterogeneity of the Internet makes it difficult to assess its overall quality. "This heterogeneity is also reflected on the level of formats and services", he states. Besides this he also argues that the studies measuring the quality of journalism on the internet from an audience perspective mostly assessed only professional representatives (Neuberger 2011a). Hence, the second way to measure quality is to ask for user's gratifications or motives, and thirdly, media credibility. His investigation (Neuberger 2014: 430) arrives to the conclusion that "people do not lose their awareness of journalistic quality when using the Internet". Neuberger (2014) considers that the majority of the respondents believed that press websites best embodied features such as credibility, regular reporting, and timeliness.

5 Common Criteria and Standards to Evaluate Media Quality

We will employ the term Media Quality to refer to the totality of the content of a media outlet. In that sense, it gathers the denomination of vertical quality (quality of programming), newspaper quality and journalistic Internet website quality. In fact,

Table 5 Matching of Media Output Quality assessment dimensions, according to every medium

Programme Quality	Newspaper content quality in the Anglo-American tradition (Gladney)	Newspaper content in the European tradition (Schönbach)	Internet Quality
Diversity	Diversity	Diversity	Diversity
Form	Form	Layout	Technical quality: * Visual design * Functional design
Content	News treatment	Content	Content quality
Interaction of form and content	–	–	Communicative quality

Source Authors

the concept of Media Quality is also suitable for crossmedia assessment, that is to say, to evaluate two different media types (i.e. television and newspapers, radio and Internet media, etc.).

Along with the goals of broadcasting announced by BBC (information, education and entertainment), Weiβ and Trebbe (1994: 41–42) included the category of "Beratung" (advice, deliberation). Accordingly, media content can be first distinguished as "news coverage" or entertainment. Then, news coverage can show three features (information, education and advice/deliberation), whereas entertainment can be classified as fiction and non-fiction. As a consequence, hybrid genres like infortainment can be assessed not only for television, but also for newspapers. All in all, a media product would be defined according to the amount of every category it offers.

Albers (1996) settled down for the first time the categorization of standards and criteria employed by prize awards. Following Albers' work, every media output could be classified under three categories: form, content and interaction of form and content. Baeβler and Kinnebrock (2004) employed a different nomenclature for similar categories: technical, content and communicative quality for Internet quality. Therefore, those three categories will frame every assessment criteria for Media Quality, named as follows: format quality, content quality and social quality (Table 6).

First of all, *format quality* includes not only technical error-free supply, but also artistic values and originality of formats. In fact, technical error-free supply becomes a "prerequisite for overall excellence", since it operates as a filter for latter errors. If prevention at an early stage has failed, content errors are thought to may have been more difficult to avoid. Moreover, format quality is aimed to show artistic values based on the originality of formats provided by technology (González-Gorosarri, 2011: 286–294).

Secondly, *content quality* is assessed according to three main criteria: diversity, usefulness and reduction of complexity (comprehensibility). The question about diversity as a criterion for quality has been long discussed. Scholars like Gladney

Table 6 Common criteria for every aspect of media quality

	Format features	Content quality	Social responsibility
News coverage	• Technical error-free supply • Artistic values • Originality	• Diversity • Usefulness • Reduction of complexity	• Monitorial Role • Facilitative Role • Radical Role • Collaborative Role
Information			
Education			
Advice or debate			
Entertainment			
Non-fiction			
Fiction			

Source Authors

(1990) and Schönbach (2004) considered diversity as a separate category from layout (here, technical quality) and content. However, most scholars agree that diversity is a criterion of content quality. Nevertheless, it has been highlight that diversity per se cannot assess content quality, since it is alleged that having more formats or genres cannot be "automatically" better. Then, diversity can only be considered a quality measure when referring to the "diversity of the relevant issues". Accordingly, diversity and relevance are taken as a single criterion to evaluate content excellence of Media Quality (Kolb 2015: 43–44).

The term usefulness has arisen to appoint a determinate information supply, which provides the interpretation key to understand deferred consequences of current events. Usefulness requirements set out the assessment of information. Applying to news, regardless of media type, usefulness has been defined as "informativeness", that is to say, the amount of information or news elements required for an elementary knowledge. For educational purposes, the acquisition of new knowledge and perspectives is applied. For entertainment, humor is required (McQuail 1992: 197–200; Mezger 2005: 73; Schatz and Schulz 1992: 696–701; Scheuer 2008: 28–32; Schirmer 2001: 97–99).

Similarly, reduction of complexity alludes to the "comprehensibility of language and context". It includes fact verification, accuracy, background reports, explaining the causes and consequences of the matter, and clarity of the language register (Bucher 2003: 24–27; Bucher and Barth 2003: 232–241; Fengler 2003: 148–158; Pöttker 2005: 127; Ruß-Mohl 2000a: 33–37, b: 252–253).

Lastly, *social quality* evaluates the pretended media's impact in a community, taking into account that journalism is eminently intentional. As a consequence, social quality can only be effectively assessed at the level of the media product and not for every media output like news reporting, for example, since media's influence is performed at the meso level of Media Quality. McQuail adopts the concept of "social responsible media" set out by the Hutchins Commission in 1947, as it was formulated on the basis that media corporations' business offers an eminent social service and not a merely commercial one. The German scholar tradition also includes social quality ("functional consciousness") as another dimension of excellence, different from that of format and content ("professional consciousness"). That is why social quality is assessed according to the roles of the media. In fact, freedom of the press states that media freely choose the social intentionality of their performance. The degree of accomplishment of such a social role is measured, but the convenience of that intentional role can only be judged at the level of System Quality. Therefore, Christians et al. (2009) distinguished four media roles: monitorial, radical, collaborative, and facilitative. The monitorial role describes the basic function of all communication as surveillance. The radical role tends to expose conflict of interests and focuses on the voiceless. The collaborative role assumes the postulates of the State. Finally, the facilitative role aims that media promote dialogue among their audience, to give them the clues to participate in society (Christians et al. 2009: 139–220; McQuail 1992: 237–300).

Table 7 Media Quality criteria for news coverage (information, education and advice or debate)

Media quality dimensions	Criteria	Standards
Format features	Technical error-free supply	Elements that obstacle supply, measured according to time or display
	Artistic values	Aesthetics
	Originality	Creativity, innovation
Content quality	Diversity	Relevant genres and formats, topics, protagonists and opinion
	Usefulness	Informativeness
	Reduction of complexity	Background reports, causes and consequences of the matter, clarity of the employed language
Social Responsibility	Monitorial Role	Journalistic objectivity
		Accountability
	Facilitative Role	Deliberation, public discussion
		Cultural context
	Radical Role	Power and legitimation
		Community
	Collaborative Role	Compliance
		Acquiescence
		Acceptance

Source Authors

Therefore, every aspect of Media Quality (news coverage or entertainment) is assessed following criteria that are gathered under three dimensions: format features, content quality and social responsibility.

In the research field of Media Studies, news coverage has been the most assessed aspect of Media Quality, since studies on fictional entertainment have been conducted in other research areas (Pujadas Capdevila 2011: 168–197). As a result, several standards deriving from the common criteria have already been employed (Table 7).

Nevertheless, the election of given indicators may have not been scientific enough, accused of subjectivity and arbitrariness. Gladney (1990) proposed the assumption of standards that had been employed at least three times in previous works. That will be the purpose of future further research. The present chapter aimed a common frame for every aspect of Media Quality, defining the criteria scholars agree to assess those dimensions.

Acknowledgement Research funded by the Basque Government (2006–2010), and the University of the Basque Country (2012–2013). Visiting fellowship at the Freie Universität Berlin granted by the Basque Country (2014–2016).

References

Abramson, J.: Sustaining quality journalism. Am Acad Arts Sci. Daedalus **139**(2), 39–44 (2010)
Albers, R.: Quality in Television from the perspective of the professional programme maker. In: Ishikawa, S. (ed.) Qual Assessment of Television, pp. 101–143. John Libbey Media, Luton (1996)
Alexander, J.E., Tate, M.A.: Web Wisdom: How to Evaluate and Create Information Quality on the Web. Lawrence Erlbaum Associates, New Jersey (1999)
Altmeppen, K.D.: Ist der Journalismus strukturell qualitätsfähig?: Der Stellenwert journalistischer Organisationen, journalistischer Produkte und journalistischer Medien für die Qualität. In: Bucher, H.J., Altmeppen, K.D. (eds.) Qualität im Journalismus: Grundlagen, Dimensionen, Praxismodelle, pp. 113–128. Westdeutscher Verla, Wiesbaden (2003)
Ang, I.: Desperately seeking the audience. Routledge, New York (1991)
Anker, H.: Qualitätssicherung im Hörfunk: das Beispiel Schweizer Radio RDS. In: Bucher, H.J., Altmeppen, K.D. (eds.) Qualität im Journalismus: Grundlagen, Dimensionen, Praxismodelle, pp. 289–307. Westdeutscher Verlag, Wiesbaden (2003)
Arriagada, E.: Buen periodismo y negocio: La ruta hacia la calidad. Cuadernos de Información **13**, 26–30 (1999)
Baeßler, B., Kinnebrock, S.: Qualitätskriterien interbetbasierter E-Learning-Systeme. In: Beck, K., Schweiger, W., Wirth, W. (eds.) Gute Seiten—schlechte Seiten: Qualität in der Onlinekommunikation, pp. 317–329. Verlag Reinhard Fischer, München (2004)
Bardoel, J.: Back to the Public? Assessing Public Broadcasting in the Netherlands. Public **10**(3), 81–96 (2003)
Barth, C.: Qualitätssicherung in Onlinemedien—Befunde zur Redaktionsorganisation und ihrer Auswirkung auf das Produkt. In: Beck, K., Schweiger, W., Wirth, W. (eds.) Gute Seiten—schlechte Seiten: Qualität in der Onlinekommunikation, pp. 203–221. Verlag Reinhard Fischer, München (2004)
Beam, R.A.: Content differences between Daily Newspapers with strong and weak Market Orientations. Journalism & Mass Communication Quarterly **80**(2), 368–390 (2003)
Beck, K., Schweiger, W., Wirth, W. (eds.): Gute Seiten—schlechte Seiten: Qualität in der Onlinekommunikation. Verlag Reinhard Fischer, München (2004)
Beck, K., Reineck, D., Schubert, C.: Journalistische Qualität in der Wirtschaftskrise. UVK, Konstanz (2010)
Blumler, J.G.: Television and the Public Interest: Vulnerable Values in West European Broadcasting. Sage, London (1992)
Bogart, L.: Reflections on Content Quality in Newspapers. Newsp. Res. J. **25**(1), 40–65 (2004)
Borges, G.: Discusión de parámetros de calidad para analizar programas de La 2 de Portugal. In: A.A.V.V, La Televisión que queremos: Hacia una Televisión de Calidad. Congreso Hispanoluso de Comunicación y Educación, pp. 1–22. Grupo Comunicar, Huelva (2005)
Bourdieu, P.: Sur la télévision. Raisons d'Agir, Paris (2006)
Brandl, A.: Hausse oder Baisse? Zur Qualität von Unternehmensinformation Print vs. Online. In: Beck, K., Schweiger, W., & Wirth, W. (eds.), Gute Seiten—schlechte Seiten: Qualität in der Onlinekommunikation, pp. 233–256. Verlag Reinhard Fischer, München (2004)
Broadcasting Research Unit (BRU) (Ed.): Quality in Television: Programmes, programme-maker, systems. John Libbey, London (1989)
Brüggemann, M.: Jetzt erst recht. Crossmedia-Strategien können die journalistische Qualität verbessern. In: Beck, K., Schweiger, W., Wirth, W. (eds.), Gute Seiten—schlechte Seiten: Qualität in der Onlinekommunikation, pp. 222–232. Verlag Reinhard Fischer, München (2004)
Bucher, H.J.: Journalistische Qualität und Theorien des Journalismus. In: Bucher, H.J., Altmeppen, K.D. (eds.) Qualität im Journalismus: Grundlagen, Dimensionen, Praxismodelle, pp. 11–34. Westdeutscher Verlag, Wiesbaden (2003)
Bucher, H.J., Altmeppen, K.D. (eds.): Qualität im Journalismus: Grundlagen, Dimensionen. Praxismodelle. Westdeutscher Verlag, Wiesbaden (2003)

Bucher, H.J., Barth, C.: Qualität im Hörfunk: Grundlagen einer funktionalen und rezipienten-orientierten Evaluierung. In: Bucher, H.J., Altmeppen, K.D. (eds.) Qualität im Journalismus: Grundlagen, Dimensionen, Praxismodelle, pp. 223–245. Westdeutscher Verlag, Wiesbaden (2003)
Bucher, H.J., Schumacher, P.: Tabloid versus Broadsheet: Wie Zeitungsformate gelesen warden. Media Perspektiven **10**, 514–528 (2007)
Buendía Hegewisch, J.M., Andonie Mena, S., Rodríguez Munguía, J., Vital, J.H.: Propuestas de Indicadores para un Periodismo de Calidad en México. México: Universidad Iberoamericana (2006). http://www.comminit.com/es/node/285965/294
Buss, M.: Qualitätsmanagement intermedial: Hörfunk, Fernsehen, Online. In: Bucher, H.J., Altmeppen, K.D. (eds.) Qualität im Journalismus: Grundlagen, Dimensionen, Praxismodelle, pp. 269–287. Westdeutscher Verla, Wiesbaden (2003)
Campàs, J.: El hipertexto. UOC, Barcelona (2007)
Chen, R., Thorson, E., Lacy, S.: The impact of newsroom investment on newspaper revenues and profits: Small and Medium Newspapers, 1998–2002. Journalism Mass Commun. Q. **82**(3), 516–532 (2005)
Christians, C.G., Glasser, T.L., McQuail, D., Nordenstreng, K., White, R.A.: Normative Theories of the Media: Journalism in Democratic Societies. University of Illinois Press, Chicago (2009)
Chyi, H.I., Yang, M.J.: Is online news an inferior good? Examining the economic nature of online news among users. Journalism Mass Commun Q **86**(3), 594–612 (2009)
Cooke, A.: A guide to finding quality Information on the Internet: Selection and evaluation strategies. Library Association Publishing, London (1999)
Costera Meijer, I.: What is quality television news? A plea for extending the professional repertoire of newsmakers. Journalism Stud. **4**(1), 15–29 (2003)
Costera Meijer, I.: Impact or content? Ratings vs. quality in public broadcasting. Eur. J. Commun. **20**(1), 27–53 (2005)
Cox, B.: Saving quality television. Br. Journalism Rev. **11**(4), 12–17 (2000)
De La Torre, L., Téramo, Mª T.: La noticia en el espejo. Medición de la calidad periodística: La información y su público. Educa, Buenos Aires (2004)
De Pablos, J.M., Mateos, C.: Estrategias informativas para acceder a un periodismo de calidad, en prensa y tv. Patologías y tabla de 'medicación' para recuperar la calidad en la prensa. Ámbitos, 11–12, 341–365 (2004)
Durand, J.: La Qualité des programmes de télévision : Concepts et mesures. Dossiers de l'audiovisuel **43**, 32–64 (1992)
Eguchi, H., Ichinohe, H. (eds.): International Studies of Broadcasting with special reference to the Japanese Studies. NHK Radio & TV Culture Research Institute, Tokio (1971)
Ellis, J.: Seeing Things. Television in The Age of Uncertainty. I.B. Tauris Publishers, New York (2000)
Esser, F.: «Tabloidization» of news: a comparative analysis of Anglo-American and German Press journalism. Eur. J. of Commun. **14**, 291–324 (1999)
Fabbro, G. (ed.): Calidad televisiva: Tendencias y valores en la programación argentina. Universidad Austral, Buenos Aires (2006)
Farré, M.: Hacia un índice de la calidad televisiva: La experiencia de Argentina. Comunicar **25**, 1–16 (2005)
Fengler, S.: Medienkritik—feuilletonistische Textsorte oder Strategie zur Qualitätssicherung. In: Bucher, H.J., Altmeppen, K.D. (eds.) Qualität im Journalismus: Grundlagen, Dimensionen, Praxismodelle, pp. 147–161. Westdeutscher Verlag, Wiesbaden (2003)
Franklin, B.: British Television Policy: A Reader. Routledge, London (2001)
García Avilés, J.A.: Periodismo de calidad: Estándares informativos en la CBS. NBC y ABC. Eunsa, Pamplona (1996)
Gladney, G.A.: Newspaper excellence: how editors of small and large papers judge quality. Newspaper Res. J. **11**(2), 58–72 (1990)

Gladney, G.A.: How editors and readers rank and rate the importance of eighteen traditional standards of newspaper excellence. Journalism Mass Commun. Q. **73**(2), 319–331 (1996)
Gleich, U.: ARD-Forshungsdienst: Beurteilung der Programmqualität im Fernsehen. Media Perspektiven **5**, 253–259 (1994)
Gleich, U.: Crossmedia—Schlüssel zum Erfolg? Verknüpfung von Medien in der Werbekommunikation. Media Perspektiven **11**, 510–516 (2003)
González-Gorosarri, M.: Albisteen kalitatea (Research on Basque Media's News Quality). UPV/EHU, Leioa (2011)
Guglielmo, G.: Vivir del aire. La programación televisiva vista por dentro. Norma, Barcelona (2003)
Gunter, B.: Media Research Methods: Measuring Audiences, Reactions and Impact. Sage, London (2000)
Gutiérrez Gea, C.: Televisión y calidad: Perspectivas de investigación y criterios de evaluación. Zer **9**, 151–184 (2000a)
Gutiérrez Gea, C.: Televisión y diversidad: Génesis, definiciones y perspectivas de la diversidad en las televisiones públicas y comerciales. Ámbitos **3–4**, 69–86 (2000b)
Gutting, D.: Zur Qualitätssicherung der Stakeholder: Kommunikation in der Unternehmenspraxis. Ein Erfharungsbericht? In: Beck, K., Schweiger, W., Wirth, W. (eds.), Gute Seiten—schlechte Seiten: Qualität in der Onlinekommunikation, pp. 287–297. Verlag Reinhard Fischer, München (2004)
Guyot, C.: Calidad editorial: Últimas noticias desde la redacción. In: Amadeo Suárez, A. (ed.). (2007). Periodismo de calidad: Debates y desafíos (pp. 63–79). La Crujía eta Fopea, Buenos Aires (2007)
Hall, P.C., Haubrich, J. (eds.): Kritik am Markt: Was kosten Qualität und Quote?. Hase und Koehler, Mainz (1993)
Haller, M.: Qualität und Benchmarking im Printjournalismus. In: Bucher, H.J., Altmeppen, K.D. (eds.), Qualität im Journalismus: Grundlagen, Dimensionen, Praxismodelle, pp. 181–201. Westdeutscher Verlag, Wiesbaden (2003)
Harbers F.: Time To Engage, Digital Journalism, 4(4), 494–511 (2016). https://doi.org/10.1080/21670811.2015.112472
Hermes, S.: Qualitätsmanagement in Nachrichtenredaktionen. Herbert von Halem, Köln (2006)
Hillve, P., Rosengren, K.E.: Swedish public service television—quality for sale? In: Ishikawa, S. (ed.) Quality Assessment of Television, pp. 231–252. John Libbey Media, Luton (1996)
Hillve, P., Majanen, P., Rosengren, K.E.: Aspects of quality in TV programming: structural diversity compared over time and space. European J. Commun. **12**(3), 291–318 (1997)
Huart, J.: Mieux concevoir pour mieux communiquer à lère des nouveaux médias: Vers des méthodes de conduite de projets et d'évaluation qualité de documents multimédias. Université de Valenciennes et du Hainaut-Cambrésis, Valenciennes (2000)
Huber, C.: Das Journalismus-Netzwerk: Wie mediale Infrastukturen journalistische Qualität beeinflussen. Studien Verlag, Innsbruck (1998)
Ishikawa, S. (ed.): Quality Assessment of Television. John Libbey Media, Luton (1996a)
Ishikawa, S.: The assessment of quality in broadcasting: research in Japan. In: Ishikawa, S. (ed.) Quality Assessment of Television, pp. 89–98. John Libbey Media, Luton (1996b)
Ishikawa, S., Leggatt, T., Litman, B., Raboy, M., Rosengren, K.E., Kambara, N.: Diversity in television programming. In: Ishikawa, S. (ed.) Quality Assessment of Television, pp. 253–263. John Libbey Media, Luton (1996)
Jiménez Piano, M., Ortiz-Repiso Jiménez, V.: Evaluación y calidad de sedes web. Trea, Gijón (2007)
Johnson, T., Kaye, B.K.: Webelievability: a path model examining how convenience and reliance predict online credibility. Journalism Mass Commun. Q. **79**(3), 619–642 (2002)
Jornet, C.: Gestión periodística: Herramientas para lograr un periodismo efectivo y de calidad. Prometeo Libros, Buenos Aires (2006)
Kleinsteuber, H.J.: Mediensysteme. In: Weischenberg, S., Kleinsteuber, H.J., Pörksen, B. (eds.) Handbuch Journalismus und Medien, pp. 275–280. UVK, Konstanz (2005)

Klüber, H.D.: Medienqualität—Was macht sie aus? Zur Qualität einer nicht beendeten, aber wohl verstummenden Debatte. In: Wunden, W. (ed.) Wahrheit als Medienqualität: Beiträge zur Medienethik, pp. 193–210. Gemeinschaftswerk der Evangelischen Publizistik, Frankfurt am Main (1996)
Kohlhaas, B.: Aproximación al Concepto de Control y Garantía de Calidad en Prensa: El caso de La Opinión A Coruña (Grupo Editorial Prensa Ibérica). Universidade de Santiago de Compostela, Santiago de Compostela (2004)
Kolb, S.: Vielfalt im Fernsehen—Eine komparative Studie zur Entwicklung von TV-Märkten in Westeuropa. UVK, München (2015)
Lacy, S.: A model of demand for news: impact of competition on newspaper content. Journalism Quarterly **66**(1), 40–48 (1989)
Lacy, S., Fico, F.: Newspaper quality and ownership: rating the groups. Newspaper Res. J. **11**(2), 42–56 (1990)
Lasagni, MªC., Richeri, G.: Televisión y calidad: El debate internacional. La Crujía, Buenos Aires (2006)
Lasagni, MªC., Richeri, G.: Televisione e qualità: La ricerca internazionale. Il dibattito in Italia. RAI-ERI, Roma (1996)
Le Masurier, M.: What is slow journalism? Journalism Pract. **9**(2), 138–152 (2015)
Lealand, G.: Searching for quality television in New Zealand: hunting the Moa? Int. J. Cult. Stud. **4**(4), 448–455 (2001)
Leggatt, T.: Identifying the undefinable—an essay on approaches to assessing quality in television in the UK. In: Ishikawa, S. (ed.) Quality Assessment of Television, pp. 73–88. John Libbey Media, Luton (1996a)
Leggatt, T.: Quality in television: the views of professionals. In: Ishikawa, S. (ed.) Quality Assessment of Television, 145–167. John Libbey Media, Luton (1996b)
Leleu-Merviel, S.: La conception en communication: Méthodologie qualité. Hermes, Paris (1997)
Lipovetsy, G.: La era hipermoderna. Anagrama, Barcelona (2008)
Litman, B.R.: The television networks, competition and program diversity. J. Broadcast. **23**(4), 393–409 (1979)
Lublinski, J.: Structuring the Science beat. Journalism Pract. **5**(3), 303–318 (2011). https://doi.org/10.1080/17512786.2010.530984
Meneses Fernández, MªD.: La selección y calidad de los contenidos mediáticos como asunto de debate periodístico. una visión diacrónica. In: Carcelén García, S., Rodríguez Wangüemert, C., Villagra García, N. (eds.), Propuestas para una comunicación de calidad: Contenidos, efectos y formación, pp. 81–89. Edipo, Madrid (2006)
Maier, S.R.: Accuracy matters: a cross-market assessment of newspaper error and credibility. Journalism Mass Commun. Q. **82**(3), 533–551 (2005)
Mbaye, S.A.A.: La perception et la représentation de la qualité des programmes de la télévision publique sénégalaise: Le "feed-back" facteur de redynamisation. Université Stendhal Grenoble 3, Grenoble (1998)
McQuail, D.: Media Performance: Mass Communication and the Public Interest. Sage, London (1992)
McQuail, D.: McQuail's Mass Communication Theory. Sage, London (2010)
Medina Laverón, M.: Calidad y contenidos audiovisuales. Eunsa, Pamplona (2006)
Meier, K.: Qualität im online-journalismus. In: Bucher, H.J., Altmeppen, K.D. (eds.) Qualität im Journalismus: Grundlagen, Dimensionen, Praxismodelle, pp. 247–266. Westdeutscher Verlag, Wiesbaden (2003)
Meier, K.: Journalistik. UVK, Konstanz (2007)
Merlo Vega, J.A.: La evaluación de la calidad de la información web: Aportaciones teóricas y experiencias prácticas. In: Zapico Alonso, F. (ed.) Recursos informativos: Creación, descripción y evaluación, pp. 87–95. Junta de Extremadura, Mérida (2003)
Meyer, P.: The Vanishing Newspaper: Saving Journalism in the Information Age. University of Missouri Press, Columbia, Missouri (2004a)

Meyer, P.: The influence model and newspaper business. Newspaper Res. J. **25**(1), 66–83 (2004b)
Mezger, B.: Publizistisches Qualitätsmanagement und regulierte Selbstregulierung: Notwendigkeit und Möglichkeiten rechtlicher Qualitätssteuerung insbesondere der Tagespresse. Dr. Kovač, Hamburg (2005)
Mirón López, L.M.: El cambio de siglo en los medios de comunicación: El futuro de la prensa pasa por la calidad. In: Carcelén García, S., Rodríguez Wangüemert, C., Villagra García, N. (eds.) Propuestas para una comunicación de calidad: Contenidos, efectos y formación, pp. 91–96. Edipo, Madrid (2006)
Mulgan, G. (ed.): The Question of Quality. British Film Institute, London (1990)
Neuberger, C.: Qualität im Onlinejournalismus. In: Beck, K., Schweiger, W., Wirth, W. (eds.) Gute Seiten—schlechte Seiten: Qualität in der Onlinekommunikation, pp. 32–57. Verlag Reinhard Fischer, München (2004)
Neuberger, C.: The journalistic quality of internet formats and services. Digital Journalism **2**(3), 419–433 (2014). https://doi.org/10.1080/21670811.2014.892742
Ojer Goñi, T.: La calidad en los programas de televisión de la BBC. Comunicación y Pluralismo **6**, 9–36 (2008)
Ojer Goñi, T.: La BBC, un modelo de gestión audiovisual en tiempos de crisis. Euroeditions, Madrid (2009)
Ornelas Ley, A., López Ornelas, M.: La calidad de la información en Internet: Un asunto de responsabilidad académica. Revista Mexicana de Comunicación **103**, 1–26 (2007)
Ornelas Ley, A., López Ornelas, M.: En búsqueda de la calidad de la información que se publica en Internet. Textos de la CiberSociedad. **12**, 1–12 (2009)
Picard, R.G.: Measuring and interpreting productivity of journalists. Newspaper Res. J. **19**(4), 71–84 (1998)
Picard, R.G.: Commercialism and Newspaper Quality. Newspaper Res. J. **25**(1), 54–65 (2004)
Plasser, F.: From hard to soft news standards?: how political journalists in different media systems evaluate the shifting quality of news. Harvard Int. J. Press/Politics **10**(2), 47–68 (2005)
Pöttker, H.: Der Deutsche Presserat und seine Kritiker Playdoyer für eine transparente Selbstkontrolle des Journalismus. In: Baum, A., Langenbucher, W.R., Pöttker, H., Schiha, C. (ed.), Handbuch Medien-selbstkontrolle, pp. 125–131. VS Verlag für Sozialwissenschaften, Wiesbaden (2005)
Prix Italia: The quest for Quality: The public at risk. Alla ricerca della Qualità: Il pubblico a rischio (The 4th International Forum on "The quest for television and radio quality: the public at risk"). RAI (Radio Televisione Italiana), Ravenna (1998)
Puente, S., Alessandri, F., Edwards, C., Pellegrini, S., Rozas, E., Saavedra, G., Porath, W.: VAP: Un sistema métrico de la calidad periodística. Cuadernos de Información **14**, 112–120 (2001)
Pühringer, K.: Journalisten Kapital und Herausforderung im Zeitungsunternehmen: Implementierung von Personalentwicklungsinstrumenten und deren Wirkung auf redaktionelles Wissensmanagement. Mitarbeitermotivation und Personalfluktuation. Lit, Berlin (2007)
Pujadas Capdevila, E.: Els Discursos sobre la "Televisió de Qualitat": Àmbits Temàtics de Referència i Perspectives d'Anàlisi. Universitat Pompeu Fabra, Barcelona (2008)
Pujadas Capdevila, E.: La televisión de calidad: Contenidos y debates. Aldea Global, Barcelona (2011)
Pujadas, E., Oliva, M.: L'evaluació de la diversitat de la programació televisiva. Quaderns del CAC **28**, 87–98 (2007)
Quandt, T.: Qualität als Konstrukt: Entwicklung von Qualitätskriterien im Onlinejournalismus. In: Beck, K., Schweiger, W., Wirth, W. (eds.) Gute Seiten—schlechte Seiten: Qualität in der Onlinekommunikation, pp. 58–79. Verlag Reinhard Fischer, München (2004)
Raboy, M.: Legal, institutional and research perspectives on Broadcast programme quality in Canada. In: Ishikawa, S. (ed.) Quality Assessment of Television, pp. 49–72. University of Luton Press, Luton (1996)
Rau, H.: Qualität in einer Ökonomie der Publizistik: Betriebswirtschaftliche Lösungen für die Redaktion. Verlag für Sozialwissenschaften, Wiesbaden (2007)

Rausell Köster, C., Rausell Köster, P.: Democracia, información y mercado: Propuestas para democratizar el control de la realidad. Tecnos, Madrid (2002)
Reiter, S., Ruß-Mohl, S. (eds.): Zukunft oder Ende des Journalismus? Medienmanagement, Publizistische Qualitätssicherung. Redaktionelles Marketing. Bertelsmann Stiftung, Gütersloh (1994)
Román Álvarez, E.: Conectividad y referencia: Estudio de la calidad informativa de los diarios digitales chilenos. In: Vera, H., Aravena, S., Pastene, M., Román, E. (eds.), Calidad de la información periodística. Investigación sobre diarios chilenos: aspectos teóricos y metodológicos, pp. 53–76.: Universidad de Santiago de Chile, Santiago (2004)
Rosengren, K.E., Carlsson, M., Tagerud, Y.: Quality in Programming: Views from the North. In: Ishikawa, S. (ed.) Quality Assessment of Television, pp. 3–48. University of Luton Press, Luton (1996)
Ruano López, S.: Cultura y Medios: De la Escuela de Frankfurt a la Convergencia Multimedia. Ámbitos **15**, 59–74 (2006)
Ruiz San Miguel, F. J.; Blanco, S.: Los contenidos televisivos y el control social de su calidad: Los weblogs, una nueva herramienta interactiva. In: A.A.A.A., La televisión que queremos: Hacia una televisión de calidad. Congreso hispanoluso de comunicación y educación. (pp. 1–16). Grupo Comunicar, Huelva (2005)
Ruß-Mohl, S.: Am eigenen Schopfe…: Qualitätssicherung im Journalismus—Grundfragen, Ansätze, Näherungsversuche. Publizistik **1**, 83–96 (1992a)
Ruß-Mohl, S.: Zeitungs-Umbruch: Wie sich Amerikas Presse revolutioniert. Argon, Berlin (1992b)
Ruß-Mohl, S.: Anything goes? Ein Stolperstein und sieben Thesen zur publizistischen Qualitätssicherung. In: Reiter, S., Ruß-Mohl, S. (eds.) Zukunft oder Ende des Journalismus? Medienmanagement, Publizistische Qualitätssicherung, Redaktionelles Marketing, pp. 20–28. Bertelsmann Stiftung, Gütersloh (1994a)
Ruß-Mohl, S.: Infrastrukturfaktor und Infrastrukturfalle: Plädoyer für ein Qualitätssicherungs-Netzwerk im Journalismus. In: Reiter, S., Ruß-Mohl, S. (eds.) Zukunft oder Ende des Journalismus? Medienmanagement, Publizistische Qualitätssicherung, Redaktionelles Marketing, pp. 243–259. Bertelsmann Stiftung, Gütersloh (1994b)
Ruß-Mohl, S.: Der I-Faktor: Qualiätssicherung im amerikanischen Journalismus—Modell für Europa?. Edition Interfrom, Zürich (1994c)
Ruß-Mohl, S.: Redaktionelles Marketing und Management. In: Jarren, O. (ed.) Medien und Journalismus, pp. 104–138. Westdeutscher Verlag, Opladen (1995)
Ruß-Mohl, S.: Berichterstattung in eigener Sache: Die Verantwortung von Journalismus und Medienunternehmen". In: Ruß-Mohl, S., Fengler, S. (eds.) Medien auf der Bühne der Medien: Zur Zukunft von Medienjournalismus und Medien-PR, pp. 17–38. Dahlem University Press, Berlin (2000a)
Ruß-Mohl, S.: Medienjournalismus auf dem Weg zur «fünften Gewalt» -die USA als Beispiel. In: Ruß-Mohl, S., Fengler, S. (eds.) Medien auf der Bühne der Medien: Zur Zukunft von Medienjournalismus und Medien-PR, pp. 252–259. Dahlem University Press, Berlin (2000b)
Ruß-Mohl, S, Fengler, Susanne (ed.): Medien auf der Bühne der Medien: Zur Zukunft von Medienjournalismus und Medien-PR. Dahlem University Press, Berlin (2000)
Ruß-Mohl, S.: Garanzia di Qualità Giornalistica nel Ciclo dell'Attenzione. Stud. Commun. Sci. **2**(1), 105–130 (2002)
Ruß-Mohl, S.: Journalismus: Das Hand- und Lehrbuch. Frankfurter Allgemeine Zeitung (FAZ), Frankfurt am Main (2003a)
Ruß-Mohl, S.: Towards a European Journalism?—Limits, opportunities, challenges. Stud. Commun. Sci. **3**(2), 203–216 (2003b)
Ruß-Mohl, S.: Qualität. In: Weischenberg, S., Kleinsteuber, H.J., Pörksen, B. (eds.) Handbuch Journalismus und Medien, pp. 374–381. UVK, Konstanz (2005)
Sánchez-Tabernero, A.: Los contenidos de los medios de comunicación: Calidad, rentabilidad y competencia. Deusto, Barcelona (2008)
Sarandos, T. http://www.revistagq.com/noticias/articulos/ted-sarandos-netflix-mas-diverso-tele-tradicional-medida-visionado/24032 (2016)

Schatz, H, Schulz, Wi.: Qualität von Fernsehprogrammen. Kriterien und Methoden zur Beurteilung von Programmqualität im dualen Fernsehsystem. Media Perspektiven (11), 690–712 (1992)
Scheuer, J.: The Big Picture: Why Democracies Need Journalistic Excellence. Routledge, New York (2008)
Schirmer, S.: Die Titelseiten-Aufmacher der BILD-Zeitung im Wandel: Eine Inhaltsanalyse unter Berücksichtigung von Merkmalen journalistischer Qualität. Reinhard Fischer, München (2001)
Schönbach, K.: A balance between imitation and contrast: What makes newspapers successful? A summary of internationally comparative research. J. Media Econ. **17**(3), 219–227 (2004)
Sjurts, I.: Medienmanagement. In: Weischenberg, S., Kleinsteuber, H., Pörksem, B. (eds.) Handbuch Journalismus und Medien, pp. 246–250. UVK, Konstanz (2005)
Smith, A.G.: Testing the surf: criteria for evaluating internet information resources. Public-Access Comput. Syst. Rev. **8**(3), 1–13 (1997)
Soler Campillo, M.: El Informe del «Comité de Sabios»: Bases para una televisión de calidad. In: V.V.A.A., La Televisión que queremos: Hacia una Televisión de Calidad. Congreso Hispanoluso de Comunicación y Educación, pp. 1–25. Grupo Comunicar, Huelva (2005)
Stark, B., Kraus, D.: Crossmediale Strategien überregionaler Tageszeitungen: Empirische Studie am Beispiel des Pressemarkts in Österreich. Media Perspektiven **6**, 307–317 (2008)
Tena Parera, D.: Una nueva propuesta metodológica en torno a la investigación científica sobre los medios impresos: El estado estético. Zer **6**, 199–217 (1999)
Thomass, B.: Interkulturelle Kommunikation und Medienethik—Interkulturelle Medienethik? In: Bucher, H.-J., Altmeppen, K.-D- (ed), Qualität im Journalismus: Grundlagen, Dimensionen, Praxismodelle, pp. 93–110. Westdeutscher Verlag, Wiesbaden (2003)
Thomass, B.: Der Qualitätsdiskurs des Public Service Broadcasting im internationalen Vergleich. In: Weischenberg, S., Loosen, W., Beuthner, M. (eds.) Medien-Qualitäten: Öffentliche Kommunikation zwischen ökonomischem Kalkül und Sozialverantwortung, pp. 53–73. UVK, Konstanz (2005)
Tur Viñes, V.: El concepto de calidad del contenido audiovisual, desde el emisor. In: Carcelén García, S., Rodríguez Wangüermert, C., Villagra García, N. (eds.) Propuestas para una comunicación de calidad: contenidos, efectos y formación, pp. 171–180. Edipo, Madrid (2006)
Van Weezel, A.: Measurement of the productivity and quality of journalistic work on the internet. In: Picard, R.G. (ed.) Media Firms: Structures, Operations, and Performance, pp. 191–203. Lawrence Erlbaum, New Jersey (2002)
Vehkoo, J.: What it QUALITY JOURNALISM and How it can be Saved. Reuters Institute Fellowship Paper, Oxford (2010)
Vehlow, B.: Qualität von Spätnachrichten-Sendungen. Reinhard Fischer, München (2006)
Vilches, L.: Convergencia y transmedialidad. Gedisa, Barcelona (2013)
Villanueva, E.: Ética de la radio y la televisión. Reglas para una calidad de vida mediática. Universidad Iberoamericana, México (2000)
Wallisch, G.: Journalistische Qualität: Definitionen, Modelle. Kritik. UVK Medien, Konstanz (1995)
Wathen, C.N., Burkell, J.: Believe It or Not: Factors Influencing Credibility on the Web. J. Am. Soc. Inf. Sci. Technol. (JASIST) **53**(2), 134–144 (2002)
Weischenberg, S.: Medienqualitäten: Zur Einführung in den kommunikationswissenschaftlichen Diskurs über Maßtäbe und Methoden zur Bewertung öffentlicher Kommunikation. In: Weischenberg, S., Loosen, W., Beuthner, M. (eds.) Medien-Qualitäten: Öffentliche Kommunikation zwischen ökonomischem Kalkül und Sozialverantwortung, pp. 9–34. UVK, Konstanz (2006a)
Weischenberg, S.: Qualitätssicherung—Qualitätsstandards für Medienprodukte". In: Scholz, C. (ed.) Handbuch Medienmanagement, 663–685. Springer, Berlin (2006b)
Weiβ, H.J., Trebbe, J.: Öffentliche Streitfragen in privaten Fernsehprogrammen. Zur Informationsleistung von RTL, SAT 1 und PRO 7. Schrif-tenreihe Medienforschung der Landesanstalt für Rundfunk Nordrhein-Westfalen, Opladen (1994)

Welker, M.: Qualitätswettbewerbe für Internetangebote: PR-Gag oder wirkungsstarke Kommunikationsplattform? In: Beck, K., Schweiger, W., Wirth, W. (eds.) Gute Seiten—schlechte Seiten: Qualität in der Onlinekommunikation, pp. 268–284. Verlag Reinhard Fischer, München (2004)
Wilke, J.: Was heißt journalistische Qualität? Auch ein Versuch zur Bestimmung ihrer Kriterien. In: Duchkowitsch, W., Hausjell, F., Hömberg, W., Kutsch, A., Neverla, I. (eds.) Journalismus als Kultur: Analysen und Essays, pp. 133–142. Westdeutscher Verlag, Wiesbaden (1998)
Wilke, J.: Zur Geschichte der journalistischen Qualität. In: Bucher, H.J., Altmeppen, K.-D. (eds.) Qualität im Journalismus: Grundlagen, Dimensionen, Praxismodelle, pp. 35–54. Westdeutscher Verlag, Wiesbaden (2003)
Wunden, W.: Medienethik—normative Grundlage der journalistischen Praxis. In: Bucher, H.-J., Altmeppen, K.-D. (eds.) Qualität im Journalismus: Grundlagen, Dimensionen, Praxismodelle, pp. 55–77. Westdeutscher Verlag, Wiesbaden (2003)
Wyss, V.: Redaktionelles Qualitätsmanagement: Ziele, Normen. Ressourcen. UVK, Konstanz (2002)
Wyss, V.: Journalistische Qualität und Qualitätsmanagement. In: Bucher, H.J., Altmeppen, K.D (eds.), Qualität im Journalismus: Grundlagen, Dimensionen, Praxismodelle, pp. 129–145. Westdeucher Verlag, Wiesbaden (2003a)
Wyss, V.: Qualitätsmanagement im Journalismus: Das Konzept TQM auf Redaktionsstufe (2003b). www.medienheft.ch/kritik/bibliothek/TQM.pdf
Wyss, V.: Qualitative Analyse der Strukturen zur redaktionellen Qualitätssicherung im privaten Rundfunk in der Schweiz 2006, ZHAW Zürcher Hochschule für Angewandte Wissenschaften (2007). http://pd.zhaw.ch/hop/2010914319.pdf

María González-Gorosarri Lecturer at the Universidad del País Vasco (Spain). Degree in Law and Journalism. European Doctor in Communication. Postdoctoral fellowship on Media Quality at Freie Universität Berlin. Research lines: quality journalism, news quality and gender equality. Durango, Spain.

Andoni Iturbe Tolosa Associate Professor of Audiovisual Communication and Advertising at the Faculty of Social Sciences and Communication, Universidad del País Vasco/Euskal Herriko Unibertsitatea. He holds a Ph.D. in Social Communication and a Degree in Journalism. He also has an extensive experience in the media (newspapers, radio and television). His research areas are new trends in communication, film aesthetics and relationships between cinema and television. Bilbao, Spain.

Information Quality and Trust: From Traditional Media to Cybermedia

Carmen Marta Lazo and Pedro Farias Batlle

Abstract The practice of journalism has suffered, in last years, an increasing discredit in the public opinion. First, we will make a review of the basis of journalistic quality, to later make and approach to information credibility as a way to assess the causes of the decline in media reputation—both traditional and cybermedia—perceived by receivers. With the aim of assessing and valuing the causes of the profession discredit among citizens, we will collect the latest reports from the Madrid Press Association on journalism, with a representative sample of total Spanish population. Therefore, we note that the main reasons why journalism is being discredited are the uncertainty concerning job security and the emergence of new non-professionalized media players on social networks.

Keywords Journalistic quality · Information credibility · Media · Professional values · Social role

1 Introduction

In last years, the journalistic profession has been discredited due to the change in the concept of information and the alteration of professional values for business profit. For this reason, information is sensationalised to achieve greater impact and to increase audience share. However, it should be not forgotten that the media exist by delegation of society, with the aim of informing citizens, with rigour and truthfulness, as established by regulation in force (art. 19 of the Universal Declaration of Human Rights, 1948; art. 20 of the Spanish Constitution) and in the various codes of deontology (Code of Principles from the IFJ, International Federation of Journalists, 1954; Professional Journalistic Standards and Code of Ethics, UNESCO, 1983; the

C. Marta Lazo (✉)
Universidad de Zaragoza, Saragossa, Spain
e-mail: cmarta@unizar.es

P. Farias Batlle
Universidad de Málaga, Málaga, Spain
e-mail: farias@uma.es

M. Túñez-López et al. (eds.), *Communication: Innovation & Quality*, Studies in Systems, Decision and Control 154, https://doi.org/10.1007/978-3-319-91860-0_12

Ethics of Journalism, Council of Europe 1993; and the Code of Ethics from the FAPE (Federation of Spanish Press Associations), collected by some authors (Jones 1980; Laitila 1995).

In this regard, as pointed out in the Code of Ethics from the FAPE, "Within the framework of civil liberties enshrined in the Constitution, which constitute the needed reference for a fully democratic society, the professional practice of journalism represents a relevant social commitment, so that the free and effective development of fundamental rights on free information and expression of ideas becomes real for all citizens" (FAPE 1993).

This research aims at identifying the reasons why of the transformation in journalism values based on public service, a new version exercised by "a modern public administration of the infotainment" (Del Rey Morató 1998).

As noted by Cebrián Herreros (2003), "At present the information, due to trade implications of the company, is subjected to commercial approaches, political propaganda, subordination to morbidity and diverse degrading approaches to reach wider audiences. The professional—journalist—performs a task of commitment to society. As a professional, he/she is a mediator between facts and audience through a corporation. The corporation is a means, not an end. The only end for every professional is to serve the audience".

On the other hand, one of the main present challenges for journalists is to face the information pollution, also referred to as info-pollution (Desantes Guanter 2004) and the information overload resulting from the digital era. "The abundance of information, the speed of communications and the increase in the quantity of media outlets highlight more than ever before the value of the social responsibility of journalists when selecting, prioritize and confirm information, together with their honesty and impartiality when presenting that information. The pillars of a journalist's credibility cannot be replaced by either the technological sophistication of present reliable communication or the overdose of information" (Bustamante Newball 2006).

Diezhandino (1994) explains that, despite changes in the digital era, journalists are the essential elements in the informative sequence, and they should regain such importance in their role of newspeople: "at the end, despite the progress made, both in technology and research, we must go back to the starting point: journalists, their daily effort, pressures, competences, their view of the world, their sound analytical approach…The important thing is always the journalist. He/she is the only, and the better tool that remains, whatever changes are coming, whatever the platform. Journalists and their invariable cornerstone: information. Technological advances do not teach how to inform, no matter how sophisticated they may be. They certainly make the task easier. There is a risk that they will facilitate so much the work that they destroy the human ability to improve their quality of their work, under the shelter of the technological quality". From this dimension, we also wonder how the technological transformation in the digital society has affected principles and values of journalistic profession.

These changes and the removal of the sense of the profession derived from the displacement towards the spectacle, as pointed out by González Reigosa (1997), force to "rethink some essential elements of the profession: the social, professional and cultural identity of the journalist; the definition of their basic qualities; the definition of professional interest and national interest; the definition of the piece of news (who should identify it? What matters the most: reality, market, pleasure, audience?". That interest in "rethinking the profession" allows us to consider the search for arguments to enhance the information quality in the practice of journalism.

In order to determine what are the pillars for the quality of information, we carry out a review of the various authors and existing regulation that reflect the different dimensions of the essential public service that the media should address, both traditional media and cybermedia. Also, we have made a review of studies on information credibility, which are extremely complex as they depend on a wide range of factors. Some of them depend on the activity of the media outlet itself and professionals that work there (care for information, corrections, degree of readiness of professionals, proper use of sources, ethical practices, and clear separation of the editorial line and information). Others focus in news perception among receivers, which depends on the training level, the degree of knowledge on the subject, and the identification of the media. As we will see, these reports enable us to have a comprehensive framework of communication in different fields, through which the deterioration of working and professional conditions in journalism has been verified, together with the emergence of opportunities and new media as a result, in essence, of technological development.

2 From the Origins of Journalism to the Most Recent Studies on Quality of Information

The initial studies on journalism arise from sociology of communication relating to the social impact of the "journalistic activity", the name of this research line. The origin lies in the sociology of Max Weber, who proposed, in the first sociological congress held in Frankfurt in 1910, to make journalism a subject for sociological analysis. In the field of public opinion and mass communication, we highlight the initial studies by Walter Lippmann addressing the effects of political propaganda and the influence of advertising. Later, different models emerge within Mass Communication Research, as communication models endorsed by Lasswell, Lazarsfeld, Schramm, Kappler and Merton. With the growth in the influence of journalism in the social life, studies on journalism arise, journalism studies through the creation of the first school by Pulitzer in Columbia, at the beginning as a training for writing and, later, as a generator of journalism studies. Thus, as pointed out by Núñez Ladevéze (2004), journalism is one of the professional specialisations, "so decisive in the training on the modern world, their social and political relevance are so significant, that one could not be understood without the other. Therefore, the study of journalism has a special interest for those who want know the complexity, the turnarounds, and the mindset of present society".

As regards the analysis of the current state of the profession, standard references exist on the context and the picture of journalism (Cantalapiedra et al. 2000;

García de Cortázar and García de León 2000; Canel and Sádaba 1999; Ortega and Humanes 2000; Escobar Roca 2002; Rodríguez Andrés 2003; Van der Haak et al. 2012; Diezhandino et al. 2012; Túñez and Martínez 2014), others show the precariousness of the professional exercise (López Hidalgo 2005; Figueras-Maz et al. 2012; Ufarte Ruiz 2012; Soengas et al. 2014), and the crisis caused by digitization (Garcia de Madariaga Miranda 2008; McChesney and Nichols 2010; Grueskin et al. 2011), and other sources make reference to the loss of information quality (De Pablos and Mateos 2004; Gómez Mompart et al. 2015) and the divergences and agreements with the audience in the way to understand journalism (Zaller 2003; Porto 2007; Belt and Just 2008; Gil-de-Zúñiga and Hinsley 2013; De La Torre and Páramo 2015).

On the other hand, analysis establishing comparatives between old and new professional profiles may be found (Túñez 2012; López et al. 2012; Agustín Lacruz et al. 2013), others focus on the change of profiles as a consequence of the implementation and the development of cyberjournalism (Beckett and Mansell 2008; Salavarría and García Avilés 2008). Other studies point out the relevance of the versatility of professional profiles (Scolari et al. 2008; Masip and Micó 2009; Gómez Patiño 2012), others suggest the use of social networks and thus a further interaction with the audience (Túñez and Sixto 2012; Noguera 2013), others propose the essence of the maintenance of traditional roles of journalists (Gómez Mompart et al. 2013; Sánchez-García et al. 2015) and other reflections demand the recovery of the inherent values of the profession towards the exercise of social responsibility (Sinova 2003; Silverstone 2006; Osuna Acedo et al. 2013; Pihl-Thingvad 2014; Soengas Pérez 2015) and the sense of being accountable to citizens (Fengler et al. 2015).

3 Analysis of Indicators of Information Quality, from Traditional Media to Cybermedia

The existing criteria when analysing the quality of news are very different, both as regards the number of indicators and the type of measurement, and depend on the author. In this regard, we find an interesting metric system of journalistic quality, to assess the coherence between the interests of the public that consumes information and the newsworthy delivery of the media, called Valor Agregado Periodístico (VAP, Journalistic Added Value), by Alessandri et al. (2001). Also, after carrying out an interesting review of the state-of-the-art, Romero-Rodríguez et al. (2016) come to define an analysis model containing 75 items distributed into three typologies of statistical descriptions: those linked to the corporate areas of the media; those linked to the socio-employment areas of the media staff; and those linked to informational content and final product. As pointed out by these authors, most of the efforts to establish taxonomies about information quality have been carried out from the German and Dutch Academia (Schatz and Schulz 1992; Pottker 2000; Arnold 2009), until the authors Urban and Schweiger (2014) until the authors compile them in six central dimensions: diversity, relevance, accuracy, impartiality and ethics.

In this regard, as regards the typologies of normative structures, we find specific proposals focused on the analysis of different types of media, both mass media and cybermedia, being the most numerous in the most recent decade.

In last years, the most interesting classifications, both as regards the analysis of quality standards and the empirical application to different media, are found in written press. For instance, the research carried out by De la Torre and Téramo (2015) concerning convergences and divergences in the criteria of information quality from the points of view of the media and the public, applied to the Argentinian newspapers Diario de Cuyo, Clarín and La Nación. It is particularly interesting to note one of the conclusions that an author draws, linked to the bet on innovation and the courage when selecting and getting into the news. According to the author, after the analysis performed, the journalistic quality goes beyond the criteria of newsworthiness and is also related to the sensitivity to take account of the impact of what is communicated, the knowledge of reasons for not forgetting the context—background and consequences, the justice to present the other side of information, the persistence to get to the heart of the matter, the vision to help audiences see beyond, and the freedom to handle pressures, especially when they come from the political and ideological fields. Also, it is worth mentioning the analysis carried out by Ramírez de la Piscina et al. (2014) on the decade-long evolution of the quality parameters, based on the concept of media performance by McQuail (1992), of five European newspapers of reference (*Financial Times, Corriere della Sera, Frankfurter Allgemeine Zeitung, Le Monde* and *El País*). The analysed items about news quality were based on three dimensions: selection process (2.5 points), development (5 points) and social contribution (2.5 points). Within the selection process, five aspects were assessed, each with 0.5 points: citation of the origin of information, character of sources, factuality of the history reported, degree of timeliness and degree of interest of the piece of news. Within the development process, each item was evaluated with one point: accuracy (correspondence between headline and body of the news), depth (5W), different perspectives within the piece of news, complementary information of other elements (images, graphics, infographics, etc.), and the appropriateness of the journalistic language. The section on social contribution had other five subsections: power control, promotion of social debate, respect to human dignity, presence of cultural references different from those of those belonging to the newspaper, and combat social exclusion. Results of the research show that the average quality rate of news in the five analysed newspapers barely exceeds the minimum to pass (5.39 points out of 10), and the main conclusion is that newspapers forget the social function historically attributed to journalism, as the watchdog of citizens concerns, which is evident in the poor results obtained in the section on the social contribution of the news.

Television studies observe quality premises in news bulletins, for instance, in national channels (Humanes 2001; Meijer 2003; López-Téllez and Cuenca-García 2005; Humanes et al. 2013), in international channels, establishing a comparative of the most significant elements about quality in some countries, such as Japan, United Kingdom, USA and Sweden (Gutiérrez-Gea 2000) or focused on the public service of broadcasters of reference, such as the BBC (Born and Prosser 2001). Among the most recent, it should be highlighted the work carried out by Aguilar-Paredes et al. (2016), which analyses the quality of prime time news bulletins of autonomous channels in Spain. The research collect previous work on the political image construction in TV news (Iyengar 1998; Semetko and Valkenburg 2000; Guerrero-Solé et al. 2013), and it is ground-breaking because it applies an index connected with political news, as regards the degree of representation of political parties and the value of newsworthy frames, reaching a deviation index of the information quality (IDCI), which enables to identify and measure the changes that should take place to improve the quality of information.

As regards cybermedia, one may find a model of the reasons behind the consumption of digital media, titles by Thorson and Duffy (2005) Media Choice Model, which includes specific variables of digital communication, such as connectivity; the interaction with the author of the piece of news and other readers; the immediacy to obtain information and the customization in the way of looking up the media, among others. It is also worth mentioning the three criteria to assess the ability of web documents to satisfy the information needs of users, which is proposed and developed by Gómez Diago (2005): credibility, coverage y novelty. Among the most complete texts on the quality of informative content on the Internet, in which the different taxonomies established by numerous authors are included and compared, we find the work made by the Australian Knight and Burn (2005). On the other hand, other interesting classification is the Structural Model of Analysis by Flavián and Gurrea (2007, 2008) on the five indicators that affect the consumption of digital newspapers: usability, reputation, credibility, privacy of readers' data and familiarity of websites.

A specially interesting study, which correlates some of the consumption criteria of online media (usability, accessibility, interactivity, news customization and extended positioning) with criteria of information quality (utility, diversity of sources, originality, impartiality, contextualization and clarity) is that carried out by Gutiérrez-Coba et al. (2012), whose goal as precisely to demonstrate whether there was a correlation between the consumption of news online, the credibility of the media in the eyes of the audience, and the quality of information issued. As a conclusion, it is proved that the main associations between consumption and quality of information are established between utility, interactivity and customization, which confirms that the user places high value on the practical use of information, likes to consult the media according to his/her interests, and expects to be able to communicate with the author of the piece of news and make comments to feedback what he/she likes and dislikes as regards the information received. Furthermore, it is worrisome the low correlation between the credibility of the media and its consumption, as there are not competences related to know the parameters of information quality. Hence the need to train critical audiences.

All these analyses serve us as examples of the need to theorize about the parameters and indicators of information quality, with the aim of exploring suitable ways to reach a quality journalism, in accordance with the premise of public service. One of the main goals is to regain citizens' credibility in the media, as in the last decades there is an increasingly marked discredit at international level (American Society of Newspapers Editors 1998; Wyss 2000; Singer 2005). The following explores studies on credibility and trust of citizens in information published by the media.

4 Studies on Credibility and Trust in Media Information

Studies on trust and credibility have historically found support in psychology and its evolution is parallel to this field. As we have seen in the former review of studies on quality of information, focus is now given to the role of culture and human diversity as factors with an impact in social behaviour and social thinking (active cognitivism). When approaching credibility and trust, we have separated traditions in research, which have described the “receivers” as passive and powerless beings behind the great powers of the media, effects that would immediately generate homogeneous consequences in a ductile mass before the persuasive action of issuers. Taking into account this different vision, we address the first research carried out for the reports on journalism from the Madrid Press Association in 2005. This vision is also shown in the Ph.D. by professor Roses (2010), the first academic research in Spain on information credibility and trust for projects of the national plan of research. From this point of view, we maintain the idea that audience is active, as they have the ability to participate in communication, interpret knowledge, “and to affect social dynamics through their nonconformity” (Roses 2010), besides having enough resources to apply a psychological resistance to the persuasive attempts by other players—such as the media—that intervene in public communication. As sustained by Diezhandino (2005), citizens should be taken into account to draw conclusions about the changes that the media should consider in their role of informing, of their journalistic role. We are at a moment when participation is greater, audience more active than ever and resistance or opposition to issuers’ persuasion is manifested dynamically through expression on social networks and, as it concerns us specially, through the trust or mistrust to the media and to their information that could set them aside as informative nutrients.

Thus, our main focus of attention is trust. The concept is understood as a kind of attitude, general psychological orientation or state of mind of citizens, related to a series of beliefs, opinion and expectations about media institutions in relation with the functional authority to perform a task—reduction of complexity, which implies a risk—i.e., being fooled, being misinformed—and that has been traditionally delegated in the media in accordance with the established commitment in their public mission—determined by the democratic culture—of each society.

Trust is an indicator of connection between social actors: it announces success and quality of its interaction. Therefore, instead of focusing the problem of trust exclusively from the point of view of the issuer and the media or the constructive techniques of discourse—that is to say, what to do to improve communication efficiency, persuasion or control; we have decided to focus it from the point of view of citizens. Thus, more light is given on the value of public use that citizens give to the media institution and on the state of the defence of people against the action of the media, that is to say, about the scepticism of those citizens in an interactive and technological scenario that allows for the measurement of their reactions.

The most recent reviews of the line of research that studies citizens’ behaviours towards the media and the credibility towards the media and their information. Also, the last scientific papers on this subject of study have focused on this research line under the name of studies on citizens trust or media scepticism, concept that are much more consistent with the theory and implied social and psychological processes, which are also studied in other related areas, making it easier the multidisciplinary transfer. This approach allows us to, among other, draw on research on behaviour, culture and social impact of institutions carried out in other areas.

After years of study by the international academia, with special emphasis after the crisis of the 80s in the American press, various methodological tools and strategies for the study of trust in journalism from a quantitative (Kohring and Matthes 2007; Tsfati 2003a, b) and qualitative perspective (Coleman et al. 2009). Also, they have been performed synchronic analysis in different countries (Lee 2005; Cooper 2008; Gronke and Cook 2007) and comparative studies between media organizations (Botan and Taylor 2005). It has been proved the relationship of trust in the media with the political behaviour of citizens (Ladd 2010); with the trust towards other social institutions (Tsfati 2002; Cappella and Jamieson 1997); with the trust in democracy (Tsfati and Cohen 2005); with the effects of mass media (Tsfati 2002, 2003a, b); and with the economy of the media and the habits of news consumption (Meyer and Zhang 2003; Tsfati and Capella 2005; Tsfati and Peri 2006; Vanacker and Belmas 2009). Likewise, research has been made on what citizens' profiles tend to be the most sceptical in the media (Tsfati and Capella 2005; Jones 2004; Jackob 2010); on trust of citizens placed into specific areas of information (Tsfati and Cohen 2005). Also, some comparative studies have been launched (Liu and Bates 2009). Despite the large amount of initiatives in last years, studies have emphasized on the analysis of new credibility deficits in journalism, associated to new media and the impact of the latter on citizens trust in journalism, and developed under the framework of the perception of credibility. Thus, the most outstanding studies have a comparative approach from the point of view of platforms and including the Internet among them (Flanagin and Metzger 2000; Schweiger 2000; Wathen and Burkell 2002; Nozato 2002; Kiousis 2001; Sundar 1999; Payne 2001; Bucy 2003; Park 2005; Lu and Andrews 2006; Choi et al. 2006; Kim and Jhonson 2009). Within this set of works, different trends of trust in the eyes of citizens are shown, depending on the kind of new media and on the characteristics and digital culture of users. Besides, there is enough literature on trust in political coverage, which is done both in traditional media and new media (Johnson and Kaye 2010). Nevertheless, while uncertainties regarding trust in traditional media have been solved by the works of the project CSO2008-05125, there is a great need for studies on these specific areas of citizens' trust. In Spain, most part of the Academia has been focused on the development of studies where the quality or impartiality of sources of information is assessed. Based on these analysis, researchers deduced whether the media had high or limited credibility, and therefore no mention of the credibility in the eyes of citizens was made.

The following pages reflect on the credibility and trust attributed to the media in the eyes of citizens.

5 Reasons for the Loss of Trust in Media Information

In the last decade, the decrease in credibility and the loss of trust in media information have been closely related with the economic crisis. We have attended a double crisis: first one, the global economic crisis, which has weakened traditional income of the media, forcing them to reduce staff, to lay off their personnel at once and to close down companies. The other one, the technological-corporate crisis, for which there was a need to restructure the media and their information in reaction to the emergence of new technologies for transmitting and receiving. And that restructuring process has not been done well nor on time, in most of the cases, in an attempt to make the most of the traditional business. The emergence of new technologies started removing market players while bringing to the surface new participants and technological platforms that became dominant in the dissemination of information, partners who were, in most of the cases, far from the information sector. Meanwhile, personal mobile devices and free information, the latter initially fostered by the media themselves, changed citizens' consumption behaviours of information, who chose to free alternatives far from professional control. They proceed to satisfy their information needs using other formula that give them what new generations demand: immediacy, agility, interactivity and transparency. This way, our consumption behaviours of information changed forever.

After outlining this scenario as a starting point, we are trying to classify the reasons-why the loss of media credibility.

As one may guess, the causes of the loss of trust may vary depending on the media system, as they have a different economic, technological and socio-cultural development. It is not the same the starting point of a North-American society than a European one, nor in consumption behaviours of the population, nor the development level and technological development of their media. Besides, the economic crisis of the last decade caused different response times on both sides of the Atlantic. Moreover, times within the European Union were also different, producing major contrasts at times of impact, economic recession and the start of recovery. We can point out, however, common elements for most of the developed countries and media systems, elements that have led to the loss of trust and that can be summarized as follows:

1. Economic crisis and late technological restructuring of the sector.
2. Changes in the habits of information consumption by citizens.
3. Appearance of new players in the media arena: more competition for the audience against players with less professional rigor.
4. Reduction in the number of traditional media and little influence of the existing ones in digital platforms.
5. Dismissals in newsrooms: less professionals to produce a greater number of information, which results in a product (information) of lower quality.
6. Major demand of immediacy: Behind the new instantaneous delivery platforms, there is less time to prepare, look for sources and compare information.

7. New landscape, more competitive, where quantitative predominates over qualitative: the search for impact in the news to the search of audiences that attract advertisers have rewarded contents able to generate impact, regardless of their quality and veracity, which become viral thanks to social networks.

To illustrate further these reasons, the Spanish case can serve as an example in which we provide data that reaffirm the reasons outlined above and that show us the decline in confidence in the information of the last decade.

6 Loss of Confidence of Spanish Citizens in Media Information

In the last ten years in Spain, we have assisted to a gradual loss of credibility in the media and the decrease in trust in information produced by them. Since 2005, the research team of the GEPUC from the University of Malaga conduct surveys to the Spanish population (more than 1000 surveys annually) on their perception regarding information quality. The first studies were included in the reports on journalism of the Madrid Press Association and, since 2012, belong to the research project SEJ067. This long-term study on trust in information enable us to have a sequence of data of a decade (period 2006–2016). This period is especially significant as it collects the previous years to the economic crisis, the period of economic crisis and the present situation of stabilisation and the slow recovery of the sector. Data reading serve us to prove citizens' loss of trust, regardless the slight improvement at industry and labour level in which we are. Also, the data serve us to validate the reasons stated before as regards the loss of trust in information.

The economic crisis broke in 2008 starts to be substantially sensed in the media from 2009. Massive layoffs and media shutdown are initiated. The turning point is reached in 2011, when the layoffs of journalists in newsrooms were doubled. According to data from the Spanish employment service, the unemployment figure was 5564 journalists and 9937 in 2011, almost doubled. At the same time, since 2008, 375 media organizations will be closed in Spain, being 2011, 2012 and 2013 the most critical years in this regard. This period also includes the three last years considered of a certain stabilization and the recovery start of the sector, still very far from the pre-crisis period before 2008. Employment slowly recovers (from 9451 unemployed journalists in 2014 to 8680 in 2015 and to 7890 in 2016), and the number of digital media created by journalists increases (200 media created in the period 2013–2016).

At the beginning of this period in 2006, the trust of Spanish in information of the media was medium-high, at least it was that way perceived by 60% of the population (see Table 1). This trend will continue over the next years, event at the height of the economic crisis. However, its turning point starts in 2011 and 2012, years of greater disappearance of the media and with greater layoffs in newsrooms. The weakness of information provided by traditional media, caused by the reduced newsrooms,

Table 1 % Trust of Spanish citizenship in the media information 2006–2016

	2006	2007	2008	2009	2010	2011	2012	2013	2014	2015	2016
Spain (%)	60	61	61	61	63	61	59	54	44	42	38

Sources APM reports 2008–2012. GEPYC. SEJ067 2012–2015. Eurobarometer 452. September–October 2016. Prepared by authors
Note Surveys from 2006 to 2012 carried out for the reports of the APM by Demométrica. In the period 2013–2015, the surveys were also conducted by Demométrica for the GEPYC. Data from 2016 correspond to the 2016 Special Eurobarometer

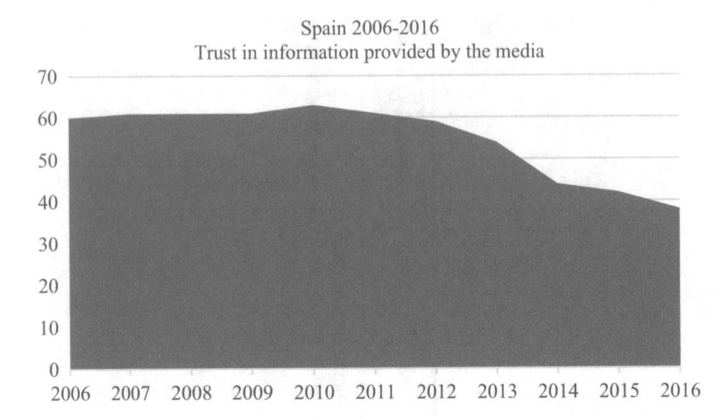

Fig. 1 *Source* APM Reports 2008–2012. GEPYC. SEJ067 2012–2015. Eurobarometer 452. September–October 2016. Prepared by authors

is added to the competition of other formulas (social networks, Twitter, Facebook, etc.), which are more agile and popular and provide concise information and easier to digest, but also less deep. The desire to position themselves and to compete as regards immediacy, transfers that superficiality to the information of generalist media, resulting in many case in inaccurate, unverified and sometimes wrong information. The new conditions of concurrence to the information market were undermining the trust in information in the eyes of society.

The following years will be the time to test new formulas to attract audiences, while traditional ones are exhausted. The interactivity of the web will facilitate access to social networks, and the media begin to create contents for this purpose. However, the reality of this kind of consumption is discouraging. An immense majority of contents that are replicated by the networks are insubstantial information, and sometimes are of low credibility. In their struggle for their positioning and survival, the media, with an exclusive mercurial vision, generate these kind of banal information—social and gossip—with great success among certain segments of population as a hook to generate impact and take them to their websites. However, the effect is not sometimes the desired one, and leads to the weakening of the media's overall image of rigor. During the following years and until 2016, the concern of population about the increase of fake information and manipulation beings. The two national electoral processes that took place in Spain, exploded the alarm over the enormous ease in manipulation of information. As happened in the US with the presidential elections of 2017, the crossroads of information from interested that were not contextualized, nor clarified or rectified, eroded the credibility of the media. As it can be appreciated in Table 1, in 2016 confidence of Spaniards in media information is the lowest of the decade (Fig. 1).

7 Causes Indicated by Citizens that Generate Distrust in the Media

What are the elements that make a citizen give more or less credibility to a media or an information? In the surveys carried out over these years, numerous causes have been identified by citizens as elements that influenced their distrust towards information. They pointed out formal aspects such as:

1. Errors in information: This information had previously been corroborated in a 1994 US study carried out by David Nelson and Paul Wang from the Northwestern University among readers in Chicago: inaccuracy was found as the first feature that influenced the credibility of information.
2. Absence of rectifications: the audience not only has the feeling that the media make many mistakes, and online media those who received more criticism. but also that the media are unwilling to correct them completely and punctually, being television and online media those which received more criticism. There was also a big difference between what the public thinks of corrections and what journalists think, who considered to be excused in their inaccuracies and, therefore, considered unnecessary to rectify, as they act under the pressure of immediacy and timing and so some errors, misunderstandings and strange interpretations must be expected.
3. Inappropriate and indiscriminate use of anonymous sources.
4. Lack of training or lack of knowledge of journalists in the issues addressed.
5. Lack of sensitivity and unethical practices in the media.
6. Negative perspective of the news and events.
7. Preconceived approach to the story and sources with which journalists are interviewed.

Together with these aspects, other were also pointed out, exceeding the mere formality of the text:

1. The excess of superficial information in any platform, especially in the media and social networks.
2. The rise of fake news.
3. The political interests of journalists and the media, which are present in information and media agendas.
4. The economic interests of media companies, which have an impact of dissemination and information processing.
5. Growing confusion between information and opinion, which also highlights the lack of independence of journalists and the media.
6. The transformation of information into spectacle in pursuing mass audiences.
7. The distance between the media agenda and the problems that citizens considered really important.

Each of the section above would deserve special attention and an in-depth reflection on the causes of these problems. Although we have reflected on some of them,

Table 2 Citizen trust in media information 2016

	Reliable %		Unreliable %		Others %	
Spain	38		59		3	
EU	53		44		3	
	Radio %			Newspapers %[a]		
	Reliable	Unreliable	Others	Reliable	Unreliable	Others
Spain	54	37	9	43	47	10
EU	66	28	6	55	38	7
	Television %			Media and social networks %[b]		
	Reliable	Unreliable	Others	Reliable	Unreliable	Others
Spain	31	65	4	26	56	18
EU	55	41	4	32	55	13

Source Eurobarometer 452. September–October 2016. GEPYC SEJ067 2016. Prepared by authors
[a]Digital and printed
[b]Social networks on the Internet, blogs, websites that host videos, etc.

the task exceeds the pretensions of this chapters, but their statements serve as elements that seek to provide greater understanding of the reasons that lead the citizen to distrust.

8 Differences Between Spain and Europe

Having seen the Spanish case, it is worth reviewing the index compared to the rest of EU countries. Table 2 shows a comparison between Spain and the EU as a whole. Data provided by the last special Eurobarometer on media pluralism and democracy show considerable differences between them. Media information is reliable for 53% of EU citizens compared with 38% of Spain. The discrediting of media information is greater in Spain, well above the set of European countries.

As for trust by media, the radio is the most reliable media in information, both EU and Spanish citizens while the less reliable are social networks in the case of EU citizens and television in the case of Spanish people.

9 The Cost of the Loss of Trust: Trivialisation of Information

The loss of media trust and the lack of confidence in their information have considerable costs that translate into a weakening of the system as a whole, generating a less informed and less free society.

Because in the new scenario in which we move, citizens have more information than ever before. But we do not need any information, but rather rigorous information that brings knowledge and reduce uncertainties. Only then we will be able to make decisions about what really happens in our environment.

But knowledge, like good information, is structured, hierarchical and complex, and so needs a rigorous treatment. And it is different from the information offered by most of the new alternative formulas, which is getting flatter and shallower. This new information leaves no room for complex reasoning that transfer the causes of things instead of the things themselves. The "trivialisation of information" makes citizens, in their new consumptions, unable to understand what really happens. And what they receive (new media, social networks) is mostly superficial and, occasionally, unverified and uncertain. New platforms have been created without professional rigor in the management of information and the drafting of traditional media has been reduced and with them the ability to produce in-depth information. Nor has it been possible to exploit the strength of the professionalism and credibility of historical media to the detriment of other less solid but more popular formulas.

The disappeared Emergent, research project by the Tow Center for Digital Journalism from the University of Columbia, monitored unverified information and rumours in the media. Prior to its closure in late 2016, the Menczer team from Indiana University monitored news shared on Facebook, classifying them into two groups: credible news and misinformation. The Menczer team was able to verify how the number of credible information was lower than the number of those catalogued as misinformation or fake news. In other words, more fake information than credible is shared on social networks.

In light of the data, one could think that the enormous effort to contrast information before disseminating it is not worth it in terms of distribution and clicks, desired elements as main economic base of the media in attracting advertisers.

The findings of the study from Indiana University are, of course, striking and confirm earlier statements on information trivialisation.

The brand does not guarantee anymore survival if it is not accompanied by rigor, and many citizens have lost their trust in brands. If they—brand—end up defrauding their audiences, as it is beginning to happen, the media are condemning themselves to irrelevance and to be replaced by other formulas of little professional rigor unless action is taken to rectify the situation.

10 Actions to Solve the Problem

Some media have already begun to seek corrective measures to restore their credibility and the trust of citizens towards their information.

As we know, it has increased the number of platforms that allow professionals to know from their office what is happening, searching via algorithms, tweets and messages on social networks that contain or may be news. These are much faster than traditional information agencies in tracking news, as the train crash in New Jersey in September 2016 and the attacks in Nice and Berlin in 2017. An example of these programs is Dataminr, a platforms used by many journalists and media to discover newsworthy elements in social networks. The platform identifies what is being trendy and allows users to check whether there are crowds of people using social networks at a specific point, as well as to know what are they tweeting. However, this carries a considerable risk if the source of the information is not well contrasted and any viral information is considered valid for being the first to give the information. The numerous cases of fake pictures and videos that did not correspond to the event with which they are accompanied have taught the media to be more cautious when using information from social networks and facilitated by individuals. The media now submit pictures, videos and information to more rigorous checks such as detecting the date and time of the file using metadata and GPS location.

One of the needed actions: greater rigor and contrast of sources, before turning it into own information and disseminating it as information from third parties. We can also point out other measures such as those taken by platforms like Facebook and Google to detect fake news after the last US elections. Facebook has started to work with CUNY (the New York's journalism school), a tool that allows detecting false news. The announcement made by Zuckerberg was followed by one by Google, which presented in March 2017 the Fact Check label. A service that started in the US and the UK but would start to be available in all countries. Although the service Google News is not available in Spain because of a disagreement with publishers, it is accessible by searching for any news that will have a fact checking label. This verification is not performed by Google, but allows third parties—not incorporated into the system of news checking—to confirm or comment the information and label them as "false" or "surely certain".

Along with these examples, we could point to many others that are being implemented as corrective measures. As we have seen in the list of reasons pointed out by citizens, action should be taken in all and each of them. However, from our point of view, there are two indispensable measures that have more to do with training than with control measures that companies and institutions can take: greater training of professionals in proper use of the enormous potential of new technologies and in their dangers when misused, but above all a better education of the new generations in the use and critical consumption of information as the best prophylaxis within a changing media and technological market of the future. The abundance and plurality of information is the best scenario when there is a selective approach. If you do not have it, you can end up burying yourself underneath.

11 Information as the Cornerstone of a Democratic Society

The standards of information quality have changed in recent years, based on the mediamorphosis that has occurred in the different communication platforms and also in the perspective in which analyses are focused. The public service and the social interest are two key criteria to follow the commitment guided by the international regulations related to professional practice. The politicization and commodification of the media have biased information and damaged the essence of journalism, affecting the discredit and criticism by citizenry.

Also, research conducted in recent years highlight the main reasons for the loss of credibility as follows: errors in information, absence of rectifications, inappropriate and indiscriminate use of anonymous sources, the lack of training and ignorance of journalists in the subjects they are dealing with, the lack of sensitivity and unethical practices in the media, the negative perspective of news and events, and the preconceived approach of stories and sources to whom journalists interview.

In the last decade, the information sector has been damaged mainly by the global economic crisis, which has resulted in labour precariousness and the technological-business crisis, which has affected the reconversion in production processes, giving less and less priority in quantity and quality to the main players, journalists, to pass the witness to new participants away from the information sector. At the same time, new generations have changed traditional demands of news consumption for others, such as immediacy, immersion and interaction, which has obliged the profession to adapt itself to other modus operandi, in which social networks and audience are gaining more and more ground.

However, the most reliable media for citizenship are traditional media—press and radio, being the less reliable television—in the Spanish case—and social media and networks in Europe.

We can affirm that having a greater diversity and plurality of media, with a more interactive character in the case of social media, does not make citizens consider information more credible, just the contrary. Hence, the witness of information should be regained by journalists, like professionals with university training for professional practice, done with rigor and contrast, and following the keys of newsworthiness criteria.

In credibility studies, the most relevant part is the study of trust of individuals. For their measurement, it is necessary to know what users of information think about it. Unlike much research tradition that gravitates in the analysis of the audience from a passive perspective, we focus on the audience as active subjects, insofar as they participate in the communication process, sometimes entering more or less that in others, even becoming controversial and nonconformists. Listening to their voice and storytelling help the media to know the pulse of their actions in favour of improving citizen interest.

As actions towards the search for information quality, it would be necessary to enhance the level of media literacy of users and prosumers of information, in order to be more critical, analytical and responsible in their use and consumption of the media, carrying out actions of literacy in digital skills, both from integration into formal teaching spaces and non-formal education environments. On the other hand, it would be interesting that the media and social networks applied accountability mechanisms in information towards the quality of the services they provide.

In essence, it is a matter of improving information as a pillar of democratic societies, as a better informed society is a freer society. This axiom is studied from the first years of the degree of journalism, and that safeguard of quality should be maintained in the media and the daily journalism routines.

References

Aguilar-Paredes, C., Gómez-Domínguez, P., Villanueva-Baselga, S., Sánchez-Gómez, L.: Propuesta de un índice de la calidad informativa de los telenoticias de las televisiones autonómicas españolas. El profesional de la información **25**(1), 75–87 (2016)
Agustín Lacruz, M.C., Marta-Lazo, C., Ubieto Artur, I. (coord.): Perfiles profesionales y espacios de empleo en Información y Comunicación. Icono 14 (2013)
Alessandri, F., Edwards, C., Pellegrini, S., Puente, S., Rozas, E., Saavedra, G., Porath, W.: VAP: un sistema métrico de calidad periodística. Cuadernos de Información **1**, 112–120 (2001)
American Society of Newspaper Editors.: Examining our Credibility, Why Newspaper Credibility has been Dropping? ASNE, Columbia (1998)
Arnold, K.: Qualitaetsjournalismus: Die Zeitung und ihr Publikum. UVK, Konstanz (2009)
Beckett, C., Mansell, R.: Crossing boundaries: new media and networked journalism. Commun. Cult. Critique (2008)
Belt, T., Just, M.: The news story: is quality a choice? Politi.Commun. **25**(2), 194–215 (2008). http://dx.doi.org/10.1080/10584600801985714
Born, G., Prosser, T.: Culture and con-sumerism: citizenship, public service broadcasting and the BBC's fair trading obligations. Modern Law Rev **64**(5), 657–687 (2001). http://dx.doi.org/10.1111/1468-2230.00345
Botan, C.H., Taylor, M.: The role of Trut in channels of strategic communication for building civil society. J. Commun. 685–702 (2005)
Bustamante Newball, J.: Desafíos para la formación universitaria en periodismo televisivo. Educere **32**(10), 115–120 (2006)
Cantalapiedra, M.J., Coca, C., Bezunartea, O.: La situación profesional y laboral de los periodistas. Zer, Revista de Estudios de Comunicación **9**, 335–366 (2000)
Canel, M.J., Sádaba, T.: La investigación académica sobre las actitudes profesionales de los periodistas. Comunicación y Sociedad **12**(2), 9–32 (1999)
Cappella, J., Jamieson, K.H.: Spiral of cynicism: the press and the public good. Oxford University Press, New York (1997)
Cebrián Herreros, M.: Análisis de la información audiovisual en las aulas. Editorial Universitas, Madrid (2003)
Coleman, S., Anthony, S., Morrison, D.: Public trust in the news. A constructivist study of the social life of the news. Reuters Institute for the Study of Journalism, Oxford (2009)
Cooper, T.W.: Between the summits: what Americans think about media ethics. J. Mass Media Ethics **23**(1), 15–27 (2008)
Guerrero-Solé, F., Pont-Sorribes, C., Palencia-Lefler, M.: Construction of the image of politics in Spanish TV news programmes. The endo- and exo-balancesof the quality of political information. Revista Latina de Comunicación Social **68**, 161–181 (2013). http://dx.doi.org/10.4185/RLCS-2013-973en
De la Torre, L., Téramo, M.T.: La calidad de la información periodística: estrategias para su observación. Coincidencias y divergencias entre los medios y el público. Estudios sobre el Mensaje Periodístico **21**, 135–144. http://dx.doi.org/10.5209/rev_ESMP.2015.v21.50666 (2015)
De Pablos, J.M., Mateos, C.: Estrategias informativas para acceder a un periodismo de calidad en prensa y TV. Ámbitos **11–12**, 341–365 (2004)
Del Rey Morató, J.: El naufragio del periodismo en la era de la televisión. Fragua, Madrid (1998)
Desantes Guanter, J.M.: Derecho a la información. Materiales para un sistema de la comunicación. Fundación COSO, Valencia (2004)

Diezhandino, M.P.: El quehacer informativo. El "arte de escribir" un texto periodístico. Servicio Editorial Universidad del País Vasco, Bilbao (1994)
Diezhandino, M.P., Aguago, G, Carrera, P., Fernández, J., Martín, O., Muro, I.: El periodista en la encrucijada. Fundación Telefónica, Editorial Ariel, Madrid (2012)
Escobar Roca, G.: Estatuto de los periodistas. Régimen normativo de la profesión y organización de las empresas de comunicación. Tecnos, Madrid (2002)
FAPE.: Código deontológico. Madrid: Federación de Asociaciones de Periodistas de España. In: http://fape.es/home/codigo-deontologico/ (1993). Accessed 11 May 2017
Farias Batlle, P.: The press crisis in Spain. Credibility and profesionality as main solutions. Infoamérica **2**, 59–68 (2010)
Farias Batlle, P.: Informe de la Profesión Periodística en España. Asociación de la Prensa de Madrid, APM, Madrid (2005, 2006, 2007, 2008, 2009, 2010, 2011, 2012)
Fengler, S., Eberwein, T., Alsius, S., Baisnée, O., Bichler, K., Dobek-Ostrowska, B., et al.: How effective is media self-regulation? Results from a comparative survey of European journalists. Eur. J. Commun. **30**(3), 249–266 (2015)
Figueras-Maz, M., Mauri-Ríos, M., Alsius-Clavera, S., Salgado-De-Dios, F.: La precariedad te hace dócil. Problemas que afectan a la profesión periodística. El profesional de la información **21**(1), 70–75 (2012)
Flavián, C., Gurrea, R.: Perceived substitutability between digital and physical channels: the case of newspapers. Online Inf. Rev. **31**(6), 793–813 (2007)
Flavián, C., Gurrea, R.: Reading newspapers on the Internet: the influence of web sites' attributes. Internet Res. **18**(1), 26–45 (2008)
García de Cortázar, M., García de León M.A. (coords.): Profesionales del periodismo. Hombres y mujeres en los medios de comunicación. CIS/Siglo XXI, Madrid (2000)
Garcia de Madariaga Miranda, J.M.: El periodismo en el siglo XXI. Una profesión en crisis ante la digitalización. Dykinson, Mexico (2008)
Gil-de-Zúñiga, H., Hinsley, A.: The press versus the public: what is good journalism? Journalism Stud. **14**(6), 926–942 (2013). http://dx.doi.org/10.1080/1461670X.2012.744551
Gómez Diago, G.: Tres criterios para evaluar la calidad informativa en Internet: credibilidad, cobertura, novedad. Global Media Journal Edición Iberoamericana **2**(4), 1–11 (2005)
Gómez Mompart, J.L., Gutiérrez Lozano, J.F., Palau Sampio, D. (eds.): La calidad periodística. Teorías, investigaciones y sugerencias profesionales. Universidades Autónoma de Barcelona, Jaume I, Pompeu Fabra y Valencia, Barcelona (2013)
Gómez Mompart, J.L., Gutiérrez Lozano, J.F., Palau Sampio, D.: Los periodistas españoles y la pérdida de la calidad de la información: el juicio profesional. Comunicar **23**, 143–150 (2015)
Gómez Patiño, M.: El (re)cambio generacional: del periodista anfibio al comunicador de la Era Amphibia. In: Túñez, M. (coord.) Viejo periodismo, nuevos periodistas, pp. 27–47. Cuadernos Artesanos de Latina, 25, Tenerife (2012)
González Reigosa, C.: El periodista en su circunstancia. Alianza Editorial, Madrid (1997)
Grueskin, B., Seave, A., Graves, L.: The story so far: what we know about the business of digital journalism. Columbia Journalism Rev. Books, New York (2011)
Gutiérrez-Coba, L.M., Salgado-Cardona, A., Gómez-Díaz, J.A.: Calidad vs. Credibilidad en el periodismo por internet: batalla desigual. Observatorio (OBS*) **6**(2), 157–176 (2012). http://dx.doi.org/1646-5954/ERC123483/2012
Gutiérrez-Gea, C.: Televisión y calidad: Perspectivas de investigación y criterios de evaluación. ZER. Revista de estudios de comunicación **9**, 151–184 (2000)
Humanes, M.L.: El encuadre mediático de la realidad social. Un análisis de los contenidos en televisión. ZER. Revista de estudios de comunicación **11**, 119–141 (2001)
Humanes, M.L., Montero-Sánchez, M.D., Molina-de-Dios, R., López-Berini, A.: Pluralismo y paralelismo político en la información televisiva en España. Revista Latina de Comunicación Social **68**, 24–16 (2013). http://dx.doi.org/10.4185/RLCS-2013-990
Iyengar, S.: Is anyone responsible? How television frames political issues. University of Chicago Press, Chicago (1988)

Jackob, N.G.: No Alternatives? The relationship between perceived media dependency, use of alternative information sources, and general trust in mass media. Int. J. Commun. **4**, 589–606 (2010)
Jones, C.: Mass media codes of ethics and councils. A compartive international study on professional standards. UNESCO Press, Paris (1980)
Johnson, T., Kaye, B.: Choosing is believing? How web gratifications and reliance affect Internet credibility among politically interested users. Atlantic J. Commun. **18**(1), 1–21 (2010)
Kim, D., Jhonson, T.J.: A shift in media credibility: comparing Internet and traditional news sources in South Korea. Int. Commun. Gaz. **71**(4), 283–302 (2009)
Knight, S., Burn, J.: Developing a framework for assessing information quality on the world wide web. Informing Sci. J. **8**, 159–172 (2005)
Kohring, M., Matthes, J.: Trust in news media: development and validation of a multidimensional scale. Commun. Res. **34**, 231 (2007)
Ladd, J.M.: The neglected power of elite opinion leadership to produce antipathy toward the news media: evidence from a survey experiment. Polit. Behav. **32**, 29–50 (2010)
Laitila, T.: Journalistic codes of ethics in Europe. Eur. J. Commun. **10**(4), 527–544 (1995)
López, X., Gago, M., Toural, C., Limia, M.: Nuevos perfiles y viejos cometidos de los profesionales de la información. In: AA.VV, Los nuevos desafíos del oficio de Periodismo. Actas del XVIII Congreso Internacional de la Sociedad Española de Periodística (SEP), pp. 55–63. SEP and Universidad Carlos III de Madrid, Madrid (2012)
López Hidalgo, A.: El periodista en su soledad. De cómo la precariedad en el trabajo condiciona la ética y la independencia del profesional, y otras respuestas sin pregunta. Comunicación Social Ediciones y Publicaciones, Sevilla (2005)
López-Téllez, J.A., Cuenca-García, F.A.: Televisión e información: análisis de los criterios de televisión de calidad en los informativos de las cadenas nacionales. Comunicar **25** (2005)
Lu, H., Andrews, J.E.: College Students' perception of the absolute media credibility about SARS-Related news during the SARS Outbreak in Taiwan. China Media Res **2**(2), 85–93 (2006)
Marta-Lazo, C., Segura-Anaya, A., Martínez Oliván, N.: Variables determinantes en la disposición al pago por contenidos informativos en Internet: Perspectiva de los profesionales. Revista Latina de Comunicación Soc **72**, 165–185. http://www.revistalatinacs.org/072paper/1159/09es.html. https://doi.org/10.4185/RLCS-2017-1159
Masip, P., Micó, J.L.: El periodista polivalente en el marco de la convergencia empresarial. Quaderns del CAC **31–32**, 91–99 (2009)
McChesney, R., Nichols, J.: The Death and Life of American Journalism: The Media Revolution that Will Begin the World Again. Nation Books, Philadelphia (2010)
McQuail, D.: Media Performance. Mass Communication and the Public Interest. Sage, London (1992)
Meijer, I.C.: What is quality television news? A plea for extending the professional repertoire of newsmakers. Journalism Stud. **1**(4), 15–29 (2003). http://dx.doi.org/10.1080/14616700306496
Meyer, P., Zhang, Y.: Anatomy of a Death Spiral: Newspaper and their credibility. National convention of Association for Education in Journalism and Mass Communication, Miami (2003)
Noguera, J.M.: How open are journalists on Twitter? Trends towards the end user journalism. Commun. Soc. **26**(1), 93–114 (2013)
Núñez Ladevéze, L.: El periodismo desde un enfoque interdisciplinar. In: Cantavella, J., Serrano, J. F. (coords.), Redacción para periodistas: informar e interpreter, pp. 17–50. Ariel, Barcelona (2004)
Ortega, F., Humanes, M.L.: Algo más que periodistas. Sociología de una profesión. Ariel, Barcelona (2000)
Osuna Acedo, S., Marta-Lazo, C., Aparici Merino, R.: Valores de la formación universitaria de los comunicadores en la sociedad digital: más allá del aprendizaje tecnológico, hacia un modelo educomunicativo. Razón y Palabra, 81. http://www.razonypalabra.org.mx/N/N81/V81/13_OsunaLazoAparici_V81.pdf (2013). Accessed 2 May 2017
Pihl-Thingvad, S.: Professional ideals and daily practice in journalism. Journalism **5** (2014)

Porto, M.P.: Frame Diversity and citizen competence: towards a critical approach to news quality. Crit. Stud. Media Commun. **24**(4), 303–321 (2007)

Pottker, H.: Kompensation von Komplexitaet: Jour - nalismustheorie als Begruendung journalistischer Qualitaets – maßstaebe. In: Loeffelholz, M. (ed.) Theorien des Journalismus, pp. 375–390 (2000). http://dx.doi.org/10.1007/978-3-322-97091-6_19

Ramírez de la Piscina, T., Gorosarri, M.G., Aiestaran, A., Zabalondo, B., Agirre, A.: Periodismo de calidad en tiempos de crisis: Un análisis de la evolución de la prensa europea de referencia (2001–2012). Revista Latina de Comunicación Social **69**, 248–274 (2014). https://doi.org/10.4185/rlcs-2014-1011

Rodríguez Andrés, R.: Características sociodemográficas y laborales de los periodistas españoles e índice de satisfacción profesional. Ámbitos **9–10**, 487–504 (2003)

Romero Rodríguez, L.M., De-Casas-Moreno, P., Torres-Toukoumidis, A.: Dimensions and Indicators of the Information Quality in Digital Media. Comunicar **49**(24), 91–100 (2016). http://dx.doi.org/10.3916/C49-2016-09

Roses, S.: Aproximación a la confianza de la ciudadanía en la institución mediática: El caso de España a través de datos cuantitativos (2006–2009). Doctoral dissertation, Universidad de Málaga, Málaga (2010)

Roses, S., Farias-Batlle, P.: Credibilidad de los medios: un análisis bivariado de las opiniones de los españoles. Revista Mediterránea de Comunicación **3**(1), 79–104 (2012)

Salaverría, R., García Avilés, J.A.: La convergencia tecnológica en los medios de comunicación: retos para el periodismo. Trípodos **23**, 31–47 (2008)

Sánchez-García, P., Campos-Domínguez, E., Berrocal Gonzalo, S.: Las funciones inalterables del periodista ante los perfiles multimedia emergentes. Revista Latina de Comunicación Social **70**, 187–208. http://www.revistalatinacs.org/070/paper/1042va/12es.html (2015). Accessed 2 May 2017

Schatz, H., Schulz, W.: Qualitaet von Fernseh pro - grammen: Kriterien und Methoden zur Beurteilung von Pro gramm - qualitaet im dualen Fernsehsystem. Media Perspektiven **11**, 690–712 (1992)

Scolari, G.A., Micó, J., Navarro, H., Pardo, H.: El periodista polivalente. Transformaciones en el perfil del periodista a partir de la digitalización de los medios audiovisuales catalanes. Zer **13**(25), 37–60 (2008)

Semetko, H.A., Valkenburg, P.M.: Framing European politics: a content analysis of press and televi-sion news. J. Commun. **50**(2), 93–109 (2000). http://dx.doi.org/10.1111/j.1460-2466.2000.tb02843.x

Silverstone, R.: Media and Morality: On the Rise of the Mediapolis. Polity, Cambridge (2006)

Singer, J.B.: The political J-Blogger: ‘normalizing’ a new media form to fit old norms and practices. Journalism **6**(2), 173–198 (2005). http://dx.doi.org/10.1177/1464884905051009

Sinova, J.: Acerca de la responsabilidad pública del periodista. La verdad, los derechos de los públicos y otras exigencias éticas del trabajo informativo. Doxa. Comunicación **1**, 171–184 (2003)

Soengas, X., Rodríguez, A., Abuín, N.: La situación profesional de los periodistas españoles: las repercusiones de la crisis en los medios. Revista Latina de Comunicación Social **69**, 104–124 (2014). http://www.revistalatinacs.org/069/paper/1003_USC/06_S.html

Soengas Pérez, X.: El pluralismo y el control de la información en las televisiones generalistas españolas. Estudios sobre el Mensaje Periodístico. Especial noviembre. Periodismo e información de calidad **21**, 199–207 (2015). https://revistas.ucm.es/index.php/ESMP/issue/archive. Accessed 1 May 2017

Thorson, E., Duffy, M.: Newspapers in the new media environment. In: NAA Marketing Conference, Orlando (2005)

Tsfati, Y.: Media skepticism and climate of opinion perception. In: World Association of Public Opinion Research Annual Conference, Italy, Rome (2001a)

Tsfati, Y.: Cynicism or scepticism? The valence of audience attitudes toward the media. In: 51st Annual International Communication Association Conference. Washington, DC (2001b)

Tsfati, Y.: The consequences of mistrust in the news media: media scepticism as a moderator in media effects and as a factor influencing news media exposure. Doctoral dissertation, University of Pennsylvania, Philadelphia, USA (2002)
Tsfati, Y.: Media scepticism and climate of opinion perception. Int. J. Publ. Opin. Res. **15**(1), 65–82 (2003a)
Tsfati, Y.: Does audience scepticism of the media matter in agenda setting? J. Broadcast. Electron. Media **47**(2), 157–176 (2003b)
Tsfati, Y., Capella, J.N.: Why do people watch news they do not trust? The need for cognition as a moderator in the association between news media skepticism and exposure. Media Psychology **7**(3), 251–271 (2005)
Tsfati, Y., Cohen, J.: Democratic consequences of hostile media perceptions: the case of Gaza settles. Harvard Int. J. Press/Polit. **10**(4), 28–51 (2005)
Tsfati, Y., Peri, Y.: Mainstream media skepticism and exposure to sectorial and extrasectorial news media: the case of Israel. Mass Commun. Soc. **9**(2), 165–187 (2006)
Túñez, M. (coord.): Viejo periodismo, nuevos periodistas. Cuadernos Artesanos de Latina, 25, Tenerife (2012)
Túñez, M., Sixto, J.: Nuevos públicos y nuevos roles en la audiencia: ¿nuevos periodistas? In: Túñez, M. (coord.) Viejo periodismo, nuevos periodistas, pp. 49–67. Cuadernos Artesanos de Latina, 25, Tenerife (2012)
Túñez, M., Martínez, M.Y.: Análisis del impacto de la función, las actitudes y las condiciones laborales del periodista en la producción de noticias: Hacia un periodismo de empresa. Zer **19**, 37–54 (2014)
Ufarte Ruiz, M.J.: La situación laboral del periodista como factor condicionante de la calidad informativa: con precariedad no hay calidad. In: Actas – IV Congreso Internacional Latina de Comunicación Social– IV CILCS– Universidad de La Laguna, diciembre 2012. http://www.revistalatinacs.org/12SLCS/2012_actas.html (2012). Accessed 7 May 2017
Urban, J., Schweiger, W.: News quality from the recipients' perspective. Journalism Stud. **15**(6), 821–840 (2014). http://dx.doi.org/10.1080/1461670X.2013.856670
Vanacker, B., Belmas, G.: Trust and the economics of news. J. Mass Media Ethics **24**(2, 3), 110–126 (2009)
Van der Haak, B., Parks, M., Castells, M.: The future of journalism: networked journalism. Int. J. Commun. **6**, 2923–2938 (2012)
Wyss, V.: Online-Journalismus in Europa: das Beispiel Schweiz. In: Klaus-Dieter et al. (eds.), Online-Journalismus: Perspektiven fu¨r Wissenschaft und Praxi, pp. 335–346. West - deutscher Verla, Wiesbaden (2000)
Zaller, J.: A new standard of news quality: Bur-glar alarms of the monitorial citizen. Polit. Commun. **20**(2), 109–130 (2003). http://dx.doi.org/10.1080/10584600390211136

Carmen Marta Lazo Professor of Journalism at the Universidad de Zaragoza (Spain). She is Director of Unizar radio station and Entremedios Digital Platform. Main Researcher on the Digital Communication and Information Research Group (GICID), which belongs to the Government of Aragon (S-115). She is Co-editor in chief of the Mediterranean Journal of Communication. She has conducted numerous scientific forums related to her research lines, based on Media Education, Digital Competence and screen consumptions and interactions. Zaragoza, Spain.

Pedro Farias Batlle Professor at the Universidad de Malaga (Spain) where he has been Vice-president. Main researcher on national and international groups, he has been for a decade the author of the yearly national report about the state of the Spanish Journalism. Author of articles about media systems and the credibility of the information, he has been member of the Spanish UNESCO commission and he is part of the UNESCO Chair at Universidad de Málaga. Zaragoza, Spain.

Citizen Influence in Social Media for Democratization of Communication

Diana Gabriela Moreira Aguirre, Santiago José Pérez Samaniego, Verónica Paulina Altamirano Benítez and Isidro Marín-Gutiérrez

Abstract The democratization of communication allows citizens accessing to truthful, contextualized and independent information, thus encouraging all actors in society to have opinions and to exercise their right to free expression. Faced with this trend, several countries in Latin America expressed their political will to transform the democratic sense of communication from the legal norm, toward enforcement of legislative initiatives that conceive of communication as a public good trying to strengthen the mechanisms of citizen participation with the aim of improve levels of access to state information. This research analyzes the moves actions and groups voluntarily constituted by civil society, as a mechanism of citizen protest, that using social media as support have been able to influence media coverage influencing as a result public opinion in order to generate social changes.

Keywords Citizen participation · Communication · Democratization Democracy · Social media · Social movements

1 Democratization and Communication

It is important to make some distinctions in relation to the terms often used as synonyms: democratization and democracy. The first describes historical-social processes determined by institutions and their processes to promote the establishment while the second term refers to a concept or category that forms part of an analytical,

D. G. Moreira Aguirre · S. J. Pérez Samaniego (✉) · V. P. Altamirano Benítez · I. Marín-Gutiérrez
Universidad Técnica Particular Loja, Loja, Ecuador
e-mail: saperez1@utpl.edu.ec

D. G. Moreira Aguirre
e-mail: dgmoreira@utpl.edu.ec

V. P. Altamirano Benítez
e-mail: vpaltamirano@utpl.edu.ec

I. Marín-Gutiérrez
e-mail: imarin1@utpl.edu.ec

M. Túñez-López et al. (eds.), *Communication: Innovation & Quality*, Studies in Systems, Decision and Control 154, https://doi.org/10.1007/978-3-319-91860-0_13

conceptual or theoretical discourse (Gallardo 1996). Democratization constitutes the activities that are carried out to settle or establish certain policies. These pursuing a goal to secure the interests and the common good of the community. This can be understood from two approaches: socialization of political and economic power and on the other hand the term participation to the extent that citizens are aware of their participation in the political life of the country.

Studies on democracy in Latin America during the last half century can be synthesized in four moments, following Javier Duque (2014): a first stage from 1960 to 1970 where studies were aimed at analyzing oligarchic domination as well as the structure of political power. Social change in this period is marked by dictatorships in the region. The second period outlines the return to democracy and to civil governments as the main characteristic. It moved from the demand of revolution to the strong exigency democratically constituted systems (Lechner 1985). This particular moment is marked by transitions from dictatorships to democracies emphasizing a lot of thinking and attentive consideration on the transition processes taking place in the region. "Transitology" was a term there coined to mainstream questions and works that had the analysis of the transitions in some countries and their modalities as central theme and topic (Schmitter 1997). The third stage was geared to analyzing the transition processes towards a model of consolidation and stabilization of democracies within a framework of globalization and global understanding of the system with the international recognition and legitimacy of the countries. And a fourth stage is aimed at understanding the quality of democracies (Munck 2002); how the states understand and construct in practice the main edges and angles that characterize a system based on democratic precepts.

During the decade of the eighties of the twentieth century the countries of this region faced processes of democratization showing different geopolitical factors as origin, as well as violations of human rights, practices of resistance, popular mobilization or wars. All of them shared as a common concern the need to identify references or patterns where traditional institutions do not exert pressure, nor domination, as usually happened in earlier historical moments. That is, the processes of democratization begin to be forged along with geopolitical factors occurring throughout the historical chronology. Facts that marked global democratic processes such as the failure of the Sandinistas in Nicaragua or the coup d'état to Allende in Chile, the fall of the Berlin Wall, the collapse of the Soviet Union and the countries of Eastern Europe. In addition to the resounding failure of the socialist model in Albania, Ethiopia and Angola. These examples of regimes inspired by ideological currents like the socialist-Marxist exerted for many years a wide influence on some countries of the region and with the collapse of these systems was losing political hegemony, with sporadic exceptions namely the Cuban experience. This led Latin American policy to become a public scenario for the defense of common national interest, within the institutional framework that would ensure the representation and resolution of conflicts of interests, rethinking policy as a body of recognized rules of institutionalized knowledge eventually recognized and shared by the political class (Ramos Jiménez 2016).

At present, the revival of currents inspired by socialist precepts such as the new system of 21st century socialism present in Latin American countries like Ecuador, Venezuela or Bolivia, as an ideological, economic and social current can be observed. Its common characteristics are replicated in the region with a number of complaints pinpointing actual threatening to individual and collective citizens' rights, communication censure, education, etc.

1.1 Democratization of the Media

Democratization of communication can be considered as the way in which citizens have access to truthful, contextualized and independent information, this way allowing all actors of society share spaces of opinion where exercise their right to free expression. Absorbed by consolidated partisan political structures Latin American societies have neglected their direct participation by committing fundamental processes such as decision making to these political parties.

In recent years, with mass media, communication becomes an evolving instrument to strengthen citizen participation, increasing public opinion spaces in the political and social agenda of Latin American governments. Its use constantly evokes not only a political sphere but a citizen view, with different purposes and objectives summed up in collective interests. Related to the democratization of the spaces of power and decision of the citizens, thus, becoming media societies. Democracy, however, far from being a perfect system, continues its construction as a means to guarantee the exercise of social justice in a framework of deep freedom. Participative democracy cannot be limited to a simple vote exercise but has to be conceived as a web of communication circuits where public affairs are discussed. We could neither conceive of a society without communication and nor a democracy without the participation of citizens.

The current and continuous technological development in the world confronts us with a scenario in which citizens change their traditional roles in relation to social media playing a leading role in the relationship with them through participation, so constructing and becoming a global village Communication. It was usual to think of both: a citizen with a passive attitude towards media and a basic classic communication model with a simple transmitter source and a passive receiver, those marking their agenda in relation to the information and knowledge that was transmitted (Borja, 2007).

A fundamental role in building bridges has been conveyed to Media to enable the exercise of social dimension of freedom of expression exercised within democracy. This democracy can only be built with political participation, a continuous process of balances and counterweights, majorities and minorities which will occur as long as citizens are properly informed to grasp the necessary elements to understand their reality (Azurmendi 2007). Information includes ideas, thoughts, theories, explanations of the world, including cosmo-visions. In other words, every element that builds an open and inclusive plural dialogue (Sánchez 2004). Citizens must evaluate, analyze and participate in decision-making in a democratic system. As well as in the public policies, programs, plans and projects that can be constituted as a means to promote and feed debate and to obtain citizen empowerment geared to the solution of the problems of his country. However, it is possible to understand that regular citizens have cultural, economic, social and technological limitations to access information about the political reality of his environment and the world (Sánchez Garrido 2004).

Under the model of modern democracies, since power is endowed by people, citizens increasingly demand to participate and be included in the processes that directly affect their lives. As well as identifying citizen participation mechanisms that can provide spaces to have their true needs and problems recognized and visible, as well as getting unsatisfied citizen demands being considered as priorities of the government when formulation new public policies. That is to say citizenship is legitimized by virtue of resolution of social demands social perception.

In considering advocacy modes used by civil society in the definition and implementation of communication policies, we assume a constructivist perspective of public policies that recognizes the role of social actors and the power struggles these processes involve. We do not, therefore, only carry out a simple approach of its formal, institutional and administrative aspects, but rather considering them as a conflictive process in which interests linked to positions and relative powers are put into play (Freedman et al. 2013). Thus, we would be talking about democratization of communications which is a necessary condition for the democratization of politics (Mata 2006).

1.2 Progress in Legislation on Democratization of Media in Latin America

Several countries in Latin America expressed their political will to transform the democratic sense of communication from the legal standard, introducing legislative initiatives that understand communication as a public good. Initiatives such as those taken in Ecuador, Venezuela, Argentina, Bolivia, Mexico and Uruguay seek promote a better and more modern government management through the introduction of citizen participation mechanisms to improve services and levels of access to official state information. It is visible hereby the increasing trend towards the use of information and communication technologies. It is also clear that many of these initiatives appear closely related to the trends of leftist government regimes in the region. This marks a debate regarding the uses and applications of these legal bodies (Table 1).

In the case of countries such as Paraguay, Peru and Brazil, they have been working during the last decade to incorporate regulations in relation to media. Paraguay is currently presenting for approval the draft of Law on Audiovisual Media that aims to ensure the democratization of communication and information; enforce communication as a right and a public good, diversity and plurality of expressions as well living as well as the promotion of democratic debate. In the case of Peru, they have been

Table 1 Legislation in the region on communication

Country	Law	Year	Object
Bolivia	General law of telecommunications, information and communication technologies	2011	Establish the general regime of telecommunications and information and communication technologies, postal service and regulatory system. It pursues well-living by guaranteeing the individual and collective human right to communication, regarding economic, social, legal, political and cultural plurality of the totality of Bolivians (male and female), indigenous people(male and female) and intercultural communities And Afro-Bolivians belonging to the Plurinational State of Bolivia
Argentina	Audiovisual communication services law	2009	Regulation of audio-visual communication services throughout the Argentine territory and the development of mechanisms for the promotion, de-concentration and promotion of competition for the purpose of cheapening, democratization and universal use of new information and communication technologies
Ecuador	Organic law of communication	2013	This law aims to develop, protect and regulate, in the administrative area, the exercise of rights to communication exercises as established by Constitution
Venezuela	Social responsibility law regarding radio, television and electronic media	2010	Establish in the dissemination and reception of messages, social responsibility of providers of radio and television services, electronic media providers, advertisers and independent domestic producers. To favor the democratic balance between their duties, rights and interests in order to promote social justice and contribute to the formation of citizenship
Mexico	Communications law/telecommunications reform	1995/2014	Its purpose is to regulate the use, progress, utility and exploitation of the radio electric spectrum, of telecommunications networks and satellite communication
Uruguay	Audiovisual communication services act	2014	Establish the regulation of the provision of radio, television and other audiovisual communication services

debating along year 2017 The Media Bill, that seeks to guarantee the right to objective and disinterested neutral information within the limits of freedom of expression and press. The last case: Brazil, started since Lula da Silva (2003–2011) and Dilma Rousseff (2011–2016) government periods, efforts and actions to regulate this sector. However, they lacked parliamentary support for their approval. Current President Temerhas stopped these former initiatives emphasizing the political situation in the country.

1.3 The Inclusion of Social Media in the Democratization of Communication

The tendency of society to interact through communities (networks) with people who share similar interests is the basis for shaping social media. Castells (2005: 403) states that "new electronic media do not move away from traditional cultures, but absorb them"; to which Scolari (2012) adds:

The emergence of new interactive and collaborative forms of communication implied major changes in the media ecosystem. The new media species modified the relationships of an ecosystem where the mass media—then called broadcasting—carried the singing and leading tone. The new forms of communication were great predators of attention: if consumers used to spend a lot of time on a handful of media (press, radio, television) in a few years they began to devote little time to many media (webs, social networks, Blogs, mobile devices, etc.). The big media strenuously struggled to adapt to the new rules of the game and survived in an increasingly hostile environment. Some media managed to survive while others, despite efforts to adapt, became extinct (Scolari 2012: 338).

In this scenario, "Social media are digital communication platforms that empower the user to generate content and share information through private or public profiles" (Interactive Advertising Bureau, 2009: 6). Which empowers the user to control the environments in which it develops (Altamirano Benítez et al. 2016). Due to the impact and penetration of social media, they have become important tools that generate a growing "democratization" of the Network from the participation of the citizens in the production of contents in contrast to information processes of the large traditional of Communication Companies (Sánchez Carballido 2008).

Social media allows the relative participation of civil society, with less power than media companies and the State, promoting changes in the rules of a game (those of the media system) where possibilities of exercising the word and opinion were quite limited and regulated. These processes are making possible the approval of the new national regulatory frameworks, as well as the incorporation of the communication issue in the governmental and social agenda of each country. Social organizations are the main drivers of this change in Latin America. Historically running in disadvantage these actors, have now managed to acquire some capacity of incidence in the processes of formulation of public policies of communication under these particular circumstances crossing through the region (Segura 2017).

When considering the modes of advocacy of civil society in the definition and implementation of communication policies, we assume a constructivist perspective of public policies that recognizes the role of social actors and the power struggles involved in these processes. We do not, therefore, only carry out an approach of its formal, institutional and administrative aspects, but rather as a conflictive process in which interests linked to positions and relative powers are put into play (Freedman et al. 2013).

As mentioned above the influence of social media on the democratization of communication, has been crucial in mainstreaming the creation of public communication policies. This influence stands out in other areas as well.

Starting from the premise democratization of media must "guarantee transparent, plural and impartial information to foster critical awareness and thinking" (Candón Mena 2013). Social media are the ideal tools to promote free access to information and regulate censorship or manipulation of information prior to its dissemination: being these the strategies that media used for a long period of years to build public opinion, a phenomenon that has been studied and described in hypodermic needle and agenda setting theories.

Democratization of communication is not a recent issue, but the emergence of Information and Communication Technologies (ICT) has made it possible and feasible to implement it. In 1980, the MacBride Commission, established by UNESCO (Díaz Bordenave 1981: 13) to study the contemporary problems of communication. Defines democratization as the process by which: A) the individual happens to become an active element, not a simple object of communication; B) the variety of messages exchanged is constantly increased. They also increase the degree and quality of social representation in communication or participation. Aspects that are currently possible thanks to the Internet and social media, which allow access to unlimited knowledge and actively participate in its construction, which allows citizens to influence the media agenda through the pressure of society and its significant changes in various political, economic, social and environmental aspects in Latin America. As Rodríguez Giralt (2002: 36) affirms, "the power of today's society operates, therefore, increasingly on the codes that regulate the flow of information." In this context, "young people have chosen their new ways and means of communication, information, debate and decision. Internet and social networks have made possible the real participation of the citizens, avoiding and modifying the dynamics of the main power groups, both political and communicative" (Fernández-Planells et al. 2013: 124).

Another aspect that influences social media regarding the democratization of communication is its own power of organization and social call and in the diffusion of ideals. Scott and Street argue that "the Internet is the organizational structure, the medium, the instrument of communication, the appropriate technology to allow new developments for collective action" (Rodríguez Giralt 2002: 33). In this scenario, young people have a leading role in shaping successful social movements. Young people familiar with the use of ICTs reflect a conflict both generational and media, becoming part of an increasingly frequent mobilization process in which young people from different countries is also being introduced. They have made use of new technologies to convene, organize and disseminate collective actions (Candón Mena 2013). But it is worth noting that most of the usual political practices of these networks are carried out in a socio-political space which is a blend between the virtual and the face-to-face spaces. Virtual and face-to-face relationships are the real scene where those political actions and their impact on decision-making centers are generated, maintained and projected. The continuous link between the virtual and the face to face spaces is where the political practices of the social movements bridge and strengthen bonds between the local and the global (Valderrama 2008).

2 From Social Media to Collective Mobilizations

Castells (2012: 111–118) states that social movements in the network show common characteristics:

They are networked in many ways, including online and off line social networks. Some are already there while many others are formed during the process, which hinders the possibility of finding an identifiable center while performing functions of coordination and deliberation through multiple nodes. This decentralized structure maximizes the possibilities of participation and reduces the vulnerability of the movement to threats of repression. They have a space of autonomy, a hybrid between cyberspace and urban space. This space of autonomy can only be guaranteed by the ability to organize in the free space of communication networks. But at the same time, transforming forces can be only exercised if the disciplinary institutional order is challenged by recovering the space of the city for its citizens. They are local and global at the same time. They start in certain contexts, for their own reasons, creating their own networks and building their own public space. But they are also global because they are connected all over the world. They learn from other experiences and hold a global debate on the Internet and sometimes convene global joint events in the network of local spaces.

They are spontaneous in origin, usually triggered by a spark of indignation related to a particular event or to the limit of disgust at the behavior of rulers. The origin of the call is less relevant than the impact of the message on the recipients. They are viral, not only in the impact of the messages but in the example of the model of the movement.

They do not have leaders, not because of the lack of candidates but because of the distrust of the majority of the participants toward any form of delegation of power. They have implicitly accepted the rule of self-government. Community is a goal, but union is the starting point and source of empowerment. The horizontality of the networks favors collaboration and solidarity, undermining the need for formal leadership. They are self-reflexive, they always question themselves, as movement and as individuals and evaluate their goals being not violent, they carry out a peaceful civil disobedience, among other characteristics (Marín-Gutiérrez et al. 2015).

Not only have social movements succeeded in transforming their agendas into public policies and expanding the frontiers of institutional politics, but they have struggled in a very relevant way, to give new meanings to inherited notions of citizenship, representation and political participation, and to democracy itself as a consequence (Escobar et al. 2011: 18).

Another feature of current social movements has been pointed out by Offe (1996: 174). The most striking feature is the fact that in their self-identification they do not refer to the established political code (left/right, liberal/conservative, etc.), or partially corresponding socioeconomic codes (such as working class/middle class, poor/wealthy, rural/urban population, etc.). Rather, the code of the political universe is codified into categories that come from the movement's approaches, such as sex, age, place, etc., or in the case of ecology and peace movements, the human race as a whole.

2.1 Movement #YoSoy132 Seeks Democratization of Information

The #YoSoy132 movement emerges within the framework of Mexico's presidential elections in 2012, in protest against election campaign of presidential candidate Enrique Peña, belonging to Partido Revolucionario Institucional and to the manipulation of information, by the media. On May 12, 2012 this candidate participates in the forum so named "Good Iberian citizen", organized and aimed by the Universidad Iberoamericana to listening to the proposals of the presidential candidates.

> Peña Nieto presented to students, teachers and authorities of the institution, the main points of his government proposal. His speech was interrupted several times by students who issued slogans against him. Some of them tried to annoy the candidate and get the attention of media by wearing former president Carlos Salinas de Gortari photo paper masks. There were posters standing in the auditorium to remind the case of police intervention in San Salvador Atenco, State of Mexico, years ago. The most critical point of the presentation occurred, perhaps, when a group of students there present questioned requesting information the PRI candidate precisely about the actions taken in that municipality. Peña Nieto changed the message assuming "the absolute responsibility of the decision that was taken, adding that when the interests of the Mexican were affected by particular interests, he took the decision to use public force to restore order and peace". "Unfortunately," he pointed in his speech, there were incidents that were duly sanctioned while those responsible for the events were registered with the judiciary system. These statements exacerbated the mood of a certain segment of the public, who literally attacked him with a greater number of offensive slogans as he left the auditorium. (Arteaga Botello and Arzuaga Magnoni 2014)

These statements aroused the student protest and the candidate left the University escorted by his security group who tried to stop the student shouting rejection slogans against him. This event recorded videos were immediately published in the social networks and were viralized immediately. However, a poor communicational strategy of the candidate's team attempting to discredit the truthfulness of the facts and the media aroused the students' indignation and a feeling of struggle.

> Leaders of PRI and the Green Ecologist Party of Mexico (PVEM) declared that those who appearing in the Ibero America University auditorium were not students, but paid agitators gathered by the leftist movements and asked the authorities of the university to investigate them. Televisa reported that Peña was received by groups of students, some supporting him and other with rejection cries and messages; at the end of his presentation, they continued, there was applause and also groups that rebuked him along with still others supporting him. Several newspapers part of the group of Editorial Organization Mexicana emphasized that the EPN visit to that house of studies had been successful in spite of the attempt of boycott. The PRI made a very cleaned up video of the event to avoid the rejection towards its candidate; however, the true event's facts had already been published using the networks of Internet. This provoked general anger against the PRI, its candidate and the media manipulators. (Alonso 2013)

Faced with the manipulation of the media and the political party attempt to dismiss the events at Universidad Iberoamericana, students realize that some other communication alternatives should be used to make their protest heard since the mainstream of media was already part of Peña Nieto's political campaign.

> By mid-May 2012, presidential campaigns were running with predictable monotony. The results of the studies carried out by certain polling houses -particularly those firms that work for Televisa- tried to convince us of the definite advantage Enrique Peña Nieto, a candidate of the Coalition Commitment for Mexico, made up of the Partido Revolucionario Institucional (PRI) And the Green Ecologist Party of Mexico (PVEM), about the other presidential candidates. Since the beginning of President Felipe Calderón administration Televisa has granted a privileged information treatment to Enrique Peña Nieto, who served then as governor of the State of Mexico during the period 2005-2011. Alejandro Quintero - one of the main directors of Televisa - managed and led a series of companies dedicated to build Enrique Peña Nieto image, being TVPromo, Radar Servicios Especializados and The Mates Group the most widely known. The governor of the State of Mexico at that time allocated considerable budget funds from the government of the State of Mexico to the promotion of Nieto's image as presidential pre-candidate. (Islas and Arribas 2012: 251)

2.1.1 The Emergence of Movement #YoSoy132

Following Castells' (2005) proposal, this movement emerges sporadically and spontaneously along with student outrage when facing the manipulation of information. In response to the PRI's disinformation campaign, which appeared empowered by Mexican media and the lack of resources to access the same information spaces. Students use the YouTube network, initially to broadcast the videos of the events raised on the university campus. But the main strategy for the constitution of the movement, is the call to people who joined the protest to record a video demonstrating their status as students simultaneously with an evidence that we were not paid by the opposition party, but participated free and voluntarily in the event, exercising their right to freedom of expression.

> On Monday May 14, a video with 131 students of the Iberoamericana University appeared on YouTube and spread as a virus. After 15 minutes the phrase "131 Alumnos de la Ibero" was the hottest topic on Twitter. This small audiovisual piece shows the protests against the PRI candidate and reproduces Peña Nieto spokesperson's audio on Televisa: "There is a group of, I do not want to say young people. They were older. I take it ranging from 30 to

> 35 years up. Inciting others. No more than 20 people. The information we are given at the end is that groups close to Andrés Manuel López Obrador were promoting and organizing this type of acts." The contradiction between what you see and what you hear becomes evident. They follow a series of close-ups of young people showing their college card, saying their name and articulating a common speech. This collective voice is not a chorus but a series of fragments that are linked: "We use our right of reply to deny them. We are students of the Iberian, we have not been driven there, and no one trained us for anything". (Rovira Sancho 2012: 442)

The video produced an unexpected effect. Thousands of young people joined the protest in social networks uploading videos and sharing their displeasure with the hashtag #YoSoy132. But one of the most important facts is that influence in public opinion allowed to disrupt the daily agenda of media and various television news programs, radio and print media began to broadcast the video in its prime time. These facts generated a massive support and allowed the consolidation of the movement #YoSoy132. The success achieved with the dissemination of the video and probably the lack of awareness of the PRI's campaign leaders and advisors about the impact of social media added to its ineffective way of dealing with the crisis, highlighted their manipulation of information through media Communication allies to Peña Nieto. As a result, the support to IberoAmerican students increased, Organization representatives from various Mexican Universities UIA, ITAM, UNAM, Tecnológico de Monterrey, Universidad Anáhuac, La Universidad del Valle de México and La Salle, among others pledged to their mates. This demonstrated the need to organize to fight for their ideals.

In this context, the organization #YoSoy132 was officially born. In its official blog (2012) a good definition of the movement appears:

> A horizontal movement formed by citizens of diverse ideological positions that seek the democratization of the country. We are citizens who do not want to live under a corrupt, repressive, authoritarian system controlled by a few ones. We want a country where the quality of life in Mexico br dignified and just for its inhabitants. A country where food, work, information, housing, security, education and health be provided by the government as a duty to the Mexican people. A country where the ecosystems, culture and other factors that represent it be respected.

Complying with the characteristics of citizen movements it declares a democratic commitment. So, it should not be linked to political action or a concrete social vision. It should encourage the free participation of all citizens (León et al. 2001: 83). The movement establishes the principles it encourages (YoSoy132 2012):

> Non-partisan: The inexistence of any organic link with any political party.
>
> Pacific: Strategic principle of rejection to the use or manifestation of violence as a resource to reach our objectives.
>
> Laical: Totally unrelated to any doctrine and religious institution.
>
> Political: Interest in public affairs and development of spaces for active citizen participation, without limiting this responsibility to the so-called political class that pretends to be the sole interpreter of the country's political affairs.
>
> Democratic: Building and decision-making in community, advocating a participatory democracy that goes beyond the representative model we now have, aimed at improving the country's cultural policy.

Plural: Inclusion of all individuals who share the principles and characteristics contained herein.

With the collected information, based on the stories of the people interviewed in the movement, Anna María Fernández Poncela (2014) states the movement objectives: "transparency in media", "democracy and informed voting", "No to fraud and injunction", "keeping the electoral process clean and "injunction prevention". Some extreme declarations such as "Change Mexico" were mentioned of course.

2.1.2 The Impact on Social Media

Despite the impact on national media agenda, students continued to use their mobile devices and social media to communicate and convene the various activities of the movement. Facebook, Twitter and YouTube were the means used to spread the ideals and add followers. They used the hashtag #YoSoy132, which became the name of the movement as a procedure to achieve their goals.

The first conquest the students got was seen in the impact of their proposals on social media transcending traditional media, as well as the introduction on the media agenda.

> In social networks such as Facebook and Twitter, the topic of Ibero students appeared to be the dominant segment of discussion about the presidential election. Most of the comments and interventions young people posts, take a stand against the candidate Enrique Peña Nieto, who is now seen as the least popular candidate within these virtual communities… With Six days of protest at the University Iberoamericana against the candidate of PRI-PVEM and three days after the first broadcast of the video "131 Students of the IBERO Respond", the issue of youth in general seems to already dominate presidential campaigns. This occurs in traditional television, print and radio media, as well as on the Internet and its social networks, where content and discussion about young people and what happened in Ibero is even greater. (Galindo Cáceres and González-Acosta 2013)

Another fundamental aspect is the visibility the movement acquires in digital surroundings. According to Portillo, the three conditions that paved the way to this visibility were: technological literacy, some educational training and an institutional belonging to the university that supported them. They showed a display of actions in networks and streets that jointly provoked a magnifying effect of their claims in the public space (Portillo 2015). This way the same afternoon the video "131 Students of the Ibero" was released, immediately became in trending topic in Mexico and the world in the social network Twitter. And the hashtag #yosoy132 was maintained as the top first for five days in Mexico and one of the top '10s worldwide.

The video of protest and identification of the students had 21.747 reproductions the first day while the organization's YouTube channel got only 2256 subscriptions. The impact of the videos is higher, achieving 1,259,384 reproductions. However, the impact on social media is even higher. Until May 2017 85,445 followers are maintained on Twitter and 123,735 on Facebook, although the accounts are no longer frequently updated and the loss of influence of the movement. It should be noted that "the total Internet users in Mexico amounted to 42 million in December 2011. That is why the IWS established Internet penetration in Mexico at 36.9%, just above the world average (32.7%), and lower than the average Internet penetration in the region (39.5%)" (Islas and Arribas 2012).

> The importance of social networks for the 132 movement lied in the fact that it appeared as an instrument of internal communication, through which information is transmitted, tasks are coordinated and topics relevant to the movement are discussed, as well as one of external communication with the environment, through which their political positions were spread, their mobilizations and activities (for example, rock concerts, etc.) were published and their relationships with affinity and solidarity groups were grown. Although the functioning of social networks accelerates the massive diffusion of information and communication, and contributes to the formation of public-political opinion, they were not by themselves a sufficient element for the constitution of the 132 as a system of protest. (Estrada Saavedra 2014)

2.1.3 From Virtual to Real Space

Castells (2012: 212–213) argues that movements tend to start on the Net, but they become a movement by occupying the urban space, either through the permanent occupation of parks or demonstrations Castells (2012: 212–213) proposes that movements tend to start on the Net, but becoming movements by occupying the urban space, either through permanent occupation of parks or planned demonstrations. This author argues that a movement can only be exercised as a transforming force if the institutional disciplinary order is challenged by recovering the space of the city for its citizens. In this context and with aiming at achieving greater impact, the students called for peaceful journeys and various cultural actions that made the movement to be identified and visible (Table 2).

Another feature of the #YoSoy132 movement is its creativity. To show its rejection to regime and media, not only marches, but cultural encounters were organized, which caused impact and increased population acceptance.

> There were no leaders, groups or speakers, but a massive irruption of various groups expressed through posters, painted bodies and other graphic forms in a convergence of horizontalities. In demanding democratization of mass media, the Movement also showed a lot of creative imagination. He repudiated a soap opera democracy, asked to turn off the television to ignite conscience. He criticized that mediated culture did not correspond to the needs of a multicultural society. Their manifestations were not solemn but festive: with music, parties of lights, concerts, poetry, painting, festivals, carnivals, costumes, jokes, performances and graphic designs in their many and varied manifestations. # YoSoy132 produced its own iconography and read books quietly in front of the PRI headquarters in repudiation of that party and its candidate. The Movement called people to participate by Twitter adding proposals to answer the question "What does each person want for Mexico?". In the marches groups of dialogue and discussion were created as multiple civil and popular spontaneous assemblies. (Alonso 2013)

The activists decided to surround the canal facilities in Mexico City for 24 h in a symbolic and peaceful action to denounce the "manipulation" of the electoral campaign in favor of the candidate of the Institutional Revolutionary Party (PRI), Enrique Peña Nieto (Seco 2012).

Table 2 Public events convened by the movement

Day	Activity	Impact
May-18	March against information bias and for "his human right to information"	Mexico City. DF 10,000 Puebla 3000 Tijuana 200 and other cities of the country with no record showed.
May-19	Anti-EPN March	46,000 people in DF
May-23	"Estela de la Luz"	Where a rally is held and a plethora of petitions are circulated about media democratization. There is no record of participants however thousands of people participated according to media reports
May-30	First University Assembly	7000 people
June-10th	Second "antiEPN" March	90,000 people
June-16th	Concert organized to support the movement by the collective Musicians #yosoy132	30,000 people Assemblies There were 9 in total from 54 initials to 136 universities

What happened in front of the main television station headquarters in the country? There were people disguised like Elba Ester Gordillo (leader of the teachers' union) and like Peña Nieto promising to broadcast more soap operas to Mexicans. A performance recreates violence in Atenco. It begins to rain but the attendants do not pay attention and dance under the rain. At night, hundreds of candles are lit and the Allied Artists continue to perform on repression, on resistance: "We are an erupting volcano seeking to rewrite its history", students tear the plastic placenta that kidnapped them and feel released "And if the earth is shaking, it is the 132 that is marching!" Recalls Isaid. A short audiovisual piece is projected over the white wall of Televisa same will later raise waves in the networks of YouTube. Images of diverse historical periods of the country are shown: the 1968 and 1971 repression against the students along with the silence this television channel kept. Under the title "LIGHT # 132" 27 this video asks: "What is being manipulated behind these walls?". On Saturday, June 23, young people organize the Cultural Festival # 132, circus workshops, children's painting, screen printing, posters, exhibition of the fascinating graphic of the movement, some works of artists. A 132 meters long sheet served for a collective mural, performances, installations, music bands, hip hop can be admired and counted. Isaíd participated in the mobile temple (a truck adapted as a theater) in a slam of poetry later performed in the main stage, where the prestigious bands Panteón Rococo, Los de Abajo, Botellita de Jerez, Natalia Lafourcade, The Damned Crocodiles, Bizarre, among others entertained people. There was no time left. Just eight days to change the country.... (Rovira Sancho 2012)

2.1.4 Achievements

If the objectives posted by the movement were the starting point namely democratization of media and an information campaign to vote consciously, the evaluation shows that the university students achieved their goals. Estrada Saavedra (2014) argues that the movement achievements could be listed as:

- First, by placing the campaigns financing and the very journalistic work at the center of the public debate, together with other voices and actors, they forced television stations to modify the contents of their informative coverage of 2012 Federal elections.
- Second, they generated sufficient social pressure on TV stations to change their original decision and transmit the second debate among the presidential candidates in their channels of greatest audience.
- Third, they organized a third unscheduled debate with three of the four presidential candidates with very low cost.
- Fourth, they effectively questioned with their public appearance the sense of inevitability of the electoral victory of Enrique Peña Nieto, succeeding in encouraging public opinion and mobilizing segments of the citizenry.

It is important to highlight the impact on the presidential debates:

> #YoSoy132 considered very relevant the widest dissemination of the presidential debates so citizens would have a better knowledge of what was at stake in the elections. He succeeded in influencing the provision for second debate to be scheduled on the broader television channels. When no acceptance from the electoral body to commit to a third debate, the Movement organized a new one that transmitted over the Internet, which broke the straitjacket of official forums, which did not allow the real confrontation of the ideas of the candidates. Three of the four candidates for the presidency met, because EPN feared to expose themselves to participate in a debate that did not take care of its figure, like the previous ones. All the political acts of this candidate had the presence of students raising protests against him. (Alonso 2013)

Transparency in information was another aspect to be sought, although the results were not positive. The two networks with the highest penetration, Televisa and TvAzteca, which formed the duopoly maintained Peña Nieto's image. They manipulated the polls and sold the candidate as an absolute winner. Nevertheless, the opinion of the people was expressed at the polls:

> Also noteworthy is the discussion generated by the alleged use of surveys as advertising, since the final result showed a smaller difference than the predicted between the PRI candidate and the PRD candidate. From a total of 41 polls published by various Mexican media between November 2011 and June 2012, an average of 44.03% of the PRI vote is extracted, compared to a 25.93% of the PRD. Considering only the last three polls conducted before the election, since the trend marks a slight reduction of the difference between the two candidates as the electoral date approaches, the PRI obtains an average of 45.3% of the votes compared to 28, 13% of the PRD. Thus, the difference predicted by the surveys shows an advantage of 18.1% in favor of Peña Nieto, 17.17% if we consider only the last three surveys. Nevertheless, the final result of the election granted the PRI of Peña Nieto 38.21% of the votes compared to 31.59% obtained by the PRD of López Obrador, showing a distance of only 6.62 percentage points. Therefore, the deviation between the results predicted by the 41 afore mentioned polls and the definitive count of the election was 11.48 points (from 10.55 on the last three polls). (Candón Mena 2013)

After the electoral results of the June 2012 elections that gave victory to Enrique Peña Nieto were proclaimed, the movement mobilized thousands of citizens throughout the country demanding the annulment of the presidential elections. But repression by the government and the usual manipulation of media forced them to abandon their struggle and concentrate on releasing the political prisoners.

> The final point of inflection of the movement was on December 2012 with the inauguration of the new president. It was the last great march of the movement, the strongest confrontation between demonstrators and police, leaving as a balance eight wounded citizen and 103 detained, which generated a strong media controversy. While television and press disseminated images of attacks by demonstrators on police stations or shops downtown, social networks showed the opposite that is, cases where the police exceeded in the use of force. So, the movement now focused on the release of prisoners, from which emerged a new hashtag: #1DMx. Under this banner, new marches were convened during December and part of January, achieving the gradual liberation of all. (Cerrillo Garnica and Lay Arellano 2013: 312)

2.2 *Student Movements in Chile*

Although their main purpose was not democratization of communication, Chilean student protests, were able to influence, through the use of social media, the media agenda and consequently social involvement. As a result, the Government listened to the students and accepted some proposals regarding the creation of public policies in favor of education. For this reason, they are analyzed in this present research. The student protests calling for the defense of public education did not begin either in 2006 (with the "penguin revolution") or in 2011. Following history and social facts, it should be stated that they have been present as a practice every year since the end of the last dictatorship. However, just like 2006 marked a milestone in terms of impact, support and conceptual clarity of the demands of a new democratic time, 2011 is considered as the year when most massive mobilizations have appeared since return to democracy (Vera Gajardo 2011).

> The student movement, led by the Confederation of Students of Chile (CONFECH), is students' union that brings together the federations of students from the country's traditional universities and some private ones. The activity of the student movement began in May, becoming the most important of its kind in recent years, and the main political problem for Piñera's government. He declared in his speech of May 21, 2011, that as the "year of higher education", adding that reforms to the quality of education approved in February 2011 along with the creation of the Quality Agency and Superintendence Of School Education, new projects for that year would be done. In this context, under the leadership of CONFECH and also the participation of high school student organizations, students began to display an intense mobilization aimed at presenting and demanding a set of reforms to the Chilean educational system to the acting government. (Vera Gajardo 2011: 291)

The student movement struggles for lack of equity in education and presents a proposal for a free, quality public education for all, without profit (Lafferte and Silva Gallinato 2014).

2.2.1 Social Media and Student Mobilization

Social dynamics have been redefined in the context of social media and ICTs, which has allowed existing social phenomena face-to-face context to be supported, as is the case of social movements and other phenomena of a broader nature such as Collective actions (Valadés García 2011). Influenced by this way mood, Chilean students used social networks to make visible their proposal visible. Although the highest incidence is evident in the fields of communication and the organization of the student movement.

Although the impact of the Internet and social media was incipient in 2006, rapidly took over to be an informative tool for young Chileans in the face of student protests against the Chilean government. According to the study "New Technologies of Information and Communication and Social Movements. Study on the Use and Perception of Internet Use in Chilean Student Mobilization in 2006" (Valderrama 2008) the behavior of students was as follows:

- 98% think that they connected to the Internet during mobilizations. The majority also stated that they spent between one to three hours on the activity (59%), whereas 24% spent less than an hour on the Internet and 15% more than three. 2% said they never connected to the Internet during this period. 41% perceive the Internet as "very important" during the mobilization, 27% consider it "extremely important" and 25% qualify it as "important".
- 69% think that "the majority" of shared content was about mobilization. Among those self-declared "active participants", most answered that at least "half" of the contents they shared on the Internet were related to the mobilization (45%) and 41% said that "most" content they shared had to do with the movement.
- Coordination of the activities of the movement, the subjects consulted stated they preferred to use "instant messaging" as their preferred medium for these purposes (78%), followed by e-mail (22%).

The study also shows that digital resources used for dissemination of campaign messages are: "avatars" (image and virtual identification) "IM nicknames" (virtual appellations of "instant messaging") to indicate their feeling of belonging or identification with the movement (58%), as communication tools 15% of young people used virtual forums and 12% blogs. Another tool used is Fotolog, a virtual community created by Scott Heiferman and Adam Seifer, in May 2002, where users have personalized public web spaces, in which one image or photo up at least can be upload a day.

The mobilized students not only visited or used Fotolog from the students' union of their school, since 73% of the consulted subjects declared to have used their personal fotologs in the mobilization with different purposes. 15% did so to "report what was happening". A 24% say that they used it to "show adherence to the movement and get more people to do so"; and 13% said: to "coordinate activities of the student movement". Messages disseminate via fotologs were informative, while the images sought to increase the sense of belonging.

In the student movement of 2011, social media were incorporated as 552 identified entries were recorded during the month of analysis (an average of 16.7 entries per day). This shows the intensive use of this digital platform by the FECH. 47,314 approvals (I like) were counted toward the published content and 8686 comments. Each entry generated an average of 15.7 comments from followers of the page (Cabalin 2014). A very low impact compared to the impact the movement provoked on the streets.

Young people have chosen their new ways and means of communication, information, debate and decision. In this context, Internet and social networks have made possible the real participation of citizens, thus avoiding and modifying the dynamics of the main political and communicative power groups. (Fernández-Planells et al. 2013). This way, social media were the tools to plan highly creative collective actions such as those highlighted by Vera Gajardo (2011):

"A Thriller for Education". Students demonstrated in the Plaza de la Ciudadanía in front of the Moneda Palace on June 24, arriving in surprise and massively disguised as zombies simulating in an extraordinarily coordinated way the choreography of the song "Thriller" by Michael Jackson. This act showed a long preparation and organizational capacity.

"Flash Mobs" or "instant crowd". One of these actions was carried out after the acting Minister of Education (Joaquín Lavín) declared an early start of the students' vacation period when the students' movement got to the peak of the student movement, with the obvious objective of disintegrating it. The "Flash Mob" designed for this occasion was called "All on Vacation!" And consisted of suddenly installing a beach atmosphere in the Plaza de Armas, in Santiago. Thus, more than 500 students wearing towels, bathing suits, playing paddles, umbrellas and suntan lotions gathered on the Plaza. It should be noted that this action was carried out in July, in the middle of the Chilean winter. This action would be replicated in different cities of the country later on.

"Besaton". It was held in July and September throughout the country with the presence of young people kissing. This action lasted 1800 s (half an hour), which represented the $1.8 billion that would be needed annually to have a qualified education.

2.2.2 Achievements

According to Garcés (2011) in a short time, the conjuncture period, it appears evident that students achieved important symbolic achievements. For example, in speech, today you can name what until yesterday was unnamable: "free education". Achievements of social nature, by succeeding in making thousands of Chileans of different ages, trades and social positions stand in solidarity with their demands. The student movement undertook a series of actions that attracted the adhesion of a good part of population. In addition, it was able to involve other actors in the educational field, and articulate as well their demands and initiatives with those of other similar social movements that broke out during 2011. The impact of the actions undertaken by the university students movement was such that succeeded in influencing government's agenda, introduced new issues in the public debate and promoted a new generation of leaders with political representation and protagonism at the national level (Avendaño 2014, 43).

3 Further Considerations

New technologies can serve as glue and bonding node, but the important thing is how links are maintained and how information and action flow through them. In short, the challenge is to create a network as flexible and solid as a web.

As we have seen, the student movement is a collective agent that intervenes in the process of social transformation, promoting, hindering, or canceling a change. It is a particular collective action, driven by the students, in a discontinuous way that provokes a confrontation that directly affects other social sectors. It is a cyclical movement that combines periods of strength with others of apathy. It is often unexpected and unpredictable, although there are pushing factors such as an increase in rates, political situation, etc. Their most common repertoires of collective action are occupations of university and street spaces, breaking academic regular activities, development of alternative forms of information and counter-information through social networks. But, above all, they use horizontal democracy (the virtual assembly) as a means of organization, elaboration and decision. As seen, the student movement is the precursor of the new social movements.

References

Alonso, J.: Cómo escapar de la cárcel de lo electoral: el Movimiento #YoSoy132. Revista Desacatos, **42** (2013)

Altamirano Benítez, V., Moreira Aguirre, G., Mahauad Burneo, M.: Comunicación pública y participación ciudadana 2.0 en Iberoamérica. De los medios y la comunicación de las organizaciones a las redes de valor. In: Simposio de la Red Internacional de Investigación de Gestión de la Comunicación. Red Internacional de Investigación de Gestión de la Comunicación, Quito (2016)

Arteaga Botello, N., Arzuaga Magnoni, J.: Derivas de un performance político: emergencia y fuerza de los movimientos 131 y YoSoy132. Revista mexicana de sociología, **76** (2014)

Avendaño, O.: Fracturas y representación política en el movimiento estudiantil. Chile 2011. Última Década, **22**(41), 41–68 (2014)

Azurmendi, A.: "La libertad de expresión en la jurisprudencia de la Corte Interamericana de Derechos Humanos: la influencia del Tribunal Europeo de Derechos del Hombre. In: Tenorio Cueto, G. (coord.) La libertad de expresión y sus fronteras contemporáneas. Universidad Panamericana, México (2007)

Borja, R.: Sociedad, cultura y derecho. Editorial Planeta del Ecuador, Quito (2007)

Cabalin, C.: Estudiantes conectados y movilizados: El uso de Facebook en las protestas estudiantiles en Chile. Comunicar **22**(43), 25–33 (2014)

Candón Mena, J.: Movimientos por la democratización de la comunicación: Los casos del 15-M y #YoSoy132. Revista Razón y Palabra **82** (2013)

Candón Mena, J.: In: http://www.redalyc.org/articulo.oa?id=199043103009 (2018)
Castells, M.: La era de la información. Economía, Sociedad y Cultura. Siglo XXI Editores, México (2005)
Castells, M.: Redes de indignación y esperanza. Alianza Editorial, Madrid (2012)
Cerrillo Garnica, O., Lay Arellano, I.: #yosoy132: redes digitales como comunicación e identidad en la acción colectiva. Revista Enfoques **13**, 294–317 (2013)
Díaz Bordenave, J.: Democratización de la Comunicación: teoría y práctica. Revista Chasqui **1**, 13–20 (1981)
Duque, J.: Guillermo O'Donnell y la democracia. Latinoamérica. Revista de Estudios Latinoamericanos **58**, 113–144 (2014)
Escobar, A., Álvarez, S., Dagnino, E.: Política cultural y Cultura política: una nueva mirada sobre los movimientos sociales latinoamericanos. Instituto Colombiano de Antropología e Historia, Bogotá (2011)
Estrada Saavedra, M.: Sistema de protesta: política, medios y el #YoSoy 132. Revista sociológica **29** (2014)
Fernández Poncela, A.: De la Red a las calles ¿y de las calles a las conciencias? El movimiento estudiantil #YoSoy132. Revista Argumentos **27** (2014)
Fernández-Planells, A., Feixa Pampols, C., Figueroas-Maz, M.: 15-M en España: diferencias y similitudes en las prácticas comunicativas con los movimientos previos. Revista Última Década **21**(39), 115–138 (2013)
Freedman, K., Heijnen, E., Kallio-Tavin, M., Kárpái, A., Papp, L.: Visual Culture Learning Communities: How and What Students Come to Know in Informal Art Groups. Studies in Art Education **54**(2), 103–115 (2013)
Galindo Cáceres, J., González-Acosta, J.: #YoSoy132 La Primera Erupción Visible. Offset Rebosán, México (2013)
Gallardo, H.: Democratización y democracia en América Latina. Revista Pasos **68** (1996)
Garcés, M.: El despertar de la sociedad. Los movimientos sociales en América Latina y Chile. LOM Ediciones, Santiago de Chile (2011)
Interactive Advertising Bureau (IAB).: Estudio de Internet e Interactividad en móviles y otros dispositivos portátiles. In: http://www.itu.int/ITU-D/ict/material/FactsFigures2011.pdf (2009)
Islas, O., Arribas, A.: Enseñanza y ejemplo de la Primavera Mexicana. Revista Razón y Palabra **80**, 249–265 (2012)
Lafferte, M., Silva Gallinato, C.: La Voz del Movimiento Estudiantil 2011. Educación Pública, Gratuita y de Calidad. Fondo de las Naciones Unidas para la Infancia, UNICEF, Santiago de Chile (2014)
Lechner, N.: De la revolución a la democracia. Leviatán **21** (1985)
León, O., Burch, S., Tamayo, E.: Movimientos sociales en la red. Agencia Latino Americana de Información, Quito (2001)
Marín-Gutiérrez, I; Hinojosa Becerra, M. y Allen-Perkins, D.: Movimientos sociales y acción colectiva. Modelos teóricos y principales movimientos sociales. EAE (2015)
Mata, M.C.: Comunicación y ciudadanía. Problemas teórico-políticos de su articulación. Fronteiras. Estudos Midiáticos **1**(8), 5–15 (2006)
Munck, G.: Una revisión sobre los estudios sobre la democracia. Temáticas, conclusiones, desafíos. Desarrollo Económico **164**(41), 579–609 (2002)
Offe, C.: Partidos Políticos y nuevos movimientos sociales. Editorial Sistema, Madrid (1996)
Portillo, M.: Construcción de ciudadanía a partir del relato de jóvenes participantes del #yosoy132: biografía, generación y participación política. Glob Media J México **12**(23), 1–18 (2015)
Ramos Jiménez, A.: Las formas modernas de la política: Estudios sobre la Democratización de América Latina. Centro de Investigaciones de Política Comparada, Mérida (2016)
Rodríguez Giralt, I.: El efecto de las TIC en la organización de la acción colectiva: la virtualización de los movimientos sociales. Universitat Oberta de Catalunya (UOC), Barcelona. In: http://www.uoc.edu/web/esp/art/uoc/irodriguez0602/irodriguez0602.html (2002)

Rovira Sancho, G.: México, #YoSoy132: ¡No había nadie haciendo el movimiento más que nosotros! Anuari del conflicte social **2**, 423–448 (2012)
Sánchez Carballido, J.: Perspectivas de la información en Internet: ciberdemocracia, redes sociales y web semántica. ZER - Revista de Estudios de Comunicación **13**(25), 61–81 (2008)
Sánchez Garrido, T.L.: El movimiento social altermundista. La nueva praxis de la acción política. El Cotidiano **20**(126), 142–158 (2004)
Sánchez, E.: Medios de Comunicación y Democracia. Grupo Editorial Norma, Bogotá (2004)
Schmitter, P.C.: Civil society east and west. Consolidating the third wave democracies. Themes and perspectives **2** (1997)
Scolari, C.: Comunicación digital. Recuerdos del futuro. El profesional de la información **21**(4), 337–340 (2012)
Seco, R.: El movimiento Yo Soy 132 cerca la cadena Televisa "contra la manipulación". El País. In: http://internacional.elpais.com/internacional/2012/07/27/actualidad/1343354496_270808.html (2012)
Segura, M.: La sociedad civil y la democratización de las comunicaciones en Latinoamérica. Íconos - Revista de Ciencias Sociales **49**, 65–80 (2017)
Valadés García, B.: Conceptualizar el papel de las redes sociales en internet en movimientos sociales y acciones colectivas. Propuesta aplicada a lo digital. Revista Razón y Palabra **77** (2011)
Valderrama, C.: Movimientos sociales: TIC y prácticas políticas. Revista Nómadas **28**, 94–101 (2008)
Vera Gajardo, S.: El resplandor de las mayorías y la dilatación de un doble conflicto: El movimiento estudiantil en chile el 2011. Revistes Científiques de la Universitat de Barcelona **1**, 286–309 (2011)
Yo Soy 132.: #YoSoy132 es… In: http://www.yosoy132puebla.com/p/acerca-de-132.html (2012)

Diana Gabriela Moreira Aguirre Ph.D. in Peace, Conflicts and Democracy by the Universidad de of Granada, Spain, Professor Titular by Universidad Técnica Particular de Loja—Ecuador, Mediator of the Center for Analysis and Resolution of Conflicts. Academic Coordinator of the UNESCO Chair of Culture and Education for Peace of UTPL, her research lines are: education and culture of peace. She is part of the Peace, Environment and Society research group. Loja, Ecuador.

Santiago José Pérez Samaniego Bachelor in Political Science mention in Globalization and Conflict by the Russian Peoples Friendship University (Moscow—Russia), Master in International Relations by the Russian Peoples Friendship University(Moscow—Russia). Ph.D. Candidate by the Universidad de Sevilla in Human Resources (Seville-Spain). At present he is Coordinator at the Interculturality Area in the UNESCO Chair of Culture and Education for Peace—Universidad Técnica Particular de Loja—Ecuador (UTPL), full-time teacher in the Department of Legal Sciences—UTPL, part of the Peace, Environment and Society research group. Loja, Ecuador.

Verónica Paulina Altamirano Benítez Ph.D. in Communication and Creative Industries with International mention and Cum Laude by the Universidade de Santiago de Compostela, Master in Communication and Creative Industries by the same university. Responsible for the Organizational Communication Section and member of the Quality Team of the UTPL. Co-author of international books, articles in scientific journals, and informative. She is part of the Digital Communication and Culture research group. Loja, Ecuador.

Isidro Marín-Gutiérrez Ph.D. BA in Political Science and Sociology from the Universidad de Granada. Doctor (Ph.D.) in Social Anthropology from the Universidad de Granada. Associate Professor at the Universidad de Huelva. Teacher-Principal Investigator of the Universidad Técnica Particular de Loja (UTPL). Member of the Technical Council of the Latin American scientific journal communication and education «Comunicar». Member of the Research Group E6, Social Studies and Social Intervention (SEJ-216). Loja, Ecuador.

A Brave New Digital Journalism in Latin America

Ramón Salaverría, Charo Sádaba, James G. Breiner and Janine C. Warner

Abstract In recent years, journalism in Latin America has been undergoing profound transformations. Confronting the stagnation or even decline of a large part of the big media industry, hundreds of digital native publications are surging up to challenge the traditional journalism landscape of this region. From the Caribbean to Patagonia, all 20 Latin American countries are witnessing the emergence of innovative online media projects that, in some cases, have already reached a high degree of consolidation. This chapter analyzes the origins, models and challenges of this emerging Latin American journalism. Firstly, it shows the historical evolution of these digital native media outlets. Then, their current characteristics are analyzed, focusing on their innovative ways to explore sustainable business models. Finally, the challenges for the future consolidation of these emerging digital media are examined.

Keywords Digital journalism · Latin America · Digital native media
Innovation · Business models

1 Digital News in Latin America

The evolution of news media industry since the end of 20th century has gone far beyond technology. Digital technologies have boosted the changes, indeed, but their effects in the media have been overarching. According to Hass (2011), digitalization

R. Salaverría (✉) · C. Sádaba · J. G. Breiner
Facultad de Comunicación, Universidad de Navarra, Pamplona 31009, Spain
e-mail: rsalaver@unav.es

C. Sádaba
e-mail: csadaba@unav.es

J. G. Breiner
e-mail: jbreiner@unav.es

J. C. Warner
SembraMedia, Los Angeles, CA, USA
e-mail: janine@sembramedia.org

M. Túñez-López et al. (eds.), *Communication: Innovation & Quality*, Studies in Systems, Decision and Control 154, https://doi.org/10.1007/978-3-319-91860-0_14

has caused the media industry to undergo more profound and structural changes than other industries, given that its value proposition is an intangible object: information. Although technology has traditionally been placed at the service of content creation, the focus has always been on how to apply creative skills and capacities. Technical teams and engineers, who for decades performed a supporting role in the media industry, are now at the heart of the creative process. Technological skills have quickly become essential to content creation, something that many managers have yet to explicitly acknowledge and do not always know how to handle successfully (Küng 2013). In sum, technological innovations have driven a general reconsideration of the overall media system.

These far-reaching changes have moved the news industry to respond with new forms of media, which were hardly imaginable just a couple of decades ago. Digital technologies have not just changed how journalism is practiced, but also who, what, where, when and even why. All five W's of journalism have transformed into something new.

Moreover, this evolution has been global (Dragomir and Thompson 2014). Not only the most technologically developed and economically powerful countries have seen a rapid expansion of digital media within their boundaries. Although at slower pace, the less developed countries of the world, especially those with emerging economies, have also witnessed the birth and growth of digital forms of media (He and Zhu 2002; Köroğlu and Tingöy 2011; Mabweazara et al. 2014). One of the regions where, despite its structural difficulties, this evolution has been more profound is Latin America (Salaverría 2016; Harlow and Salaverría 2016; Mioli and Nafría 2017).

Composed of 20 countries—all the 19 Central, Caribbean and South American countries where Spanish is the main language, plus Portuguese-speaking Brazil—, this extensive region with 600 million inhabitants has seen a significant shift in its media landscape. That change has had more to do with the increasing albeit uneven penetration of the internet (see Table 1). By 2015, the median usage of internet among the 20 countries was 52.7%, although there were differences of more than 50 points between the leading and the last ones. Not surprisingly, the Latin American countries with higher internet penetration rates are those where the digital media industry has developed faster (Salaverría 2016).

1.1 Digital, Digital-Only, Digital Native

One of the spheres in which the pervasive digital technologies have boosted innovations is the typology of news media. The multiplication of online media publications has had not only quantitative effects but also qualitative. The digital "fourth media" (Gang 1998), which joined the print press, radio and television since mid-1990s, has enjoyed in recent years such a fast process of diversification, that today it has a broad variety of digital media models.

Indeed, today it is no longer enough to consider all digital media under just one comprehensive category. On the contrary, it is becoming evident that, depending

Table 1 Percentage of individuals using the internet in Latin America (2000–2015)

	2000	2005	2010	2015	Population (1 July 2015)
Argentina	7.0	17.7	45.0	69.4	43,416,755
Bolivia	1.4	5.2	22.4	45.1	10,724,705
Brazil	2.9	21.0	40.7	59.1	207,847,528
Chile	16.6	31.2	45.0	64.3	17,948,141
Colombia	2.2	11.0	36.5	55.9	48,228,704
Costa Rica	5.8	22.1	36.5	59.8	4,807,850
Cuba	0.5	9.7	15.9	37.3	11,389,562
Dominican Republic	3.7	11.5	31.4	54.2	10,528,391
Ecuador	1.5	6.0	29.0	48.9	16,144,363
El Salvador	1.2	4.2	15.9	26.9	6,126,583
Guatemala	0.7	5.7	10.5	27.1	16,342,897
Honduras	1.2	6.5	11.1	20.4	8,075,060
Mexico	5.1	17.2	31.1	57.4	127,017,224
Nicaragua	1.0	2.6	10.0	19.7	6,082,032
Panama	6.6	11.5	40.1	51.2	3,929,141
Paraguay	0.7	7.9	19.8	48.4	6,639,123
Peru	3.1	17.1	34.8	40.9	31,376,670
Puerto Rico	10.5	23.4	45.3	79.5	3,683,238
Uruguay	10.5	20.1	46.4	64.6	3,431,555
Venezuela	3.4	12.6	37.4	61.9	31,108,083

Source ICT Facts and Figures 2016 (ITU 2017) and World Population Prospects (UN 2016)

on their origin, structure and publishing platforms, the types of digital media are increasingly different from each other (Salaverría 2017). Digital media outlets begin to show differences at multiple levels: from very specific aspects, such as their different degree of adoption of innovations (García Avilés et al. 2016), to much more general aspects related to their disparate mode of understanding and practicing journalism (Suárez Villegas 2015; Arrese and Kauffmann 2016; Harlow and Salaverría 2016).

The typologies of digital media can be built over many different theoretical criteria: platform, temporality, topic, reach, ownership, authorship, focus, economic purpose, and dynamism (Salaverría 2017). But apart from these categories, it is also possible to see differences in relation to the origin of the media outlets. In fact, research is beginning to find contrasts between digital publications derived from traditional media and those born directly on the internet (Nicholls et al. 2016). Digital media created as internet versions of printed publications or broadcast media follow patterns and structures that clearly resemble their parent media. On the contrary, internet-born publications show forms and models that are increasingly specific, not inspired anywhere else (Kilgo et al. 2016).

It seems therefore necessary to distinguish between digital media in general and digital native media, and even digital-only media. The label "digital media" refers to all those publications that, regardless of their origin, are published in online networks. For its part, we use the term "digital native media" for those that are born directly on the internet, without being the alter ego of any previous offline publication. Finally, by "digital-only media" we mean those publications that are published solely in digital format, although they may have been originated either outside or inside the network. This label applies, for instance, to those newspapers and magazines that, due to their economic losses, need to stop their printed edition, but keep their digital edition running. In this case, we have a digital-only medium, but not a digital-born publication.

This triple distinction is important because, as we will explain later, within the Latin American news market digital publications, in general, are gaining momentum. But, more especifically, digital native publications are shaking the market with a new wave of fresh journalism.

1.2 Origins of Digital Native Media in Latin America

In Latin America, the history of digital native media begins as early as 1995. That year, when many large Western media companies still had not even launched their first online publication, in Nicaragua a humble online news bulletin called Notifax started to be distributed exclusively through the internet (Solórzano 2016). This bulletin never became more than a simple collection of news, but it inaugurated the history of the digital-born publications in the region.

A few years later, in 1998, the first truly relevant digital native medium of Latin America appeared: El Faro of El Salvador, a publication still working today (Harlow and Salaverría 2016; Tamacas 2016). Its launch as a medium exclusively on the internet was more out of necessity than out of conviction: its promoters, a group of independent journalists, wanted to publish a print newspaper but lacked the financial resources to do so. In these circumstances, they decided to provisionally launch a news website only on the internet, at least until they could raise funds to publish a "true" newspaper. The economic difficulties, however, continued, to the point that, during the first years, the founders of El Faro worked practically without being paid. However, their good journalism led them little by little to receive funds from international foundations and to win recognition, in international journalism contests and prizes. This support allowed them to consolidate the project. Today, twenty years after its launch, El Faro is a fully consolidated medium and an essential reference among digital media in Latin America.

Many other digital native media have followed El Faro's trail. In May 2017, the Observatorio de Nuevos Medios (New Media Observatory; www.nuevosmedios.es), a comprehensive directory of Spanish-language digital native media, recorded 1,678 publications, of which 875 were distributed among the 19 Spanish-speaking Latin American countries. Although the list of Latin American digital native media is very long, not all of them have reached the same degree of consolidation and recognition, of course. However, today in all the Latin American countries digital native media outlets can be found, some of which are thriving and have achieved remarkable longevity and influence (Salaverría 2016: XXIX–XXXI).

Some publications can be highlighted, thanks to their consolidated editorial structure and high social media reach. In Argentina, among others, Infobae (launched in 2002), Minutouno.com (2007) and MDZ Online (2008) stand out. In Brazil, Agência Pública (2011). In Chile, El Mostrador (2000) and Ciper (2007). In Colombia, La Silla Vacía (2009) and Las 2 Orillas (2013). In Costa Rica CRHoy (2012). In Cuba, 14ymedio (2014), an independent publication promoted by intellectuals and anti-government journalists. In Mexico, Animal Político (2010), Sin Embargo (2011) and Aristegui Noticias (2012). In Nicaragua, Confidencial (2010). In Peru, besides older digital native media such as La Encuesta (1996) and Pueblo Continente (1996), there are also more recent projects such as IDL-Reporteros (2010) and Ojo Público (2014). In Puerto Rico, NotiCel (2011) should be highlighted, whereas in the Dominican Republic, in addition to the pioneering Diario Electrónico Dominicano (1996), which soon closed, another closed publication stands out, Clave Digital (2004–2010); after its closure, that news website was followed by Acento. Venezuela, a country with a especially troubled political life in recent years, has also been a hotbed of digital native media: among many others, Noticiero Digital (2009), La Patilla (2010), Run Run (2010) and Efecto Cocuyo (2015).

The list of digital native media in Latin America is growing fast. Some of them were launched in the first decade of the 21st century, and a few of them even in the 1990s. However, the blossoming of digital natives began after 2010 mainly. It seems that the decline or, at least, stagnation of the legacy media has opened up the opportunity for the emergence of many digital native publications in all countries of the region. And they have done so with an undeniable commitment to regenerating journalism (Harlow and Salaverría 2016).

It seems, indeed, that these new digital media are managing to do quality journalism. Two digital native publications, Connectas of Colombia and Aristegui Noticias of Mexico, were among the reporting partners on the Panama Papers investigation that won the Pulitzer Prize (ICIJ 2017). Also, for the first time, all five of the winners of the prestigious Gabriel García Márquez Awards in fall 2016 were digital natives (FNPI 2016). The increasing impact of these digital native publications can also be measured by the fact that dozens of them have seen their original research picked up by the national and international press, including The New York Times, BBC, Al Jazeera, and The Guardian (Warner et al. 2017).

2 Models of Sustainable Digital Native Media in Latin America

The rise of digital news media over the past 25 years has prompted various efforts to catalogue and categorize them as a way of understanding them. Journalists, scholars, and media development organizations around the world have created lists with various goals in mind (Breiner 2017a). Many of the studies attempt to show how the new digital media ecosystem is functioning and what strategies are showing the most promise for achieving sustainability (Harlow 2017).

Some of the research has focused on the unique characteristics of media markets in Latin America. As noted by Warner et al. in a white paper (2017) by the digital media research and training platform SembraMedia, the 400 million Spanish speakers and 200 million Portuguese speakers (Brazil) in the region's 20 countries share much history and culture. They also share similarities in the structure of their media markets, where journalists live under the threat of violence and the media's financial state is often precarious:

> News ownership is highly concentrated in these countries, and government advertising is frequently used to reward compliant media outlets. Even in the face of these legal, financial, and physical threats, entrepreneurial journalists are building sustainable businesses around quality journalism. The advent of social media and easy-to-use web design tools has made it possible to launch a digital media venture almost entirely on sweat equity. (Warner et al. 2017, p. 7)

The cutbacks by traditional media have created opportunities for new digital news organizations to fill the gaps in coverage, or to broaden and deepen topics long-neglected by the established media. Many of the new sites were founded by laid-off journalists or those frustrated with censorship, low pay, and poor working conditions. Some were founded by people with no journalism experience whatever but with an interest in informing the public and serving their community. The result is a dynamic ecosystem of digital media that is changing public discourse in the region and challenging the traditional power oligarchies.

2.1 *Studies of Latin America's Digital Media Ecosystem*

Besides the aforementioned Observatorio de Nuevos Medios, various organizations and academic researchers have done studies of the digital media landscape of Latin America, with a variety of aims and purposes (Mioli and Nafría 2017; Harlow 2017). Rey and Novoa (2013) studied 650 digital media in Colombia, partly to determine whether digital media were available in all geographic areas of the country. Most of the 650 were focused only on digital (no print or broadcast link) and just over half were journalistic (p. 8). The media organizations, however, were reluctant to cooperate with researchers. Only 61 of the media organizations (9%) responded to a questionnaire, so the data had to be assembled from a variety of public and private sources. The report did not give clues about sustainability. Of the respondents, only three had more than one million monthly users at the time. As for revenue sources, 23 listed advertising, 18 private capital, and six donations. Seventeen carried no advertising (p. 26).

In 2015, Poderopedia, whose software allows users to visually map connections among people, undertook a study of the connections among media owners and the political and business elite of Colombia and Chile. The aim of the studies was to document the level of concentration of media ownership. The studies looked at 509 media organizations in Chile and 220 in Colombia (Mioli 2015). It was financed by the Open Society Foundations and received support from universities Alberto Hurtado in Chile and La Javeriana in Colombia. Among Poderopedia's findings was that in Chile, the media groups El Mercurio and Copesa dominated the newspaper and digital media sectors. In Colombia, the researchers found that the media were controlled by large business conglomerates. The studies validated the close connection among power elites and their influence over the information available to the general public. However, the databases had no information about revenues or business models of the websites, and spotty information on traffic and audience size.

One of the main sources on digital media entrepreneurs in the region has been the annual Colloquium of Latin American journalists, conducted by the Knight Center for Journalism in the Americas at the University of Texas. In 2013, the Colloquium topic was sustainability, and 23 digital news organizations participated. At the time, many reported that they were dependent on foundation grants and lacked the marketing and sales skills to build an audience and attract advertisers and sponsors (Quesada 2013). In addition, participants said the business models of digital media in the U.S. did not translate well to Latin America, where private foundations and venture capital investment play less of a role (Nelson 2013). Among conclusions drawn were that digital media needed to develop multiple revenue sources beyond advertising to survive.

At the 2017 edition of the Colloquium, some 20 digital natives and traditional media discussed their innovations, including some in revenue generation. Among them was Juanita León, CEO of La Silla Vacía, which focuses on politics and power in Colombia. She described how the publication offers brands the opportunity to sponsor sections of the site dealing with topics such as leadership, education, innovation, and gender. She has had no problem selling these sponsorships at USD $10,000 each (López Linares 2017a; Warner et al. 2017).

Among other digital entrepreneurs at the 2017 Colloquium was Silvia Ulloa, director of the digital news site CRHoy of Costa Rica. She said the publication is financed by a banker who favors independent journalism. CRHoy has ten journalists who Ulloa has trained to be multi-skilled, each a "one-man band" in her words. The site was founded in 2012, but has already achieved more traffic than all the other news sites in Costa Rica by focusing on investigative journalism, according to Ulloa.

In recent years, journalism in Latin American countries has substantially changed. New players have emerged, and small start-ups have gained popularity and attention in a short period of time, creating a very dynamic sector still to be systematically analysed. SembraMedia is attempting to do just that. As an organization whose mission is doing research and sharing best practices on new digital media, it has developed a directory of these media. As of May 2017 they had 600 organizations listed (López Linares 2017b). To deepen this data, Omidyar Network, a philanthropic investment organization, partnered with SembraMedia to conduct an in-depth study of 100 digital natives in Latin American—25 each from Argentina, Brazil, Colombia, and Mexico. This is the largest study ever done of the digital media market in Latin America (Warner et al. 2017). The research, under the supervision of researcher and consultant Mijal Iastrebner, was conducted by a team of seven who spent five months interviewing the founders and directors of the digital news startups. In selecting the 100 media for this study, SembraMedia asked the researchers to seek out the top digital players in each country, while also being careful to include a diverse mix of geographic coverage areas: 44% identified themselves as providing international coverage, 63% national, and 23% identified themselves as not national or international but local or provincial, as some media identified themselves in more than one category. Some of the main results of this research are further explained in the following pages.

2.2 *Independence and Credibility as Economic Assets*

A recurring theme in the mission statements and value propositions of the 100 organizations was that they positioned themselves as independent of the traditional oligarchy of shared media-political-business interests that control public discourse in the region. The following mission statement from the Facebook page of Aristegui Noticias, an investigative website in Mexico, was typical:

> In societies such as ours, where the control of mass media rests in a few hands, the old cultural inertias that restrict free expression, debate, and accountability continue to define the public life of our country. So we are convinced that journalists and independent professionals should be given the task of developing new projects, creating new spaces, and creating alternatives. Our purpose for this digital age is to exercise, from here, as in other spaces, a liberated journalism.

This commitment to independent journalism is consistent with the findings of Harlow and Salaverría (2016) in their study about the level of "alternativeness" of digital native media in Latin America. According to this research, where content of 69 online native news sites was analyzed, many of them "committed themselves to covering news regardless of pressures from owners or advertisers" (p. 1009).

In various ways, the mission statements of the organizations researched by Warner et al. (2017) expressed dissatisfaction with the traditional media's collusion with special interests, their failure to report on sensitive topics, and their neglect of impoverished rural areas. These digital natives used terms such as "independent", "different", "human rights", and "investigative journalism" to describe their content. Their manifestos described their products as more interactive, conversational, explanatory, accessible, and user friendly than the top-down product of their traditional rivals. Their positioning reflected a global, industry-wide reawakening to the importance of credibility as a media asset that can be monetized by focusing on readers rather than advertisers as the most important clients of a media organization. The trend was manifested recently in the increase of subscriptions to trusted media brands such as The Guardian, The New York Times, The Washington Post and others (Newman 2017; Davies 2017; and Doctor 2017).

Many journalists interviewed in the SembraMedia study were inspired to develop new media ventures because they were frustrated by the polarized political discourse in their countries. They aimed to create credible media voices independent of political organizations. In line with the rapid increase in fact-checking initiatives in Europe, the U.S., Southeast Asia, and elsewhere—113 at last count (Graves and Cherubini, par. 1)—11 of the publications in the SembraMedia study said they employ the technique. Chequeado of Argentina is the leader among them. It has shared its fact-checking methods with many other news organizations, including several in this study: Detector de Mentiras (Lie Detector) of La Silla Vacía in Colombia, Truco (Trick) of Agência Pública in Brazil, and El Sabueso Verificador (Bloodhound) of Animal Político in Mexico. The data showing the lack of credibility of news media in Latin America suggest there is a huge opportunity for independent media. Latinobarómetro has been doing surveys of press credibility in 18 Latin American countries since 2004. Consistently over the last dozen years, two-thirds of those surveyed agreed with the statement that the news media "are frequently influenced by institutions or powerful people" (2016, p. 41). In the same survey, people were asked if they agreed with the statement that "the news media are sufficiently independent", and the results were equally grim for journalists. Only a quarter of survey respondents thought that journalists were independent, and the four countries of this study were on the low end—24% in Argentina, 22% in Mexico, 21% in Brazil and 20% in Colombia.

2.3 *Revealing Corruption Has a Cost*

The work some of these journalists are doing poses real risks for their own lives. Almost half (45%) of the organizations surveyed said they have been subject to blackmail, threats, or violence because of their journalistic work. Mexico is one of the most dangerous places in the world for journalists because they can be murdered with relative impunity—nine in 2016 alone (Reporters without Borders 2017, p. 5). However, the most commonly mentioned method of reprisal against a media organization was economic, namely withdrawal of government advertising. Law enforcement was another. After Congresso em Foco of Brazil published a list of government officials' whose salaries exceeded the legal limit, they were subjected to an orchestrated attack of 50 lawsuits by those officials. The publication won 48 of the suits, and two others were pending. In Mexico and Argentina, a favorite government tactic has been to initiate a tax audit of a publication that seems to never end. Half of the organizations have suffered cyber attacks because of their news coverage, ranging from hacked email and social media accounts, to denial of service (DDoS) attacks, to digital smear campaigns.

The most important job of the media is to be a watchdog for the public interest and reveal corruption and self-dealing among powerful political and business interests. Many of the media in SembraMedia's study have done just that. At times, they have paid a high price for it in the form of economic reprisals (advertising boycotts by government and business), cyberattacks, smear campaigns, threats, violence, and murder. Still, these organizations have made a tremendous commitment to public service and are having an impact. Their work has been picked up and redistributed by national and international news organizations. They have been honored nationally and internationally for their work.

2.4 Strong in Journalism, Weak in Technology and Business

A solid, sustainable digital media operation needs at least three legs of a stool—journalism, technology, and business. However, the focus of the founders of these digital natives is tipped heavily toward journalism. When asked to define their specialty, in terms of experience or education, the founders replied: 53% journalism, communication, content production; 20% business, marketing, or administration; 12% humanities, literature, social sciences (political science, sociology, etc.); 11% web technology; and 4% audio visual production, design.

Not including the founders, these organizations had a median of 10 people involved in producing content (many of them unpaid collaborators), one working on technology and none in business activities such as sales, marketing, and administrative (To be clear, the sites with the most revenue had a median of two people in sales.). One-fifth of the publications had no employees working on technology and instead outsourced web development and other technical tasks to consultants.

Given the lack of business experience on the teams, it is no surprise that they have weak revenue structures. The most common revenue streams are banner ads on the website (31%), followed by creation of native ads or branded content (28%), consulting services (28%), training services (19%), grants (16%), crowdfunding and donations (15%), and Google AdSense (15%). Only 9% reported revenue from events and 5% from subscriptions.

Creating content for third parties, either as an editorial or advertising product, is proving to be an important way for these publications to support independent journalism. Native ads and branded content are especially important on mobile devices, as they evade ad blockers. Training in best practices—particularly ethical standards—could be a promising way to build native advertising as a revenue source. It may also be advisable to build a network of native ad content creators, to take advantage of the increasing demand for this advertising product.

Small-scale Businesses. These media are mostly very small businesses, and only a few have potential to scale. Their revenues are modest. When calculated in local currency, and converted to dollars at the year-end exchange rate, the 2016 revenues of the 90 organizations that shared financial information totaled USD $15.1 million—an average of USD $168,000. To put that USD $15.1 million in perspective, the combined revenues of all the media in our study were less than the 2016 revenues of ProPublica (USD $17.2 million), the Pulitzer-prize-winning digital news organization in the U.S., which has a staff of 50 journalists. Although comparing figures is imprecise because of inflation and exchange rates, the gross numbers for the Latin American media are inarguably small. This is not the kind of revenue stream that can create a strategically important media sector.

Few Audiences Are at Scale. The average number of Twitter followers was 174,000 for all the media, but that number is skewed high by some publications that have community managers dedicated to developing the audience and a strategy for these channels. The median number of Twitter followers was just 10,000, meaning half had more, half less. With Facebook, the average number of followers was 456,000, the median 16,800; YouTube average 16,700, median 350; Instagram average 18,200, median 1,800. All the sites surveyed except one use both Twitter and Facebook; 70% use YouTube, and 62% Instagram. LinkedIn, Snapchat, WhatsApp, and Telegram all are used by 12% or less.

A small number of the publishers had developed significant audiences: 18% had 1 million or more unique users monthly; however, 30% had less than 10,000 unique users. Many of the publishers evidently did not know how to measure their audiences: 12% did not answer the question about the number of unique users, and 20% did not answer the question about traffic from mobile devices. The last data point was particularly troubling given that the majority of publications in SembraMedia's survey were getting at least half their traffic from mobile. In addition, 38% of respondents said they did not have a database or mailing list of their users, another indication of a weak knowledge of how to connect with their users. Email newsletters are an important way for digital organizations to monetize their audience. Links in a newsletter carry users directly to the publication's own content and advertisers and thus avoid competing with Google and Facebook for digital advertising. In this way the publication can own its audience and advertisers, deepen the relationships, and groom potential donors and sponsors.

Survivors of Multiple Crises. An explosion of new technologies and tools, many of them free, has aided the development of new digital media: 51% of those in the study were founded since 2013. However, it is notable that despite volatility in the digital sector, a global recession, and various national economic crises, half of the 100 have survived more than four years. Older and more experienced does not exactly correlate with higher revenues or readership.

Nationally Focused Sites Do Best. The survey respondents could identify themselves as serving more than one geographical audience—hyperlocal, provincial, national, and international—and many did so. The locally focused media had the most precarious financial position and were the most vulnerable to attacks by the local political and business powers. If they were in poor regions providing information to underserved populations, there was almost no chance of getting business or government support if they revealed corruption and self-dealing. Most of the 10 highest revenue generators in the SembraMedia study, those who are bringing in $500,000 or more per year, "are blending entertainment and political coverage to drive millions of visitors (the median was 3.7 million sessions per month)" (p. 29). In the second tier of revenue, 14 media with revenues from $100,000 to $499,000, "the types of journalism being produced varied widely, but general news and political coverage were the most common, with a smattering of culture, science, environmental, and human rights coverage."

Women Take Charge in Digital Media. One of the most notable pieces of data to emerge from the SembraMedia research was the role of women in these startups. In traditional media in Latin America, women make up a tiny percentage of the management teams (Vega Montiel 2014). In the survey, more than 60% of the media organizations had women among their founders. Women also accounted for 35% of the directors named in the survey: 23 were executive directors or CEOs, 13 were editorial directors, 14 were directors of sales/marketing/commercial, and 13 were in administration, human resources, and finance. There was only one woman director of technology.

These numbers are more remarkable when compared with Vega Montiel's study of Mexican media, possibly the most detailed study of women leaders in Latin American media. Among her findings:

- Owners: Women made up only 1% of the owners of major television properties in Mexico, 13% of radio, and 0% in major newspaper groups. Many of the women owners in radio inherited their shares.
- Directors: There was not one woman among the 52 board members of Grupo Televisa and Televisión Azteca, the two dominant television groups, nor among two other prominent TV groups. In radio, women have 8% of the board seats, and in newspapers, 11%.
- Executives: In television, no women had any senior executive positions—not in administration, finance, or editorial. In radio, they had 11% of these strategic positions and in print media 13% (2014, pp. 198–205).

Vega Montiel observed that women's lack of access to positions of authority in the media in Mexico "leaves them marginalized in one of the most important sectors of global capitalism: the cultural industries" (p. 205). This has other implications as well for Mexican society, she observed: Gender inequality is perpetuated, the rights of women are not a subject of public discourse, and sexual stereotypes are reinforced in news and entertainment (p. 206).

2.5 *Collaboration as a Solution*

Traditional media and digital natives have begun to join forces on transnational projects as a way of multiplying the impact of their scarce resources of money and time. The most obvious of these was the aforementioned Panama Papers project, which revealed how businesses and individuals were using offshore corporations and bank accounts to avoid taxes and scrutiny, sometimes illegally. However, there is a larger trend in Latin America of cooperation on cross-border investigations. López Linares (2017c) provides an excellent overview of the many projects under way across the region, such as this one:

> "Memoria Robada" (Stolen Memory) is another emblematic project of collaborative transnational journalism based on huge databases which required the use of special tools that facilitated the sharing and visualization of information. This project –which presented a database on the illicit trafficking of cultural pieces in Latin America – is an investigation of Peruvian site Ojo Público in collaboration with La Nación of Costa Rica, Chequeado of Argentina, Plaza Pública of Guatemala, and Animal Político in Mexico.

Connectas of Colombia, one of the organizations in the SembraMedia study, has positioned itself as a coordinator of journalists, NGOs, publishers, and other organizations from across the region to produce in-depth reports on topics such as the plunder of the Amazon region's natural resources in the name of economic development (Torres 2017). Univision, the Spanish language broadcast and digital outlet in the U.S., has begun to partner with media outlets throughout the region—including many digital startups such as El Faro in El Salvador and Animal Político in Mexico—to expand its audience and impact (Mioli 2016). This also increases the reach, impact, and influence of the digital media.

3 Future Developments and Challenges: The Road Ahead

During the last twenty years media companies worldwide, but especially in Latin American countries, have lived through massive economic and technological disruptions. The effects of digitalization are still going on but are not the only ones being noticed: the economic development of most of these countries in recent years and the profound societal changes have substantially modified the general and the particular context where media, and journalists, work. More and better educated citizens are getting more active roles as audiences, and the public demands to control the traditional political-business-media oligarchy are also growing. Digital technology offers new and creative ways to deal with these demands: as some of the examples depicted in this chapter show, a lot of the initiatives started using the internet as the only possible way to use a real journalistic approach to serve the information needs of citizens.

In this context, five elements could be decisive to understand the future of media in Latin American countries as well as to define the main challenges media companies will face in the coming years: (1) women, (2) sustainability, (3) training, (4) market orientation and (5) collaboration.

Digital media have opened up a reservoir of talent in Latin America by bringing previously unheard voices to public attention. And women are clearly eager to make their mark in a sector where many of the traditional barriers to their advancement are lower, as the figures on women founders and directors indicate. From the point of view of creating democratic societies that are more fair and just, these new media seem to be in the vanguard. That should make them more attractive as an investment vehicle for organizations that support gender equality and human rights, such as social impact investors like North Base Media and the Media Development Investment Fund, which have made some investments in Latin America (Breiner 2017b).

At the same time, there is still a long way to go in order to fulfill this goal. While women have a more central role in starting and leading some of the new journalistic initiatives, their presence still has to be consolidated in the top leading positions in the traditional media companies. According to the "Global Report on the Status of Women in the News Media" published in 2011 by the IWMF, women in the region were 43.7% of those in the senior professional level, and their representation was even higher, 45.8%, in the junior level. They were also doing well in the two management occupational levels: 46.4% in senior management and 40.5% in middle management. But the report also found women's representation was particularly low at the top of companies, where they were only 21.5% of those in governance, and less than a third (30.5%) of those in top-level management. Women's marginalization was also clear in the creative and technical aspects of news production, where women held less than 25% of the positions (Byerly 2011). Some organisations, such as IWMF, are focusing their efforts on training women in journalism.

Sustainability is one of main worries for both traditional and digital media. In recent years, many media organizations have chosen to deal with the economic crisis by reducing costs, which has had an impact on production (fewer pages printed, reduced services, broadcasters and publications closed) and the workforce (salaries reduced, early retirements and layoffs). Although this restructuring may have been necessary in certain cases, many organizations have made these cuts with the aim of holding out until the economic situation improves. Other organizations, in contrast, have understood that the recession can also be an opportunity to redesign the future and develop more sustainable business models. But at the same time, the publications clearly are not capitalizing on all of the assets that they have, given the market opportunities created by the shrinking traditional media and the lowered barriers to entry made possible by digital tools for publishing and distribution. Either because of neglect or lack of know-how, they are not taking advantage of digital measurement tools to identify opportunities to increase the size or loyalty of their audiences. By missing those opportunities, they are of course missing revenue opportunities.

But digital media are suffering from a different set of problems regarding sustainability: when evaluated on a purely economic basis, very few of these digital startups have the scale or the profit possibilities that would interest traditional investors looking for a monetary return. Their greatest value can be calculated not in the value they create for advertisers but for readers, users, and members of the general public.

The key to assessing the value of these organizations is to take into account not only their economic value as centers of profit and loss, assets and liabilities. Their value lies in the information service they provide their communities, in the horizontal networks of communication they provide to strengthen those communities, and the democratic values of participation that they encourage. When that value is taken into account—the social value, or the social capital, to use the language of sociologists—then these organizations can be attractive to the social-impact investors discussed above. Helping these organizations achieve economic stability to support their public service will be the key for transforming the media ecosystem of the future and re-establishing the Fourth Estate as a pillar of modern democracies.

Working towards a sustainability model would first require rethinking what is sustainability. According to Nieto (1973), media companies' profitability should take into account the very particular nature of their core business: to attend a public need, which implies an effort of constant market adaptation that makes it difficult to obtain high profits. Business models are usually badly understood, focusing only on how to generate more revenue and not on a broader picture, including the definition of the whole value chain.

Both journalists and media managers have to be trained differently now. The biggest missing element in the digital natives was in the skills related to business, marketing, sales, and administration. With the right support and training, these organizations could move relatively quickly from surviving to thriving. Key to this training would be helping digital media develop multiple revenue streams, the lack of which was illustrated earlier. The second biggest need was training in better use of technology, particularly to generate multimedia, expand into new platforms, and measure and track audience response and growth. Many of the startups had simply migrated their old journalism products, production methods, and distribution strategies to the digital format, where they did not function as well. At the moment, these abilities, skills and capabilities are being gained through a varied set of courses offered by institutions and professional centers, including the media organizations: one-fifth of the organizations in the SembraMedia study generate revenue from training. Providing grants to support their training efforts could have a double impact, on the revenue of the media organization and on improving journalism standards in the region. Universities and Schools of Communication, with a long tradition in the Latin American region (Ferreira and Tillson 2000), are now starting to adapt their curricula to these new needs, but there is still a long way ahead for most of the countries (Sádaba 2016).

Market opportunities and changing consumer behavior also have a growing influence over the innovation strategies of companies in the media industry. In terms of digital products and services, the internet constitutes a global marketplace in which the range of content available to users is multiplied. This also increases the level of competition faced by domestic press organizations, as they are not only competing with rivals within their own countries, but also with media outlets in other countries and with content that is offered in other languages and even in other formats. Thus, competition also helps set the pace of innovation.

In addition, in terms of consumer/audience behavior, the increase in mobile phone usage has radically changed the way people access content, including news content. Market orientation is a must in this new scenario, and the ability to read and understand trends and behaviors should also be addressed in the training of future journalists and media managers.

Finally, collaboration with international and national media organizations appears to be a promising opportunity for startups. Helping small companies build content-syndication relationships with larger media outlets could represent a lucrative new revenue stream and help them build their audiences. If the digital natives collaborate more administratively, they might also be able to reduce costs via shared services for accounting, legal, and other professional services. But this collaboration also makes strong sense at an editorial point of view: cases such as the Panama Papers (ICIJ 2017) show that different media outlets, big and small ones, could work together to exploit the big data the internet makes possible, to offer both a global and a local picture of the news.

4 Disclosure

As indicated in the bylines, one of the authors of this chapter, Janine C. Warner, is co-founder of SembraMedia, a nonprofit organization dedicated to increasing the diversity of voices and quality of digital media in Spanish in the Americas. Another one, James G. Breiner, visiting professor at University of Navarra, is a board member.

References

Arrese, Á., Kaufmann, J.: Legacy and native news brands online: do they show different news consumption patterns? Int. J. Media Manag. **18**(2), 75–97 (2016)

Breiner, J.: Mapping the world's digital media ecosystem: The quest for sustainability. In: Berglez, P., Olausson, U., Ots, M. (eds.) What is sustainable journalism? Integrating the environmental, social, and economic challenges of journalism. Peter Lang, New York (2017)

Breiner, J.: Investor sees 'great returns' from new digital media. News Entrepreneurs (2017b)

Byerly, C.M.: Global report on the status of women in the news media. International Women's Media Foundation, Washington, DC (2011)

Davies, J.: Road to 1 million: guardian has gone from 15,000 to 200,000 'paying members' in the past year. Digiday (2017)

Doctor, K.: Newsonomics: Trump may be the news industry's greatest opportunity to build a sustainable business model. Nieman Lab (2017)

Dragomir, M., Thompson, M. (eds.): Mapping digital media. Global Findings. Open Society Foundations, New York (2014)

Ferreira, L., Tillson, D.J.: Sixty-five years of journalism education in Latin America. The Florida Commun. J. **27**, 61–79 (2000)

FNPI: Estos son los ganadores del Premio Gabo (2016). https://premioggm.org/2016/09/estos-son-los-ganadores-del-premio-gabo-2016/

Gang, G.: Facing a coming era of the fourth media. Journalism & Commun 3 (1998)

García-Avilés, J.A., Carvajal-Prieto, M., De Lara-González, A., Arias-Robles, F.: Developing an index of media innovation in a national market: the case of Spain. Journalism Stud. 1–18 (2016)

Graves, L., Cherubini, F.: The rise of fact-checking sites in Europe. Reuters Institute for the Study of Journalism (2016)

Harlow, S.: Quality, innovation, and financial sustainability. Central American entrepreneurial journalism through the lens of its audience. Journalism Pract. 1–22 (2017)

Harlow, S., Salaverría, R.: Regenerating journalism: exploring the 'alternativeness' and 'digitalness' of online-native media in Latin America. Digit. Journalism **4**(8), 1001–1019 (2016)

Hass, B.H.: Intrapreneurship and corporate venturing in the media business: a theoretical framework and examples from the German Publishing Industry. J. Media Bus. Stud. **8**(1), 47–68 (2011)

He, Z., Zhu, J.: The ecology of online newspapers: the case of China. Media Cult. Soc. **24**(1), 121–137 (2002)

ICIJ.: Panama Papers wins Pulitzer Prize. https://www.icij.org/blog/2017/04/panama-papers-wins-pulitzer-prize/ (2017)

ITU.: ICT Facts and Figures 2016. http://www.itu.int/en/ITU-D/Statistics/Pages/stat/default.aspx (2017)

Kilgo, D., Harlow, S., García Perdomo, V.M., Salaverría, R.: A new sensation? An international exploration of sensationalism and social media recommendations in online news publications. Journalism (2016). http://journals.sagepub.com/doi/abs/10.1177/1464884916683549

Köroğlu, O., Tingöy, O.: Online journalism in Southern East Europe. AJIT-e: online academic. J. Inf. Technol. **2**(3), 1–15 (2011)

Küng, L.: Innovation, technology and organisational change: legacy media's big challenges. An introduction. In: Storsul, T., Krumsvik, T. (eds.) Media innovations, a multidisciplinary study of change, pp. 10–12. Nordicom, Göteborgs Universitet, Sweden (2013)

Latinobarómetro: Informe Latinobarómetro. http://www.latinobarometro.org (2016)

López Linares, C.: Journalists from Ibero-America share innovation projects at the Knight Center's Colloquium on Digital Journalism. Knight Center for Journalism in the Americas. https://www.isoj.org/journalists-from-ibero-america-share-innovation-projects-at-the-knight-centers-colloquium-on-digital-journalism/ (2017a)

López Linares, C.: SembraMedia reveals digital media growth in Latin America, but says organizations still face challenges. Knight Center for Journalism in the Americas. https://knightcenter.utexas.edu/blog/00-18315-sembramedia-reveals-digital-media-growth-latin-america-still-faces-challenges-must-be (2017b)

López Linares, C.: Journalistic investigations without borders: Latin American journalists innovate with transnational projects. Knight Center for Journalism in the Americas. https://knightcenter.utexas.edu/blog/00-18166-journalistic-investigations-without-borders-latin-american-journalists-innovate-transn (2017c)

Mabweazara, H.M., Mudhai, O.F., Whittaker, J. (eds.): Online journalism in Africa: trends, practices and emerging cultures. Routledge, London (2014)

Mioli, T.: "Media map" project reveals ownership concentration in Chile and Colombia. Knight Center for Journalism in the Americas. https://knightcenter.utexas.edu/en/blog/00-16302-media-map-project-reveals-ownership-concentration-chile-and-colombia (2015)

Mioli, T.: Univision and Latin American journalists collaborate to fact-check final U.S. presidential debate. Knight Center for Journalism in the Americas. https://knightcenter.utexas.edu/blog/00-17580-univision-and-latin-american-journalists-collaborate-fact-check-final-us-presidential (2016)

Mioli, T., Nafría, I. (eds.): Innovative journalism in Latin America. Knight Center for Journalism in the Americas, Austin, TX (2017)

Nelson, A.: 6th Ibero-American colloquium on digital journalism: region's budding news sites discuss sustainability. Knight Center for Journalism in the Americas. https://knightcenter.utexas.edu/00-13959-6th-ibero-american-colloquium-digital-journalism-region's-budding-news-sites-discuss-sustai (2013)

Newman, N.: Fake news, algorithms and guarding against the filter bubble. Journalism, media and technology predictions 2017. Reuters Institute for the Study of

Journalism. http://www.digitalnewsreport.org/publications/2017/journalism-media-technology-predictions-2017/#fake-news-algorithms-and-guarding-against-the-filter-bubble (2017)
Nicholls, T., Shabbir, N., Nielsen, R.K.: Digital-born news media in Europe. Reuters Institute for the Study of Journalism, Oxford (2016)
Nieto, A.: La empresa periodística en España. Eunsa, Pamplona (1973)
Premio Nacional de Periodismo. http://www.periodismo.org.mx/g2014.html (2015)
Quesada, J.D.: El boom de la prensa digital latinoamericana. El País (2013)
Reporters without Borders: Roundup 2016 of journalists killed worldwide (2017)
Rey, G., Novoa, J.L.: Medios digitales en Colombia 2012. Consejo de Redacción y Centro Ático de La Universidad Javeriana, Bogota (2013)
Sádaba, C.: Epílogo. Innovación en el sector de los medios. In: Salaverría, R. (ed.) Ciberperiodismo en Iberoamérica, pp. 423–433. Fundación Telefónica & Editorial Ariel, Madrid (2016)
Salaverría, R. (ed.): Ciberperiodismo en Iberoamérica. Fundación Telefónica & Editorial Ariel, Madrid (2016)
Salaverría, R.: Typology of digital news media: theoretical bases for their classification. Mediterr. J. Commun. **8**(1), 19–32 (2017)
Sin Embargo. http://www.sinembargo.mx/30-09-2015/1504434 (2015)
Solórzano, R.: Nicaragua. In: Salaverría, R. (ed.) Ciberperiodismo en Iberoamérica, pp. 255–272. Fundación Telefónica & Editorial Ariel, Madrid (2016)
Suárez Villegas, J.C.: Nuevas tecnologías y deontología periodística: comparación entre medios tradicionales y nativos digitales. El Profesional de la Información **24**(4), 390–395 (2015)
Tamacas, C.M.: El Salvador. In: Salaverría, R. (ed.) Ciberperiodismo en Iberoamérica, pp. 145–167. Fundación Telefónica & Editorial Ariel, Madrid (2016)
Torres, F.: La carretera que corta el corazón de la Amazonia. Connectas (2017)
UN: World Population Prospects, the 2015 Revision. https://esa.un.org/unpd/wpp/DataSources/ (2016)
Vega Montiel, A.: Igualdad de género, poder y comunicación: las mujeres en la propiedad, dirección y puestos de toma de decisión. Revista de Estudios de Género. La Ventana, **5**(40), 186–212 (2014)
Warner, J., Iastrebner, M., LaFontaine, D., Breiner, J., Peña Johannson, A.: Inflection point: impact, threats, and sustainability: a study of Latin American Digital Media Entrepreneurs. http://data.sembramedia.org/ (2017)

Ramón Salaverría Ph.D. and Associate Professor of Journalism at the School of Communication, Universidad de Navarra (Pamplona, Spain), where he serves as Associate Dean of Research. He was Chair of the Journalism Studies Section at ECREA (2010–2012). He is the author and editor of several books and research articles on digital journalism, mainly about Europe and Latin America. Pamplona, Spain.

Charo Sádaba Ph.D. and Associate Professor of Digital Marketing at the School of Communication, Universidad de Navarra (Pamplona, Spain), where she is also the Dean of the School. Since 2001 she has been involved in several international and national research projects about the impact of the Internet on the media sector and particularly on media business models, innovative management practices, and users' role. Pamplona, Spain.

James G. Breiner Digital media consultant specialized in entrepreneurship and innovation. He previously worked as newspaper editor and publisher with business publications in the U.S. Recently he has done consulting and training in Europe, Asia, and Latin America on behalf of the Poynter Institute, International Center for Journalists, and Garcia Marquez Foundation. He is visiting professor of Communication at the Universidad de Navarra (Pamplona, Spain). Pamplona, Spain.

Janine C. Warner Co-founder of SembraMedia and the International Center for Journalists Knight Fellow. She is an online instructor, the author of more than a dozen books, and numerous articles and publications about digital journalism and the Internet. She has worked with thousands of digital media entrepreneurs in the U.S., Latin America, and Europe and travels extensively to present at conferences and universities. Los Angeles, CA, United States of America.

The Spanish Digital Media Industry's Transition

Xosé Rúas-Araújo, Iván Puentes-Rivera and Ana Culqui Medina

Abstract This chapter shows the results of a survey that was carried out among the managers of a 106 SMEs and other large well-known Media companies based in Spain, whose turnover exceeded the six million € a year quote, taking into account the Data and the Trends of the economical development that the Spanish Media Industry is ongoing. The survey was conducted on a random sample, using Simple Random Probability, with a Maximum Margin of Error on the Total Sample of 9.0%, and a 95% Confidence Interval. The two major challenges we face underlie on the managers' answers to questions related to the digital transition and on knowing how to adapt to the changes brought by Big Data. The results throw a very pessimistic prospect for the Printed Press and Traditional Advertising Payments, with a greater optimism when it comes to Private TV and Radio, where all the hopes are set on the Digital and Mobile Media as well as on Corporate Communications.

Keywords Media · Advertising · Business · Press · Radio · Television
Digital methods

1 Introduction: A New Horizon for Media Management

In order to understand, the results of the survey about the Spanish Media Industry, carried out by XESCOM—A network that was brought about by the New Media (the University of Santiago) in 2014, NECOM Research Groups (from the University of Vigo) and iMARKA (the University of Coruña)—First of all, we have to discuss, the data and the trends of the World Association of Newspapers

X. Rúas-Araújo · I. Puentes-Rivera (✉)
Universidade de Vigo, Vigo, Spain
e-mail: ivanpuentes@uvigo.es

X. Rúas-Araújo
e-mail: joseruas@uvigo.es

A. C. Medina
Pontificia Universidad Católica del Ecuador Sede Ibarra, Ibarra, Ecuador
e-mail: amculqui1@pucesi.edu.ec

M. Túñez-López et al. (eds.), *Communication: Innovation & Quality*, Studies in Systems, Decision and Control 154, https://doi.org/10.1007/978-3-319-91860-0_15

(WAN-IFRA 2015, 2016), World Newsmedia (WNMN 2015), and the Spanish White Press Book (AEDE) (2016), the Spanish Audiovisual Report of the Spanish National Commission for Competition Markets (SNCCM 2016), and also a study on Advertising in Spain (carried out by Infoadex 2016), (SABI 2016), we also took into account an Analysis of the Balance of the Iberian Systems, from the European Audiovisual Observatory (2016), the World Yearbook TV Trends (Eurodata 2016), the European Broadcasting Union (EBU), the International Communications Market Report, 2015 (OFCOM 2016), the Entertainment and Media Outlook (EMO 2016), Analytics Trends, by Deloitte (2016), and the European Communication Monitor (2016), as well as studies of the Public Relations Consulting Companies and the Spanish Telecommunications Association (PRCCSBTDA) and also of the Board of Telecommunication Directors Association (DIRCOM).

In the survey conducted by the World Newsmedia (WNMN 2015), and the World Association of Daily Newspapers (WAN-IFRA), the results threw that among the 170 interviewed leaders coming from 50 different countries; most of them prioritized the use of big data, the promotion of social networks, and the integration between various departments, as well as training in new skills, such as specific aspects of star management for long periods of time. The automation of Big Data, Applied Intelligence, Management, Commitment Algorithms (Engagement), which are used amongst most users of media and social networks and the trends that were revealed by European Communication Monitor (ECM 2016), who carried out another global survey interviewing of 2.710 professionals from 43 different countries also.

Traditional Media are facing disruptive innovations (Christensen et al. 2015) caused by new operators who have emerged from an ongoing technological revolution in communications, the 2.0, 3.0, and 4.0 Webs, the Social Networks, and the so-called Internet of Objects and Artificial Intelligence, respectively. Two main industrial trends (Evans 2015) emerge today in the National and Global Media Industry. On the one hand, a mega-cluster of global corporations—the Big Glass (Google, Apple, Facebook and Amazon) present nowadays in the majority of the links in the communication's value ecosystem (Miguel de Bustos 2017)—and on the other, a hyper-fragmentation of small or medium-sized operators, who depend on those because they control the software or the content distribution platforms.

The audiovisual sector is also facing a continuous and ongoing transition, going from analogue to digital into linear TV (reducing their consumption on twelve EU countries) but also experiencing growing consumer demand (streaming), in the middle of a great fragmentation of audiences, the crisis of the traditional system for assessing and measuring value, increased Internet Advertising (in seven European Countries it already exceeded TV Investments), Growth of Programmed Advertising based on data, loss of competitive power of public broadcasting, concentration of traditional private audiovisual groups and strong entry of new Telecommunications Operators and digital News-Diaries (Netflix, Amazon, YouTube, etc.).

Factors such as the recent economic crisis,[1] and the issues that advertising has experienced led to a rebound in the Public Relations Sector and in all sorts of integrated communications, in addition to the need to incorporate General Managers (Costa 2001) in Communication Companies & Organizations, with wider functions, both from the "Organizational" point of view and form an "Strategic Executive Planning" point of view. All this, on the basis of a Public Relations definition, when understood as the direction and management of all communications between an organization and its public, seen here as an open system to which it corresponds the management of all formal informative processes, that occur inside and outside the institution (Grunig and Hunt 1984).

On the other hand, the lack of resources to invest on advertising investment is an incentive for the practice of public relations, as an investment in the medium and long term, in search of credibility, given that it has, as they say at the and Ries (2003), that the products do not create free publicity, but people do.

From there the power of third-party, that people speak for the brands and their companies, tell their stories (Salmon 2007) and generate conversation value, taking into account that the marketing is not just products, but also perceptions, which responds to the definition of company's image as a stereotype of the Organization mental representation publics are formed as a result of the information that perform about the same (Caprioti 1992: 30).

In the perception of the image of a company come into play components cognitive, emotional and behavioral, defined through dichotomous traits as reliable/non-reliable, modern/outdated; warm/cold; efficient/inefficient; powerful/weak; big/small to help measure the credibility, confidence and attitude of the public towards the organization.

In addition to the financial capital, that includes all monetary and physical assets of the companies and their value in the market, there is intellectual capital, comprising assets invisible, structural, and human, the thinking soul of the company that transcends the ownership of it (Roos et al. 2001) and where talent is essential.

Keys the good reputation and the intangible business value based on attributes such as visibility, differentiation, authenticity, transparency, which in addition to others such as the commercial offer, the innovation, quality and responsibility, as well as being emotionally attractive (both for employees as for those who aspire to be), as well as dialogue and related to their public internal and external (Villafañe 2004). A positive social value, which implies community recognition, but we also need to express a good goal-driven behavior and goodwill (goodwill) (Sotelo 2001: 99).

The last of the reports published by the Spanish Public Relations & Communications Consulting Companies Association (SPRCCCA 2008) on the General View of the sector, precisely pointed out the interest that 85% of the Companies Commu-

[1]The advertising crisis of 1993, predicted the end of traditional advertising mentality and the loss of the role of agencies in the communication strategy of the advertisers and the incorporation of new communication services and the assumption of the corporate, understood as the strategic management of all factors that influence the image of an organization (Villafañe 2004: 22).

nication Departments of consulting firms have to improve their corporate reputation and, at the same time, their credibility and corporate image were at stake, and these are worrying factors in which concern has grown in recent years.

However, at the same time, the forecasts of growth in the sector still remained optimistic, specifically, 69% of those polled pointed to that you increase, compared with 21% indicated that it would be equal and with respect to future trends, the professionals of the sector were more ways out in specialization by sectors, although he also began to sign up a tendency towards offering full services (26%) and integration into large groups of communication (22%).

More recently, the Association of Directors of communication of Spain, entity that analyzes corporate communication and its professionals, points in its last report (DIRCOM 2015), four in five companies (81%) exerted a "communication with a conscience", by listening proactively to the different stakeholders and communicating in a manner transparent, responsible and truthful.

In addition, 85% of the respondents belonging to this institution says that the importance of communication in companies has increased in recent years, although, as a result of the crisis, 65% indicated that the budgets devoted to communication has been reduced and 56% said that, similarly, declined the number of people who work in communication in companies.

With regard to the training aspect, DIRCOM of 2015 report also shows that the number of graduates in journalism in communication management positions has increased in the last five years, going from 35% in 2010 to 49% in 2015. It also stresses that training in digital skills is in the first position in terms of areas of interest, that half of the respondents (51%) spent between 1 and 50 h of additional training last year and 29% between 50 and 100 h. For his part, European communication Monitor 2016 highlights the need for more and better training in the management of media, Big Data, and social networks (Zerfass et al. 2016).

Precisely, the training needs of new communicators, as consequence of the Bologna process, is among the issues for years (Vázquez and Fernández 2012), intending to observe the possible divergences between the academic world and the professional world and giving account both need to understand that the consumer is not a mere receiver of content, but that it is also a transmitter (prosumer) and, as such, calls for participation and consideration.

Responses given by communication professionals, who demand more training on creativity related issues, the search for ideas and the production of innovative and original content, as well as learn to adapt and respond to the changes in attitudes and ways of understanding the life of the new audiences and its relationship to the digital environment.

1.1 The Media Industry's Evolution

The Media world Industry, which by 2015, billed about 1.72 billion dollars (1.61 billion), went through five years of moderate growth of which 4.4%, was in nominal terms, up until 2020, the year in which the already mentioned turnover reached 2.14 billion in U.S. currency (well over 2 trillion €), representing a slowdown due to developments in the last five years, according to the forecasts done by the EMO'S (2016), Entertainment and Media Outlook 2016–2020. Also, the average growth of the Media Industry in Spain over the next five years was a 2.7%, significantly lower than the Global Evolution.

These Communication Activities that are deemed to exceed the Global Average Growth on a 4% in Spain in the coming years (2016–2020) are Pay-TV (4.9%), TV Advertising (4.8%), Billboards (4.17%) and Internet (8.5%), the latter doubling the percentage rate of other businesses. The daily press, magazines and printed paperbacks were found at negative margins on the horizon of the last five years, on the second decade of the 21st century (Table 1).

The five major trends that stand out in the study on the EMO (2016) Outlook, that was conducted last year, that normally comes to light every five years, and looks over a momentum of media spending activities which are aimed at adults of over 35, and also at the impact of Local Contents, in Global Markets, potential offerings of packages of different services (Multiple Telecommunication Operators Packages (i.e. TV, Mobile Phone Companies, Internet, etc.), and the fact that a lot of hybrid and integrated companies have popped up, increasing in this manner their ability to respond to all the different market's needs, faster. And as a natural consequence of this, the appearance of the 2.0, the 3.0, and the 4.0 Webs, which have also increased a general automation of things along with the rise of Artificial Intelligence, a new technological revolution is coming about, which will lead us to the loss of jobs within traditional roles in favour of new but also slightly more autonomous external hires.

This "mega-cluster of companies" that are developing in the global media industry highlights the fact that one of our main concerns is the relationship between media and the distribution of social media. This was a major concern and it prompted a survey, which was carried out by Nic Newman in the Digital News Project (2017), for the Reuters Institute on trends and predictions. As these concerns grew, the reactions forced Google and Facebook to seek agreements with major traditional content providers, so that they could both share the advantages of these more evenly, also through branding and by setting up quality control procedures on journalism, due to the lack of credibility of networks and social media, which is right in the eye of the debate on the controversy of false news in the age of beyond-the-truth.

Seventy per cent of managers surveyed by the Reuters Institute were concerned about the distribution of false or inaccurate news, 46% are concerned about the role of social media, 33% of them are more sensitized about the financial sustainability of their companies in relation to previous years and most of them want to take better advantage of the possibilities that social messaging brings about (56% rely a lot on Facebook, 53% on WhatsApp and 49% use Snapchat). Their priorities are; enhanc-

Table 1 Economical outlook of media in Spain from 2011–2020

	2011	2012	2013	2014	2015	2016	2017	2018	2019	2020	2015–2020 (%)
Pay TV	1808	1810	1798	1804	1950	2092	2217	2319	2405	2472	4.9
Public TV	2251	1833	1727	1919	2047	2196	2339	2459	2546	2593	4.8
Press	2207	1875	1693	1645	1613	1570	1536	1509	1478	1447	−2.1
Magazines	938	875	837	844	848	850	850	848	846	837	−0.3
Radio	525	454	404	420	450	467	488	506	523	539	3.7
Internet access	5979	6426	6871	7557	7706	8076	8356	8586	8820	9057	3.3
Video games	866	820	814	857	877	906	936	966	996	1026	3.2
Cinema	662	637	524	548	607	590	609	629	649	671	2
Publishing	3056	2791	2503	2364	2260	2185	2191	2200	2204	2228	−0.3
Music	493	483	441	467	474	482	489	500	510	518	1.8
Public. internet	875	894	967	1168	1359	1555	1706	1832	1942	2041	8.5
Billboard publishing	422	351	310	323	338	347	76	380	398	415	4.17
B2B	1906	1904	1865	1877	1894	1915	1949	1983	2018	2058	1.6
DVT grants	2335	2141	1978	1785	Public TV Grants from the Telecommunications and Audiovisual CNMC's Report 2015						

ing digital video facilities (89%), focusing on the profitability of Pay-Walls (45%), creating new revenues, and streaming beyond digital advertising, enhancing memberships (14%), experimenting with virtual reality (38%), forcing previous access records, taking advantage of data using algorithms, digital management robots and applications to interact with as the user's personal assistants.

Amid this wave of changes and emerging trends, the opinion of Deloitte's 2016 puts emphasis on the integration of strategies, people, processes and data. His chapter on trends in the media focuses on the situation of short-term advertising blockers of mobile devices, niche which has a market of 1 billion $, estimated on virtual reality, the exploitation of European Football by 30,000 million $ (nearly 29,000 million €), emerging Videogame Sports Competitions (eSports), Mobile Phone Apps, Digital Payment methods and the downfall of traditional TV Marketing with the income of new competitors. Audiovisual On-Demand (SVOD), and with Netflix already present in more than 130 countries, which revenues have increased by 77% in the last two years.

Netflix's highly successful business model is based, actually, on a wide range of global audiovisual entertainment catalogue, along with a very flexible pricing system and an optimized management of the users' digital data. Another argument over why it has been so successful is that a 72.3% of its Communication Managers consulted by the European Communication Monitor (Zerfass et al. 2016) believed that Big Data will change their careers and 75% of them considered that they have to adapt their activities to external search engines and algorithms, although only 29.2% admits to using them. From these 55.3% use data for planning, a 45.9% for measurement, and a 36.5% to generate content. Spain, Italy, France, United Kingdom, Ireland, Finland and Poland are the European countries that stand out the most in pioneering with the use of large data, adapting it to algorithms, and basically paving the way for the automation of many communicational activities.

The mixed payment model of the Communication Industry is being re-thought of, but the essence of it is not, as you can see by the manager's response, advertising revenues are not enough to ensure sustainability of both; businesses and their activities. Much of the Global Media Advertising Industry pie's share has been already appropriated in recent years by Social Networks and Digital Communication Channels, at the expense of traditional media. The damage that the loss of advertising caused hence, the revenues that faced traditional media, were one of the main concerns and this was the basis of another piece of research and a survey carried out by Network World Newsmedia (WNMN 2015).

There is also the World Newsmedia Research, linked to the World Association of Newspapers (WAN-IFRA) that states that only 5.8% of the surveyed Official Media had generated profits above a 20% in the last financial year, percentage actually considered very ambitious, but which nonetheless, was a reality in this industry a decade ago. In relation to Stagnation Indicators or Low Growth in the sector, the tendency is that (83% of the polled people) think that it is was amenable to make further adjustments and reductions in costs. And in that sense, four priorities for renewal or restructuring were came to show the relevance of these areas, the printing

Table 2 Global evolution of the printed press

Years	Broadcasted printed press				Daily headers			
	World	Europe	Spain	Portugal	World	Europe	Spain	Portugal
2010	551.309	89.227	3.777	533	12.696	2.292	134	18
2011	519.302	85.011	3.520	480	12.926	2.296	116	23
2012	615.257	69.228	3.008	413	7.515	2.146	113	18
2013	644.757	61.180	2.550	300	4.582	1.526	110	17
2014	685.158	54.261	2.350	223	4.430	1.337	110	17

processes, as well as the measures to be taken, either if that means focusing more on industrial processes or else if it means suppressing paper-back issues completely.

Press is an ever extending, ever developing mixed-media Industry, which is nowadays facing the decline of print but also going through a digital revolution, in the midst of a strong conversion, loss of diffusion and impact, plant closings, mergers and headers and the dismissal of workers. The data that the World Association of Newspapers (WAN-IFRA 2015) holds shows a recovery in 2014 of the Broadcasting Industry globally—driven by Asian countries of most recent literacy—but an important decrease in North America, Europe, Spain and Portugal, even to a much smaller degree on the headers of the greatest employer known as the World Press. And that is due to a decline of closed traditional media and an emergence of new digital which we have already listed, and that are not part of the WAN-IFRA, Institution that was set up by a merger of the Association of Publishers, also formerly known as Graphic & Printing Industry's Organization (Table 2).

The Worldwide Press revenue amounted to 168,000 million $ (157.8 million €) in 2015, 90,000 million in the U.S. currency (53%) corresponded to the sale of printed issues or the sale of digital subscriptions compared to 78,000, which came from advertising using the same currency. Around 92% of revenues came from the Written Press formats although the sale of Digital Payments and Digital Circulation (Paywall) rose 30% in 2015, a staggering 547% in the past five years. Digital Press Payments revenues amounted to 3000 billion dollars, starting to compensate already for some of the newspapers falls and the overall decline experienced by the demand of printed issues. On the whole, the revenues of newspapers went down a 1.2% in 2015 and a 4.3% in the last five years.

The Press Business is still based on printing, despite the loss of broadcasting, because the marketing of advertising was largely solely based on issue dissemination, which was what made the figures and it was deeply rooted in a traditional methodology or model that has not been able to be replaced by digital means as yet nowadays, making it lack in credibility and consensus by the metrics of any potential advertisers. Although, this has been changing, soon the tables will slightly turn due to Programmed Advertising (which will be processed by algorithms) which is starting to be the norm anyway.

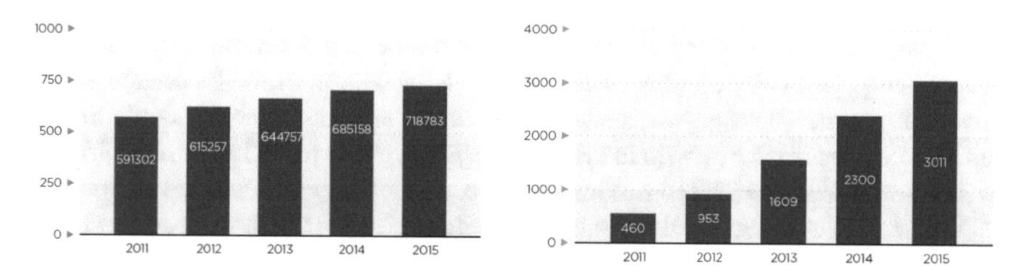

Fig. 1 Dissemination of the World press and online income

Therefore, one of the most important challenges for the Press in the coming years is the set up of a system for a mixed type of measurement for the paper & digital media, being able to access it across all social networks (Fig. 1).

The Newspapers World Association estimated that 1400 million people in the world usually read a newspaper in printed issue version or in any means of digital support. This Association estimates that 40% of Internet users read the digital press and that the surveys indicate that one in five would be willing to pay for it. The digital transformation and the increase of online readers is appreciated more in the mainstream media. The Washington Post announced by the end of 2016, a growth of 50% of readers and 40% of digital advertising in the last year, to beat the record of 100 million visitors in the United States and 30 million out of the country. On the other hand, the strategy of many other newspapers and local and regional of the world was to resist and more focused on larger numbers of paper-based editions.

The Spanish Press, however, suffers a loss of 8.7% diffusion although it has managed to increase its revenue on advertising (816 million on a set of 107 titles), according to data on the White Book of Press in Spain (AEDE 2016) in a 2.4%. A recovery of the revenues from Advertising and operating economic results of the whole of the Spanish Media Companies went up by 55% (29.9 million positive) with respects to 2014. The Spanish Digital Press is the second means of online access after online TV. AEDE highlights a slight tendency to increase on readers' digital access, which can forecasts a recovery of the balance in 2018, as well as 20% of access through mobile devices (Table 3).

Analyzing the data on the whole European market of 28 countries, daily newspapers and printed issues and magazines are only areas of the media industry that

Table 3 Advertising revenue from European media, 2010–14 (in millions of euros)

Years	Daily	Magazines	Television	Radio	Cinema	Outdoors	Internet	Total
2010	21.361,8	10.010,4	29.196,0	4.813,4	626,4	5.461,5	16.326,4	87.795,9
2011	20.609,4	9.485,5	29.239,8	4.797,7	616,4	5.454,5	18.985,7	89.189,0
2012	18.788,4	8.504,1	28.053,7	4.741,4	677,9	5.432,2	21.768,4	87.966,1
2013	17.123,4	7.764,5	28.035,7	4.676,3	600,6	5.302,4	24.320,2	87.823,1
2014	16.236,6	7.382,2	29.415,6	4.827,5	648,6	5.502,8	27.200,5	91.213,8
2014–13 (%)	−5.2	−4.9	4.92	3.2	8.0	3.8	11.8	3.9

are missing the rise of advertising effects. On the other hand, revenues from Internet Advertising practically already equals to television. Seven of media broadcast and advertising areas (newspapers, magazines, television, radio, cinema, outdoors and the Internet) recorded a growth in the last five years, going from 3.9% to 87.795 over to a billion in 2010 to 91.213 million by 2014.

With respects to advertising, data from Infoadex report that, on the investment of the Spanish advertising market a growth is shown of 4.7% on all areas in 2015 compared to the previous year, both in conventional and non-conventional means, but the percentage of the total amount of the market was higher in the latter with respect to the first, specifically, 42.7% on the conventional media and 57.3% on non-conventional media. Within mainstream media, the largest volume of advertising spending in 2015 was made for television (40.1%), followed, in second place, by the Internet (24.9%) and, further away, Newspapers (13.1%), Radio (9.1%), Billboards (6.5%), Magazines (5.1%), Sunday Supplements (0.8%) and Cinema (0.4%).

With regards to the investment in non-conventional media advertising, the largest percentage was an increase recorded in the year 2015–2014, which recorded in market fairs and exhibitions and mobile marketing (Messaging, Advergaming, Apps), with an increase of the 15.6 and 15%, respectively, over the previous year. For its part, acts of sports sponsorship investment registered an increase of 8.9% and acts of patronage in other areas of social marketing and CSR experienced a growth of 2.3% in 2015 to 2014. On the other hand, highlight the falls in the off-line environment: promotional games (−15.1%), advertising gifts (−9.5%), and loyalty (−9.1%) cards.

If we talk about the European Audiovisual Market in general, we can see a clear economic recovery of 2.8% on revenues in 2014, with regards to 2013, but practically the same figures s the ones we had five years ago. On the 2011 the European Audiovisual recorded revenues of 101.968 million € and in 2014, it all amounted to 105.790 billion, 0.9% higher. But differences between the different activities, such as growth notable, close to 30% growth on the audiovisual on-demand, which passes from 919 million € in 2010 to 2.501 million by 2014. Second highlights too, albeit with a more modest increase of 4.3%, revenues from pay television.

The economic evolution of the conventional television between 2009 and 2013 was positive in Europe, through revenue from 74.875 to 83.632 million €. Although, with a slight increase in the years 2012–13, according to Fontaine (2015), of the European Audiovisual Observatory. The change from linear television to pay On-Demand television is strong and accelerated in some media. And the average expenditure per household varies markedly between countries also, you would pay 724 € in Denmark, to a 107 € in Romania and 116 € in Lithuania. That spending per household is assessed based on advertising contributions, pay-TV's and allocation of public funds from the canon or approved by each countries' Governments. The European average spending per household on all kinds of television is 396.8 € and Spain is fairly below, exactly at 279.3 €.

The television market in Spain, recorded an increase of an 11.5% over 2014, for a total of 4.2224 million €, by 2015 is dominating by a double duopoly between Mediaset and A3 Media Television, content of open access with Movistar and Ono Vodafone in the convergent broadcasting of payment. Added to the revenue private television makes and the addition of public, State and Regional Grants, the figures rise up to 5.9404 million €, 8.7% more compared to 2014, according to the annual report of telecommunications and Audiovisual of the National Commission of Markets of the competition (CNMC 2016). Pay television was the business area that grew up the most in Spain in 2015, with a turnover of 2.064,6 million €, 18.5% more compared to 2014.

The economic impact of global data and trends underscore the importance of the media industry both at a global, and a national or European level. Market European information and communication generates seven million direct jobs, 40% of them in the small and medium enterprises, at an annual volume of 1,340,000 million Euros, of which global business represents just a 9.5% of the GDP of the EU. Of the seven main sectors that make up the Telecommunications' & Broadcasting Markets are the ones who recorded a greater productivity by employee, according to data from Eurostat (2015). The Global Press Industry generates three million direct jobs, according to WAN-IFRA (2016).

The total contribution of the activities cultural to the internal gross product of Spain in 2012 amounted to 26.031 million € compared to 28.224 billion in 2011 and, with regards to the audiovisual sector, this contribution to the GDP was 6310 million compared with 6.953 million, respectively.

The audiovisual and multimedia represents 24.2% of cultural activity in Spain, according to the Yearbook of cultural statistics (2015) published by the Ministry of Education, culture and sport, preceded the Book Industry and the Press (39.1%). The per capita expenditure was reduced from 266 € per year, in 2013 to 260 € a year in 2014. And expenditure per household in 2014 was 653, 6 billion €, the 44.3% on Internet and audiovisual activities, to 35.2 in 13.9% both in press and cultural services as books. The total employment in the cultural sector by 2014 amounted to 511.800 workers, 3% of the Spanish Total.

1.2 The Challenge of Ongoing Changes

What precisely this digital environment and its consequent revolution has doomed the sector of the media to a deep change (home, 2012), in a scenario where the distribution of online content, posed a scenario of great potential for new forms of communication (left-Castle, 2012) emerges. As the industrial bases of the traditional sector of the communication were resent, they seek new avenues for value creation for citizens and society (Picard 2012). It is in this context that the media industry has intensified its presence in recent years in social networks (Campos-Freire 2010) and has diversified strategies.

The hyper-fragmentation and clusters which we noted above are present in the model of global information and the industry network, in which Start of self-employed entrepreneurs compete with traditional companies and large content distribution social media, operating under a variety of economic, cultural, political, and material contexts with very little or next to no means, but hopefully create trends that help the future of journalism (Deuze 2017).

In order to face the new challenges, the communication industry has undertaken in the past and in the present restructuring plans to offer a response to processes of convergence technology, business, professional and content to meet the challenge of the multi-platform production (Boczkowski 2004; Dailey et al. 2005; García-Avilés and Carvajal 2008; López and Pereira 2010; Larrondo et al. 2012).

The media industry has had to establish new organisational structures to interfere with certain guarantees of success in a society (Castells 1996) network where the digital adventure requires business models that monetize content (Álvarez et al. 2011). Communication companies, on the digital stage, assume that the value of information increases when it has more relevance for the consumer, who behaves and interacts (Carpentier 2011), feeding the field of digital sociability.

2 The Future According to the Communication Companies

The study shown here is based upon a sample of companies in the Communications Sector in Spain, distributed as follows: Printed Press (15.1%), Radio (45.3%), Television (18.9%) and Native Digital Media (7.5%), Advertising Agencies (5.7%), Broadcasting (6.6%) and Press (0.9). All of them are companies with a minimum of 6 million € of annual turnovers, divided between SMEs and large enterprises. In particular, 49.1% are companies with less than 10 employees, 37.7% of 10 to 49 employees, a 9.4% who have among 50 and 250 employees and a 3.8% of them with over 250 employees.

For this reasons, a total of 106 interviews with companies were carried out (publishers, managers, C.E.O's and the responsible for issuing technical qualifications) were now contacted by phone and sent a questionnaire by mail. There was a two month dateline for the study, so between October and November of 2016, coinciding with the margin provided to each company and the people that took part, so that they had enough time to study and reflect on the questions of quite an extensive questionnaire.

The survey was conducted on a sampling with simple random probability, with a Maximum Margin of Error on the Total Sample of 9.0% and a 95% confidence interval (2 Sigma), a quite a high response rate on the universe of the existing media in Spain.

One of the first questions posed to the sample of people that took part on the survey and to companies was what were their estimates and forecasts for all industries and developments when it came to Spanish Communications for the next three years. The data reflected a pessimistic picture, in terms of the printed press and for general information and traditional advertising, with estimates of a negative growth of a 49% of a total of 26.3% of poll takers, respectively.

Happier forecasts when it came to the growth of the Printed Press, which lied between 0 and 1% (an 18.8% of polled people) and a 2 and a 5% (17.7% of those polled), if it is true we have to assume that, there was a forecast of growth for traditional advertising there, between 20 and 21% of those surveyed thought there would be growth forecasts of between at least a 1 and a 5%. One small caveat within the newspaper industry is that the Sports Media and, above all, the Economic would be it. In the case of the Business Press, 22.2% of polled people considered that their growth will be negative in the coming years, while 36.7% projected that just 0 and a 1%, as in the case of the Sports Media, which stands at 35.5%. I.e. that we can predict a fall on the Economic Section and also that Sporting Press will decrease in size, while the forecasts for growth are very slim. In addition, 20% of the polled people forecast a rise of between a 1 and a 5%, in traditional advertising.

On the other hand, most expectations lie on digital media; digital advertising and mobile media, where polled people estimated a slight negative growth and, on the other hand, these were the only media where over 30% of polled people expected a growth of over a 10%.[2]

In the case of audiovisual media, the prospects are rosier than the press, with less predictions of negative growth in the next three years judging by the people polled. In particular, in the case of the radio, a quarter of polled people expected positive growth of between 1 and 5%, with one slightly higher with respect to the one of television, an average of 27.4% of polled people foresee a rise of between 2% and 5%. Also News Agencies also recorded very negative forecasts, with respect to other sectors; in particular, almost 30% of polled people expected a growth on the sector of between a 1 and a 2%.

Another significant fact also, is the relative growth considered by the people polled in the case of the Corporate and Political Communications.

Consequently, data show a tendency and forecast an evolution and transfer of business from traditional to online media and corporate communication journalism and advertising. The data of everything indicated above is shown below (Table 4).

Below the average growth envisaged for the heads of Spanish communication companies in different sectors, led by digital media and followed by Digital Media and Mobile Applications Advertising, also used for Business and Political Communications. What we can see here are the values of the General press lie between 0 and a 1% growth, different tot those of the economic and sports section of a 1 and a 2% (Table 5).

With respect to the prediction of the importance that will have digital advertising and mobile media against traditional advertising, more than half of entrepreneurs, managers and managers of Spain media companies consulted believe that advertising will migrate in the coming years, largely to social networks and, in particular, to Youtube (51.1%) and Facebook (50%) (Table 6).

Very similar responses to these were the ones indicated by Media Headings, people interviewed when asked about the degree of strategic importance they attribute

[2]The Interactive Advertising Bureau (IAB), Association of advertising, marketing and communication in Spain, shows an increase of 24.4% of advertising investment in digital media during the first half of the 2016, compared with the same period of last year, going from 606.7 investment registered between January and June of 2016 to 754.7 million € in the same months of 2015.

Table 4 Forecast of the industry's developments in the communication sector for the next three years

Response	Negative growth	Between 0 & a 1%	Between a 1 & a 2%	Between a 2 & a 5%	Between a 5 & a 10%	More than a 10%
General printing data (%)	49.0	18.8	10.4	17.7	3.1	1.0
Sports press (%)	16.1	35.5	16.1	18.3	12.9	1.1
Economic press (%)	22.2	36.7	20.0	15.6	5.6	0.0
Radio (%)	5.1	27.3	24.2	25.3	13.1	5.1
TV (%)	6.3	13.7	26.3	27.4	13.7	12.6
Digital media (%)	1.0	6.1	6.1	17.3	34.7	34.7
Press agencies (%)	7.9	29.2	29.2	18.0	10.1	5.6
Traditional advertising (%)	26.3	21.1	21.1	20.0	9.5	2.1
Digital advertising (%)	2.1	9.4	13.5	18.8	26.0	30.2
Mobile media advertising (%)	2.1	13.7	8.4	20.0	24.2	31.6
Business communication (%)	2.2	24.4	16.7	23.3	18.9	14.4
Political communication (%)	8.9	16.7	20.0	26.7	14.4	13.3

Table 5 Forecast of growth average for the next three years

Response	Average growth (%)
General printing data	0.3
Sports press	1.9
Economic press	1.0
Radio	3.0
Tv	4.2
Digital media	8.5
Press agencies	2.7
Traditional advertising	1.6
Digital advertising	7.3
Mobile advertising	7.4
Business journalism	4.7
Political journalism	4.2

Table 6 Evaluation on the evolution of social media advertising

Response	Nothing (%)	Little (%)	A lot (%)
Facebook	7.3	42.7	50
Twitter	10.4	44.8	44.8
LinkedIn	19.6	45.7	34.8
Instagram	15.2	44.6	40.2
YouTube	10.9	38.0	51.1
Pinterest	27.3	45.5	27.3
Other	16.9	48.1	35.1

Table 7 Assessment of the importance of social networks for business management of and political communications

Response	Nothing (%)	Little (%)	A lot (%)
Facebook	2.1	21.9	56
Twitter	2.1	18.8	79.2
LinkedIn	7.6	40.2	52.2
Instagram	5.4	33.7	60.9
YouTube	2.1	22.1	75.8
Pinterest	17.4	38.4	44.2
Other	13.2	32.9	53.9

to social networks when it came to business management & to political communication, more than 75% of entrepreneurs and of media managers gave a fair degree of importance, particularly to Twitter (a 79.2%), Facebook (76%) and YouTube (75.8%) (Table 7).

In contrast to the forecasts of growth in the industry of communication in the next few years, we have asked managers of media companies if they believed that the current media business model was in crisis.

In this sense, areas where the majority of polled people indicated that they were deeper in crisis were, on the one hand, the press printed payment (a 66.3%) and free (a 48.9%), and, on the other hand, the regional public television channels (a 63.5%) and at a National level, RTVE, model that 54.8% of poll takers deemed to be in crisis. Conversely, areas that poll takers consider that are little or not at all in crisis are the digital press, all of them, payment and the free ones (paywall), digital magazines, themed radios and generalist TV.

Models in crisis, therefore, face the Digital Printed Media and, also, public TV faces the general and Private Channels; open and encrypted TV payment channels, of which nearly 70% of poll takers (69.8%) think that their businesses model are a little or not at all in crisis, an even higher proportion and as a result of this, a more optimistic, in the case of more general and specialized radios (Table 8).

In relation to the aforementioned aspect, we have asked the C.E.O's of Spanish Communications Companies what they thought of it, as a business models facing the immediate future, the financing of Media, on a scale of 1 to 10, taking into account that 1 is the lowest and 10 the highest.

The most outstanding formulas, with a scoring average of over 5, were advertised, with an average of 6.4, followed by sponsorships (a 6.3), the realization of micropayments (5.6) and the Crow Funding (5.5) or participatory financing of ideas or projects. Other alternatives such as the signing off Pay TV, paying per viewing (pay per view) and freemium,[3] received, each of them, an average score of 5.2, ahead of other formulas as the total paywall (5.1) and mixed paywalls (5.0).

[3]Business model designed for large volumes of customers, offering a free shuttle service (free) or low cost to all other more advanced (Premium) payment for response to specific demands or broader service offerings.

Table 8 Assessment of the crisis in the media business models

Answers	Nothing (%)	Little (%)	A lot (%)
Printed issues payments	2.0	31.6	66.3
Free printed press	13.8	37.2	48.9
Free digital press	54.2	33.3	12.5
Payment (Paywall) digital press	25.3	47.7	27.5
Printed journals	6.2	42.3	51.5
Digital magazines	28.7	53.2	18.1
Generalist open TV	28.4	52.6	18.9
Coded pay TV	15.6	54.2	30.2
General radio	23.8	50.5	25.7
Specialized radio	16.7	56.3	27.1
Press agencies	15.2	48.9	35.9
RTVE	12.9	32.3	54.8
Regional public TV	10.4	26.0	63.5

Table 9 Importance attributed to the following business models of the future

Answers	Average score
Issue sales	4.6
Paywall total	5.1
Mixed paywall	5.0
Freemium	5.2
Micropayments	5.6
TV subscription payment	5.2
Pay-per-view	5.2
Advertising	6.4
Sponsorship	6.3
Crow funding	5.5
Fee for public TV	4.1
Return of advertising to RTVE	4.7

Only 5 of them were assessed, the direct sale of issues, with an average of 4.6 rating, and some alternatives to public television, such as the return of advertising to RTVE (4.7) and the establishment of a set canon for public television (4.1).

Some have hopes in the establishment of a fee or a canon for public television, but they have greater confidence in advertising and the search of other forms of social and collective involvement through sponsorships and patterns of small payments, in search of greater involvement and higher levels on satisfaction of audiences (Table 9).

Table 10 Valuation according to the degree of interest to your company

Response	Average score
Investing in technology	8.3
Developing new products	8.0
Reducing jobs	2.5
Internal reorganization	7.3
Improve working processes	7.8
Outsourcing processes	6.4
Searching income in other industries	7.1
Searching new business models	7.6
Digital tools training	8.0
New networks	7.9
Mobile communication	7.9

Precisely, in line with the previous question and the possibility of finding new ways of funding and participation, we consulted many communication companies in Spain who were responsible for giving great importance, today, and has an image and visibility of the averages (8.1 out of 10), (8) hearings and content quality control (7.7), but also many more biased opinions such as credibility (8), talent (7.9), training (7.4), reputation (7,8), creativity and innovation (7,8), freedom of expression (7.7), honesty (7.6) and transparency (7.3) and Corporate Social responsibility (6.9).

Entrepreneurs and Media Managers also have valued as issues of great interest to organizations the investment in new technologies (with a 8.3 average rating), training in digital tools (8), new product development (8) and the new communication networks and mobile applications (7.9), together with the improvement of work processes (7.8), the search for new models of business (7.6) the internal reorganization (7.3), or even the pursuit of income outside the sector (7.1). In addition, other aspects such as the interest in Bigs Data are also valued with a 6.7 average by consulted representatives of Spanish Communication Companies. All of these values are believed to be well above other aspect, such as the outsourcing processes (6.4) or reduction of employment (2.5) (Table 10).

Ultimately, we asked about what were the three main priorities of their organization for the next three years, we consulted the leaders of Spanish Communication Companies and they indicated the following:

1. Billing (keeping and making new customers and advertising, fidelity).
2. Audience (keeping/making new customers) and
3. Technology (technical improvements, creation and adaptation of new technologies and digital networks and platforms).

The following issue identified by poll takers, also related to this third aspect, the renewal and innovation of social networks.

3 Future Challenges

As indicated at the beginning of this paper, the scenario described by the media industry shows a pessimism that faces the printed press of the ways of payment with traditional advertising, with our hopes placed in the digital and mobile media and corporate communication.

In particular, almost half of those surveyed expected negative growths in the printed press for general information in the next three years and more than a quarter in the case of traditional advertising. Media makers are also new forms of financing (micropayments, paywall) and participation through the sponsorship and patterns like (crowfunding), maintenance and increase the number of hearings through the enhancement of intangible and with a little investment that need put into discovering new networks and technologies, communication and advertising through social networks, especially Facebook and YouTube.

The results of this survey are in line with some other results shown by earlier studies, considering the attention put on social media and the digital revolution's management, they also put special attention to the volumes and the information flow, actually, one of the most important challenges in order to achieve the leadership, has to be through the redefinition and location of key public, forms of monitoring conversations that occur in social networks so that we can have a faster response to information needs (Moreno et al. 2014).

On the other hand, most future hopes are placed on a more corporate, business and political communication level, along with the search for new business opportunities, showing the way towards a broader view of journalism and traditional advertising and to the management of digital surroundings, the diversification of supply and demand services, throughout a thoroughly comprehensive means of communication.

Similarly, it proposes an adaptation to a new public relations paradigm, the idea is studying the idea that the relationship between the public and groups may affect or be affected (stakeholders), above the idea of communication (Míguez and Bahamonde 2011).

A two-way process in which communication companies try not only to meet but also to listen to their audiences, where public relations form part of the address (Dircom), as a way of communicating to the public and with the public, with the help of new technologies.

In the light of these findings arises the need that media companies have to improve training in new technologies and adapting to new digital surroundings, as well as improving communication for themselves, incorporating a more comprehensive and versatile view of its activity, so that they do not only serve the audience (readers, listeners and viewers), but also society and the environment in which they are registered.

Acknowledgements This work was made in the framework of the programming of activities of the XESCOM network (networks 2016 XESCOM GI-1641), the Department of Culture, Education and Management, from the Xunta de Galicia's University, with reference ED341D R2016/019, as well as R&D Projects, State Promotional Scientific Research and Technical Programming of Excellence,

State sub-programming, knowledge generation for the Ministry of Economy and competitiveness of Spain on *Indicators related to broadcasters' governance, funding, accountability, innovation, quality and public service applicable to Spain in the digital context* (reference CSO2015-66543-P) and *Uses and informative preferences in the new media map in Spain: journalism models for mobile devices* (reference CSO2015-64662-C4-4-R).

References

ADECEC: La comunicación y las relaciones públicas en España. Radiografía de un sector. http://prnoticias.com/images/stories/comunicacion/ARCHIVOS/presentacin_del_estudio_adecec.pdf (2008)

AEDE: Libro Blanco de la Prensa de España. Asociación de Editores de España. www.aede.es (2016)

Álvarez, I., Benamou, J., Fernández-Boch, J.M., Solé, C.: España CONecta. Cómo transformar internet en la economía española. The Boston Consulting Group, Boston. http://www.espanaconecta.es/pdf/Spanish_Executive_Summary.pdf (2011)

Boczkowski, P.: The processes of adopting multimedia and interactivity in three online newsrooms. J. Commun. **54**, 197–213 (2004)

Campos-Freire, F.: Los nuevos modelos de gestión de las empresas mediáticas. Estudios sobre el Mensaje Periodístico **6**, 13–30 (2010)

Caprioti, P.: La imagen de la empresa Estrategia para una comunicación integrada. El Ateneo, Barcelona (1992)

Carpentier, N.: Media and participation. Intelect, Bristol (2011)

Castells, M.: The Information Age: Economy, Society and Culture, vol. I: The Rise of the Network Society. Blackwell, Oxford (1996)

Christensen, C.M., Raynor, M.E., McDonald, R.: What Is Disruptive Innovation? HBR. https://hbr.org/2015/12/what-is-disruptive-innovation (2015)

CNMC: Informe del Mercado de las Telecomunicaciones y el Audiovisual de España. www.cnmc.es (2016)

Costa, J.: El Director de Comunicación. La nueva figura central en la empresa del siglo XXI. In: Benavides, J., Costa, J., Costa. P., Fajula, A. (eds.) Dirección de Comunicación Empresarial e Institucional, pp. 47–66. Gestión 2000, Barcelona (2001)

Dailey, L., Demo, L., Spillman, M.: The convergence continuum: a model for studying collaboration between media newsrooms. Atlantic J. Commun. **13**(3), 150–168 (2005)

Deloitte: Analytics Trends The Nest Evolution. https://www2.deloitte.com/us/en/pages/deloitte-analytics/articles/analytics-trends.html (2016)

Deuze, M.: Considering a possible future for digital journalism. Revista Mediterránea de Comunicación/Mediterr. J. Commun. **8**(1), 9–18 (2017). https://www.doi.org/10.14198/MEDCOM2017.8.1.1

DIRCOM: El estado de la comunicación en España y América Latina. Fundación Canal de Isabel II, Madrid (2015)

Eurodata: One TV Year on the world. Eurodata TV Worldwide, Paris (2016)

EUROSTAT: Statistics Explained. Information and communication service statistics—NACE Rev. 2. http://ec.europa.eu/eurostat/statistics-explained/index.php/Information_and_communication_service_statistics_-_NACE_Rev._2#Further_Eurostat_information (2015)

Evans, P.: De la deconstrucción a los big data: como la tecnología está transformando las empresas. Reinventar la empresa en la era digital. https://bbvaopenmind.com (2015)

Fontaine, G.: Trends in linear television revenues. European Audiovisual Observatory. http://www.obs.coe.int/ (2015)

García-Avilés, J.A., Carvajal, M.: Integrated and cross-media newsroom convergence: two models of multimedia news production: the cases of Novotécnica and La verdad multimedia in Spain. Convergence **14**(2), 223–241 (2008)

Grunig, J.E., Hunt, T.: Managing Public Relations. Fort Worth. Holt, Rinehart and Wilson (1984)

IAB: Estado de la inversión publicitaria en medios digitales en España. Primer semestre de 2016. http://www.iabspain.net/wp-content/uploads/downloads/2016/09/Estudio_Inversion_Medios_Digitales_S12016.pdf (2016)

INFOADEX: Estudio de la inversión publicitaria en España 2016. http://www.asociacionmkt.es/sitio/wp-content/uploads/Estudio-InfoAdex-2016-Resumen-2.pdf (2016)

Larrondo, A., Larrañaga, J., Meso, K., Airreazkuenaga, I.: Convergencia de medios y redacciones: el caso de la Radio Televisión Pública Vasca (EITB). El Profesional de la Información **21**(4), 347–353 (2012)

López, X., Pereira, X.: Convergencia digital. Reconfiguración de los medios de comunicación en España. Servicio de Publicaciones de la Universidad de Santiago, Santiago de Compostela (2010)

Miguel de Bustos, J.C.: Big Data y Big GAFA (Google, Facebook, Amazon y Apple. Reflexiones sobre economía de los datos. Universidad del País Vasco (2017)

Míguez, M.I., Baamonde, X.M.: La evolución de las relaciones públicas hacia la dirección de comunicación: aproximación histórica en el contexto académico español. Razón y Palabra **16**(75) (2011)

Moreno, M.A., Navarro, C., Humanes, M.L.: El liderazgo en relaciones públicas y gestión de la comunicación. Análisis cuantitativo de los factores de liderazgo en el sector en España. Palabra Clave **17**(3), 946–978 (2014)

Newman, N.: Journalism, Media, and Technology Trends and Predictions 2017. Digital News Project. Reuters Institute. http://www.digitalnewsreport.org (2017)

OBS: Yearbook 2015: Key Trends. Television, cinema, video and On-Demand Audiovisual Services. The Pan-European Picture. European Audiovisual Observatory (2016)

OFCOM: International Communications Market Report 2015. https://www.ofcom.org.uk (2016)

Picard, R.: La creación de valor y el futuro de las empresas informativas. Por qué y cómo el periodismo debe cambiar para seguir siendo relevante en el siglo XXI. Media XXI, Porto (2012)

PWC: Entertainment and Media Outlook 2016–2020. www.pwc.es (2016)

Ries, A.L.: La caída de la publicidad y el auge de las relaciones públicas. Ediciones Urano, Barcelona (2003)

Roos, J.G., Dragonetti, N., Edvinsson, L.: Capital Intelectual. El valor intangible de la empresa. Paidós, Barcelona (2001)

SABI: Sistema de Análisis de Balances Ibéricos. In: www.informa.es (2016)

Salmon, Ch.: Storytelling. La máquina de fabricar historias y formatear mentes. Península, Barcelona (2007)

Sotelo, C.: Introducción a la comunicación institucional. Ariel, Barcelona (2001)

Vázquez, M., Fernández, A.B.: Las necesidades formativas de los nuevos comunicadores según los profesionales y su reflejo en los Grados de Comunicación. Estudios sobre el Mensaje Periodístico **18**, 889–897 (2012)

Villafañe, J.: La buena reputación. Claves del valor intangible de las empresas. Ediciones Pirámide, Madrid (2004)

WA-IFRA: World Trends Database. Asociación Mundial de Diarios. WAN-IFRA. http://www.wan-ifra.org (2015)

WAN-IFRA: World Press Trends 2016. www.wan-ifra.org (2016)

WNMN: World Newsmedia. Innovation Study 2015. The seventh annual survey chronicles media company priorities to manage current challenges and ensure future success. Global Digital Media Trendbook 2015. www.wnmn.org (2015)

Zerfass, A., Verhoeven, P., Moreno, A., Tench, R., Vercic, D.: European Communication Monitor: Exploring Trends in Big Data, Stakeholder, Engagement and Strategic Communication. Results of a Survey in 43 countries. EACD/EUPRERA, Brussels. http://www.zerfass.de/ECM-WEBSITE/media/ECM2016-Results-ChartVersion.pdf (2016)

Xosé Rúas-Araújo Tenured Lecturer of Techniques of Electoral and Institutional Communication at the Faculty of Social Sciences from the Universidade de Vigo. He is the principal investigator of the neuro-communication area of the research group CP2: Persuasive Communication from the Universidade de Vigo. Pontevedra, Spain.

Iván Puentes-Rivera Degree in Advertising and Public Relations, Post-Graduate course in Research on Communication and Ph.D. in Communication. Member of the research group CP2: Persuasive Communication from the Universidade de Vigo. He has lectured at the Graduate Course in Advertising and PR. His research focuses on political communication, public relations and social networks. Pontevedra, Spain.

Ana Culqui Medina MBA in Communication and Creative Industries from the Universidade de Santiago de Compostela, Chair Professor in Pontificia Universidad Católica del Ecuador Sede Ibarra of the areas of Image Theory and Audiovisual. She is a coordinator of the School of Social Communication at the same University. Journalist and radio producer in Ecuador. Ibarra, Ecuador.

Context, Investigation and Condemnation: Los Ángeles Press

José Luis González-Esteban, Carmen María López-Rico and Manuel Ortiz-Marín

Abstract Journalism has become a high-risk activity in some areas of Mexico. The number of assassinated journalists grows and grows, and the way this 'war' is reported is vastly conditioned by the pressure that professional journalists are under. This situation has led some Mexican journalists to search for innovative solutions within new media and narrative formats that make a complete break with 'la nota roja' (red news). The Mexican journalist Guadalupe Lizárraga has opted for long tail journalism from the other side of the border. Context, investigation and condemnation of corruption and the violation of human rights are the hallmark of *Los Ángeles Press*; the case we are going to analyze in this chapter.

Keywords Investigation · Human rights · Mexico · Context · Violence · *Los Ángeles Press*

1 Journalism in Mexico

According to Reporters Without Borders (RWB 2016), Mexico, where almost 130 reporters have been assassinated in the last 20 years, is the third most dangerous country in the world to work as a journalist, it is only surpassed by Syria and Afghanistan, both war zones. It has been at the top of this ranking in Latin America since 2004, and was only surpassed by Brazil and Honduras in 2013. Spring 2017 (just as this chapter was completed) was a particularly dark time for journalism with the murder of seven journalists: Cecilio Pineda (Guerrero), Ricardo Monlui (Veracruz), Miroslava

J. L. González-Esteban (✉)
Universidad Miguel Hernández, Alacant, Spain
e-mail: jose.gonzalez@umh.es

C. M. López-Rico
IMEP, Alacant, Spain
e-mail: carmen.lopezr@umh.es

M. Ortiz-Marín
Universidad Autónoma de Baja California, Mexicali, Mexico
e-mail: mortiz@uabc.edu.mx

M. Túñez-López et al. (eds.), *Communication: Innovation & Quality*, Studies in Systems, Decision and Control 154, https://doi.org/10.1007/978-3-319-91860-0_16

Breach (Chihuahua), Maximino Rodríguez, Filiberto Álvarez, and more recently Javier Váldez and Sonia Córdova between 14th and 16th May 2017. There was also a failed attempt on Armando Arrieta (Veracruz). This makes a total of 33 journalists killed during Enrique Peña Nieto's term of office from 2012 to now, as reported by the organization "Artículo 19" in their report 'Estado de Censura' (2017) and updated for this research.

Mexico ranks as the second country with most impunity in the world, preceded by the Philippines, according to the "Índice Global de Impunidad de México" (Le Clercq Ortega and Rodríguez Sánchez Lara 2016). Likewise, RWB (2016) warns that 99.75% of these attacks and assassinations are still unresolved.

1.1 The Media Ecosystem in Areas of Conflict

To be able to understand the conditions for working as a journalist on the northern border of Mexico, we need to be aware that this region covers more than three thousand kilometers from Tijuana, Baja California as far as Matamoros, Tamaulipas. It contains six states: Baja California, Sonora, Chihuahua, Coahuila, Nuevo León and Tamaulipas. Demographically speaking, the most important cities on this border are: Mexicali, the state capital of Baja California, with a population of more than 988,000 inhabitants and Tijuana, where more than 1641,000 people live (the fourth most populated city in Mexico). The other city of significant importance is Ciudad Juárez, Chihuahua with 1391,000 inhabitants (Encuesta Intercensal 2015, INEGI).

These three cities together have the highest demographic density and comprise 80% of the border population. There are border cities with a lower number of inhabitants like: Nogales, Sonora; Piedras Negras, Coahuila; Nuevo Laredo, Reynosa and Matamoros, the latter three belonging to the Tamaulipas state. The total number of inhabitants on the border is 7.5 million (3.77 million men and 3.73 million women). Between 2000 and 2010, the population in the border area increased by about 18 percent on the Mexican side (Organización Panamericana de Salud, Organización Mundial de la Salud y Salud en Las Américas 2015). It is in this context that we will describe the most important newspapers in the leading border cities in the states of Baja California, Sonora, Chihuahua, Coahuila and Tamaulipas, with the exception of Nuevo León which does not have an important border newspaper to document.

1.1.1 Baja California

This border state is home to four daily newspapers and one weekly newspaper, which are important for their circulation and their journalistic background. Although this state has 23 printed newspapers recognized by the "Padron Nacional de Medios Impresos de la Secretaría de Gobernación" (2017), the most important papers are: *El Mexicano, Frontera* and the weekly paper *Zeta* in Tijuana, and *La Voz de la Frontera* and *La Crónica* in Mexicali. They are described below.

El Mexicano, belonging to the publishing company Tenedora Elcoli, S.A. de C.V., was first established as a cooperative and is now a limited company. It is one of the most important and oldest newspapers in Baja California, whose first issue was published on 21st September 1958 as a broadsheet paper, which it still is today. It has six editions: Ensenada, Mexicali, Tijuana, Rosarito, Tecate and a special section for the city of San Diego, California. This paper was a cooperative affiliated to the "Confederación de Trabajadores de México", associated with the "Partido Revolucionario Institucional"—PRI (Trujillo 2000), and still maintains an organic relationship and identifies with the federal government or with the municipals whose governors belong to this political party. The printed version of the newspaper indicates that the president of the board of Directors is Eligio Valencia Roque and the managing director is Eligio Valencia Alonso (El Mexicano 2017), it has a circulation of 36,448 copies (Secretaría de Gobernación 2017), and is a member of the "Asociación de Editores de los Estados"—AEE (AEE 2016).

Frontera first appeared on 25th July 1999, joining its sister newspapers *El Imparcial de Hermosillo*, Sonora and *La Crónica* in Mexicali, Baja California. It belongs to the publishing company Impresora y Editorial, S.A. de C.V. as part of the Healy Newspaper group, which currently publishes 37,710 copies (Secretaria de Gobernación 2017), *Frontera* like *La Crónica* maintains an independent and even tense relationship with some state governments and on occasions with groups that have business and political power. Its website indicates that it has the same managing director as *La Crónica*, Luis Alejandro Bernal García but a different local director, Jesús Manuel López for Tijuana, while Juan Fernando Healy Loera is president of the board of directors. It has four permanent sections in the standard printed version and one section for news about San Diego, California and Ensenada (Frontera 2017).

Zeta has a slogan "free as the wind", it is a weekly paper founded in April 1980 by Jesús Blancornelas and Héctor Félix—both deceased. This weekly paper stands out because in the first 10 years of being published "it went from 32 pages in 1982 to an average of 80 pages in 1986. Halfway through this year, the number of pages increased to 96" (Trujillo 2000: 15). It currently has an average 80 pages with a circulation of 18,579 copies and belongs to the company Choix, Editores, S.A. de C.V. (Secretaria de Gobernación 2017). The current directors are Adela Navarro and César René Blanco Villalón and it has a website (Zeta 2017) where it publishes all the printed information in seven sections.

The weekly newspaper *Zeta* is a unique example of an independent newspaper within the context of Mexican newspapers, since it is managed by journalists. Its co-founder, Jesús Blancornelas is known for his journalistic work condemning corruption among local politicians and the narco-trafficking cartels. This work was acknowledged by the University of Colombia with the María Moors Cabot award in 2002. Through its editorial policy, *Zeta* is characterized for taking an independent journalistic stance which is usually made tense by political groups and/or strong interests from certain business sectors, and more recently by organized crime. This independent stance, therefore, makes *Zeta* a unique paper within Mexican journalism.

La Crónica de Baja California, first appeared on 7th November 1990, at first as a broadsheet. Its predecessor was the newspaper *Novedades de Baja California*, which appeared in 1982 and was bought years later by the Healy Newspaper Group. Apart from the standard print version, this paper also has an online version (Crónica 2017). *La Crónica* indicates that it has a daily circulation of 25 thousand copies. Its current president of the board of directors is Juan Fernando Healy Loera and the general director is Luis Alejandro Bernal García, who is also director of *Frontera* in Tijuana, *Frontera* in Ensenada and *el Imparcial* in Sonora. The director of *La Crónica* is Javier A. Lozano. This group of newspapers along with another three newspapers in different states belongs to the Healy Newspapers chain, which is a member of the "Sociedad Interamericana de Prensa"—SIP (2017). The Healy Newspaper Group website indicates that its printed version has 100,000 readers a day and the digital version—La Crónica.com—has 35,000 cybernauts a day (Media Kit 2014).

La Voz de la Frontera, directed by Cristóbal Garcilazo, started its circulation on the 20th September 1964 as a broadsheet newspaper, which it still is today. The first owners of the paper were Mexicale businessmen, and in 1990 it was acquired by the "Organización Editorial Mexicana" (OEM) which it still belongs to (Ortiz 2006). The publishing company is called Compañías Periodísticas del Sol del Pacífico, S.A. de C.V.,—it comprises newspapers from the cities of Acapulco, Mazatlán, Tijuana, La Paz and Culiacán—, and according to the "Padrón Nacional de Medios Impresos de la Secretaría de Gobernación" (2017) its daily circulation is 50,702 newspapers. This newspaper has a strong government bias, whatever political party is in power in the state, but it is consistent with the federal government when it is in line with the PRI. The President of the OEM was Mario Vázquez Raña, now deceased, and the current president and director is Paquita Ramos de Vázquez with Francisco Edgardo Leal Corrales as local director.

On its website this organization describes itself as "a media company which has a presence throughout nearly all Mexico. It has a group of 70 newspapers, 24 radio stations and 43 websites on the Internet in its name. It also has companies from other sectors, such as paper production or high definition television shows" (OEM 2017). On average it produces between 80 and 90 pages and its digital version comprises the following sections: local, latest news, financial, republic, column writers and various sections (La Voz de la Frontera 2017).

1.1.2 Nogales, Sonora

The majority of the important newspapers are in the capital of the state, Hermosillo, and in the border town of Nogales, the newspaper with the greatest circulation and tradition is *Nuevo Día* from Nogales belonging to the company Periódicos Nuevo Día, S.A. de C.V., which according to the "Padrón Nacional de Medios de la Secretaría de Gobernación" (2017) has a circulation of 28 thousand printed copies. The current director is Jaime Juaristi Santos and it has a website (Nuevo Día 2017) where part of the printed information is published and where it indicates that the paper was founded in 2003.

1.1.3 Ciudad Juárez, Chihuahua

Most newspapers are published in the capitals of the border states. However this is not so in Chihuahua—in this case, because of its social, cultural and historical importance, as well as its economic and geographical significance, several newspapers are published in Ciudad Juárez. Two of the most relevant papers are: *Diario de Juárez*, member of the "Asociación Mexicana de Editores de Periódicos, A.C." and SIP; and *el Norte de Ciudad Juárez*, also a member of SIP. *El Diario de Juárez*, published by the company Publicaciones e Impresos Paso del Norte S. de R.L. de C.V, is one of the most important newspapers in this city and has a circulation of 46, 255 copies. Its printed version has an average of 40 pages with five sections, and it is also circulated in the neighboring town of El Paso, Texas (Secretaría de Gobernación 2017).

Its current director is Osvaldo Rodríguez Borunda and its website (Diario 2017) indicates that it was founded in 1976 and belongs to a newspaper chain in the state of Chihuahua. Some of these newspapers are: *El Diario de Chihuahua, El Diario de Nuevo Casas Grandes, El Diario de Delicias, El Diario de Parral* and *El Diario de Cuauhtémoc*, and the weekly paper *Regional*—altogether they publish 200 thousand papers a day. In addition, as indicated on its website, through its journalistic prestige it has received the "Premio Nacional de Periodismo 2010", based on an articles condemning the attacks on journalists from this publishing house. For this reason, it also received the María Moor Cabot award from the University of Columbia in New York, USA.

El Norte de Ciudad Juárez, belongs to the publishing house Omega Publicaciones, S.A. de C.V., it has a circulation that has reached 34,155 copies and it was founded in May 1990. Its current director is Óscar Cantú Murguía and it is a member of SIP according to the "Padrón Nacional de Medios" (Secretaría de Gobernación 2017). This prestigious journalistic medium shut down both its print edition and digital edition (Norte Digital de Ciudad Juárez 2017) as from 2nd April because of the assassination of their collaborator Miroslava Breach Velducea on 23rd March as she was leaving her house to go about her work as a journalist. She worked for papers like *El Heraldo de Chihuahua, el Diario de Chihuahua* and was also a correspondent for the national newspaper *La Jornada*. In a letter published on the first page of the newspaper, its director Óscar Cantú makes his condemnation:

> Dear reader, today I'm addressing you about our decision to shut down this morning paper, since, among many things, there are no guarantees nor security to carry out critical and counteractive journalism. The paper you have in your hands will be the last printed edition that Norte de Ciudad Juárez will publish. In the last 27 years, with few exceptions, we were left to fend for ourselves. We fought against the current, receiving attacks and punishments from individuals and governments for having brought to light their bad practices and acts of corruption. (Cantú 2017: 1)

This decision was made after both state and federal authorities ignored the continual threats to journalists and the media, especially in Ciudad Juárez.

1.1.4 Piedras Negras, Coahuila

This town has more than 163 thousand inhabitants and the newspaper *Zócalo de Piedras Negras* is published here. It appeared in June 1965, its publishing company is Editorial Piedras Negras, S. A. de C.V. and it has a circulation of 12,970 papers a day. It is also a member of the "Asociación Mexicana de Editores de Periódicos, A. C." (2016), and its president is Francisco Juaristi Santos, as indicated by the "Padrón Nacional de Medios Impresos" (Secretaría de Gobernación 2017). This paper publishes various editions for the cities of Saltillo, Monclova and Acuña, all of them in the state of Coahuila. The printed version of this paper has six sections and contains between 20 and 30 pages.

1.1.5 Nuevo Laredo, Tamaulipas

According to the 2010 census (Inegi 2015), this border city has 399,431 inhabitants, and several newspapers are published here. The most important, regarding circulation, are *El Bravo* and *El Mañana de Nuevo Laredo*. The former was founded in August 1951 and belongs to the Compañía Periodística del Bravo, S. A. de C. V. It has a circulation of 31,684 papers a day, and its website (El Bravo 2017) indicates that it has six sections and contains from 40 to 45 pages. Without being professedly pro-government, its stance is consistent with the state governments and municipals. The other newspaper, *El Mañana de Nuevo Laredo*, belonging to Casa Editora Argos, S.A. de C.V., was founded in 1932, that is to say, it has an 85-year journalistic tradition of informing people from both sides of the border, both Mexico and United Sates. This newspaper maintains a stance that identifies with the federal government when it belongs to PRI. According to its website (El Mañana 2017), the founding director was Heriberto Deándar and the current director is Ninfa Cantú Deándar, and it publishes a section dedicated to the neighboring city of Laredo, Texas. The printed edition is made up of seven sections with an average of 40 pages and 14,402 copies are printed a day (Secretaría de Gobernación 2017).

1.2 *Organization of the Profession at Risk. The Journalists Network of Ciudad Juárez*

The Journalists Network of Ciudad Juárez was established in April 2011. The Network is a group that brings together working journalists, there is no membership, and therefore, it is comprised of the founders—five female journalists: Rocío Gallegos, Araly Castañón, Sandra Rodríguez, Luz del Carmen Sosa and Gabriela Minjares, and recently joined by Alicia Fernández as a collaborator. Except for Sandra Rodríguez, who is the editor of *Sin Embargo*, the rest work for *Diario de Juárez*. The average age of these journalists is 40 and most of them have a degree and master normally in

Communication and/or Political Science. It has a unique gender component which aims to vindicate the role of the professional woman in an area where they have been brutally punished. *El Diario de Juárez* has been around for 38 years and throughout this time the only women to reach management of the paper is the journalist Rocío Gallegos, also a founding member of the Journalists Network of Ciudad Juárez. Gallegos has a degree in Communication from the University of Monterrey and is one of the few women to direct a news medium in Mexico. (González and López 2016):

> As a journalist it's an important challenge for me to be able to demonstrate that a woman can have a very different impact on the management of a news medium. The structures of power are very complex in Mexico, and for women it is especially difficult to reach positions of responsibility. Besides, there is also the aggravating circumstance that we live in a very male dominated society and many men cannot accept that a woman should have the last word in decision making.

The Journalists Network of Juárez primarily aims to consolidate a professional, ethical, committed and quality association for journalists in Juárez who are aware of their rights. All this in the interests of a society that increasingly demands to be informed so as to counteract the many problems it is faced with. The Network considered it necessary to unite as a platform for all the working journalists who through an exchange of experiences and concerns find ways to carry out journalism in an increasingly safer way and to offer more quality and benefits to the community. It aims to generate a learning dynamic in various senses, especially in techniques for safe journalistic coverage in situations of conflict. As we have already mentioned, journalism is a high risk profession in Mexico and particularly in Ciudad Juárez. In this sense, the Network programs feedback and training workshops as a space where knowledge, experiences and concerns can be exchanged among journalists. Another important strategy is to contact and organize sessions where experiences can be shared with journalists from other parts of Mexico and the world. "We need to communicate our problems to international journalists so that in their countries they can inform about the extreme situation we are in here" declares the journalist, Luz del Carmen Sosa, from the newspaper *El Diario* and also member of the Journalists Network of Ciudad Juárez.

The work of a journalist is different in every region of Mexico, it is complex to conceptualize a general state of the profession by territories because it depends on the economic, social and political circumstances of each region, conditions that are as diverse as the country itself. As far as Ciudad Juárez is concerned, it stands out because of the persistence with which "a troop of reporters, photographers and camera crew from about thirty different media have maintained the coverage of the fight against narco-trafficking that we have experienced for years, and of the many events that sustain it and other events that converge on the border", explains journalist, Gabriela Minjares also from *El Diario* and founding member of the Journalists Network of Ciudad Juárez.

This professional group, therefore, acts in three areas—labor, training and condemnation, all in relation to the uniqueness of the territory where they carry out their work as journalists. With respect to labor, Rocío Gallegos, stresses that "many colleagues work in unprotected conditions regarding social security because companies fail to comply with their fiscal responsibilities". They also vindicate better salaries, since the minimal monthly salary is no more than 200 euros. Another serious problem, as Gabriela Minjares explains, is that the majority of journalists work in a controlled environment, where the lack of regulation in Mexico regarding official or governmental publicity means it is used as a ways of awarding or punishing different media depending on their editorial viewpoints and, at least in the case of Ciudad Juárez, this is more decisive for censorship than for the very violence attributed to organized crime.

One of the main tasks of the Journalists Network of Juárez is continuous training, part of which involves self-defense courses and survival in high risk coverage. Experienced reporters like Héctor Saavedra, who has years of experience with *El Diario de Juárez*, at radio stations like *860 Radio Noticias* and as a news correspondent for *Agencia EFE* in the most violent periods of Ciudad Juárez, emphasizes the need to know all the possible resources to protect lives and to be able to work in a profession with some minimal guarantee. Even so, he recognizes that the pressures and threats become so strong that many journalists are forced to self-censorship or directly search for new professional challenges outside Mexico. Saavedra values the work of the Journalists Network of Juárez because he considers that only by being united and through the coordination of the group of journalists in this city, will it be possible to minimize risks and make a broader condemnation that has an international projection.

In this sense, the Network presents itself as a group whose core aim is education and the constant professionalization of the journalists in Ciudad Juárez. This is done through the organization of workshops, strategy development and other activities that can reinforce skills for investigation, writing, narrative, approaching victims, and equally important, for preventing risks to the journalists themselves. Its proponents summarize the essence of the Network as a space where all the journalists in Ciudad Juárez can above all find support and a place to exchange ideas.

In short, the Network's activity is managed in two areas: on the one hand, training including the prevention of risks arising from violence, and on the other hand social condemnation. This strategy is evident on their website, and especially on their social networks. Workshops for writing, locution or data journalism are alternated with a systematic condemnation of the abuses and aggressions that journalists suffer in this border area.

2 'La Nota Roja' as the Journalistic Genre of Violence

2.1 Origin and Evolution of 'La Nota Roja'

'La nota roja' genre aims to disseminate information related to violent events like assassinations, natural disasters, accidents, sex, etc. Its maxim is sensationalism, which is what most appeals to the readers' emotions. All of this contrasts with the lack of commitment and context.

A very newsworthy event, Sofía Ahumada's suicide in 1899, published in *El Chisme: diario de la tarde joco-serio ilustrado con noticias*, was related as follows:

> From the force of the impact on the lower right part of her face, her eyes had come out of their sockets, especially the left one, which was hanging by an optic nerve. Her lower jaw, smashed, came out of her mouth, and brain mass came out of her nose and out of her ears. One of her legs, the left one, was half bent and the right one was completely straight. (Castro 2008 cited in Melchor 2012)

There are several hypotheses about the origins of the term: one is that during the printing of a newspaper in Guadalajara in 1889, the print operator stamped his hand with red ink on the news article about a murder to give it more realism; or another is that it could be due to a stamp used by the "Tribunal del Santo Oficio" in their sentences (Op.Cit. 2012).

In the 19th century, when this genre appeared in Mexico, engravings were already included to illustrate events. However, it is in the 20th century when the spectacular nature of the news increases because of the inclusion of real photographs of the corpses, blood, and bullet casings, etc. Image is fundamental in this genre, and in the twenties they would include dramatized images of the event so that the audience could visualize the violent episode, although it was a recreation (Op.Cit. 2012). Currently, the images are hyper-realistic, they show the corpses and the consequence of the event. "This genre enforces its own esthetics. The photographer must underpin the event, not the situation that caused it" (Castellanos 2003: 37).

The first magazines to commercialize violence were *Alerta* and *Alarma*, relating mainly murders committed by citizens. However, rising drug consumption in the seventies and the spread of drug trafficking in the eighties meant that a large part of the general press and 'la nota roja' press became occupied with narco-trafficking and the violence related to it. It should be noted that in the period 2006–2012, during the presidency of Felipe Calderón, who declared open war on narco-trafficking, the executions, murders, mutilations, etc. appeared almost daily in the media.

Nevertheless, in 2009, the authorities issued a press release asking the press to stop abusing sensationalism. This led to a considerable reduction in news related to narco-trafficking (Rodríguez Morales 2010 cited in González-Esteban and López-Rico 2016) although it did continue to be a genre which was in great demand in the popular press (Alfaro Víquez 2014). This sector of the population has always been identified with 'la nota roja', because in its initial stages, they were the ones who committed most crimes according to the authorities, and besides this type of reading was accessible entertainment for them (average cost of a paper was 7 Mexican pesos).

'La nota roja' had its highpoint on television in the nineties, especially on TV Azteca and Televisa, with specialized programs about these news stories. However, at the start of the 21st century, almost all these programs disappeared, when several companies threatened to withdraw their advertising if they did not stop broadcasting them (Op.Cit. 2014).

In the press, this genre is doing well, "the newspapers with the highest circulation in the Federal District are the popular newspapers whose front and back pages feature

AÑO 5 NUMERO 295

MAGAZINE DE

POLICIA

SALE LOS LUNES

COMPLETAMENTE AJENO A LOS CUERPOS DE SEGURIDAD PUBLICA

SEÑALAR LAS LACRAS DE LA SOCIEDAD ES SERVIRLA

Asesinado por un maton

Fig. 1 Front page *Magazine de Policía*, 1940. *Source* Jesse Lerner, The shock of modernity, cited in Op.Cit. (2012)

'la nota roja' and the sports news" (Op.Cit. 2014: 636). The most widely read newspapers in the country are *La Prensa*, a newspaper that specializes in 'la nota roja', followed by the sports paper *Esto*. Other widely disseminated papers belong to the so-called popular press like *El Gráfico*, *El Metro* and *Extra*. The target population of 'la nota roja' is men between 26 and 55 years old, working class, middle-low income and primary or secondary studies (Álvarez 2016) (Fig. 1).

2.2 *'La Nota Roja' in Mexican Press Today*

Although at the end of the 19th century 'la nota roja' or crime news used to be mixed with news about other subjects, nowadays, we can find it either in a specific section within the general press or in specialized papers.

The high demand for this type of information has led to the creation of different online sites about these news stories: *Notaroja.mx*, *Lapoliciaca.com*, *Gacetamexicana.com*, etc. Thanks to the multimedia provided by the internet, we can find videos as well as photos "whose only interest is blood and injury, the text is secondary because the image is worth more than a thousand words" (López Peña 2011: 87).

This genre is characterized by a text focused on describing the place where the event took place and how it happened and the state of the victim, without considering the causes or the perpetrator (Freeman 2016). In this case, the characters are more important than the people themselves, nobody will remember the victim since the important thing is the facts and to show the human condition (Arriaga Ornelas 2017).

The front page and the back page are fundamental—they have major headlines, explicit photographs and use red in their design. In some cases, as in *Lapoliciaca.com*, they create a top 10 of the most important news stories of the week to help the reader choose the most shocking events.

The style is based on short headlines, with verbs of action which emphasize the event. The text is written in colloquial and crime-related language, and although they sometimes include literal judicial texts, the use of figure of speech, which had defined the style of this genre in its initial stages, has disappeared. It is important to point out the excessive use of adjectives for describing the facts in detail and even morbidly. "'La nota roja' of today no longer resorts to figure of speech to shape a narrative; some testimonies about the facts are given and even the titles introduce ambiguous information about what is referred to in the content" (Op.Cit. 2011: 20). In this way they are able to transmit an emotionally charged text to the readers as if they were direct witnesses, but the information is presented out of context and stereotyped, constructing a biased social representation of delinquency/violence (Op.Cit. 2014).

> Unlike almost any other genre, in 'la nota roja', its quality as a witness impregnates the discourse constructed in an impersonal style. However, in strict terms, this type of story usually "misleads" about the person of the narrative (…). It doesn't matter if the journalist was not there when everything happened; when structuring their story not only do they situate themselves at the scene as an attentive spectator, but they also manage to take the reader with them. (Op.Cit. 2017)

According to López Peña, "over the years, 'la nota roja' reporters have resorted to the structure of informative news without completely adopting it, without adjectives and the inverted pyramid guide; (…) some of the original characteristics of the genre have disappeared: making judgements within the narrative of the event. It loses its essential element by trying to eliminate morals. It is not clear which genre they are using. 'La nota Roja' and informative news are complements" (2011: 22).

The subject—violence related to narco-trafficking—is of great interest to part of the general public. In some social strata narco-traffickers are seen as heroes and their position of power and money is idealized. Even in music the 'narco-corridos' eulogize the biographies of criminals. Consequently, Mexican society absorbs these stories through different media, and the rise of violence is stamped on the reality of today; organized crime is the industry, the media markets it, so 'la nota roja' becomes leading news in the media agenda.

On the other hand, "natural phenomena alluding to the end of the world have always appealed to the public—explosions, spreading disease and floods have always had popular appeal. 'La nota roja' has a stylistic presentation that converts misfortune into something accessible, comprehensible and fascinatingly disturbing." (Op.Cit. 2011: 18).

Ultimately, with its advocates and opponents, there is no doubt that 'la nota roja' forms part of the Mexican identity; it is, indeed, a reflection of the violence that the country is experiencing, as are the stories of this genre that construct the identity and social consciousness of the country. It is also true that hiding behind 'la nota roja' and this lack of context, and very often no signature for this type of reporting, is self-censorship through fear of reprisals (Fig. 2).

3 The Case of *Los Ángeles Press*

3.1 Justification and Genesis of This News Medium

In this context of violence, self-censorship and a media which merely reports this violence in the most sterile way possible ('la nota roja'), apart from some exceptions like *Proceso*, *Zeta* or *La Jornada*, some interesting and brave reporting initiatives are appearing. These actions are based on investigation and condemnation of cases of corruption and the systematic violation of human rights. Only for this reason, for its philosophy, for choosing long tail journalism, 'slow journalism', opting for new narratives and for being a digital native, *Los Ángeles Press* is an innovative media in a context of extreme difficulties for working in journalism. *Los Ángeles Press* was founded in 2011 by the journalist Guadalupe Lizárraga and has its headquarters in the border city of San Diego (United States). Lizárraga is a journalist with recognized prestige in Mexico, with an extensive professional career and an outstanding academic education. She was born in Mexicali (Baja California), she completed her first professional studies at the Escuela Normal Urbana Federal Fronteriza, and then she did a master in Journalism in Spain at the School of Journalism of El País-UAM. She has a diploma in Public Policy and in Political Logic Analysis from the UNAM and has had specific training in working with women who are victims of violence at the Mental Research Institute in Palo Alto (California). She has received several awards and distinctions and published numerous research papers in journalism, with special

POLICIACA
Martes 18

FRONTERA
FRONTERA.INFO
Tijuana, B.C.

Editor:
Víctor Magdaleno
vmagdaleno@frontera.info

Coeditor Gráfico:
Víctor M. Aranda Reyes
varanda@frontera.info

Juez dicta prisión por homicidio en grado de tentativa

Acusan a imputado de disparar a otro hombre en cuatro ocasiones

REDACCIÓN/GH

Un sujeto señalado como presunto responsable del delito de homicidio en grado de tentativa recibió prisión preventiva como medida cautelar, informó la Procuraduría General de Justicia del Estado.

El imputado es identificado como Omar Orlando "N", quien fue denunciado por un habitante de la calle Presidente León de la Barra, en la colonia Presidentes.

Según la declaración del denunciante, escuchó detonaciones cerca de su domicilio, por lo que salió a la calle y vio como Omar Orlando "N" corría en dirección cuya, blandiendo un arma en la mano derecha.

SE TROPIEZA

Al ir corriendo, Omar Orlando "N" se tropezó y cayó al suelo, golpeándose la cabeza y quedando semiinconsciente.

A la llegada de los agentes de la Policía Municipal, se descubrió que Omar Orlando "N" había disparado en contra de un hombre minutos antes en cuatro ocasiones.

El lesionado proporcionó la descripción de su atacante y esta coincidió con la de Omar Orlando "N", por lo que fue detenido y turnado al Ministerio Público.

En la audiencia inicial, el Juez de Control dictó prisión preventiva como medida cautelar, con lo que el imputado continuará con su proceso en el Centro de Reinserción Social.

• Omar Orlando 'N', preso.

• Juez ordena el arresto del acusado como medida cautelar.

• Una de las víctimas de los asesinatos recientes yace en la vía pública, abatido por disparos.

SUMA 3 VÍCTIMAS LA OLA CRIMINAL

Tijuana está por acumular 100 homicidios en poco más de 4 meses y medio de este año

ÁNGEL GONZÁLEZ
agonzalez@frontera.info

Tres hombres fueron asesinados a diferentes horas y puntos distintos de la ciudad, en el transcurso del domingo, confirmó la Procuraduría General de Justicia del Estado.

En la calle San Juan Bosco de la colonia Lomas de San Martín, se reportó la presencia de lo que parecía un cuerpo envuelto en cobijas, a las 7:00 horas.

Agentes de la Policía Municipal revisaron el hallazgo y constataron que era un cadáver, por lo que notificaron a elementos de la PGJE.

Los peritos llevaron a cabo el proceso correspondiente y se detalló que el cuerpo correspondía a un hombre de alrededor de 35 años, el cual sufrió golpes contusos y presentó un surco en el cuello.

A las 9:30 horas, un hombre fue ejecutado en vía pública con arma de fuego, en la avenida Melchor Ocampo, en la colonia del mismo nombre, de la delegación Centenario.

A la llegada de los oficiales, el hombre, de alrededor de 25 años, ya se encontraba sin vida, siendo trasladado a las instalaciones de Servicio Médico Forense.

En otro hecho, a las 19:03 horas, fue encontrado el cuerpo sin vida de un hombre de aproximadamente 40 años, en la calle Chichén Itzá, en la colonia Mariano Matamoros Sur.

El fallecido sufrió lesiones de arma de fuego, siendo llevado a Semefo, en espera de que sea identificado.

Estadísticas de la Procuraduría General de Justicia del Estado señalan que en abril se han registrado 67 muertes en Tijuana, con las que se llega a 399 en 2017.

67
HOMICIDIOS SE HAN COMETIDO EN LOS DIAS QUE LLEVA ABRIL

399
ASESINATOS ACUMULA TIJUANA EN LO QUE VA DEL PRESENTE AÑO

Transeúnte agrede a chofer de taxi libre

ÁNGEL GONZÁLEZ
agonzalez@frontera.info

Un conductor de taxi libre fue agredido por un peatón en la avenida Centenario, frente al Palacio Municipal.

El hecho ocurrió a las 12:00 horas y según una de las pasajeras del taxi, se encontraban en el alto, esperando turno para avanzar.

Cuando el taxi hizo el alto, un hombre que iba a acompañado de dos mujeres le reclamó su tardanza para cederle el paso, lanzándose una botella de agua en el parabrisas.

El taxista se detuvo y, al tratar de bajarse de su vehículo, el hombre, de unos 30 años, no le permitió salir, golpeándolo en repetidas ocasiones hasta causarle lesiones en la cabeza.

"Parecía que estaba de mal humor porque el chofer hizo su alto bien y en ningún momento se portó grosero con el joven", explicó la pasajera.

Oficiales de la Policía Municipal que se encontraban cerca aseguraron al agresor; el conductor, identificado como Alfredo, fue atendido por paramédicos de la Cruz Roja.

Fuentes policiales confirmaron que el agresor no fue turnado al Ministerio Público, al llegar a un arreglo con el conductor.

Fig. 2 Front page of 18/04/2017 crime section. The newspaper *Frontera de Tijuana*

attention to cases like that of Isabel Miranda Wallace or that of Nestora Salgado or, especially her investigative journalism about feminicides in Ciudad Juárez. In this sense, we should mention her latest book titled: *Desaparecidas de la morgue* (2017).

This news medium is presented to its readers as a channel of information for learning about 'the news that isn't reported in your country' and it has the personality and the fighting spirit of its founder. Lizárraga (2016) herself defines *Los Ángeles Press* as:

Fig. 3 Front page of 21/05/2017, the main news item is the assassination of the journalist Javier Valdez

> A service of well-prepared news in Spanish from the United States, whose purpose is to break down the media circus and the censorship that exists in a country like Mexico, and other countries in the Central American region.

Working on context, therefore, becomes a way of innovating, as is having to transfer journalistic activity to the United States and organize a network of investigative journalism of an international nature (Fig. 3).

3.2 Context, Investigation and Condemnation

Three key concepts serve to define this innovative medium about Mexico with its headquarters in United States—context, investigation and condemnation. The references that have been the inspiration behind setting up and developing this communication medium are related to the founder's journalistic training itself. She recognizes how she was influenced by the school of investigative journalists and international information from the newspaper *El País* in the eighties and nineties of the previous century. Lizárraga also refers to the influence of the Spanish digital media natives like *Eldiario.es* or others like *La Marea*, both fifth and eleventh respectively (2015) in the ranking for innovation in Journalism, developed within the Master of Innovation in Journalism at the UMH.

> In Mexico, with very few exceptions, journalism is commercial, business and corporate. For this reason, we look to the trends in investigation that feature the great North American

media like *The New York Times* or *The Washington Post*, and also the innovative projects appearing in Spain, which are far-reaching and have many readers.

The consolidation of a journalism of context, investigation and condemnation in five years is a result of the depth of the sources Guadalupe Lizárraga works with and her team of collaborators. In order to extend her network of sources, mostly victims of abuses of power and violence, and to achieve their credibility and trust, this Mexican journalist's work has gradually developed a reputation with her personal stamp on it, making use of fundamental social networks like twitter and Facebook: @gpelizarraga y https://www.facebook.com/guadalupe.lizarraga. On Twitter, Lizárraga introduces herself as "a writer and independent journalist, founder of *Los Ángeles Press*", having more than 11,300 followers at present (around 3895 followers of the news medium itself).

For many victims Los Ángeles Press represents the hope that they can become known and can condemn the conflicts which they are immersed in. Many of these victims reach us through social networks and from that moment we dedicate our time to them, helping them and guaranteeing independent and rigorous investigative journalism. This is the only way we can gain their trust and help the reputation of our media to continue to grow.

In this context of violence, conflicting interests, threats etc. many of these sources reach the media through social networks, and they do so anonymously. In some cases this is from fear, in other cases it is from a desire to intoxicate, generate confusion, manipulate and try to hinder a journalistic investigation. "We have a verification process for anonymous sources that finally leads to establishing direct contact, either by telephone or in other cases in person", indicates the editor of this news medium.

3.3 Business Model and Growth Prospects

Los Ángeles Press defines itself as a non-profit news medium which emerged to fill the gap in cross-border media. It is organized as a civic association and has a strong international philosophy, with collaborators from Chile, El Salvador, Spain and other Spanish speaking countries. After five years' experience, the foundations of this news medium have been built on its contents and on the investigative journalism it practices. Its growth prospects are addressed through alliances with other similar news media with the common nexus of the Spanish language. The editor of LA Press foresees that in five years it will be financially solid because it covers informative needs and is generating a following which is helping them grow exponentially.

One of the collaborators on this journalistic project, the Spanish photojournalist José Pedro Martínez, is studying the development of the *Los Ángeles Press* business model (2016):

Funding for this news medium is supported through four channels of income: donation campaigns and sales of annual print editions; producing journalistic products for third persons; association with other investors, corporations and organizations; and diversification of activities. The latter will lead to book publications and documentaries using their own business model and holding photographic exhibitions, workshops, seminars, courses and lectures. In this sense, the social condemnation and informative nature of this medium takes on a differential factor.

One of the hallmarks of this project is financial independence, so at the expense of growing less or more slowly, *LA Press* foregoes advertising which could reduce or compromise this independence, whether it comes from business or fundamentally from institutions. Although they have managed to develop their activity thanks to altruistic and voluntary work and the micro-donations from readers, the project aims for a diversified business plan which allows them to combine income with different activities. This plan consists of strengthening the network of users who get involved in the donation campaigns, involvement of users through this sponsorship will be compensated through the different actions launched from the news medium. The diversification of activities also involves the sale of their own products, like a quality yearbook.

Currently, the average audience for *Los Ángeles Press* is a round 50,000 users a month, with more than 8000 followers on Facebook. Those responsible for the medium aim to promote loyalty among the Mexican public and increase their audience in the United Sates, where more than eleven million Mexicans reside. In this sense, they aim to continue networking and incorporating professional groups, associations, universities, NGOs, etc.

The project tries to incorporate other types of alliances and synergies by offering more than one advertising space. *Los Ángeles Press* offers journalistic products for organizations from other areas like NGOs and public entities, that is to say, strategic synergies and alliances that allow them to grow but at the same time remain independent.

3.4 Slow Journalism and New Narratives in Contrast to 'La Nota Roja'

Los Ángeles Press presents itself to the public as "a permanent exercise in new digital journalism which reinforces new public spaces created through networks and through the intellectual ability of an active and aware general public". It promotes online journalism that defends human rights and democracy, reporting the facts calmly despite being a digital medium, marked by its immediacy. For this reason, they prefer an in depth investigative and monitored journalism, where public condemnation is listened to and society is considered a source of information about the violation of rights in Mexico.

In this sense, genre is of special importance in this medium when covering issues that reflect the concerns of the Mexicans. So, we can observe how the 2448 posts published to date are classified under the tags and categories referring to the most relevant issues in Mexico and on the border with the USA.

Periodistas asesinados en México en 2017

18 de Mayo del 2017 Mexico Violento

Imagen original de La Jornada.

En México, el periodismo es profesión de alto riesgo. Hasta el 16 de mayo, van seis periodistas asesinados tan solo en el 2017. Javier Valdez es el nombre más reciente en una lista que crece rápidamente.

Fig. 4 Report from 18/05/2017 in the section *Violent México, Los Ángeles Press*

Out of the 426 tags, we can appreciate that among the 35 most important issues are human rights (451), extreme unrest in the state of Guerrero (183), President Enrique Peña Nieto (66 + 63), narco-trafficking (61), threatened journalists (36) and murdered journalists (30), violence against women and Ciudad Juárez (30) or information about PRI (35). These are issues which are difficult to deal with in depth in the Mexican press because of the serious consequences for these journalists and the media. In this sense, and given the violence against journalists, especially in the first half of 2017 (just as we completed this chapter), *Los Ángeles Press* is busy reporting and giving context to this scourge, with examples and by updating data and sources, as illustrated in Fig. 4.

In the analysis of the content of *Los Ángeles Press*, we can verify the in-depth treatment of events supported by many documented sources, as in the Wallace case, where *Los Ángeles Press* discovered proof that Wallace's son (presumably assassinated) was still alive, revealing a false kidnapping. Another case receiving thousands of visits to the different posts on the website was the kidnapping of the regional coordinator of the Community Police of Olinalá (Guerrero), Nestora Salgado who was imprisoned in 2013 under false charges of kidnapping for denouncing the involvement of local authorities in narco-trafficking (Table 1).

Likewise, in the categories that function as sections, we observe how *Violence in México* is the most used (18%), followed by *Our voice* (13%) and *The network*

Table 1 Number of posts on *Los Ángeles Press*, tagged according to subjects. Drawn up by authors

Most used tags			
Human rights	451	2012 Elections Mexico	43
Guerrero	183	Indigenous Rights	39
Los Angeles Press	134	Olinalá	38
Report	109	Educational reform	37
Sergio Ferrer	105	Threatened journalists	36
Mexico	92	USA	35
Nestora Salgado	83	PRI	35
Ayotzinapa	77	Tlapa	34
Enrique Peña Nieto	66	Francisco Bedolla Cancino	33
Feminicides	66	Ricardo Santes	31
Peña Nieto	63	#YoSoy132	31
Narco-trafficking	61	Emmanuel Ameth	30
Alberto Buitre	54	Ciudad Juárez	30
Corruption	52	Violence against women	30
Ramsés Ancira	49	Assassinated journalists	30
Community Police	49	Isabel Miranda de Wallace	28
Freedom of expression	44	Wallace case	28
AMLO	43		

fights on (10%). With respect to *The network fights on* it divides information by geographical areas: Asia, Europe, USA, Latin America and the Middle East, which clearly illustrates the transversal journalism of *Los Ángeles Press*, which is engaged in condemning the impunity and violation of human rights, not just in Mexico but everywhere in the world.

However, throughout Mexico and on the border with the USA, their commitment to the problems of violence has a greater presence because the information is close to its readers, and especially in the neighboring country. These issues are presented in the section *Violent Mexico*, which includes the sections *Pending Justice*, *Narco-politics, Corruption and more of the same* and *Violence against women*, all of which are present in the most significant categories. Another of the main sections deals with the United States and contains *Straight from Los Angeles*, *Migration* and *What's going on in the USA*. The section *Our voice* is the section that brings *Los Ángeles Press* collaborators together to write about different matters like *Mexican Intelligence*, *Women today*, *Contempt of the century*, *Audio report America* and *The people's blog*, which is about Hidalgo and is included on their website. Apart from these sections, there are others focusing more on the information than subject matter like *Latest news* and *Reports*, which are mixed in with the categories and tags mentioned above. Also, there is a section *Arteleaks* about Arte and Culture as a means to condemn (Table 2).

Table 2 Most used subject categories in *Los Ángeles Press*. Drawn up by authors

It is important to point out that most of the information is in Spanish, and although they have a section in English, they only translate some specific posts, which makes it obvious that this medium is aimed at Mexicans and the Hispanic public in general.

Social networks are fundamental to the dissemination of this medium. On Facebook, they have 8145 likes, while on Twitter they have 3888 followers, which added to the 11,300 of its founder @gpelizarraga, constitute a very active channel. This is used to either publish the information they have or for the public to be able to contact them to condemn injustices or to provide some feedback.

Los Ángeles Press presents itself as an alternative to 'la nota roja', it tries to provide information that is verified and responds to the causes of violence in Mexico. It especially aims to ensure accountability in most of the areas that affect the human rights of everybody and especially of the Mexicans. This distancing from the usual coverage of issues of violence in the Mexican media, beyond structure and contents, is also appreciated in the treatment and writing of texts, in the type of headline, construction of phrases, use of adjectives, attributing sources, etc. Within the framework outlined above, the narrative, which favors context and in-depth contents, is especially innovative in comparison with the traditional Mexican media. To conclude and according to the words of Guadalupe Lizárraga:

> We aren't interested in 'la nota roja', so fleeting, superficial and uncommitted that it dies a few minutes after being published online or on paper, our hallmark is long-term journalism and validity. If a user reads one of our reports in two years' time, it will still provide knowledge and context about what is happening in a failed state like Mexico, where the rule of law has collapsed. Our best investigative journalism is innovative in this sense, it doesn't lose validity, which is our challenge.

Acknowledgements This chapter has been written in the framework of the research work by MediaFlows, their group and their national projects: *Communication flows in the processes or political mobilization: media blogs and leaders of opinion* (reference CS02013-43960-R). Project financed by the Ministry of Economy (2014–2016); and the current (2017-2020), continuation of the previous R+D project: *Strategies, agendas ad discourse in election cyber-campaigns: communication media and the general public* (reference CSO2016-77331-C2-1-R).

References

Alfaro Víquez, E.G.: Los estudios del periodismo de nota roja. In: AMIC (ed.) La investigación de la comunicación en México ante la reforma constitucional en materia de telecomunicaciones, *ra*diodifusión competencia económica, pp. 630–642. Memoria XXVI Encuentro Nacional AMIC, San Luis de Potosí. http://amic2014.uaslp.mx/g5/g5_19.pdf (2014)

Álvarez, R.: La Nota Roja mexicana: más de cien años del periodismo más escabroso que puedas imaginar. Magnet. https://magnet.xataka.com/en-diez-minutos/la-nota-roja-mexicana-mas-de-cien-anos-del-periodismo-mas-escabroso-que-puedas-imaginar (2016)

Arriaga Ornelas, J.L.: "Colombianización" o "mexicanización" periodística. La nota roja en los noventa. Razón y palabra, 26. http://www.razonypalabra.org.mx/anteriores/n26/jarriaga.html (2017)

Asociación de Editores de los Estados AEE: http://www.aee.com.mx/?p=periodico-mexicano-tijuana (2016)

Artículo 19: México: Article 19 lanza informe anual "Estado de Censura". https://www.article19.org/resources.php/resource/37906/es/m%EF%BF%BD%EF%BF%BDxico:-article-19-lanza-informe-anual-%E2%80%9Cestado-de-censura%E2%80%9D (2017)

Cantú, O.: "¡Adiós!". El Norte de Ciudad Juárez, p. 1. http://nortedigital.mx/adios/ (2017)

Castellanos, U.: Manual de fotoperiodismo: retos y soluciones. Universidad Iberoamericana (2003)

Diario: http://diario.mx (2017)

El Bravo: Versión impresa. http://www.elbravo.mx (2017)

El Mañana: Directorio. http://www.elmanana.com.mx (2017)

El Mexicano: Directorio, p. 2. http://www.el-mexicano.com.mx/directorio (2017)

Freeman, L.: Estado bajo asalto: la narcoviolencia y corrupción en México. Las consecuencias indeseadas de la guerra contra las drogas. Washington Office on Latin América (WOLA). http://www.wola.org/sites/default/files/downloadable/Mexico/past/Estado%20bajo%20asalto.pdf (2016)

Frontera: Directorio. http://www.frontera.info/Empresa/Directorio.html (2017)

González-Esteban, J.L., López-Rico, C.M.: Cobertura de la violencia en zonas de riesgo: el caso de la Red de Periodistas de Ciudad Juárez. Index Comunicación **6**(1), 225–248. http://journals.sfu.ca/indexcomunicacion/index.php/indexcomunicacion/article/view/267/232 (2016)

Instituto Nacional de Estadística, Geografía e Informática INEGI: Encuesta Intercensal. http://cuentame.inegi.org.mx/poblacion/habitantes.aspx?tema=P (2015)

La Crónica: Principal. http://www.lacronica.com/Home.html#Principal (2017)

La Voz de la Frontera: Portada. https://www.lavozdelafrontera.com.mx (2017)

Le Clercq Ortega, J.A., Rodríguez Sánchez Lara, G.: Índice Global de Impunidad México. Fundación Universidad de las Américas Puebla, Puebla. https://www.udlap.mx/igimex/assets/files/igimex2016_ESP.pdf (2016)

Lizárraga, G.: Personal Interview in San Diego. United States (2016)

Lizárraga, G.: Desaparecidas de la morgue. Editorial Casa Fuerte, California (2017)

López Peña, M.: El estilo de la nota roja. Thesis, Escuela de Periodismo Carlos Septién García, México. http://quijote.biblio.iteso.mx/catia/CONEICC/cat.aspx?cmn=browse&id=315378 (2011)

Los Ángeles Press.: Periodistas asesinados en México en 2017. http://www.losangelespress.org/periodistas-asesinados-en-mexico-en-2017 (2017)

Martínez Pérez, J.P.: Los Ángeles Press, las noticias que no se ven en tú país. Master's thesis, Universidad Miguel Hernández, España (2016)
Melchor, F.: La experiencia estética de la nota roja. Los orígenes del periodismo sensacionalista en México. Replicante. Cultura crítica y periodismo digital. http://revistareplicante.com/la-experiencia-estetica-de-la-nota-roja (2012)
MIP: Ranking de Innovación en Periodismo. Universidad Miguel Hernández, España. http://mip.umh.es/ranking/ (2015)
Norte digital de Ciudad Juárez: Principal. http://www.nortedigital.mx (2017)
Nuevo Día: Edición impresa. http://www.nuevodia.com.mx (2017)
Organización Editorial Mexicana: Inicio. https://www.oem.com.mx/oem (2017)
Organización Panamericana de Salud, Organización Mundial de la Salud, Salud en Las Américas: Frontera Estados Unidos-México. http://www.paho.org/salud-en-las-americas-2012/index.php?id=63:united-statesmexico-border-area&option=com_content (2015)
Ortiz, A.: Los medios de comunicación en Baja California Universidad Autónoma de Baja California, México (2006)
Reporteros Sin Fronteras RSF: Veracruz: Los periodistas frente al estado de miedo. https://rsf.org/sites/default/files/rapport_veracruz_es-2.pdf (2016)
Rodríguez Morales, Z.: Entretejidos comunicacionales. Aproximaciones a objetos y campos de la comunicación. Editorial CUCSH, México (2010)
Secretaría de Gobernación: Padrón Nacional de Medios Impresos. http://pnmi.segob.gob.mx (2017)
Sociedad Interamericana de Prensa: List of members per country. http://www.sipiapa.org/contenidos/socios-pais.html?text=&c=435 (2016)
Trujillo, G.: La canción del progreso. Larva y Instituto de Cultura y Arte y cultura de Tijuana, Tijuana, México (2000)
Zeta: Portada. http://zetatijuana.com (2017)

José Luis González-Esteban Ph.D. in Journalism. MBA in Community Law from the European Union. Professor of Journalism and Mass Communication at the UMH (Spain). His lines of research focus on the study of the processes of transformation of politics and journalism. He has been guest lecturer at several North American, Mexican and European universities. Elche, Spain.

Carmen María López-Rico Ph.D. in Journalism. Professor of Journalism and Audiovisual Communication at UMH and IMEP (Spain) and guest professor at the UABC (México). As for her research, she has focused on political communication, with special attention in the electoral periods in Spain and US elections. She is currently researcher in Mediaflows group at Universidad de Valencia. Elche, Spain.

Manuel Ortiz-Marín Research Professor at the Faculty of Human Sciences from the UABC (Mexico). Degree in Journalism and Collective Communication from the UNAM; professor in Teaching and Educational Administration from the UABC; doctor in Social Communication by the University of Havana and postdoc at the Universidad Nacional de Cordoba, Argentina. Member of the National System of Researchers. Baja California, Mexico.

Digital Infographics: The Key to Information Paradigm

Ana Gabriela Nogueira-Frazão and Yolanda Martínez-Solana

Abstract The issues of memory continuity and informative credibility present themselves to Media Online as one of the main everyday obstacles. The search for structures and mechanisms for informative condensation and qualitative progression is the challenge that arises for this second stage of the current Technological Paradigm of Information and Communication. In a process where, due to the demands of a Net Generation, already digital native, visuality is in command, Digital Literacy is a crucial and undeniable necessity and, to Journalism, the Digital Infographic presents itself, as the structure that embraces all the Media competences, conferring to information and time, cohesion, longevity and efficiency.

Keywords Online journalism · Digital infographics · Digital literacy Newsblending · Informative cohesion

1 To the Continuation of Informative Memory

The truth is known and insurmountable: the events are endless and the Web's informational skills allow the simultaneity of its dissemination. In fact, the reticular memory structure, multiplex and also multiple and cumulative (Palacios 2003) articulates the competences of an immersive journalism, which is activated upon navigability.

From gatekeeper to gatewatcher, from newsmaking to newsblending, by media circumstances, the role of the journalist remains fundamental and irreplaceable. A deep knower, this professional takes on the role of a compass in the face of content's profusion. Thus, speech decoder (Alsina 1996: 18) in the union of the techniques of this hyper-productive and multi-dialogic platform, the journalistic 'tribe' is the

A. G. Nogueira-Frazão (✉)
Universidade Fernando Pessoa, Porto, Portugal
e-mail: ana@ufp.edu.pt

Y. Martínez-Solana
Universidad Complutense de Madrid, Madrid, Spain
e-mail: mymartin@ucm.es

M. Túñez-López et al. (eds.), *Communication: Innovation & Quality*, Studies in Systems, Decision and Control 154, https://doi.org/10.1007/978-3-319-91860-0_17

guardian of the efficacy of the symbolic ecologies (De Oliveira 2011) that builds cyber-identity and, also, the transnational interpretive community (Traquina 2002). In fact, the journalist knows that only the use of all the digital tools will increase his narrative competence because, in view of the journalistic self-service (Salaverría 1999), the pillars of the cyberjournalist should be the intelligibility, the visibility and the legibility (Charaudeau and Maingueneau 2006: 233) of the contents, as well as their critical and co-relative capacity of raw material.

Let's consider immersion, complementarity, speed and immediacy. Sancho (2008) writes that digital infographics should be considered as one of the resources of cyber-journalism, but perhaps the difficulty for a consensus in considering the infographic structure as a journalistic gender lies in its structural versatility, antagonistic to the classicism of journalistic structures that, even online, still comprise rules of composition which, altered, may give rise to informational noise.

But in the course of this information's speed buzz, the Producer's action over the content is not sufficient for the effectiveness of each one of those innumerous paths. The codex is Visual literacy, and the interpretation skills of visual messages are obligatory not only for their understanding, but also for the establishment of meaningful co-relations in (and from) each one of them. That is, through its characteristic of translinguistic hypertextuality Digital Infographic (because interactive and hypertextual) is, in a recursive way, a structure of paths and senses so that the integrative relationship that the infographic presents between the topic and the drawing turns out to be the most chameleonic and plurimorphological content of informative genders (Sancho 2008), characteristics which translates into an informative complex most likely complete and plurinarrative.

In addition, transmissibility, edition, diffusion and storage (Peltzer 1992: 90–93) are the pillars that define the infographic composite as cyberjournalistic and that makes Infographics the format to retrieve the informative gender on the Web (De Pablos 1999: 13).

2 Within the Scenario of Digital Information

Information is, today, a broadband concept. In fact, after the advent of New Communication Technologies, information assumes a wide range of concepts, techniques and configurations. From this multiplex narrative scenario, despite the fact that information is not dependent of a Media to exist, it is dependent on the Media to express itself and overcome the issues of informational entropy, despite one (or as a consequence) interactive, comprehensive hypertextual memory, diversified, immediate and cumulative.

Lévy (1998: 34), conceptualizing the questions of knowledge engineering, assumes that all human relations, as well as intelligence (individual, by social convergence), depend on the permanent and uninterrupted metamorphosis of the mechanisms, also assuming the technique as a fundamental dimension of the development and sustainability of these cognitive spaces, of a society of the mind and of the system of memory (Dividino and Faigle 2004).

But it is not possible to forget that the cognitive process unfolds through a simultaneous dynamics between considering and disregarding the information that, whatever its format, the network offers us. And assuming that by, the principle of Levy's metamorphosis the networked text is in a process of continuous (re)creation, if the question of informative utility leads us to choose, the question of the informational quantity (which, in itself, distances itself of the Media) leads, in face of processing and manipulation, to the user's entropy and deductive blocking.

However, hypertext is a multiple (for its different media) and multiplex structure (for the fusions of its different content), by interdependent and non-linear tree association, where the link is the measured value. Thus, the breakdown of the physical limits (Palacios 2003) serves not only for the question of the potentiation of the informative characteristics vis-a-vis the format, but also for the question of the 'infoxication' (Torres 2004), synonym, among others, for 'data asphyxiation' (Pettigrew *apud* Huber and Van de Ven 2005) or 'data smog' (Shenk apud Juskalian 2008): an informational overload that is associated with the concept of cognitive stress (Marcuschi 1999) and which Hari (2011) assumes as the reason Wired is not only explained "*with Internet connection*" but also "*high, frantic, unable to concentrate*".

In fact, too much information can be as harmful as the reverse. David Lewis (*apud* Buchanan and Kock 2001) certifies the term 'information fatigue syndrome', which explains that information overload can enhance the difficulty in embedding information—and, consequently, knowledge—as it leads to a blockage by clogging the cognitive process of informational perception.

However, revealing characteristic of the paradigm of informative individualization, is the fact that, from the unique stimulus-objective relationship, the interactivity stimulates the communicative process. Under this scenario, from the sender (the information architect), there is a need, during the processing of the event, to position itself analytically in the face of the identification of the system that best enhances the perception and informational interpretation. This emanating dynamic of projection of do-know and do-feel intends that, because of the utility of the information, that information will be reintegrated into the same information-communicative system, in a systematization that, within a tacit contract of communication, duplicates the process of the informative discourse to advance through the indispensable mutual ratification of the restrictions of the informative act. That means that, from the circumstances of the realization of an informative exchange (Charaudeau 2006: 68) the process that translates the world described and commented from the informative speech to the "News" built event, of the communications contract: structurally, a transformation—transaction—interpretation process (Nogueira-Frazão 2016: 220).

In fact, in this relationship between communicative and linguistic competences, visual information is supported by a conventional meaning controlled by all the communicative actors what allows a heterogeneous translinguistic decoding (because individual), but consensual (because social) of the multiplex narrative, a communication contract that, in this case, refers to the norms conveyed by the acceptance of the media access requirements (Damásio 2001a, b: 150–151). And it is from this conception of a cognitive ecology that Lévy (1998) assumes that the space of the sensory-intellectual interactions that fits the mental life of individuals must be habitable within a space where knowledge is in permanent metamorphosis. In fact, an online text is not read, but explored in a scanning reading (Gradim 2011) on the reading spaces, through a cognitive process of Measurement, Mapping, Exploration and Navigation (Dade-Robertson 2011) unfolded by information-seeking (Ellis 1989a, b, Wilson 1999).

The interpretation is individual and comes from the sum of the chosen sequences of informative cells. New technologies now activates the expansion of forms of knowledge, and a single informational entry becomes informative only if a certain degree of usefulness is achieved (whatever it may be) and is always the result of processing, systematization, and manipulation of data in order to achieve (or make achieve) knowledge.

3 Infography in the Face of Digital Literacy and Interactivity

The online is inseparable from interactivity: this ability to add functionality to a polysemic and hyper-adaptive multiplex narrative (Lipovetsky 2010: 95), that adds itself to the space and to the informative dimension with unparalleled individualization and documentation skills. This process drives the information—assuming it to be precise, timely, complete and concise (Gouveia 2006)—to be a tool that allows us to make conscious choices, but the web's ability to produce understanding is lost when not only the amount of information and its ergonomics, but also the arrival interval does not permit the cognitive pause that allows the interpretation and thought laterality that adds knowledge.

However, interactivity is inseparable from the path, the choice and the search for knowledge in a technological advent that joins instantaneousness to gratuitous information. In journalism, the behavior of information professionals has had a great impact on the productive and value-information routines. Also the behavior of the public has adjusted recharacterizing itself not to the format, like in the other advents, but to the Media.

Interactivity assumes the principle of heterogeneity and the hypertextual metamorphosis that projects path and meaning. The digital is multiplex and the narration is built by the sum of informative cells and the new multi-platform and multi-format journalistic model, impregnates of an informational consumption that requires an informative feed increasingly faster, clear and precise, less descriptive and more contextualized (Predoza et al. 2013: 4). Applied to the Infographics, the link is assumed to be a connection between lexias (Luna 2009), permissive of the process, but not of the content instantaneousness, and what permits the infographic structure to be a mechanism that enhances the perceptual capacity (Cairo 2011: 25), thus, cognitive stimulus to knowledge increase (Mayer and Moreno 2002).

However, it has already been realized that increasing information does not necessarily imply an increase in knowledge and may even imply a parallel increase in the difficulty of accessing what is actually intended (Nogueira 2013: 29). And that's why the choreography between interactivity and content has such a fundamental role in what is usability and why information architecture must plan the information structure translating its objectives to a very high degree of effectiveness in the relationship between the effort and the result over access to the information intended by the user. Today, the product is already beginning to be prepared for the future, but the user is still a digital immigrant, rather than a digital native (Prensky 2001), a generation that is only now beginning to manage an informative awareness. But how much is prepared for it?

The truth is that the transition has been slow and the technologies have not been seen as a means, but as the purpose. The advances have been unmistakable and the transition has been massive and transversal and it has already been realized the need of specialized areas in the Media for the potentiation of the content and the media platform. It has also been learned that the individualization of content facilitates the way of the user. There is, in fact, a new type of reader, but there is also a percentage of readers who are not prepared for the new journalism (Pereira 2015: 81). The return to the information cell is inescapable and information is increasingly visual, but media education issues are only now gaining credibility.

While Digital Immigrants learn and adapt with degrees of usability from such success that even merge their origins, Digital Natives have steeped themselves in technological everyday, in information speed, in multi-tasking, in non-linear reading, in image before reading. They work best when networked and take instant gratification as the best reward: a generational shock taking into account the need for time to tangibilize reality…

But to what extent does the ability to know how to interrogate the Internet fit the understanding of this process? To what extent is scanning reading of the Net Generation a mechanical act in which the search is by impulse and not conscious and/or deductive? What is the degree of web-functional illiteracy and what impact does it have on digital information reception?

The study carried out by Pereira (2015) is an important step in the perception of this reality, in Portugal. In an investigation that aimed to explore the difficulties that are behind the process of expanding interactive infographics concerning the Web as an information platform, this study, had 302 valid surveys with 50% of respondents between the ages of 21 and 30, a significant group that chose Internet as their preferred update Media, and that showed easiness in finding, while perceived process of Web content. About Infographics, Pereira (2015) reveals that, although the sample evaluated the infographic structure in a positive way, it was surprising the considerable percentage of respondents who were unable to interact with it, but, above all, the general lack of knowledge about what is an Interactive Infographic, a particularly worrisome issue since this percentage was in the active age groups that present close proximity to the Web. Also the focus group had some initial fear

Table 1 List of values presented by Pereira (2015) (own authorship)

	Internet as 1° media (%)	Easy to find/perceive content online[a]	No knowledge of interactive infographics (%)
<21	65.5	65.4% (máx. 24.1%)	69.0
21–30	70.1	75.4% (máx. 19.5%)	70.1
31–40	40.0	60% (máx. 25.0%)	65.0
41–50	25.0	43.8% (máx. 6.3%)	87.0
51–60			100
61–70			75.0

[a]Sum of a positive scale of 7–10 (as totally ease)

of exploring an unknown subject: those who were completely lost (about 25%) and those who interacted, but with uncertainty and hesitation (about 20%), but in none, the response on which the greatest difficulty in the interaction overlapped with the interactive action itself (Table 1).

Technological advances have facilitated information management as well as the elaboration of interactive infographics, not to mention making it possible to cross data in a more practical way (Predoza et al. 2013), however, this visual literacy not only presents a set of information and experiences co-participated, but gathers the intelligible competences for their understanding (Dondis 2007: 227): knowing how to organize, relate and extend the meaning of a visual message is essential for the perception of information ergonomics (Chammas and Moraes 2007) that provides and stimulates content delivery.

But due to the lack of incidence of this educational practice, the task is dense and vast, however visual literacy certifies and concludes the process of informative-technological interpretation since it encompasses, shares and systematizes the meaning of each information body. Deciphering these identities makes possible, even with different degrees of usability, the decoding of the hypertextual narrative and the advance to a process of an emergent evolutionary *continuum*, which, assuming Potter (*apud* Damásio 2001a, b) corresponds to a state of literacy that, visual, depends directly on the trinomial information-environment-technological expression, for the Media, which unfolds, for the user, in knowledge-understanding-technology, respectively (Damásio 2001a, b: 136–137).

4 The Productivity Scenario

Sartori (1997: 59) considers the audiovisual culture, uncultured and, in fact, in the face of the Web's advent, the lack of interactivity, scope and heterogeneity of Media's information dynamics—today, already traditional—end up assuming a behavior of *narrowcasting*. But in the informational input, the transmission of messages by images is still the fastest.

Perceived the importance of the digital infographic composite as a structure that simplifies information, that holds the reader and benefits the immediacy of content while preventing the deterioration of its value by mutualism of information cells, in an evaluation derived from the exploratory study presented by Nogueira-Frazão (2016: 241), and with a parallel strategy, through the analysis of the components Periodicity, Interactivity, Staticity, Integration and adding Participation, we found the indicators that relate the positioning of the professional, as well as the journalistic enterprise, concerning the digital infographic composite.

In short, the infographic styles are maintained from the earlier analysis. In Portugal, Diário de Notícias (DN) keeps absent from infographic production since 2014 and Correio da Manhã (CM), currently publishes images without date and that only make you notice some informative updating, from the titles. But as characteristics of the 118 infographics analyzed -seven days per sample, all in sight as infographics of the day, of that day or in list—did not differ much through the years: 45 in 2014, 30 in 2015 and 43 in 2017. Although very similar to 2017, of the three years, 2015 is the one that maintained the highest level of Periodicity (Fig. 1), a record only possible because El País (ELP) remains as the online newspaper that maintains a, sometimes extensive, daily production, much probably supported by its Editorial Design that can help to accelerate the constructions process (Nogueira-Frazão 2016: 275). However, the high rate of production of this Media, is sustained from an equal degree of Staticity, tendentiously free of Interactivity (Fig. 2). On the other hand, with the exception of 20 min (20M), in Spain, and Público (PUB), in 2017, Interactivity is also a criterion that has been decreasing in Digital Infographics. In summary, of the 118 infographics only nearly over 1/3 of the sample (50 units) is Static and Interactive.

But before continuing, lets locate the digital infographics in each of the sample elements. In 20M, the link to the infographics is very accessible in "Gráficos", but although El País presents different national and international publications, it is a journal that does not include any direct link to the Infographics section. In Portugal, the Público includes the infographic composites within the "Multimedia" link located in "Mais" and Correio da Manhã does the same, only, at the end of the page. For its part, Jornal de Notícias, includes Infografics in 'tag/infografia.html'. Concerning Diário de Notícias, awarded in 2011 with the Favorite Website Awards (FWA) and, in 2010, with the recognition of the best design project for informational site, for its aesthetic quality that enhances the news and its perception[1] is the only which, since 2014, does not have infographic journalistic structures.

And if the external link, that intersects, that could be a tool for assimilation and habituation to Digital Infographics and of impulse for literacy, does not exist or is difficult to find such (as much as composite Interactivity, a factor of appeal to the structure and to information dynamics), also the issue of Integration (Fig. 3), particularly in Portugal (and except for Público), is absent. Spain has a different

[1] http://www.dn.pt/tv-e-media/media/interior/amp/dnpt-ganha-premio-de-melhor-site-de-informacao-1549184.html.

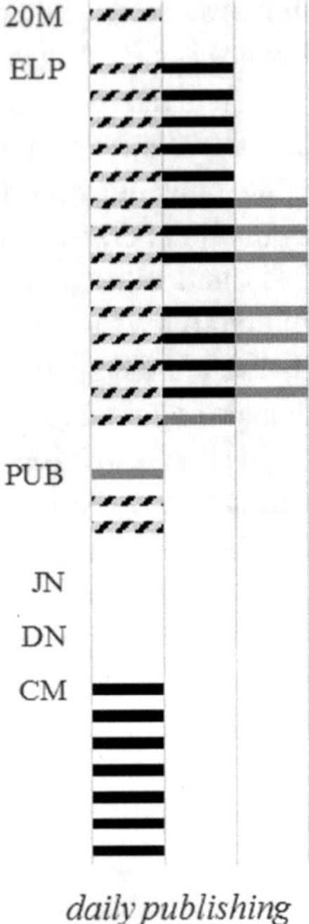

Fig. 1 Periodicity

scenario, however, with a major decrease from 14 infographics in 2015, to 4, in 2017. And if Integration is a visibility tool, between 'Comments' (on page) and 'Shares' (in special on Facebook), the values for Participation correspond immensely to the very weak projection verified: much more shared than commented, only 14 of the infographics have the participation of the users and, majority, in Portugal (Fig. 4).

On another front, the Producers efforts, here represented by Infographics Society (https://sociedadeinfografica.wordpress.com), so far, the only professional association in the Iberian Peninsula.

As for some authors (Alonso 1998; De Pablos 1999; Sojo 2002), also for the respondents,[2] there is no doubt that the Infographic can be considered a Journalistic genre (Nogueira-Frazão 2016: 259). In this sense, on the side of the Producer, it is necessary to strengthen the Media in favor of the format. Thus, essential to the construction of an Infographic composite applied to Online Journalism, the use of Image, Interactivity and (much) of Navigability potentiates and drives the storage and retrieval of data, which, for its most important confluence between the structure and the Media, should be the elements with greater incidence in the composite Digital Infographic (Nogueira-Frazão 2016: 259–260).

[2]29 members, 12 interviewed all over 35 years of age, of Portuguese nationality, with proven specializations in the areas of Design, Information, Journalism and Infographic and professionals from various national and international media.

Fig. 2 Structure

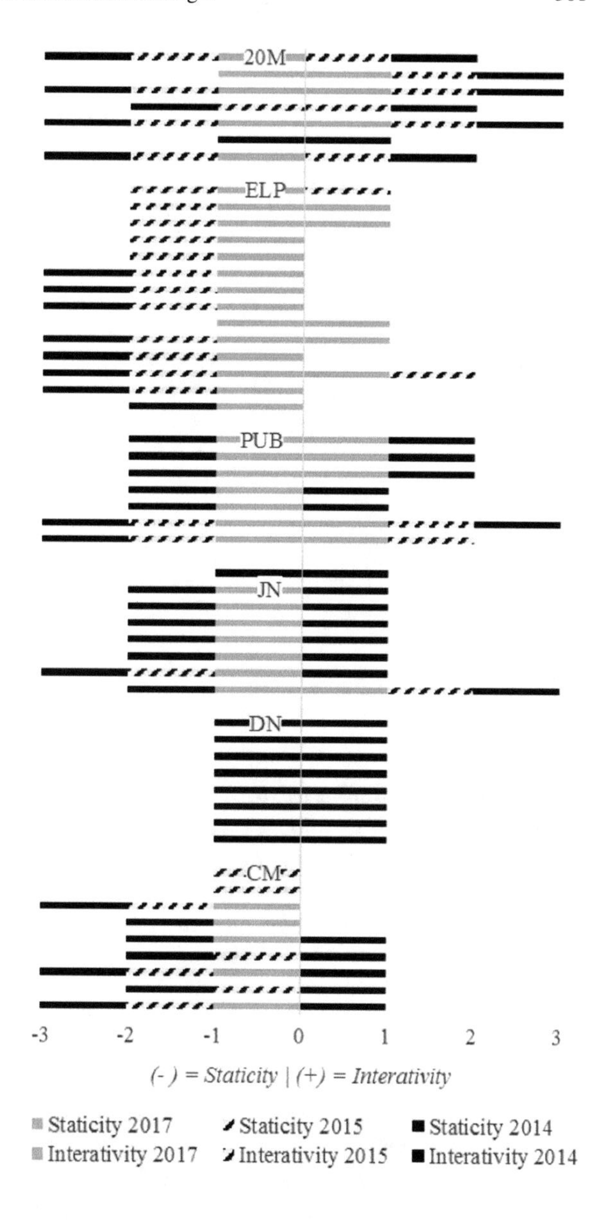

But, knowing this, who should be responsible for the composite infographic? How many levels of execution is the development of an Infographic? Contrary to what one might think, the tasks are not as segmented (Nogueira-Frazão (2016: 263), as we may believe, with the exception of writing and data management, the responsibilities of creating a Digital Infographics tend to be both. However, in the question of building the journalistic infographic structure, which is what it says in the organization of its

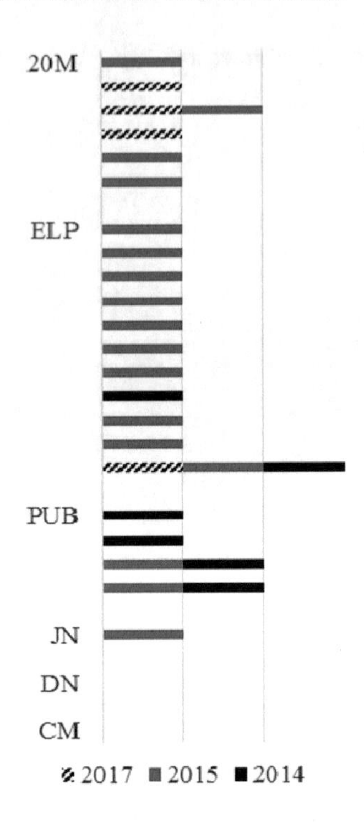

Fig. 3 Integration

informative cell multiplex network, the responsibility should be shared between the journalist and the infographics producer.

And because cyberspace is multidimensional, in the creative process applied to infographics, the skills that are favorable to a reading that allows this multiplex platform to flow in favor of the perceived ergonomics are of extraordinary importance. Thus, layering information, its systematization and organization at the different levels of the online journalism pyramid, assumes, in advance, and as justify the data of the conception (Fig. 5), a good systematization and knowledge of the contents as well as an understanding careful and accurate on the stimuli that hold the attention of the reader to the infographic composite.

In other words, in addition to the importance of the relationship between the infographic compound and the graphical presentation of the information, and the way in which this presentation captures and maintains the reader's attention, the relation of informational and visual synchronicity to the content that obviously intensifies the dialogue of the user with the information (which includes Interactivity, Navigability and Content Levels) is of the most importance to the Producer, curiously, more than the question of structure, organization and visual fluidity of the composite.

And the truth is that digital infographics, deriving from the Media as an informative model that offers the user the empowerment of the Media itself, is the structure to which the functional aesthetics of online journalism converges. From here, the analysis must be made on what are the main reasons for the use of Digital Journalis-

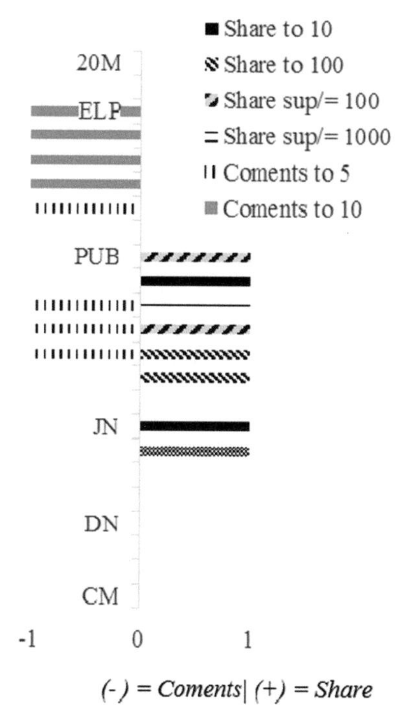

Fig. 4 Participation

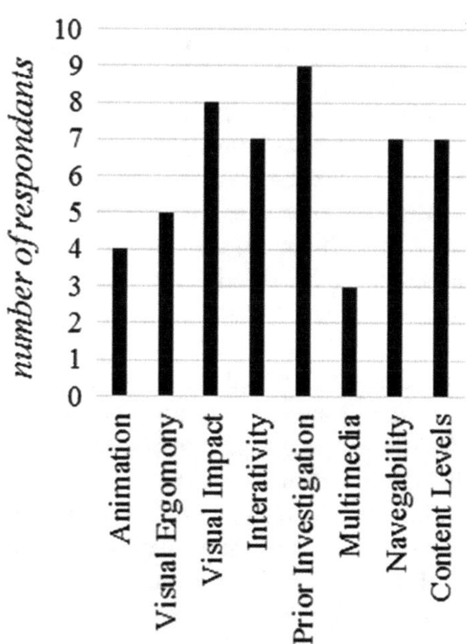

Fig. 5 Elements of conception

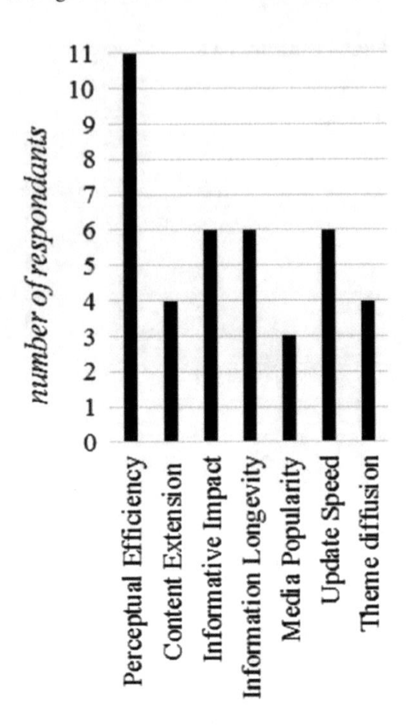

Fig. 6 Reasons to use infographics

tic Infographics (Fig. 6). As expected, perceptual efficiency is an heavy-weight as a reason to deepen the use of the Infographics in Online Journalism, a reason that, supported by de information impact, longevity and update speed, comply Instantaneity, something that, as known, Net Generation is used to and even demands; a concept that proposes contents depth and laterality of the theme.

Thus, realizing that the visualization of information is a tool to increase cognition (Cairo 2008: 34) and that the simplification of the information is completely discordant to the deterioration of the data (Cairo 2008: 33), it is assumed that the questions of speed of novelty its actualization, of memory as knowledge, of the physical and meaningful nature of the message, as well as of the efficacy of the smallest size (Drucker 1989: 223) transform, through an Objectivist Rhetoric (Casasús 1991: 121), the Qualities of Information (accurate, timely, complete and concise) in the Qualities of Language, moving from a Newsmaking to Newsblending Paradigm, through a format that stands out in the probabilities of minimizing the effort in the information retention process, amplifying the perceived result, by arousing the attention and conquering the concentration of the reader.

However, this digital structure of interpretation and organization of journalistic content, which processes, through Interactivity and Levels of Contextualization and Information Exploitation (Nogueira-Frazão 2016: 149), a sensorial and affective effect of permanence and narrative, may not correspond to this expectation, as well identified in the data about Infographics' levels of Interactivity. In this scenario, the other side of the exaltation of the use of Digital Infographics is the Structural Faults

(Fig. 7), elements that may jeopardize the reception and perception of the message, which is, as in the long run, the question of Memory and Scope.

For SI respondents, the greatest danger is that infographic composite does not take into account the amount of competencies in each one of the elements that verify the information perception, nor if it doesn't include the perceptive objective in the information strategy. And the fact is that the change from the classic informative structure to a multiplex narrative with a reading of Contextualization and Exploration can allow a professional who does not master his complexity of lexias and interconnections, not to fulfill the informative objectives.

And perhaps the value of the visual impact as well as the care with perceived effectiveness certify some security concerning the visualizations, the certainty of readability as well as navigability (a plus for interactive) as an intuitive process are Reasons for care and, in absence, for failure. However, if we consider the importance that Interactivity has for the Producer as an element of conception, if we consider the issue of interest as registered visualizations -which is like saying, sharing or commenting- and taking into account that the quality of the overall experience of the User is going to evaluate the usability of the product (a process that, through the link as Measurement Unit of visualizations, gives the value of the structure) the data referring to 2017 are not the most favorable…

Less important seems to be the issue of information continuity, but assuming the versatility and independence of a web structure, is also assumed its evolutionary autonomy. For this same reason, the issue of failure due to lack of convergence Multimedia presents itself as a failure, above all because, journalism will only survive if takes advantage of the digital platform (Palacios *in* La Capital 2012). But although Interactivity should be taken as a very important item, to the SI respondents, its absence may not be a fault, probably considering, in advance, the editorial criteria of the Media as is not, for certain, the compliance of being a way for topic diffusion in the Web: it´s there, it is what it is, so use it if you want.

Nevertheless, above all, it is necessary to evaluate the scenario of inclusion of Digital Journalistic Infographics in the daily media as a process that has its own time, just as the paradigm transition, in some cases, still has it. Each step must be certain and resilient, and the fact is that, for the producers, if it was undeniable that the Infographics is a journalistic structure, it is also indisputable that the Infographic is a rising area (Nogueira-Frazão 2016: 264) or because is more often used in the traditional areas or because it starts to leave these limits…

But the care that the Producer takes on issues of informational and visual efficiency, attesting to the efficacy of the composite, the data tends to indicate that the question of Utility it may not be so linear. That is, the little use and projection that periodicals make of the informational composite does not allow the trust between product and user to be established (Túñez and Nogueira 2017). This means that the percentage of effort that exists does not correspond (yet) to a sensorial and affective effect of permanence and narrative, specifically, by the transverse absence of continuity (1) Productive, in the convergence of formats, (2) Informative, at the confluence of contents, (3) of Usability, in the relation between the Media and the User and (4) of Integration, in the relationship between the Media and the Product.

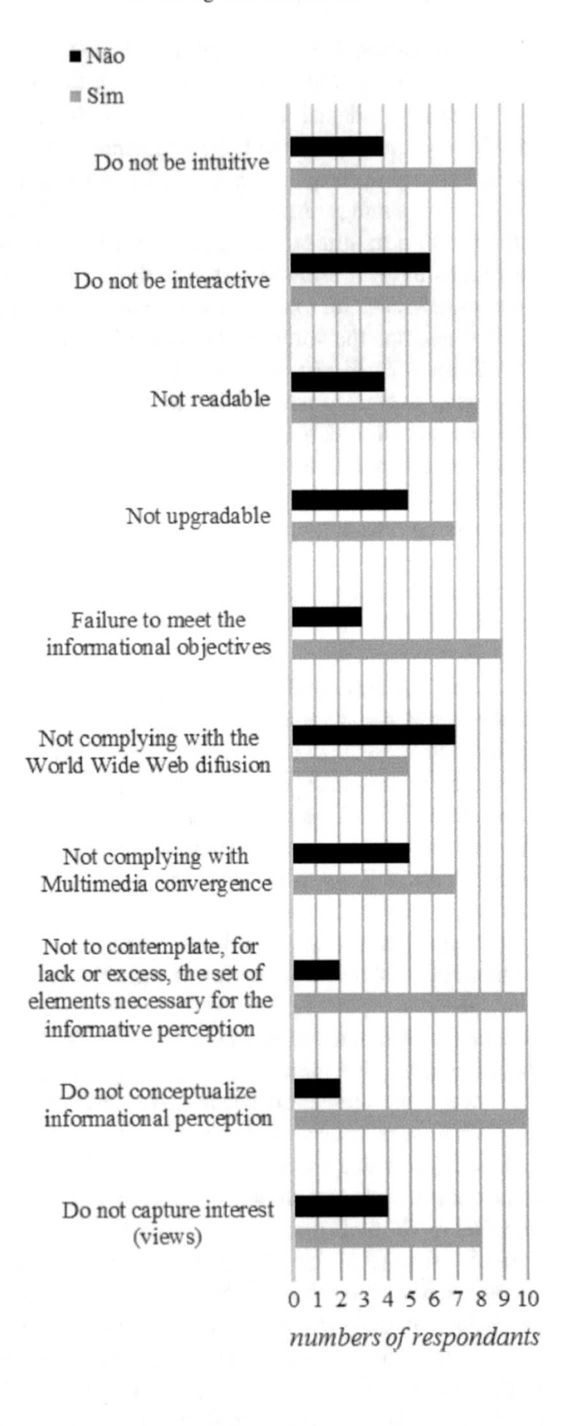

Fig. 7 Failures of online infographic structure

In short, despite its innate potential, the full Efficiency and Usefulness (as Content Confront and Progression) of infographic's content fall short on the Media's information function, a requirement that ultimately interferes with the proportional decrease of content's Utility either to the Media either to the Reader, which are confronted with the absence of cohesion for Continuity and by Continuum: the first, for the need to recover, in an effective and integrated way, the continuity of the topic and the interest of the reader, which, in this case, covers all ages, socio-cultural and economic segments; the second, because of the speed between the event and its instantaneousness, that is, the widening of distance and information diffusion.

On the other hand, the deficit of Usability (as Content Integration and Adaptability) also interferes with the retrieval of the infographic composite, above all, by distancing itself from the nucleus of the informative flow on the Web—informative dexterity, density and instantaneousness—that goes alongside a versatile, comprehensive and transverse, eclectic, hybrid and multiplex language that, through a composite, connective and meaning-producer, that facilitates the readability and enjoyment of the discourse(s) and allows, to the reader, this informative honesty to totally understand (Costa and Mole 1991: 21) and to be oriented in the totality.

Online Infographic is an emerging format (Nogueira-Frazão 2016: 294). The absences are of integration and the style of the composite does not interfere in the final perceptive result. The format, even that isolated, is accomplished. On the other hand, the Generation Net profile privileges visual information to text (Sousa 2012). Therefore, it is not the role of the producer, but the owner to make time for the editors with (new) publication strategies prepared to redefine the current concepts of informative journalism.

The online is inseparable from interactivity: this ability to add functionality to a polysemic and hyper-adaptive multiplex narrative (Lipovetsky 2010: 95), which adds itself to the space and to the informative dimension with unparalleled individualization and documentation skills, drives it—assuming to be precise, timely, complete and concise (Gouveia 2006)—to be a tool that allows to make conscious choices. But the Web's ability to produce understanding is lost when not only the amount of information and its ergonomics, but also the arrival interval does not permit the cognitive pause that permits the level of interpretation and thought laterality that adds knowledge.

However, interactivity is inseparable from a path, from the choice and the search for knowledge in a technological advent that joins instantaneousness to gratuitous information. In journalism, the behavior of information professionals has a great impact on the productive and information-value routines, also the behavior of the public has adjusted, recharacterizing itself not to the format, as in other advents, but to the Media.

5 Digital Infographics: *Stage II* of the Communication Paradigm

In its top exponent, Online Infographics brings together the structural competencies to support informative impact and effective information management, thus translating into the epitome of thought and codification technology (Costa and Mole 1991: 29). In fact, the assertive, directed and individualized requirement of *Homo Videns*, for which the image is secondary to the word, allows the resumption of a *Homo Sapiens* positive integration (Sartori 1997: 9) and the interactive and polysemic potential of this online structure, as well as the competent use of Media through the empowerment of its content, transforms the message into a hyper-adaptive platform (Sabbatini and Betania 2004). For the information architect, the challenge of, through the Infographics, to give intelligible access to data that would not be in another format, allows the internet user to perceive memory and knowledge.

But to this discursive inventory (Traquina 2005: 43–44) effectively allow a simulacrum of contact (Dalmonte 2008)—which, through the speed of interpretation, intensifies the dynamic skills of visual intelligence—it is essential to consider the value of Utility and Efficiency, since the degree of Usability applied to a infographic, multiplex narrative allows the association of paths and senses, and potentiates lateral and depth knowledge on time and in time. And if the visual syntax does not prescribe (Calado 1994: 49) this informative, continuity mise-en-scene ends up diluting, with effectiveness and credibility, sustained by detail and comprehensiveness, the issues of lack of space and time, in a process of functional profitability of the information, so that Digital Infographics can be considered the structure that ensures, through correlative inclusion, the information longevity (Freitas 2008) by survival and utility.

However, in spite of the transmedia simultaneity (APDSI 2012), the convergence of the communicated senses occurs in a singular way and is reflected within the interrelations of the individuals. And if, on the production side, the definition of an infographic project should satisfy the characteristics of the Net Generation (which systematizes the time for the effectiveness of the click), on the other hand of this communicative competence in the face of the linguistic multi-functionality, a socio-pedagogical dynamics of media literacy must be born to translate and assure Media critical competences and its hand-to-hand responsible communication content phenomenon, therefore, of citizenship.

From this point of view, the characteristics of the reader (immersive) and of the reading (by scanning) converge where the multi-predictability of the multi-narrative assumes a timeless and timelessness hierarchy over the spaces of instantaneous information. When used, because of its cellular structure of assertive lateral and deep information, Digital Infographic identifies itself as an informative magnet and keystone, as the origin of readability, longevity and information resilience, but mostly, as a discourse fluidity informative structure that allows the user to orient themselves towards the whole: an effective interface of citizenship service (Sancho 2010).

And this digital and informative availability (Donas 2004) breaks with a continuous cycle of obsolete, meaningless and disaggregated information, which allows the creation of a techno-informative and self-instrumental rationality, but social validator, because interpretive. The infographics may then be the epitome of an ecology of knowledge (Donas 1999) and of the perceived citizenship that could be the first stone of Ramonet's "5th Power" (Ramonet 2003), of a recovery of news value to the web and, consequently, of the news-value for the Media.

And the truth is that within the issues of compensation and cognitive stimulation, if a Media offers sources in quantity, quality, and diversity simultaneously and to the information, set and dimension, that Media is, tendentiously, more credible therefore, the most influential (Edo 2003: 111). In this scenario, if to the producer, Utility and Efficiency are the conceptual basis, to the online journalistic business, the time-actuality-information-credibility dynamics on which it moves, still does not present, in a confirmed way, a maximization strategy of resources to information understanding (development) and sustainability (facilitation).

But the future looks promising. Given the results of Infographic Society's data and some Infographics seen in the research[3] the degree of differentiation of a digital infographic composite—which we characterized as (1) speed of content integration, (2) accessibility, (3) time savings (4) popularity, (5) adaptability to audience needs and of content levels, (6) dexterity and intelligibility, and (7) longevity—becomes capable of suppressing the risk of entropy in the perceptive quality of the message. On the other hand, since only perceptive reading stimulates memory and communication, to the effect of design that ensures usability, to the success of the discursive coherence of the multimedia body, is indistinguishable the aesthetic of ergonomics of the information structure.

In resume, if it is the Legibility and Readability (IBM, n/d) that materialize digital infographic dynamics, it is the Usability of an intelligible content that confers deepening and fragmentation to the complementary and convergent informational infographic composite. However, parallel to the socio-informative requirements, the technological advance is shaped and, with it, the habits of cyberjournalistic consumption are molded, a process that includes reconfigurations not only in the hierarchy but also in the narrative coherence, as well as in the value of its structuring elements and it will be an integrated newspaper room that will be able to respond more rapidly and effectively to the progressive demands of an increasingly multiplatform reader

[3]As examples: (1) "Para lá do Marão já não vão só os que lá estão" (JN—29.03.2016)—animated digital infographics in a tripartite scenario, with interaction by instruction and horizontal structure that explores and explains the all about an important road tunnel in Portugal; (2) "Um Planeta em exaustão" (PUB—25.03.2017)—static, site-based, interactive infographic with instructional interaction that explains the scenario of the ecological footprint and biocapacity; (3) "Virus y bacterias: pequeños, pero matones…" (20M—22.10.2013)—static infographic, manipulated by interaction and manipulation of a horizontal scroll structure that presents the relative dimensions between the human being and dangerous viruses and bacteria; (4) "As equipas e os pilotos do Mundial de Fórmula 1" (JN—18.03.2014)—animated scrolling, bipartised, fostered activated by interactive interaction with instruction and horizontal dynamics that explores the teams and circuits of the F1 World Championship; (5) "50 años del asesinato de Kennedy" (20M—19.11.2013)—animated, interactive interaction by manipulation and scroll digital infographic in a vertical structure that explains and explores the JFK assassination.

(Salaverría 2010), who demands an Intelligent Speed, Summative Memory, Effective Credibility and an Informative laterality and depth.

In fact, being information online the front line of the Information Society, 'Phase I' of the advent of the New Technologies of Communication and Information is complete. It is urgent now for an emerging literacy to act on this hybrid narrative, stratified, non-linear, in information overload and greedy of informative immediacy, transforming it into a phenomenon of communication-therefore of citizenship—so that through Digital Infographics, 'Phase II' of the Online Journalism scenario of the current Paradigm of Communication begins.

References

APDSI—Associação para a Promoção e Desenvolvimento da Sociedade da Informação: Cultura e Arte na SI: indústrias criativas: 14ª Tomada de posição do GAN (grupo de alto nível). http://www.apdsi.pt/uploads/news/id604/14tpgan_apdsi_vfinalrev3.pdf (2012)
Alonso, J.: Grafía, el trabajo en una agencia de prensa especializada en infográficos. Rev. Latina de Comunicación Social, 8. [Online]. https://www.ull.es/publicaciones/latina/a/49inf6.htm (1998)
Alsina, M.: La construcción de la noticia. Paidós, Barcelona (1996)
Buchanan J., Kock N.: Information overload: a decision making perspective. In: Köksalan, M., Zionts, S. (eds) Multiple Criteria Decision Making in the New Millennium. Lecture Notes in Economics and Mathematical Systems, p. 507. Springer, Berlin, Heidelberg (2001)
Cairo, A.: Infografía 2.0: visualización interactiva de información en prensa. Madrid, Alamut (2008)
Cairo, A.: El arte funcional: infografia e visualização de informação. Alamute, Madrid (2011)
Calado, I.: A Utilização Educativa das Imagens. Porto Editora, Porto (1994)
Casasús, J.M., Ladevéze, L.N.: Estilo y géneros periodísticos. Ariel, Barcelona (1991)
Charaudeau, P., Maingueneau, D.: O discurso das mídias. Contexto, São Paulo (2006)
Chammas, A.S., Moraes, A.D.: Ergonomia Informacional: fundamentos e aplicação. Encuentro Latinoamericano de Diseño. http://fido.palermo.edu/servicios_dyc/encuentro2007/02_auspicios_publicaciones/actas_diseno/articulos_pdf/ADC089.pdf (2007)
Costa, J., Mole, A.: Imagen didáctica. CEAC, Barcelona (1991)
Dade-Robertson, M.: The Architecture of Information: Architecture. Interaction Design and the Patterning of Digital Information. Routledge, New York (2011)
Dalmonte, E.F.: Efeito de real e jornalismo: imagem, técnica e processos de significação. FAMECOS/PUCRS, (20), 41–47. http://revistaseletronicas.pucrs.br/ojs/index.php/famecos/article/viewFile/4832/3688 (2008)
Damásio, M.J.: Práticas educativas e novos media: contributos para o desenvolvimento de um novo modelo de literacia. Minerva, Coimbra (2001)
Damásio, M.J.: Novas literacias, novas ferramentas educativas. Caleidoscópio. Rev. de Comunicação e Cultura. **1**, 59–69. http://revistas.ulusofona.pt/index.php/caleidoscopio/article/view/2184 (2001b)
De Oliveira, A.R.: O que é cibercultura@ - e como ela pode inverter o sentido do vetor tecnológico. Rev. Eletrônica do Programa de Pós-graduação da Faculdade Cásper Líbero. **1**, 1–7. http://www.revistas.univerciencia.org/index.php/comtempo/article/viewFile/7580/7232 (2011)
De Pablos, J.M.: Infoperiodismo: el periodista como creador de infografía. Madrid, Síntesis (1999)
Dividino, R.Q., Faigle, A.: Distinções entre memória de curto prazo e memória de longo prazo. http://www.ic.unicamp.br/~wainer/cursos/906/trabalhos/curto-longo.pdf (2004)
Donas, J.B.: Saber e dizer na era do ciberespaço: em direção a uma ecologia do conhecimento. HIPNOS **4**(5), 355–364 (1999)

Donas, J.B.: ¿Qué puede esperar la democracia de internet? Una reflexión sobre la crítica de Langdon Winner al poder político transformador de la tecnología. Argumentos. **7**, 13–49. https://dialnet.unirioja.es/ejemplar/106876 (2004)

Dondis, D.A.: Sintaxe da linguagem visual. Martins Fontes, São Paulo (2007)

Drucker, P.F.: As novas realidades: no governo e na política, na economia e nas empresas, na sociedade e na visão do mundo. Livraria Pioneira Editora, São Paulo (1989)

Edo, C.: Periodismo informativo e interpretativo: el impacto de Internet en la noticia, las fuentes y los géneros. Comunicación Social, Sevilla (2003)

Ellis, D.: A behavioral approach to information retrieval system design. Journal of Documentation. London, **45**(3), 171–212, Sept. (1989a)

Ellis, D.: A behavioral model for information retrieval system design. J. Inform. Sci. London, Cambridge, **15**(4–5), 237–247 [Online]. https://journals.sagepub.com/doi/pdf/10.1177/016555158901500406 (1989b)

Freitas, H.: Infografias de imprensa, digitais e televisivas: Comunicar seduzindo o olhar. Rev. JJ (jul-set). http://pt.scribd.com/doc/55437100/%C2%ABInfografia-um-novo-genero-joranalistico%C2%BB-35-da-JJ (2008)

Gouveia, L.: A Gestão da Informação: um ensaio sobre a sua relevância no contexto organizacional. In: CXO Media. Information Lifecycle Management: criar a empresa centrada Informação, pp. 174–180. CXO Media, Lisboa (2006)

Gradim, A.: Os géneros e a convergência: o jornalista multimédia do século XXI. Agora Net–Rev. sobre Novos Media e Cidadania, (2). www.labcom.ubi.pt/files/agoranet/02/gradim-anabela-generos-convergencia.pdf (2011)

Hari, J.: Johann Hari: how to survive the age of distraction. Independent. http://www.independent.co.uk/voices/commentators/johann-hari/johann-hari-how-to-survive-the-age-of-distraction-2301851.html (2011)

Huber, G.P., Van de Ven, A.H (eds).: Longitudinal field research methods: studying processes of organizational change. Sage Publications, Thousand Oaks (2005)

IBM: Conseguindo boa legibilidade e "leiturabilidade" na Web. iMasters Blog. http://www.ibm.com/developerworks/community/blogs/fd26864d-cb41-49cf-b719-d89c6b072893/entry/conseguindo_boa_legibilidade_e_leiturabilidade_na_web1?lang=en (n/d)

Juskalian, R.: David Shenk on Data Smog. Columbia Journalism Rev. http://archives.cjr.org/overload/david_shenk_on_data_smog.php (2008)

La Capital: El periodismo que sobrevivirá es el que aproveche la plataforma digital para contextualizar la noticia - La Ciudad. [Online].www.lacapital.com.ar/la-ciudad/El-periodismo-que-sobrevivira-es-el-que-aproveche-la-plataforma-digital-para-contextualizar-la-noticia-20120605-0071.html (2012)

Lévy, P.: As tecnologias da Inteligência – O futuro do pensamento na era da Informática.http://wp.ufpel.edu.br/franciscovargas/files/2015/03/LEVY-Pierre-1998-Tecnologias-da-Intelig%C3%AAncia.pdf (1998)

Lipovetsky, G., Serroy, J.: O Ecrã Global. Cultura mediática e cinema na era hipermoderna. Edições 70, Lisboa (2010)

Luna, D.D.: Links como articuladores do sentido na narrativa do hipertexto jornalístico do portal JC Online. Anais Hipertexto, 1–8. www.ufpe.br/nehte/hipertexto2009/anais/g-l/_links-como-articuladores-do-sentido-diogenes.pdf (2009)

Mayer, R.E., Moreno, R.: A cognitive theory of multimedia learning: implications for design principles. http://lowellinstitute.com/downloads/Research/Cognitive%20MultimediaTheory.pdf (2002)

Marcuschi, L.A.: Linearização, cognição e referência: o desafio do hipertexto. Coloquio Latinoamericano de Estudios del Discurso, 3. http://web.uchile.cl/facultades/filosofia/Editorial/libros/discurso_cambio/17Marcus.pdf (1999)

Nogueira-Frazão, A.G.: Infografando: análise das práticas infográficas das edições online de jornais diários generalistas da Península Ibérica. Ph.D. dissertation, Universidade de Santiago de Compostela, Santiago de Composela (2016)

Nogueira, P.: Infografia digital: da produção à receção. Masters' thesis, Universidade do Minho, Braga (2013)
Palacios, M.: Ruptura, continuidade e potencialização no jornalismo on-line: o lugar da memória. In: Machado, E., Palacios, M. (orgs.) Modelos do Jornalismo Digital, pp. 13–36. Edições GJOL, Salvador. www.facom.ufba.br/jol/pdf/2003_palacios_olugardamemoria.pdf (2003)
Peltzer, G.: Jornalismo iconográfico. Planeta Editora, Lisboa (1992)
Pereira, A.R.B.: Dificuldades de expansão da infografia interativa: um problema de produção e receção. Masters' thesis, Universidade do Minho, Braga (2015)
Predoza, N., Lima, P., Nicolau, M.: Tratamento da informação no webjornalismo: a infografia e o uso de ferramentas digitais. Rev. Temática, **9**(8), 1–13 (2013)
Prensky, M.: Digital natives, digital immigrants, On the Horizon, MCB University Press, **9**(5) [Online]. https://www.marcprensky.com/writing/Prensky%20-%20Digital%20Natives,%20Digital%20Immigrants%20-%20Part1.pdf (2001)
Ramonet, I.: O quinto poder. Observatório da Imprensa. http://www.observatoriodaimprensa.com.br/artigos/jd211020032.htm (2003)
Sabbatini, M., Betania, M.: Infografias interativas: novos suportes de informação para o jornalismo científico digital. In: 8° congresso Braliseiro de Jornalismo Científico [Online]. www.sabbatini.com/marcelo/artigos/2004sabbatinimaciel-abjc.pdf (2004)
Salaverría, R.: De la pirámide invertida al hipertexto: hacia nuevos estándares de redacción para la prensa digital. Novatica, (142), 12–15. http://dadun.unav.edu/bitstream/10171/5186/4/de_la_piramide_invertida_al_hipertexto.pdf (1999)
Salaverría, R.: Estructura de la convergencia. In: López-García, X., Pereira Fariña, X. (eds.) Convergencia digital: reconfiguración de los medios de comunicación en España, pp. 27–40. Servicio de Publicacións e Intercambio Científico, Santiago de Compostela (2010)
Sancho, J.L.V.: Algunas consideraciones sobre la infografía digital. Portal de la Comunicación. https://www.portalcomunicacion.com/download/58.pdf (2010)
Sartori, G.: Homo videns: la sociedad teledirigida. Tauros, Madrid (1997)
Sojo, C.A.: Periodismo Iconográfico (y XI): ¿ Es la infografía un género periodístico?. Rev. Latina de comunicación social, **51**. [Online]. https://www.ull.es/publicaciones/latina/2002abreujunio5101.htm (2002)
Sousa, A.F.T.: A infografia é jornalismo? Rev Comunicando, **1**(1), 43–56. https://www.revistacomunicando.sopcom.pt/ficheiros/20130108-infografia.pdf (2012)
Torres, M.B.P.: B: El periodista online: de la revolución a la revolución. Comunicación Social, Sevilla (2004)
Traquina, N.: Uma comunidade interpretativa transnacional: a tribo jornalística. Media & Jornalismo **1**(1), 45–64 (2002)
Traquina, N.: Teorias do jornalismo: porque as notícias são como são, vol. 1. Insular, Florianópolis (2005)
Túñez, M., Nogueira, A.: Infographics as a mnemonic structure: analysis of the informative and identity components of infographic online compositions in Iberic newspapers. Rev. Commun. Soc. **30**(1), 147–164 (2017)
Valero Sancho, J.L.: La infografía digital en el ciberperiodismo. Rev. Latina de Comunicación Social, (63). www.ull.es/publicaciones/latina/08/42_799_65_Bellaterra/Jose_Luis_Valero.html (2008)
Wilson, T.D.: Models in information behaviour research. J. Documentation, **55**(3), 249–270. http://informationr.net/tdw/publ/papers/1999JDoc.html (1999)

The Public Image of Book Translators in the Digital Press

Silvia Montero Küpper and Ana Luna Alonso

Abstract Our work seeks to analyze how media coverage has approached the news regarding book translation in recent years. We study the discourse in the digital press starting from a quantitative assessment with the aim to analyze how the translator's image is represented and how visible translators are. We ground our theory on Bourdieu's concept of cultural, economic, social and symbolic capital (Bourdieu 1979, 1992). Thus, we take into consideration several aspects related to editorial and cultural policies. The starting point for our investigation are the findings made to date regarding the treatment of the figure of translators and interpreters. We note how the press hardly gives importance to book translators, despite the fact that translation is essential for the development of society.

Keywords Translation studies · Book translation · Visibility of the translator in media · Analysis of press and online media · Spanish press

> In a world full of translations, the translator's profession remains invisible, the great unknown that is, paradoxically, everywhere: at work, in cinema, on the Internet, in advertising, in the media, in the streets. The figure of the translator should be better known to the general public, as they need to be explained what makes a literary translation, for example, reach us and also with quality. (Delgado 2016/06/02)[1]

[1] All quotations are translations.

S. Montero Küpper (✉) · A. Luna Alonso
Universidade de Vigo, Vigo, Spain
e-mail: smontero@uvigo.es

A. Luna Alonso
e-mail: aluna@uvigo.es

M. Túñez-López et al. (eds.), *Communication: Innovation & Quality*, Studies in Systems, Decision and Control 154, https://doi.org/10.1007/978-3-319-91860-0_18

1 Introduction

The press can turn an anonymous person into a well-known one. As stated by Russo (2013: 2), the press not only develops and consolidates the image of famous people but also acts as a springboard for some professions about which little or nothing is known and, in turn, enables us to learn about the social image of a professional group.

> Newspaper corporations, although informative, pursue persuasive and ideological goals; they seek to influence public opinion. [...] the intention is not only to share a message with the recipients but to provoke an attitude in them. (Hernández Guerrero 2009: 103)

We agree with Nour El Islam Sidi Bah that "newspapers are an inexhaustible source of information, opinions, views and judgements about the activity [of translation and interpreting] and those who perform it" (2015: 9, 16). In this sense, during his presentation at the 22nd International Conference of AESLA, El Madkouri (2005) described, from the viewpoint of critical discourse analysis applied in translation, the Arabic linguistic and/or cultural phenomena which were "translated or not" in the Spanish press; and he confirmed that the role of the media in forging the images and the representation of the news object is unsurpassed by any other, for promptness and outreach (El Madkouri 2005: 541). To this, we ought to add the credibility still enjoyed by the media, despite all the fake news we witness on a daily basis.

Translation in general, and book translation, in particular, is an industry that generates significant profits and yet it lacks the media presence to position it where it deserves to be. Neither economic, nor political, nor social factors are alien to translation. These phenomena are a part and parcel of the idiosyncrasy of individual translators, of their cultural background, and of the social reality dictating ideological and linguistic rules and conventions. Likewise, we should not underestimate the discourse that accompanies translated texts; who is producing them; how particular themes or styles, ideology, values, cultural codes, etc., are being sold, that are introduced, sometimes very subtly, in the explanations in the introduction, in the footnotes, in the front or back cover of the texts (Montero Küpper 2011). These are all elements or agents (in possession of symbolic and/or economic capital) which must be considered when undertaking a scientific study.

Lefevere (1992) introduced the concept of 'patronage' as an element that exerts pressure (power systems such as publishers, the educational system or the press) and which is capable of influencing the promotion of works and authorship. Bourdieu (1992) analyses the literary field in which writers interact in their capacity as culture-producer agents. For sociologists, translators may hold symbolic power and supposedly may or may not follow the rules of the cultural field where they interact, as well as considering any restrictions inherent to the translation as commissioned. Awards granted to translated works, for example, add prestige and symbolic value to the texts and the authors.

The place for any translation in the literary field will mainly depend on the positions occupied by the author and the translator, as well as on the prestige of the languages or cultures of the source and target literary fields. According to Casanova

(2002), the literary value accorded to literary texts is "at least partly" relevant to the language in which they were created. The vitality of a language is measured by the strength it has in its own territory, but also by its international recognition.

The four basic criteria that enable us to measure the different capitals a publisher may have or may mobilise are the "economic capital", i.e. business volume and company size; "cultural capital", i.e. the prestige of the titles in the publisher's collection; "social capital", i.e. contacts and circle of influence; and "symbolic capital", i.e. the position of the publisher and authors and works published in the national and international literary and publishing fields (Fukari 2005: 144–145).

To a certain extent, economic, political and cultural factors govern cultural exchanges, more precisely as a function of the space, time and areas or sectors that are the object of study, which in turn are represented by different institutions and individuals that bring about the diversity of social functions involved in translation (Sapiro 2008).

In general, translation is performed in two well-differentiated fields: major productions (commercial publishing), which plan their catalogues by focusing on best-sellers or the so-called fast-turnover texts, and limited productions (small, independent publishers looking for a niche in the market), which normally publish the so-called long-sellers.

2 The Book Market and Translation—Figures and Facts

Analysing the volume of creation and translation provides a lens through which to understand how markets function, both internally and in relation to the spaces that surround them. Calvo describes a literary world that has changed since the beginning of the economic crisis through

> a fragmentation of the publishing market, in which the two major groups are being resized to adapt to declining sales, and smaller publishers also pay less due to their size. [...] added to the situational issue, the economic crisis, there is another which is structural: literature used to account for 50% of the cultural consumption and now it is down to 20%, surpassed by the Internet and series downloads". (Calvo quoted in EFE 2016/03/13)

The economic crisis of 2008 did not bypass the European book market: according to the global data by the Federation of European Publishers (FEP 2017: 2–4) for the European Economic Area (EEA) in reference to the period 2006–2016, the turnover in this sector started to decline in 2008, especially from the year 2011, until 2014. However, the sales figures for the years 2015 and 2016 point to the beginning of a slight recovery with a timid upward trend. According to the absolute invoicing data provided by the aforesaid Federation, the average year-on-year change in the recovery period is 1.25% (FEP 2017). Among the five EU countries with the highest turnover in the field of book publishing are Germany, followed by Great Britain and France, together with Spain and Italy. These same countries are also the ones where most new titles are published each year (FEP 2017: 2–4).

FEP (2017) itself estimates that the book publishing industry was not so gravely affected by the crisis compared to other sectors. This is, of course, a global assessment, without differentiating the situation in countries more or less affected by the financial and economic crisis. In fact, the book industry in Germany, whose economic situation in recent years has been much different to that of countries like Greece, Italy and Spain, shows little changes in annual turnover since the beginning of the crisis. The slight decrease by 4.3% recorded for the five-year period from 2011 to 2015 does not affect the overall trend of Germany's turnover (Börsenverein des Deutschen Buchhandels 2016).

By contrast, in Spain, there has been a marked decline in 2015 with respect to the turnover in 2008 (FGEE 2016: 114). We must remember that the Spanish publishing industry had experienced significant growth since the 1980s,[2] but the evolution of turnover (and the number of books sold) in the last ten years reveals a clear regression. After peaking in 2008 in economic terms, the Spanish domestic book market started to fall markedly. However, as from 2014, the process seems to be reversed with a positive year-on-year variation, at current prices, by 1.7% (2014) and 2.8% (2015). According to the report for 2015 by the Spanish Association of Publishers Guilds (FGEE), the turnover at current prices fell by 25.1% in the decade from 2006 to 2015[3] (2016: 58) down by 30.8% compared to the turnover for 2008 (2016: 116). In fact, during the period 2008-2014, there is a continued decline in the contribution of this sector to the GDP, that being 1.05% in 2008 and a mere 0.85% in 2014. Nevertheless, book publishing still ranks as the first cultural industry in Spain (Francí Ventosa 2016: 117),[4] which for the period from 2008 to 2014 added up to an average of 2.6% of the total GDP.[5]

Figures 1 and 2 show the turnover for Spain, and the number of copies sold from 2006 to 2015.

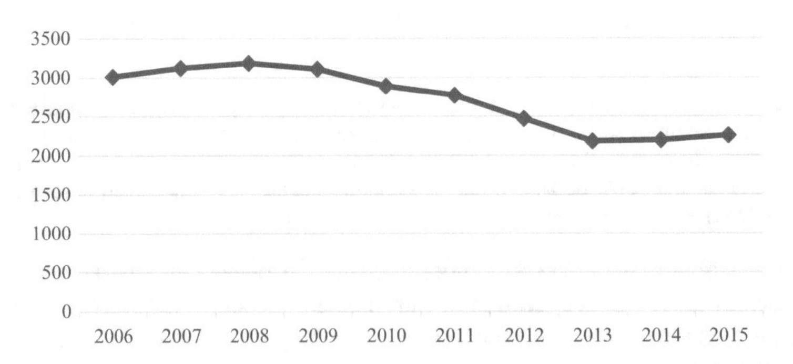

Fig. 1 Gross Turnover from Book Sales in Spain, 2006–2015 (€ million). Compiled by the authors from FGEE (2016)

[2]Cf. for further detail Francí Ventosa (2016).

[3]This equals a decrease by 35.2% of the average turnover at constant prices (FGEE 2016: 58).

[4]36.5% on average for the period 2008–2014 for the overall cultural sector.

[5]Contribution to the GDP in the five-year period prior to the crisis, 2003–2007, is of 3% (MCU 2011).

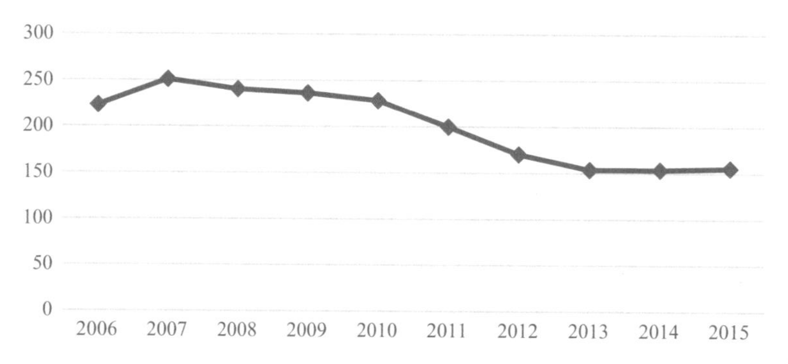

Fig. 2 Number of Book Sales, 2006–2015 (in million). Compiled by the authors from FGEE (2009, 2013 and 2016)

The figures show a clear decline in turnover, in parallel to the decrease in the number of books sold from 2008 to 2014. However, things seem to pick up in 2015 with a positive year-on-year change (2014/2015) by 1.2% (FGEE 2016: 113). Nevertheless, the number of copies sold continues to be lower than during the years of the crisis.

Regarding new titles published, in what refers to the overall European context, FEP (2017: 2) reports the opposite trend, although the number of new titles was more moderate after 2008 than in the years preceding the start of the crisis.

In Spain, however, the number of new titles grew each year until this trend was reversed in 2011. With a delay of 4 years regarding the onset of the decline of the number of copies sold in 2008, there is a marked downward trend in new titles published. From 2014, in parallel to the sales figures, a slight recovery apparently started which, incidentally, has had no repercussion on the number of books translated. However, the number of titles published in 2016 is lower than in 2008.

Table 1 summarises the data on the evolution of the number of titles registered with ISBN in Spain since 2001, which we deem appropriate to relate to the number of translations registered. The data show that the number of translations published in Spain experienced a significant decline as from 2009. In 2015 and 2016 they accounted for 16.15% of total publishing, compared to an average of 22.08% (including reprints) in the period 2009–2013. During the previous five-year period (2004–2008), translations accounted for one-quarter of titles published.[6] In any case, the percentage of translations in Spain's book market is higher compared to their weight in the UK (approx. 2.5%, Donahaye 2012) or Germany (around 12% in the years 2008–2015, Börsenverein des Deutschen Buchhandels 2016).

The data referred include all kinds of texts, but we know that literary translation occupies an important position in Spain's publishing landscape. In 2016, literary translations of texts for adults as well as for children and youths account for more than half of the total translations (MECD 2017). Moreover, it must be noted that

[6]Percentages calculated from CULTURABase statistics and MECD (2017) forecasts for 2016.

Table 1 Titles published, translations, the percentage of the weight of translations (the numbers in brackets do not include reprints) and their year-on-year variation 2001–2016. Compiled by the authors from MECD (2017) and CULTURABase

Year	Pub. titles w/r	Pub. titles wo/r	Pub. translations	% translation	% YOY variation of translations
2001	71,345	57,709	17,030	25.4	+12.9
2002	73,624	59,459	17,149	24.5	+0.7
2003	82,196	67,675	20,124	25.8	+17.4
2004	82,207	67,373	19,516	23.7	−3.0
2005	84,335	69,427	20,618	24.5	+5.6
2006	87,440	71,827	23,752	27.2	+15.2
2007	96,770	81,491	23,955	24.8	+0.8
2008	104,223	86,949	25,851	24.8	+7.9
2009	110,205	91,768	25,223	22.9	−2.4
2010	114,459	95,559	25,236	22.1	+0.1
2011	111,907	96,862	(19,005) 23,559	21.1	−6.6
2012	104,724	90,978	(19,221) 23,063	22.0	−2.1
2013	89,130	76,803	(16,237) 19,865	22.3	−13.9
2014	90,802	79,224	(16.301) 19,233	(20.06) 21.2	(+0.4) −3.2
2015	81,323	79,397	(12,858)	(16.2)	(−1.1)
2016	99,867	86,000	(13,867)	(16.1)	(+7.8)

From 2015 the Ministry of Education, Culture and Sport stopped counting reprints in its statistics. From CULTURABase we took the data on books published in Spain from previous years for re-editions and first editions (published titles without reprints: cfr supra Pub. Tit. wo/r), as well as the total publishing including reprints (Pub.Tit. w/r) for the years 2015 and 2016. However, we have no data which include reprints for translations in 2015 and 2016 (Pub.Trans.), nor data without reprints for the years prior to 2011 (such data are expressed in the table between parentheses). In order to offer comparable information in all years and since we include the variable "translation" that accounts for reprints until 2014, the table shows all the available data on published titles, including reprints as well as single prints. The figures include books published in all types of media (paper, digital…)

a part of this is due to translations from and into the official languages of Spain,[7] where translations play a significant role in the linguistic and cultural normalisation process, as well as in the process of reconfiguration of the literary systems proper

[7]The large percentage of book translations in Spain is due precisely to the co-official status of Spain's languages. Between 1996 and 2007, translations between such languages rose to 24.3% of total translations in 2007, though the following years the figures fell back to those of 1996 (18%). Percentages in 2006: 19.8; 2007: 24%; 2008: 18.5%, and in 2009 down to 17.4% (MECD 2010: 13 f.; Montero Küpper 2016).

to the Communities with co-official languages. This implies a certain responsibility towards the quality of the translations and texts and is also the reason for translators to deserve a special position in a society in which the accrual of cultural capital involves their commitment to a delicate aesthetic and stylistic effort.

For this reason and for the volume of translations generated every year, involving a large number of translation professionals in the publishing industry, the figure of the translator should enjoy more visibility in society and hence in the media.

3 Previous Research

The question of translator visibility is something that concerns translatologists since the existence of translatology as a discipline (Venuti 1995; Von Flotow 1991; Kellman 2010; Olshanskaya 2011; Vanderschelden 2000 or Bielsa and Bassnett 2009). Translation Studies are interested in knowing the role of translators and their status in society.

The works consulted show that although the translation and interpreting profession has achieved more recognition and social visibility, partly thanks to the incorporation of the T&I Degree in university education, the media, in general, do not actually reflect such a standing. This chapter seeks to test whether this hypothesis corresponds to reality through the image offered in recent years by the media in general and the digital press in particular. Professor Baigorri points out that:

> the written press in our days has been a source of information as well as a means to create public opinion, so what is published influences decisively in the way the recipients of information build an image of the protagonists of the news. (Baigorri 2012: 229)

We are interested in learning whether the press helps to consolidate the idea that we are facing a recognised profession, or if, on the contrary, it is still difficult for T&I professionals to guarantee their rights as other colleagues have done in their own fields.

The thesis defended by El Islam Sidi Bah (2015) analyses 109 press articles from four Spanish newspapers during the period 2004 to 2014,[8] and summarises that although

> in recent years, the issues of PSIT [Public Service Interpreting and Translation] have attracted the interest of a growing number of researchers; the works published so far, while still valuable, hardly address media issues in relation to this activity and the image that society reflects with regard to those who perform such tasks. (El Islam Sidi Bah 2015: 21)

El Islam Sidi Bah (2015: 21–22) journeys along what he calls "the few works interested in researching on translators and interpreters in the light of news publications" and quotes the study by Ebru Diriker (2005) focusing on the analysis of the image of conference interpreters in different Turkish newspapers, magazines, television news and programmes from 1988 to 2003. This researcher offers data on

[8]The digital edition of the dailies: *El País*, *El Mundo*, *ABC* and *La Razón*.

the reasons why the interpreting profession draws the attention of the media and compares the opinions reflected by communication professionals with those of the interpreters themselves, on four recurring themes:

1. Big Events (in which interpreters play a crucial role),
2. Big Money (which interpreters allegedly earn),
3. Big Mistakes (which interpreters allegedly make),
4. Big Names (which some interpreters have), and
5. Big Career (which some items promise SI can be).

Without dismissing the recognition of simultaneous interpreting professionals, according to the study, journalists show some lack of knowledge about the profession when they speak of "fidelity" to the word. Interpreters, in turn, try to explain that the active role they play during their work does not imply any absence of loyalty to the speaker's discourse.

El Islam Sidi Bah focuses on the figure of translators, but especially on interpreters working for public services (PSIT), and in his review of the known scientific literature, he cites a historical study (late 19th–early 20th Centuries) by Baigorri (2012) who rescues articles of the British and American press describing situations involving interpreters, acting as professionals or as ad hoc resources. For the researcher, the fact that interpreters are mentioned in the press "symbolises the progressive awareness of their function, although the way in which it is perceived is often out of focus or arguable, according to the parameters we use today" (Baigorri 2012: 229).

The research work by Russo reviewed by El Islam Sidi Bah is also diachronic. It is an article by the former analysing the image of professional interpreters. She emphasises that the figure of the conference interpreter has lost prominence in the media in favour of "everyday interpreters":

> Over time, the interest of the press towards interpreters has changed its focus. At first the aim was to emphasise interpreting as a myth [...] Then, as the profession spread and stepped out of the prestigious headquarters of international organisations to meet the needs of linguistic and cultural mediation in less glamorous though equally important environments, such as hospitals, courthouses, police stations, social services and schools, the attention has focused more on news about the interpreting profession and its functions or dysfunctions, when the service is not guaranteed or when the person performing this role does not meet the standards required by the communicative situation. (Russo 2013)

The research by El Islam Sidi Bah (2015: 51–32) follows this comparative line between interpreters working for public services and conference interpreters, as well as the role of literary translation professionals. Some highlights of the results of this work are:

- there are few articles on translation and interpretation in the press compared to other professional activities,
- the concept of "translation" is better known, more used and widespread than that of "interpreting", but there is certain confusion and ignorance when choosing to use one or the other,

- interpreting is the speciality most usually addressed, with almost 70% of the articles, compared to 21% which approach translation, and slightly less than 10% addressing both specialities together,
- the three areas of translation and interpreting most frequently dealt with by the press are PSIT (legal matters and in conflict zones), conference interpreting and literary translation,
- conference interpreting is the most valued one, though the fees are considered to be excessive. Literary translation ranks next, with some reference made to the low pay received by these professionals,
- articles on translation tend to be positive towards translators, especially literary translators; whereas, in the news on interpreting, the image of interpreters is a largely negative one (El Islam Sidi Bah 2015: 247–250).

Thus, the studies conducted to date show that the discourse of the press continues to portray conference interpreting and literary translation as the most prestigious fields.

4 Methodology

Newspapers contain many different types of journalistic texts inserted in their various sections. In general, and although the genres have evolved and no longer act as watertight compartments, the name "journalistic genres" is given to the different groups in which the diverse textual typologies may be included. Hernández Guerrero (2009) speaks of three broad categories: informative, interpretative and argumentative genres.

The main features of this textual type are simplicity (using short sentences with a simple structure), economy (saying things in the shortest and most concise way possible) and the continuity that the discourse must achieve (organised according to relevance, argument and style).

Both the reading audience and the industry professionals -editors and publishers- know what kind of relationships are established between headlines, stand-firsts and the body of the news, as well as those established between the news and the rest of the newspaper in which it appears: front page, different sections and pages, position on the page, size, etc. (Hernández Guerrero 2009: 33–34).

The headline determines how the editor and the publisher view the event, obliging the reader to address the news from such a point of view. It may act as a summary, but also as a sample of more relevant information stated in the stand-first and in the body of the news. The headline is especially interesting because of its multiple functions. It invites to read the text, providing a brief overview of the information, and it has a clear identifier function. One of the most interesting aspects is that "it may be interpreted on its own through the contextual knowledge of the recipient, without the need to resort to the textual development of the information" (Hernández Guerrero 2009: 36).

The informative density contained in a headline is very useful to disclose the topics covered in the news. According to Vella Ramírez and Martínez López (2012), this part (the headline) of the news "must summarise the content of the news or, at least, the most important part and, at the same time, try to capture the reader's attention" (Vella Ramírez and Martínez López 2012: 187). This idea is also shared by Lozano Ascencio et al. when they affirm that "the analysis focused on the headlines provides us with the most accessible and cared for discursive environment of written journalism" (Lozano Ascencio et al. 2010: 8), since here is where the summary of the content of the document lies, therefore, reading and assessing a headline could replace reading the full text.

At the beginning of this chapter, we reviewed some studies on the visibility of translators and interpreters in the press. We come forward in time, approaching our days, to continue the work applying a similar methodology. Once having analysed the texts compiled from the press, we observe that in many cases a translator is not mentioned in the headlines though is actually part of the news. Therefore, and bearing in mind our goals, we considered it necessary to focus not only on such headlines in which a translator is news, but also in those where translation is news or a translator is mentioned in the body of the text, regardless of whether they are news or not: "In order to identify the frame, the information content of the news is less important than the way in which the researcher may interpret it" (Águila-Coghlan and Gaitan Moya 2012: 7).

Thus, we base the methodology on a primarily qualitative approach. To learn about the translator's image in the media an ad hoc textual corpus has been compiled. On this occasion, we opted for a case study, which will be extended to other cultures, based on the general-interest digital press: *El País*, whose current tendency could be situated between social-democratic and liberal; *El Mundo*, liberal-conservative; as well as the dailies of conservative ideology *ABC* and *La Razón*.

In order to delimit the corpus, we have decided to include only such articles (news, information, opinion…) for the years 2015 and 2016 that contain the word "traductor*" (translator),[9] thus completing the research by El Islam Sidi Bah (2015), although our interest is exclusively addressed to book translators. The analysis also entails reading in detail the surrounding context to determine the semantic and pragmatic approaches and values which relate to the translator.

5 Results

Gómez Patiño reminds us that the form is the content and insists on that the attitude a medium of communication may have towards a topic becomes manifest when it prioritises, selects and treats the text in a specific manner: "in one or another page, with one or another space, with one or no photograph, by creating a monograph

[9] We include the masculine and feminine form, as well as singular and plural. We exclude, however, the words: 'interpreter', 'interpreting', 'translation' and 'translating'.

booklet, etc." (Gómez Patiño 2011: 123). In our study, we distinguish the articles in which translation is news from those in which the news is the translator. In addition, we differentiate cultural supplements (*Babelia, ElCultural.es, ABC Cultural*) from cultural sections or social news, etc. in the newspapers, since the articles in which a translator is news may appear anywhere, that is to say, it is not an exclusive subject-matter to cultural supplements. The following are the most significant results, to which we add a few examples.

As we have already pointed out, the field of translation most treated by the press in the four newspapers is the literary field, highlighting positive aspects such as prizes, tributes, and good reviews:

Ramón Buenaventura, Premio Nacional de Traducción 2016
EFE Ramón Buenaventura (Tánger 1940) ha sido galardonado con el Premio Nacional a la Obra de un Traductor correspondiente a 2016, y Luis Baraiazarra ha sido el ganador del Premio Nacional a la Mejor Traducción por su versión al euskera de las «Obras… completas de Santa Teresa de Jesús» (ABC, Libros, 3/11/2016).

La Escuela de Traductores de Toledo, premio internacional de traducción
ABC La Escuela de Traductores de Toledo ha recibido el octavo premio internacional Rey Abdullah Bin Abdulaziz de Traducción, en un acto que por primera vez en los ocho años de trayectoria de este galardón se ha celebrado en España (Abc, Castilla La Mancha, 7/05/2016).

Jordi Fibla obtiene el Premio Nacional a la Obra de un Traductor
Arnau Pons el Premio Nacional a la Mejor Traducción por la obra Cristall d'alè/Atemkristall de Paul Celan, que ha traducido del alemán al catalán
Jordi Fibla (Barcelona 1946) ha sido galardonado con el Premio Nacional a la Obra de un Traductor correspondiente a 2015, un premio que tiene por objeto distinguir el conjunto de la labor realizada por un traductor español, sea cual sea la lengua o lenguas utilizadas en el desarrollo de su labor, y que haya dedicado especial atención a la traducción de obras extranjeras a cualquier lengua del Estado español (ELCULTURAL.es, 5/11/2015).

Ana María Bejarano gana el premio Nacional a la Mejor Traducción por «Gran Cabaret», de David Grossman
ABC Semítica de la Universidad de Barcelona, Bejarano Escanilla coordinadora los seminarios de traducción literaria hebreo-español y los cursos de lengua hebrea de la Escuela de Traductores de Toledo de la Universidad de Castilla la Mancha (ABC, 28/11/2016 17:31:48).

The presence of translators is rare, even in reviews where the authorship of the translation is identified. Precisely in this context, where the figure of the translator should be news, there are few mentions of good translation works and even fewer comments on lapses.

El club del crimen
Weldon Kees
Traducción de Ezequiel Zaldenwerg. [...]
Los poemas de *El club del crimen*, en versión bilingüe, fluyen con naturalidad en español. Es mérito del traductor, Ezequiel Zaidenwerg (El Cultural, Irazoki, F. J. 31/03/2017).

Many reviews still tend to determine the quality of a translation based on the degree of (in)visibility of the translator, even when the latter is named if only out of courtesy. Thus, in an interview, the writer and translator Javier Calvo summarises the dichotomy of (in)visibility as is usually demanded:

Javier Calvo
"Evito el contacto con los autores que traduzco como si fueran la peste"
[...] La invisibilidad del traductor literario es connatural a su tarea, debemos aspirar a ella. Pero, paradójicamente, para conseguir que una traducción sea más fiel, más invisible y, en definitiva, más efectiva, hay que darle un poco más de espacio al traductor para que cree, para que sea más artesano, como antes. Pero hoy la traducción se ha devaluado al convertirse en un proceso industrial (Calvo entrevistado por Díaz de Quijano. *Elcultural*.com 21/03/2016).

The approach to the discourse diverges when we hear the voice of media professionals or that of translation professionals:

Doce visiones sobre la traducción
Cees Nooteboom, Bernardo Atxaga, María Teresa Gallego, Miguel Sáenz, Luisa Gutiérrez y otras voces autorizadas opinan sobre la realidad y los retos del oficio (Culturaelpaís.com, Babelia, En Portada, 26/08/2016).

It is precisely the interviews with prestigious literary translation professionals that reveal, in a more or less evident way, what the working conditions, training, and translation practices are, and to what extent they have changed in recent times:

La escritora y académica Carme Riera, nueva presidenta de Cedro
[...] En la misma línea se manifestaba Riera, que considera necesario que los escritores, traductores, periodistas y editores sean "justamente" retribuidos por la utilización digital de sus obras, como ocurre en la mayoría de los países de Europa (ElCultural.com, 24/06/2015).

In the news, there is usually no room to mention professional groups, nor the ethical or conduct codes agreed to between the parties, which in the case of book translation are provided for in the Law on Intellectual Property.[10] In a way, this contributes to a greater ignorance about the profession and underlines its lack of social consideration.

---

[10]Law 23/2006, of July 7th, which modifies the recast text of the Law on Intellectual Property, adopted by Royal Legislative Decree 1/1996, of April 12th. This aspect is expounded in Luna Alonso (2016).

Although it is true that headlines synthesise the message of the news, and although in Spanish they are characterised by being more elaborate and extensive (Vella Ramírez and Martínez López 2012), we have detected that sometimes even in the subtitle the word "translator" is omitted, giving preference to other attributes:

Carlos García Gual: "Vivir sólo en el presente es vivir en una prisión intelectual"
El filólogo, Premio Nacional de Traducción, interviene en los diálogos de la Fundación Juan March (El Cultural, 1/04/2015).

When the translator is news, but for being also a writer, publisher or any other mediator agent, the journalist's discourse might become patronising and the fact of being a translator becomes a secondary issue.

Kathryn Woolard, Alfred Konijnenbelt y Kirsten Brandt, premios Ramon Llull de promoción de la cultura catalana
En esta edición del certamen se ha galardonado el director de un festival de artes de la calle holandés, una profesora de antropología lingüística americana y la traductora al alemán de «Incerta Glòria» (ABC.ES, Cataluña, 11/11/2016).

The image of the translator is that of a figure subordinated to another, which is usually that of the original author whose work is being translated:

De la finitud
Günter Grass
[...] Este libro supone, pues, un adiós tranquilo, que relata sus afinidades y fobias, y muestra sus talentos, el de artista, escultor, pintor, que complementaron su lado de escritor. Grass reunía una vez al año a sus traductores, y todos ellos recuerdan que nunca regresaban a casa sin llevarse un recuerdo inesperado del talento del artista, una edición especial, un grabado, etcétera. Y al igual que Hemingway, su presencia física tendía a expandirse con la charla [...] (El Cultural, Gullón, G., 27(05/2016)

The translator is usually defined through other, often intellectual conditions, such as those of a literary expert, writer, philosopher, philologist, etc.:

Juan Arnau
"Si fallan los lectores es porque fallamos los escritores"
Marinero, astrofísico, filósofo, traductor... Juan Arnau (Valencia 1968) está revolucionando el pensamiento español con su *Manual de filosofía portátil* y con *La invención de la libertad*, ambos publicados por Atalanta (El Cultural, 25/11/2016).

Sometimes, a situation of subsidiarity arises towards the authorship or even the work itself that may lead to absolute anonymity:

La traductora Anita Raja es Elena Ferrante
Así se desprende de una exhaustiva investigación, centrada en sus ingresos y posesiones inmobiliarias, llevada a cabo por el diario económico « Il Sole 24 Ore » y hecha pública, de forma simultánea, en Italia, EE.UU., Alemania y Francia. (ABC.ES, Cultura, Ángel Gómez Fuentes, Corresponsal en Roma, 02/10/2016).

6 Closing Remarks

Taking into account the weight of translation in today's publishing landscape, and especially in the literary field, the relevance attained by the translator is indisputable. Translation and creation constitute a cultural asset with an impact on intellectual development, and they contribute to the sustainability of society. To this, we must add the associated economic value of being a cultural industry. In this chapter, we do not refer to translation solely as an accumulation of symbolic capital, as an accrual of aesthetic facts and knowledge to be transmitted and incorporated. We must remember that translation can also contribute to the coexistence of different cultures, revealing modes of action and fostering mutual understanding and rapprochement between communities.[11]

The media, in addition to acting as the notaries of social reality, are, to a large extent, a reflection of popular sentiment in general, and of the reading population in particular, constituting a very valuable thermometer to get to know social reality in a documented way (Gómez Patiño 2011: 120).

Knowing the positioning of the Spanish press practically amounts to knowing the national public opinion, given the close relationship between public opinion and what is published (Gómez Patiño 2011: 121). Although it is well known that one of the essential tasks of the media is not only to please but to inform while instructing without departing from the deontological framework. From the articles consulted we draw a vision adapted to the social reality of the object studied, which confirms that the figure of the translator is not profiled with enough entity to grant it the category obtained by other professions that achieve to have autonomous social capital. The printed media offer an image of the translator that remains static over time. In view of the themes and contexts analysed, the subordinate position of the translator in relation to other entities is evident, even in the news in which the translator is news. In texts in which the translator is not news, the translator subject as a service provider becomes blurred, eventually to become incidental.[12]

References

Águila-Coghlan, J.C., Gaitan Moya, J.A.: El encuadre (framing) de las noticias de la televisión española sobre la cumbre del cambio climático en Cancún. Sociedad Latina de Comunicación Social p.1–20, III Congreso internacional Latina. http://www.aeic2012tarragona.org/esp/abstracts3.asp?id_seccion=6&id_slot=64 (2012)

Baigorri Jalón, J.: La imagen caleidoscópica del intérprete: Algunos ejemplos de la prensa anglosajona de los siglos XIX y XX. In: Cruces, S., del Pozo, M., Álvarez, A., Luna Alonso, A. (eds.) Traducir en la frontera, pp. 229–244. Granada, Atrio (2012)

[11]It is important to note that only now, on May, 24 2017, the United Nations recognises the role of professional translators and interpreters in connecting nations, and fostering peace, understanding and development. http://www.fit-ift.org/itd-adopted/?lang=fr.

[12]Translated from Galician by Michael Skinner.

Bielsa, E., Bassnett, S.: Translation in Global News. Routledge, London (2009)
Börsenverein des Deutschen Buchhandels: Der Buchmarkt in Deutschland. https://www.boersenverein.de/de/182716.pdf (2016)
Bourdieu, P.: La Distinction. Critique sociale du jugement. Les Editions de Minuit, Paris (1979)
Bourdieu, P.: Les régles de l'art. Genèse et structure du champ littéraire. Éditions du Seuil, Paris (1992)
Casanova, P.: Consécration et accumulation de capital littéraire. La traduction comme échange inégal. Actes de la recherche en sciences sociales, Sept., 144, 7–20. Seuil, Paris (2002)
CULTURABase: Ministerio de Educación, Cultura y Deporte. Estadística de la Edición Española de Libros con ISBN (s.d.). http://www.mcu.es/culturabase/cgi/um?M=/t16/p16&O=culturabase&N=&L=0
Delgado, P.: El escritor fantasma. Según Javier Calvo. ABC Cultural. http://www.abc.es/cultura/cultural/abci-escritor-fantasma-201606020931_noticia.html (2016, June 2)
Díaz de Quijano, F.: Javier Calvo. Evito el contacto con los autores que traduzco como si fueran la peste. Elcultural.com (2016, May 3). In: http://www.elcultural.com/noticias/buenos-dias/Javier-Calvo/9082
Diriker, E.: Presenting simultaneous interpreting: discourse of the Turkish media, 1988–2003. aiic.net. (2005, March 21). http://aiic.net/p/1742
Donahaye, J.: *Three percent? Publishing data and statistics on translated literature in the United Kingdom and Ireland.* Aberyswyth: Literature Across Frontiers. http://www.lit-across-frontiers.org/research/literary-translation-data-and-statistics-sources/December2012.pdf (2012)
EFE: Javier Calvo da visibilidad al traductor en "El fantasma en el libro". Eldiario.es. Barcelona. http://www.eldiario.es/cultura/Javier-Calvo-visibilidad-traductor-fantasma_0_495201449.html (2016, March 13)
El Islam Sidi Bah, N.: La prensa y la traducción e interpretación en los servicios públicos. Ph.D. thesis, Universidad Autónoma de Madrid, Madrid. https://repositorio.uam.es/bitstream/handle/10486/669535/el_islam_ould_sidi_nour.pdf?sequence=1 (2015)
El-Madkouri, M.: La traducción del Otro en la prensa española. In: Carrió Pastor, M.L. (ed.) Perspectivas interdisciplinarias de la Lingüística Aplicada, T. 2. Análisis del discurso, lenguas para fines específicos, lingüística de corpus y computacional, lexicología y lexicografía, traducción e interpretación, pp. 541–552. Asociación Española de Lingüística Aplicada, Valencia (2005)
FEP Federation of European Publishers: The Book Sector in Europe: Facts and Figures—2017. http://www.fep-fee.eu/-Publications-.pdf (2017)
FGEE Federación de gremio de editores de España: Comercio Interior del Libro en España 2008. http://federacioneditores.org/documentos.php.pdf (2009)
FGEE Federación de gremio de editores de España: Comercio Interior del Libro en España 2012. http://federacioneditores.org/documentos.php.pdf (2013)
FGEE Federación de gremio de editores de España: Comercio Interior del Libro en España 2015. http://federacioneditores.org/documentos.php.pdf (2016)
Francí Ventosa, C.: La traducion literaria en España 1980–2015. In: Galanes Santos, I. et al. (eds.) La traducción literaria. Nuevas investigaciones, pp. 115–132. Comares, Granada (2016)
Fukari, A.: Les maisons d'édition- Freins ou moteurs du processus de traduction? In: Peters, J. (ed.) La traduction. De la théorie à la practique et retour, pp. 141–150. PUR, Paris (2005)
Gómez Patiño, M.: Análisis del tratamiento de la mujer en la prensa española. Día Internacional de las Mujeres. Estudios sobre el mensaje periodístico, **1**(17), 119–140. file:///C:/Users/Ana%20Laura/Downloads/36949-39122-1-PB.pdf (2011)
Hernández Guerrero, M. J.: Traducción y periodismo. Peter Lang, Bern (2009)
Kellman, S.: Alien autographs: how translators make their marks. Neohelicon **37**(1), 7–19 (2010)
Lefevere, A.: Translation, Rewriting, and the Manipulation of Literary Fame. Routledge, London & New York (1992)
Lozano Ascencio, C., Piñuel Raigada, J.L., Gaitán Moya, J.A.: Las verdades implantadas en los titulares de prensa sobre los temas de comunicación. Análisis de la construcción de discursos

hegemónicos a partir de las auto-referencias hacia la comunicación. Razón y Palabra **15**(74), http://www.redalyc.org/articulo.oa?id=199516111051 (2010)
Luna Alonso, A.: Códigos de conducta y calidad en traducción. In: Ferreiro Vázquez, O. (ed.) Traducir e interpretar lo público, pp. 153–174. Comares, Granada (2016)
MCU Ministerio de Cultura: Cuenta satélite de la Cultura en España. Resultados detallados 2000–2009. Secretaría General Técnica. Subdirección General de Publicaciones, Información y Documentación (ed.). http://www.mcu.es/culturabase/pdf/resultados_detallados_2009T22P22.pdf (2011)
MECD Ministerio de Educación, Cultura y Deporte: La traducción editorial en España, Madrid: Servicio de estudios y Documentación, S.G. de Promoción del Libro, la Lectura y las Letras Españolas D.G. del Libro, Archivos y Bibliotecas, Madrid. http://www.mecd.gob.es/cultura-mecd/dms/mecd/cultura-mecd/areas-cultura/libro/mc/observatoriolect/redirige/estudios-e-informes/elaborados-por-el-observatoriolect/TRADUCCION_2010.pdf (2010)
MECD Ministerio de Educación, Cultura y Deporte: Avance de Panorámica de la Edición Española de Libros 2016. Observatorio de la Lectura y el Libro. http://www.mecd.gob.es/mecd/cultura-mecd/areas-cultura/libro/mc/observatoriolect/redirige/destacados/2017/abril/observatorio/PEEL2016-avance.html (2017)
Montero Küpper, S.: Loyalty of publishers and quality: Galician literary translation peritexts (2000–2009). In: Luna Alonso, A., Montero Küpper, S., Valado Fernández, L. (eds.) Translation Quality Assessment Policies from Galicia/Traducción, calidad y políticas desde Galicia, pp. 235–252. Peter Lang, Bern (2011)
Montero Küpper, S.: Políticas para la traducción de textos literarios. In: Galanes Santos, I. et al. (eds.) La traducción literaria. Nuevas investigaciones, pp. 53–72. Comares, Granada (2016)
Olshanskaya, N.: De-coding intertextuality in classic and postmodern Russian narratives. Transl. Interpret. Stud. **6**(1), 87–102 (2011)
Russo, M.: El intérprete en la prensa. inTRAlinea Special Issue: Palabras con aroma a mujer. Scriti in onore diAlessandra Melloni. Edited by María Isabel Fernández García & Mariachiara Russo. http://www.intralinea.org/specials/article/2007 (2013)
Sapiro, G. (dir.): Translatio: le marché de la traduction en France à l'heure de la mondialisation. CNRS, Paris (2008)
Vella Ramírez, M., Martínez López, A.B.: Análisis de estrategias y procedimientos traductológicos utilizados por "El País" en la traducción de títulos de prensa del español al inglés. Sendebar **23**,177–206 http://revistaseug.ugr.es/index.php/sendebar/article/view/35 (2012)
Vanderschelden, I.: Quality assessment and literary translation in France. Eval. Transl. **6**(2), 271–293 (2000)
Venuti, L.: The Translator's Invisibility: A History of Translation. Routledge, Londres (1995)
Von Flotow, L.: Feminist translation: contexts, practices and theories. TTR **4**(2), 69–84 (1991)

Silvia Montero Küpper MA degree in Romance Philology, Modern German Literature and Comparative Literature (Universität Bonn) and Ph.D. in Translation and Interpreting (Universidade de Vigo), Montero Küpper lectures on Translation Studies at the Universidade de Vigo since 1995 and has as areas of research Galician literary translation (history and sociological viewpoint), German-Galician contrastive stylistics and lexicography, currently focusing on the translator's visibility in media and translation policies. Dortmund, Germany.

Ana Luna Alonso Degree in Romance Philology and Ph.D. in French Philology by the Universidade de Santiago de Compostela, is currently senior lecturer at the Department of Translation and Linguistics at the University of Vigo. Luna Alonso research areas, devoted to Translation Studies, include socio-cultural research from and into minority languages; the history of translation and translators; the study of translation flows and policies for publishing translations. Vigo, Spain.

Transparency, Online Information and Pre-election Surveys

Carlos Pío Del Oro-Sáez, Berta García Orosa and Nuria García Guillín

Abstract The diffusion of pre-election studies should respect national and international standards of transparency of opinion studies. Compliance with these codes of dissemination of pre-election opinion studies guarantees readers' access to accurate and truthful information. Based on these standards, this paper analyses the digital media dissemination of pre-election polls in the 2015 elections in Spain, subject also to a proper electoral law. The methodology used was the analysis of the published content information and the in-depth interviews with those responsible for the selected opinion studies.

Keywords Transparency · Dissemination · Pre-election surveys
Political communication · Digital journalism

1 Introduction

Media are often the link between voters and political parties, regulating the main issues and approaches during the election campaign and trying to respond to the voters' political information demands.

The pre-electoral surveys are a particular case of opinion studies that take on special importance during this period. Voters access this information for data that helps them decide their vote and allows them to understand the differences between political parties or candidates, and their programs. Of particular relevance is the role of these studies in the decisions of the undecided, and their importance proceeds from the scientific character, based on statistical techniques, which is presupposed to the surveys.

C. P. Del Oro-Sáez (✉) · B. García Orosa · N. García Guillín
Universidade de Santiago de Compostela, A Coruña, Spain
e-mail: carlospio.deloro@usc.es

B. García Orosa
e-mail: berta.garcia@usc.es

N. García Guillín
e-mail: nuriagarciaguillin@gmail.com

M. Túñez-López et al. (eds.), *Communication: Innovation & Quality*, Studies in Systems, Decision and Control 154, https://doi.org/10.1007/978-3-319-91860-0_19

In dissemination work, the journalist's role is relevant. However it is not the only agent that is usually involved in the process because the study is often performed by a firm, outside the media. The independence between the two institutions, the method of work and transparency will mark the level of professionalism of the surveys by the audience.

The guidelines that the journalist should follow when it comes to disseminating the results and presenting their reports are governed by codes of good practice and professional ethics. But once the study is publicly disseminated, the organization responsible for the elaboration of the research may be obliged to another level of transparency in the dissemination. In this way, the moment a survey is published two systems of regulations come together for the information that must be made public. This situation occurs when the media and the company conducting the study belong to national or international associations that subject them to carry out a diffusion framed within the recommendations in force for each association. The aim of the principles governing the public dissemination of results should be to ensure that potential consumers of this information have sufficient means to judge their validity and reliability.

These standards of action are particularly relevant in the case of pre-election voter intentions surveys, since the electorate consumes the information collected in the media and rarely recurs to the original source, that is, the firm or entity contracted to perform the study, trusting the message transmitted by the media. It is necessary to insist on the scientific nature of these surveys that can generate a deceit in the electorate not sufficiently informed about its limitations. Proper professionalism will result in voters deciding their choice based on reliable information enabling the formation of other effects (underdog and bandwagon) undesirable to the vote (Maarek 2011). But, in addition, the use of language and its presentation in the media are fundamental elements that should not motivate changes in the intention to vote and raise doubts about possible manipulation.

In the new digital dimension of the media, journalists must, in addition to informing, attract the audience interest and correctly use the digital tools for the production and circulation of their messages. In this environment where activities and responsibilities are changing, where the search for truth prevails and the receivers become the originators of theoretically journalistic texts, professional ethics becomes even more important.

2 Diffusion of Opinion and Pre-election Studies

The dissemination of market, opinion and social studies is widely regulated at the international level. Research institutes and companies that conduct pre-electoral studies often belong voluntarily to associations that demand different levels of transparency in the public dissemination of their results. This degree of transparency is reversed only if the study has been commissioned by an entity that does not make a public dissemination of the results.

Table 1 NCPP Standard Disclosure

Level	NCPP
1	Sponsorship of the survey Fieldwork provider (if applicable) Dates of interviewing Sampling method employed Population that was sampled Size of the sample Size and description of the subsample Margin of sampling error Survey mode Complete wording and ordering of questions Percentage results of all questions reported
2	Estimated coverage of target population Respondent selection procedure Maximum number of attempts to reach respondent Exact wording of introduction Complete wording of questions in any foreign languages Weighted and unweighted size of any subgroup cited in the report Minimum number of completed questions to qualify a completed interview Whether interviewers were paid or unpaid Details of any incentives or compensation provided for respondent participation Weighting procedures used to generalize data to the full population Sample dispositions to compute contact, cooperation and response rates
3	Release anonymized raw datasets (ASCII, SPSS, CSV format) Post complete wording, ordering and percentage results of all publicly released survey questions to a publicly available web site for a minimum of two weeks Publicly note their compliance with these Principles of Disclosure

Source Own elaboration

A standard of requirements for the results dissemination of the polls is the proposal of the National Council of Public Polls (NCPP) of the USA. In Principles of Disclosure (NCPP 2006) it presents three levels of transparency for the presentation of the results of the electoral studies (Table 1). The first level is mandatory for research institutes and includes the need to recommend to the media to disseminate this information or provide it, in this case, upon request of an interested party. The second level of transparency refers to the obligation for research institutes to respond to certain requests for information from potential interested parties in the event that they carry out the study diffusion. The third level is a recommendation that includes the public diffusion of data sets. The British Polling Council (BPC) follows the NCPP recommendations and makes it public in its Statement of Disclosure (BCP n.d.) understanding that it is a responsibility shared by customers and companies. It does not reach level 3 of transparency.

The World Association for the Market, Social and Opinion Research (ESOMAR) and the World Association for Public Opinion Research (WAPOR) jointly publish in the context of opinion and electoral surveys a guide that has as one of its objectives to ensure that the public has sufficient information to make an unbiased and adjusted interpretation of the results presented. The Guideline on Opinion Polls and Published Surveys (ESOMAR and WAPOR 2014) further understands that electoral surveys are subject to the same requirements as any type of social and market research techniques and highlight the role journalists have to inform the public about possible electoral results. Equally, BPC and NCPP affect the need to train journalists in the knowledge of the methodology of opinion polls. The text includes obligations for research institutes aimed at maintaining audience confidence in electoral studies that are publicly disseminated at the same level 1 of NCPP transparency. They also make explicit indications so that readers can clearly distinguish between the direct results of the survey and those produced from the primary information. However, the guide itself points out the difficulties of including this information in the journalist's practice, which is governed by its own ethical and practical codes and limits the examples presented to a brief technical file of the survey. On the other hand, they make recommendations similar to level 2 of transparency of the NCPP for the research institutes, emphasizing that the means of communication should facilitate the obtaining of this information in its own website or facilitating the website address of the company that carried out the survey. Level 3 transparency only includes it as a recommendation in countries where possible. The guide is complemented by specific guidelines for a good realization of the pre-electoral and voting intentions surveys highlighting the importance they have for the public. Among them it must be enhanced that it recommends that research institutes indicate the percentages of undecided, non-voters and those who do not respond to the question about the meaning of their vote in pre-election polls.

Other codes of good practice such as ICC/ESOMAR International Code on Market, Opinion and Social Research and Data Analytics (ICC-ESOMAR 2016) and the American Association for Public Opinion Research (AAPOR) Code of Professional Ethics and Practices Research (AAPOR 2015a) adapt these recommendations in a context of more general work adapting in their latest revisions to the growing use of new technologies and digital formats. These codes mainly refer to the standards of transparency that should be followed by research institutes and their relationship with clients and participants in the study. They also include the basic norms of ethical conduct when it comes to publicly disseminating the work. They generally insist on the obligation for the diffusion of their studies to contain sufficient basic information to determine the validity of the results presented. They recommend that technical information can be obtained in addition to the published information and ensure the quality of the information disseminated. The AAPOR Code of Professional Ethics and Practices also adds a guide to good professional practice for organizations for both the conduct of the research and subsequent disclosure. These principles are similar to those included in the NCPP and ESOMAR/WAPOR codes and are binding on researchers. They include the need to reflect clear specifications on the statistical models and indicators used in the study in order to replicate the research. However, it allows research institutes to justify the inability to disseminate datasets. The code is complemented by the Transparency Initiative Disclosure Elements (AAPOR 2015b) which supports organizations to systematically practice good transparency practices in disseminating their research.

In the context of working in the media, journalists are mediators between reality and citizens. They select a part of the events that occurred for publication. This process of selection between some and other facts is marked by the professional honesty, the general interest of the citizens and veracity of the published information. The media must select truthful facts, contrasting all possible sources of information and not hiding in the piece of information no important element for their correct compression. The International Federation of Journalists (IFJ) and many other national associations of journalists such as the Federation of Press Associations of Spain (FAPE) insist on the commitment to the search for the truth that will always lead to reporting only on facts of which the origin is known, without falsifying documents or publishing false, misleading or deformed information material. The IFJ Declaration of Principles on the Conduct of Journalists (IFJ 1986), initially adopted in 1954, emphasizes among the duties in the transmission of journalists' information respect for truth, honesty in publication and non-suppression of essential elements of this information.

More specifically, the European Code of Ethics of the Journalistic Profession of the European Union or Ethics of Journalism (Council of Europe 1993), adopted by European countries, insists that the dissemination of news must be carried out truthfully without being mediated by an increase of the audience and demanding an adequate professional formation for the journalistic activity. However, this resolution by the Council of Europe is really a recommendation for countries rather than an obligatory practice for the profession. Many country codes of ethics, press and media associations include similar recommendations (FAPE 1993; Australian Press Council 2014; The Guardian News & Media 2011; The New York Times n.d.; Grijelmo 2014). In some cases they are a statement of intent that contains recommendations that are not an obligation.

Most of these codes emerge as the response of the professional journalism sector to the fundamental right to freedom of information contained in United Nations General Assembly Resolution 59 (I) (United Nations 1946) and article 19 of The Universal Declaration of Human Rights (United Nations 1948) which includes the freedom to transmit and disseminate without obstacles. UNESCO is the agency in charge of promoting actions to guarantee this freedom of the press (UNESCO 1983). Freedom of information has historically been recognized, alongside freedom of expression, in other international documents such as article 19.2 of the International Covenant on Civil and Political Rights (United Nations 1966) or the American Convention on Human Rights 1969). The fulfillment of these general principles is regulated and developed through international organizations, professional associations of national and international journalists and in the constitutions of democratic countries.

In recent years, UNESCO has continued its work by promoting initiatives on the right of access to information and its role as an integral element of the development of societies. In relation to the media, it continues to insist on fair and professional information (UNESCO 2005), in order to highlight the role of the Internet as a means of expression and diffusion in the face of traditional media supports (Dutton et al. 2011) and the self-regulation of media as a means of transparency and accountability to society (UNESCO 2011). Many communication groups endorse these general recommendations by adapting them to the different formats of their publication, either audiovisual, printed and/or online.

It is undeniable that the new technological trends on the Internet and its possibilities have caused important changes in the media (López and Campos 2015) that have been outlining the current hybrid media system (Chadwick 2013; Hamilton 2016). Hypertext, multimedia, updating and the possibility of interactivity (Bolter and Grusin 1999; Díaz-Noci 2001) characterize this new scenario full of possibilities that, at the same time, looms as a threat of the traditional values of journalism and requires the search for new ethics in the online publication (Ward 2016). The real-time circulation of information establishes an uninterrupted cycle of news in which journalists must combine the speed of updating information with the essence of journalism, reflection and deepening (Bradshaw 2007). However, the media insist on the need to continue with traditional journalistic standards (Reuters 2016) and respect traditional ethics at all times, even if the inclusion of new technologies for production and circulation of informative material are positively valuated (Hohmann 2011). The loss of control of the message in the digital age (in which the audience can already participate) and the emergence of the so-called "aftertruth" leads to the delimitation of new ethical codes. An informative dynamic marked by instantaneity and interactivity, the prudence in the use of information obtained in social networks and the insertion of hyperconnections in their articles to sufficiently verified websites are some of the ethical challenges that the journalist faces in digital media (Cruz-Álvarez and Suárez-Villegas 2017).

Knowing the impact of the publication of opinion studies in the media, many associations have published guides for journalists that allow them to distinguish between scientific and non-scientific surveys and to help them professionally inform the public (Kernel n.d.; Spanish Association of Market Studies, Marketing and Opinion—AEDEMO 1989). Some of them contain specific examples on electoral surveys (Gawiser and Witt 2004), which shows the interest in the correct dissemination of these results. The extensive literature and existing works related to pre-electoral surveys analyze their goodness and validity from different perspectives. The accuracy of the estimates (Caballé et al. 2013; Udina and Delicado 2005; Traugott 2005) and the transparency of these studies (Mateos and Penadés 2013) are also recurrent themes.

Section 10 of the Editorial Guidelines of the British Broadcasting Corporation—BBC (n.d.) contains the broadcasting standards in the group of electoral surveys, including the type of minimum information to be disseminated, style indications for the presentation report in the means and the limitations of the different types of surveys. The importance of meeting these standards for the publication of results leads to the publication of specific guides for the electoral processes held in the United Kingdom since 2015 and to documents on the presentation of statistical data in the medium (BBC 2017). This type of documentation aims to meet the objective of accuracy and impartiality in the presentation of information but also helps the training of the journalist in the line suggested by Ethics of Journalism (Council of Europe 1993). In the same direction can be included the guidelines of The New York Times (2008) that comes to limit the type of studies that can be disseminated and the Australian Press Council (2001). Other associations and organizations have also developed initiatives to encourage the specific training of journalists in the dissemination of statistical data. The Statistics for Journalists online training course of the Royal Statistical Society or A guide to writing stories about numbers (United Nations Economic Commission for Europe—UNECE 2009) are two valid examples.

In general, the treatment received by pre-electoral press surveys differs from country to country. It obeys, in part, the diversity of regulations and casuistic of each country and its tradition as an element in electoral campaigns, even within those belonging to the European Union. On one hand, it should be expected to be similar or respect existing regulations for the dissemination of social studies and market. On the other hand, its diffusion in the media must follow the guidelines of good journalistic practices. And finally, comply with the possible regulation developed by each country in this, regard that can limit or regulate the behavior of the press in its dissemination activity. It is for these reasons, in addition to the importance they have for the population, that the regulation of the publication of this type of surveys contains differentiated elements in comparison with other types of social studies and opinion.

The electoral regulations of each country generally regulate the electoral process (candidates, voters, voting, duration of the campaign, financing of the electoral campaign, electoral propaganda and publicity in the public or state media, …). However, in a few countries, these laws regulate activity and dissemination in printed or online media during the election campaign. The activity of the press in this period is governed by its own codes of good practice and by the fundamental principle of freedom of the press. Only in a few cases, such as France, Spain, Mexico and Costa Rica, electoral norms provide temporary limitations to the media when disseminating pre-election polls before elections day or ballot box or partial results during the voting. These last restrictions are currently under debate in many countries due to the use of the internet and social networks as an information vehicle, since the regulation affects exclusively the media of each country and are not effective.

But in addition to this recent debate on the difficulties of restricting the dissemination of pre-electoral studies during a certain period of time, it must be pointed out that social networks and online media have become a proven system of dissemination during the electoral campaign. The political parties' communication cabinets use them to launch messages that reach the electorate, the electorate uses them as a more current and plural information procedure than traditional media and has been a source of introduction of campaign distorting elements as in past US presidential election and in France. Although the difficulties to legislate or regulate the contents on the Internet are indisputable, it is necessary that the information of vote intention studies arrives adequately, maintaining the standards that are needed for users to access useful and truthful information.

3 Dissemination of Pre-election Surveys in Spain

The Spanish legislation on elections (Organic Law 5/1985 of the General Electoral Regime—LOREG) is one of the few electoral laws that includes a chapter dedicated to the dissemination of pre-electoral studies. Article 69 of section VIII specifies what elements must be transmitted during the electoral campaign (Table 2), restricting the possibility of publishing, disseminating or reproducing pre-electoral studies from 5 days before the time of the elections. This article obliges both the company conducting the study and the media without expressly distinguishing the type of support. It attributes to the Central Electoral Board the task of ensuring that published polls do not contain forgeries, concealments or deliberate changes and establishes sanctions and penalties for non-compliance with the regulations. Although the legislation seems very clear on these issues, in practice it presents limitations on the control of the accuracy and goodness of the studies published (Gálvez Muñoz 2000) and the existence of a register of organisms that includes the companies that can carry out studies of intention of voting with a minimum level of guarantees and objectivity for its later diffusion, as it is realized in Portugal and Belgium. It also does not include important elements on the methodology followed in the process of obtaining results and the composition of the sample which, in the Spanish case, becomes an essential issue to be able to carry out the attribution of seats applying the electoral legislation itself (Diez Nicolás 1996).

The general elections of December 20th, 2015 can be considered peculiar in comparison to the ones realized in the last years due to the change that lived the policy of Spain during those elections. Two new parties were faced with the traditional two supported by a large number of votes in a situation not lived in the recent democratic history of the country. The intensive use of social networks designed a novel scenario as an element of diffusion of messages and information by the new parties. The dissemination of pre-electoral surveys was a system to capture the interest of the population. Data from the post-electoral Elections General 2015 (Centro de Investigaciones Sociológicas—CIS 2016) show that although the Spanish

Table 2 Information to be disseminated according to Spanish legislation

LOREG Spain
Organisation which conducted the poll and its sponsor (and address)
Dates of fieldwork Universe represented Sample size Sampling method employed Margin of sampling error Survey mode Full question wording Number of respondents who have not answer each question

Source Own elaboration

electorate was not interested in the campaign (50.8%), 62.8% were aware of a pre-election survey and 22.5% of them took them into account when deciding their vote. In addition, 45.7% of voters are informed of the campaign through newspapers and 68.8% of those using the internet, 47.1% have followed campaign information on online press pages and a 54.1% in other digital media and social networks. For this reason, in this study, an approximation has been carried out to determine whether the dissemination in the online media of electoral surveys complies with current regulations and international transparency standards.

Although the space available in traditional media is limited, it is not limited in the digital version of printed newspapers because the advances of internet technology allow to include all the necessary information in the own media or through links to other own pages or of the firm responsible for conducting the study. This type of solution would meet the international publication standards (ESOMAR and WAPOR 2014).

According to the data consulted electronically and published by the Office of Justification of the Diffusion—OJD (www.introl.es), the headlines with greater diffusion in Spain, of daily character and on general or economic information, thus discarding Sports information in the written press, are *El País* (http://elpais.com/) and *El Mundo* (www.elmundo.es). Between them, they covered 21.41% of the print run and 20.72% of the total press release in Spain, in June 2015. In their digital editions, in December 2015 (El País 2016), these two media presume more than 50% of the total number of unique users of the five most widely read digital newspapers in Spain. These media usually work with the same body, on a permanent basis, in relation to pre-election surveys: Sigma Dos (http://www.sigmados.com/) for *El Mundo* and Metroscopia (http://metroscopia.org/) for *El País*.

The Spanish electoral campaign officially began on December 4th and ended on December 18th, however the selected media, in that period, published some of their information based on previously published studies of voting intention that were not selected for the analysis following the definition of pre-election study of Benavente (2013). The news included were those containing information on pre-election studies. During that time frame, 41 news items were published between the two media related to this pre-election online survey (48.8% in *El País*) that were revised. The contents of the online edition vary with respect to the printed version of the same medium because they do not collect all their pieces and make other summaries with links that lead to more general data, sometimes outside the domain of the medium itself. The three studies that gave rise to this information were located in their original source, in the web pages of the organisms that made them. One of these studies (CIS 2015) was broadcast the day before the official start of the election campaign but it was decided to include it because most of the first news of the two media during the campaign contained references to its results and is the only study elaborated by a state agency. The two private entities released their surveys on December 14th, the last possible day to do so, according to Spanish legislation. The analysis was completed by telephone interviews with representatives of the private institutes that conducted the investigations, in order to obtain their opinion on general aspects of the studies of vote intention in which they gave some information that allowed to complete the information not published in its web page or by the means of communication. The CIS refused to respond to the interview. To the transparency analysis it was used a basis from the Spanish electoral legislation in which it expressly includes the mandatory information that must be disseminated.

Table 3 shows the percentages of the analyzed news that fulfills LOREG in the media selected for each of the issues required by the legislation. The data show that, on average, the regulation is not met even to indicate the domicile of the entity that conducted the survey, except in a news from *El País* that indicates the Metroscopia website. In many cases the information is not strictly reflected in the news and the media use links to other pages to access the information not collected. It is important to note that the news corresponding to the first day of diffusion of the pre-electoral study specifies almost all the data including a technical file of the study but when subsequent analyzes are done, in many cases, it does not provided the necessary information. In total, 6 news from *El Mundo* (23.8%) includes all the necessary elements compared to 2 news from *El País* (10.0%). Many of the news published by *El Mundo* (52.4%) contain holdings of results by Autonomous Communities and, in this case, the news is more deficient in compliance with the legislation, since they only include the institute responsible for the elaboration of the study, without a link to it. Something similar happens in 30% of the *El País* news. The two media include, in one of the days, a link with the results of the CIS (2015) survey that supports the first news of the campaign regarding pre-election polls, although in the case of *El País* is held on December 10th, almost at the end of the legal period to broadcast results, while *El Mundo* does so on the first day. However, none of the media acts in this way in the case of studies commissioned by them from private companies. This is why the percentage of news that meets the requirement to have the complete questionnaire is so low. It is also important to notice the less caution into reflect the percentage of people who do not answer each question. The percentages present in the charts that accompany the news do not add up to a hundred in some presentations and, in other occasions, are added in epigraph of other minority parties.

The ban gathered in LOREG regarding the dissemination of polling results five days before the election has not been fulfilled. A total of 5 news items containing results of polls appeared (4 in *El Mundo*), although they do not correspond to new pre-election polls, include partial information from previous studies. Table 3 further includes a relevant issue for the end user of the information that is the clear differentiation in the news writing between the direct vote and the estimated vote. The table shows that it can be an improbable element in the texts.

The transparency of the agencies in the diffusion of the studies was made in comparison with the standards of ESOMAR and WAPOR (2014). As shown on its website, Sigma Dos is associated with ESOMAR and AEDEMO, while Metroscopia and CIS do not include any information on membership of associations. The major difference with respect to level 1 of NCPP transparency (2006) is the need to diffuse the margin of error of the study if the sample is probabilistic. Although the responsibility for dissemination belongs to the client (the press), both codes indicate that the accuracy and correct dissemination of the results should prevail. These two codes state that the data in Table 4 should be available and that, in general, any results

Table 3 Percentage of digital news on pre-election polls that comply with LOREG

LOREG Spain	*El País*	*El Mundo*	Mean
Organisation which conducted the poll and address	95.0(1)	100.0(1)	97.6
Organisation which sponsored the poll and address	95.0(1)	100.0(1)	97.6
Dates of fieldwork	70.0(2)	47.6(2)	58.5
Universe represented	60.0(2)	47.6(2)	53.6
Sample size	70.0(2)	47.6(2)	58.5
Sampling method employed	30.0(2)	47.6(2)	36.6
Margin of sampling error	65.0(2)	47.6(2)	56.1
Survey mode	65.0(2)	47.6(2)	56.1
Full question wording	15.0(2)	23.8(2)	19.5
Number of respondents who have not answer each question	35.0(2)	47.6(2)	41.5
Limitación temporal de presentación de sondeos	95.0(2)	80.9(2)	87.8
Mean	63.2	58.0	60.5
Direct and Estimated voting	35.0	4.8(2)	19.5

(1) Do not include address. (2) Most information is accessed with links to other pages of the same medium. *Source* Own Elaboration

report should contain this information even if it is accessible through web links. The study is easily accessible on the CIS website. For the two private organizations, the information gathered in three of their own news items published in their pages related to the study (2 of Metroscopia and 1 of Sigma Two) has been used.

The CIS publishes all the information required in several documents, allowing the download of the microdata. The information disseminated by Sigma Dos of the study in its own web site is the most limited, with almost no elements related to the fact sheet of the survey carried out. In the case of Metroscopia the news that spread in its web page is more complete but partial with respect to the code of ESOMAR and WAPOR. In that entity, for the questions included in the news, the information corresponding to the percentages of people who "do not vote" or "do not respond" is added to the minority parties. This requirement is only fully complied with by the CIS since the two companies only provide this information and the text of the questions disclosed in their news. Only the CIS in its press release of the day of the study's presentation provides a telephone contact, to request information. In other cases, it may be assumed that the general contact of the company may serve as a mechanism for requesting additional information. As the two private entities have disseminated their own news about their studies, LOREG forces them to include all the necessary elements in them. Only Metroscopy (except for the complete questionnaire) publishes the information indicated by the Law. Although it is not necessary to publish the

Table 4 Information disseminated in the web page of the organisms in news or own links

ESOMAR and WAPOR	CIS	Metroscopia	Sigma Dos
Organisation which conducted the poll	Yes	Yes	Yes
Organisation which sponsored the poll	Yes	Yes	Yes
Dates of fieldwork	Yes	Yes	Yes
Universe effectively represented	Yes	Yes	
Sample size Geographical coverage	Yes Yes	Yes	
Sampling method employed	Yes	Yes	
Margin of sampling error (NCPP)	Yes	Yes	
Survey mode	Yes	Yes	
Full question wording	Yes	Yes	Yes
Question order and context	Yes		
Percentages of respondents who give 'don't know' and 'not vote' answers	Yes	Yes	Yes
General weighting variables (if used)	Yes		
If the questions form part of a more extensive survey	Yes		
How and where additional details can be obtained	Yes	Yes	Yes
Estimation process (not required)	Yes	Yes	
Microdata (not required)	Yes		

Source Own elaboration from the web pages of CIS, Metroscopia and Sigma Dos. Accessed 08 may 2017

methodology used for the estimation of the vote and to differentiate it from the direct vote, both the CIS and Metroscopia make clear indications in the presentation of the information that allows to differentiate between both, adding that other procedures can lead to different results.

The interviews carried out by telephone to José Pablo Ferrándiz, principal investigator of Metroscopia, and Manuel Mostaza, operations director of Sigma Dos (Guillin 2016) allow to complete the previous analysis taking into account the perspective of the private entities that carried out the studies, approaching the problematic of its correct diffusion and methodological differences. Regarding compliance with LOREG, the principal investigator of METROSCOPIA pointed out that its article 69 is of a lax character and that it does not seem necessary to subscribe all its points. "In the law you say one thing and then you systematically do another and no one complains about it, so it is understood that there is flexibility. Although it is important for reliability and for dignifying work" (Ferrándiz, J. in Guillin 2016, pp. 52).

During the interviews the two institutes kindly provided the full questionnaire used in their research (Guillin 2016). The total non-disclosure of the questionnaires could be understood as the responsibility of the client as the agencies deliver all the information to the client and the client is in charge of publishing the information. “Our control with the media is only that we do not touch the data delivered” (Mostaza, in Guillin 2016, pp. 85).

The fundamental differences between the three questionnaires are their extension and the precision of the questions (7 and 5 questions in the Sigma Dos and Metroscopia questionnaires respectively *versus* the 38 questions in the CIS). In addition, only the CIS publicly offers the complete ordered list of questions as they were formulated. When interviewed in this regard, the responses by the institutes’ researchers were different. Sigma Dos understands that this is a media error since they give them the information and are not responsible for not spreading it. In Metroscopia they recognize the problem and affirm that they are initiating measures to solve it: “The topic of the space in the written press is complicated. Space is taxed and they eliminate things that they consider less necessary, something that with the Internet is stop happening, little by little” (Ferrándiz, in Guillin 2016, pp. 87). Currently some institutes, such as Metroscopy are beginning to solve this problem through its own website. Although the studies analyzed are not available, it is possible to find the questionnaires, data sheets, results and microdata of two studies: Pulso Electoral I and Social Climate. The latter, of monthly periodicity, although *El País* does not link its web with that of the organism in its diffusion so they can be hidden. Concerning Sigma Dos, they do not perform this type of diffusion.

The precision in the writing of the questions can be explained through the type of telephone interview that the two private entities use in which the speed is high. The concision of the questionnaires also influences the percentage of interviews covered entirely by Sigma Two, indicating that the percentage of interviews not completed in the study (Mostaza in Guillin 2016, pp. 86) is not expressly indicated. A more detailed study that the questionnaires writing provided, allows to appreciate the existence of some biases that can go unnoticed for the people who respond to the survey of Sigma Dos: Is it alright that …?, Do you think … or, on the contrary, consider positive …? Do you think that pensions in Spain can be at risk, that is to say, that in the future they have to be reduced?

The three agencies use methodologies conditioned by the hypothesis used, the number and type of interviews. “The process of elaboration begins in all cases with an interview between the client and the institute, where one limits the information that is required,” explains Manuel Mostaza (Guillin 2016, pp. 36). The public entity CIS performs 17,615 personal interviews distributed provincially, which corresponds with the structure of the Spanish electoral constituencies. In that line Ferrándiz also manifests that, on the part of Metroscopia, conducts a survey with almost 2800 telephone interviews expanded in 18,000 to estimate the vote: “There are 52 elections, one for each constituency. To give a result of the division of seats you must have samples of the fifty-two provinces.” (Guillin 2016, pp. 88). In the case of Sigma Dos, they perform 8350 telephone interviews but this issue is not mentioned. The distribution by provinces seems adequate for the Spanish electoral system, otherwise the results may reflect the general tendency of the vote and not allow to make reliable assignments of the number of elected political representatives of each party.

The number and type of interviews may influence the speed of the study. However, Metroscopia performs the fieldwork much faster with a volume of interviews than the other two surveys. This issue is important because during the electoral campaign can produce rapid movements of the intention to vote motivated by elements that can distort the electoral message (news, crossings of accusations between candidates, cases of corruption,…). It is true that conducting telephone interviews and a shorter questionnaire are also factors that allow the field work to be carried out more quickly, compared to the CIS that performs personal interviews and employs a large questionnaire. The existing debate on the telephone interviewing is a recurring theme in the literature on opinion surveys. Although no one doubts the scope by telephone, there is a greater presence of mobile phones *versus* fixed telephones, in Spain, although this trend is decreasing with the years. Data presented by the National Institute of Statistics of Spain indicate that 96.7% of households had a mobile phone compared to 78.4% of households with landline (INE 2015). For now, the two private entities conduct pre-election surveys using fixed telephone numbers with CATI support. The representative of Metroscopia supports the efficiency of using mobile phones in Spain, the over the landline "… we do some studies with mobile phones … to see if the difference influences" (Ferrándiz, in Guillin 2016, pp. 88) while in Sigma Dos Support the validity of this procedure "is not a problem today, it may be in the future, but not now" (Mostaza, in Guillin 2016, pp. 86).

In the Spanish case, the two companies defend the importance of the process of vote estimation in obtaining electoral results. Both indicate that it is an indispensable tool, result of the experience and the habitual work with social opinion studies in which is used the declared vote and the sympathy or affinity with other parties, to make the estimate. In this process, Mostaza acknowledges that "information is reasonable, but it can always be improved" (Guillin 2016, pp. 85), while Ferrándiz points out that the results are not manipulated to favor any political option and that the degree of accuracy of its estimates affect the business prestige of the entity.

The effects of these studies on the electorate's intention to vote are minimized from the perspective of the two research institutes. "It influences citizenship but in a positive sense because it allows to generate democracy. They provide the citizen with information that, if not, they would not have access" points Sigma Dos (Mostaza in Guillin 2016, pp. 86). "They influence the same as an electoral campaign, a debate… What has to be demanded is that the quality of the polls is correct" (Ferrándiz in Guillin 2016, pp. 90). These statements coincide with the results of the Postelectoral General Elections 2015 (CIS 2016), where it is indicated that only 4.2% of the people who had knowledge of pre-election polls changed their initially expected voting decision in view of the results presented.

4 Exploring the Way Forward

The analysis presented above does not attempt in any way to analyze the quality of the studies used or their accuracy according to the degree of precision in relation to the final results but to know if the disclosure of the same allows the online reader to determine the reliability and accuracy of their results. The vast majority of the population perceives these studies as a global picture of the intention to vote. In that sense, the three studies have, regardless of their differences, the same validity. However, the transparency and correctness of its presentation in both the online media and on the websites of private organizations can be improved. Only the CIS complies with international dissemination standards and Spanish electoral legislation.

The available space is an obstacle to the publication of information by the traditional media but, in online media, this does not seem too relevant. Some institutes currently solve this problem through their own web page as recommended by leading market research and opinion associations. The media must also make this effort and take advantage of the possibilities offered by digital media in which space is not a restriction and can introduce more visual elements and links.

The Spanish electoral legislation delimits the characteristics that must have the dissemination of pre-election surveys as a rule of obligatory compliance for both companies and media. It also attributes to the Central Electoral Board the power to ensure that data and information from the polls reach the public in a truthful manner, without concealment and complying with the law. There does not seem to have been a mechanism of superior control that has been effective during the electoral campaign because the two online media and the private entities failed to comply with the rule regarding the necessary dissemination content and the corresponding rectifications have not taken place. This effort of external control, either by the Central Electoral Board or by some national or supranational association of market research and opinion firms, would result in certification of the independence of private entities and opinion institutes, adding transparency and credibility to their studies, and would dignify their work. A public register of companies certified to carry out pre-election studies would guarantee a scientific and professional activity within the current regulations.

The prohibition period for disseminating survey results during a pre-election timeframe established by some legislations is not effective since some media publish online news on polling results at that time. The arguments on this aspect of the legislation are opposed. Advocates mainly point to the possible direct effects on the voters (underdog and bandwagon) while detractors point out that it does not favor free access to information. In view of what has happened both in Spain and in other countries with the dissemination on the internet of results of voting intentions in foreign publications during the prohibition period, it seems that this particularity of the legislation is not effective and must be taken into account for the future, as claimed by different associations and entities. Surveys are a tool more than online press consumers and there is no clear consensus that its diffusion makes change the meaning of the vote and even do so, nor would it necessarily be bad if the information presented possesses the necessary guarantees of precision and transparency.

Perhaps for the general public it is less important to know in depth the methodological mechanisms related to the estimation of the vote that differentiate it from the declared vote, which, in Spain, is known as "electoral kitchen". However, for more specialized readers and political analysts it would be advisable for research institutes to collect, at least on their web pages, a somewhat more detailed explanation of the approximation mechanisms for this calculation, which is usually the one with the most presence in the information disclosed in the online media. This concealment makes the term "electoral kitchen" have a pejorative translation in the reception of information for the public and does not manifest the important and indispensable technical effort of specialized social analysis that is carried out in the agencies and companies in charge of elaborating the pre-election studies. It should be taken into account that the distribution by provinces of the electoral system requires, in many cases, large samples and that the estimation process is necessary to make a faithful estimate of the intention to vote. The mystery surrounding this estimation process does not seem to help public opinion recognize the credibility and performance of an independent work of the entities in relation to the means of communication in which their studies are disseminated.

With the degree of technological advance and the importance of the internet and social networks, it is difficult to understand that there are still deficiencies in the dissemination of pre-electoral studies, since the media and the organizations are aware of the interest that their results arouse during the electoral campaign and have the mechanisms to solve the observed deficiencies. It is a shared responsibility in which each actor should be followed by his codes of good practices but guaranteeing access to information true to reality presented without concealment. A well-conducted and well-publicized pre-election poll is an image of the electoral pulse and an instrument of value to the electorate that can help maintain interest in the campaign and increase the active participation of citizens. Public opinion during the election campaign is changing and the dissemination of online media surveys is one of the factors that can motivate it. If the proliferation of surveys developed without the necessary rigor adds an incorrect subsequent presentation, the consequences can be unpredictable and perhaps not the most beneficial for each country at any given time.

The current technology helps the speed of information transmission and it is expected that both the media and the entities actively collaborate so that the information of the pre-election surveys reaches the consumers of online press with the necessary guarantees of good journalistic practice. The efforts of the associations of market and opinion studies that cover research institutes and those carried out by media associations are many by publishing, updating and disseminating codes, guides and guidelines for action. The co-responsibility in the diffusion of the two main actors of the process should not be a difficulty to ensure that users get a proper dissemination of the surveys.

References

American Association for Public Opinion Research: AAPOR Code of Professional Ethics and Practices. http://www.aapor.org/Standards-Ethics/AAPOR-Code-of-Ethics.aspx (2015a)
American Association for Public Opinion Research: Transparency Initiative Disclosure Elements. https://www.aapor.org/Standards-Ethics/Transparency-Initiative/Educational-Materials.aspx (2015b)
American Organization of States: American Convention on Human Rights "Pact of San Jose, Costa Rica" (Treaties_B-32). http://www.oas.org/en/topics/human_rights.asp (1969)
Australian Press Council: Standards of Practice. Advisory Guidelines: Opinion Polls. http://www.presscouncil.org.au/advisory-guidelines/ (2001)
Australian Press Council: Standards: Statements of Principles. http://www.presscouncil.org.au/statements-of-principles/ (2014)
Benavente, L.: Apuntes sobre encuestas electorales. Apuntes de Ciencia & Sociedad **3**(2), 173–179. https://doi.org/10.18259/acs.2013021 (2013)
Bolter, J.D., Grusin, R.: Remediation Understanding New Media. MIT Press, Cambridge (1999)
Bradshaw, P.: Un modelo para la redacción del siglo XXI. Cuadernos de Periodistas **12**, 59–74 (2007)
British Broadcasting Corporation: Reporting Statistics. http://downloads.bbc.co.uk/rmhttp/guidelines/editorialguidelines/pdfs/ReportingStatistics.pdf, (2017)
British Broadcasting Corporation: Editorial Guidelines. http://www.bbc.co.uk/editorialguidelines/ (n.d.)
British Polling Council: Statement of Disclosure. http://www.britishpollingcouncil.org/statement-of-disclosure/ (n.d.)
Caballé, A., Grima, P., Marco-Almagro, L.: Are Election Polls Right? Analysis of the Accuracy of Election Poll Predictions. Revista Española de Investigaciones Sociológicas **142**, 25–46. http://dx.doi.org/10.5477/cis/reis.143.25 (2013)
Centro de Investigaciones Sociológicas: Preelectoral Elecciones Generales 2015. Panel (1ª Fase). Estudio 3117. Octubre–Noviembre 2015. http://www.cis.es/cis/opencm/ES/1_encuestas/estudios/ver.jsp?estudio=14245 (2015)
Centro de Investigaciones Sociológicas: Postelectoral Elecciones Generales 2015. Panel (2ª Fase). Estudio 3126. Enero-Marzo 2016. http://www.cis.es/cis/opencm/ES/1_encuestas/estudios/ver.jsp?estudio=14258 (2016)
Chadwick, A.: The Hybrid Media System: Politics and Power. Oxford University Press, Oxford (2013)
Council of Europe: The Ethics of Journalism. Parliamentary Assembly Resolution 1003/1993. Strasbourg. http://assembly.coe.int/nw/xml/XRef/Xref-XML2HTML-en.asp?fileid=16414&lang=en (1993)
Cruz-Álvarez, J., Suárez-Villegas, J.: Pautas deontológicas para el periodismo digital. El profesional de la información **26**(2), 249–254. https://doi.org/10.3145/epi.2017 (2017)
Díaz-Noci, J.: La escritura digital: hipertexto y construcción del discurso informativo en el periodismo electrónico. Universidad del País Vasco, Bilbao (2001)
Díez Nicolás, J.: Predicción de escaños electorales mediante encuestas. Revista Española de Investigaciones Sociológicas **74**, 264–289 (1996)
Dutton, W.H., Dopatka, A., Law, G., Nash, V.: Freedom of Connection, Freedom of Expression: The Changing Legal and Regulatory Ecology Shaping the Internet. UNESCO, Paris (2011)
El País: EL PAÍS cierra 2015 como líder digital en España y en castellano. http://politica.elpais.com/politica/2016/01/30/actualidad/1454185717_321005.html (2016)
ESOMAR and WAPOR: ESOMAR/WAPOR Guideline on Opinion Polls and Published Surveys. http://www.esomar.org (2014)
Federación de Asociaciones de Periodistas de España: Codigo Deontológico. http://fape.es/home/codigo-deontologico/ (1993)

Gálvez Muñoz, L.A.: Organismos de sondeos, encuestas electorales y derecho. Revista de Estudios Políticos **110**, 97–122 (2000)

Gawiser, S. R., Witt, G.E.: Twenty questions a journalist should ask about poll results. In: Babbie, E. (ed.) The Practice of Social Research, pp. A25–A30. Wadsworth/ Thomson-Learning, Belmont, CA (2004)

Grijelmo, A.: Libro de Estilo de El País. Aguilar, Madrid (2014)

Guillín, N.: Una aproximación al análisis de la transparencia de la publicación de los sondeos pre-electorales en las elecciones generales del 2015. Estudio de El Mundo y El País. Final Degree Project [unpublished manuscript], Universidade de Santiago de Compostela, Santiago de Compostela (2016)

Hamilton, J.F.: Hybrid news practices. In: Witschge, T., Anderson, C.W., Domingo, D., Hermida, A. (eds.) The SAGE Handbook of Digital Journalism, pp. 164–178. http://dx.doi.org/10.4135/9781473957909. SAGE, London (2016)

Hohmann, J.: ASNE 10 Best Practices for Social Media. http://asne.org/content.asp?contentid=77 (2011)

ICC and ESOMAR: ICC/ESOMAR International Code on Market, Opinion and Social Research and Data Analytics. https://www.esomar.org (2016)

Instituto Nacional de Estadística: Encuesta sobre equipamiento y uso de tecnologías de información y comunicación en los hogares. http://www.ine.es/prensa/np933.pdf (2015)

International Federation of Journalists: IFJ Declaration of Principles on the Conduct of Journalists. http://www.ifj.org/about-ifj/ifj-code-of-principles/ (1986)

Kellner, P.: A Journalist's Guide to Opinion Polls. http://www.britishpollingcouncil.org/a-journalists-guide-to-opinion-polls/ (n.d.)

Ley Orgánica 5/1985, de 19 de junio, del Régimen Electoral General. Texto Consolidado 01/11/2016. Boletín Oficial del Estado núm. 147, de 20 de junio de 1985, pp. 19110–19134. http://www.boe.es/buscar/act.php?id=BOE-A-1985-11672 (1985)

López, X., Campos, F.: Journalism in Change. Media XXI, Porto (2015)

Maarek, P.J.: Campaign Communication and Political Marketing and Communication. Wiley-Blackwell, Chicester, United Kingdom (2011)

Mateos, A., Penadés, A.: Las encuestas electorales en la prensa escrita (2008–2011). Errores, sesgos y transparencia. Metodología de Encuestas 15, 99–119 (2013)

National Council on Public Polls: Principles of Disclosure. http://www.ncpp.org/?q=node/19 (2006)

Reuters: Handbook of Journalism. Specialised Guidance. Reporting from the Internet and Social Media. http://handbook.reuters.com/index.php?title=Reporting_From_the_Internet_And_Using_Social_Media (n.d.)

The Guardian News & Media: Editorial Guidelines. Guardian New & Media Editorial Code. In: https://www.theguardian.com/info/2015/aug/05/the-guardians-editorial-code (2011)

The New York Times: The New York Times Polling Standars. In: http://www.nytimes.com/ref/us/politics/10_polling_standards.html (2008)

The New York Times: Standards and Ethics. http://www.nytco.com/who-we-are/culture/standards-and-ethics/ (n.d.)

Traugott, M.W.: The accuracy of the national pre-election polls in the 2004 presidential election. Public Opinion Quarterly, **69**(5), 642–654. https://doi.org/10.1093/poq/nfi061 (2005)

Udina, F., Delicado, P.: Estimating Parliamentary composition through electoral polls. Journal of the Royal Statistical Society: Series A (Statistics in Society) **168**, 387–399 (2005)

UNESCO: International Principles of Professional Ethics in Journalism. Fourth Consultative Meeting of International and Regional Organizations of Journalists. Paris. http://www.unesco.org/new/en/communication-and-information/freedom-of-expression/professional-journalistic-standards-and-code-of-ethics/ (1983)

UNESCO: Dakar Declaration on Media and Good Governance. http://www.unesco.org/new/en/unesco/events/prizes-and-celebrations/celebrations/international-days/world-press-freedom-day/previous-celebrations/worldpressfreedomday200900000/dakar-declaration/ (2005)

UNESCO: Professional journalism and self-regulation new media, old dilemmas in South East Europe and Turkey. UNESCO, Paris (2011)
United Nations: Calling of an International Conference on Freedom of Information. General Assembly Resolution 59 (I). http://www.un.org/documents/ga/res/1/ares1.htm (1946)
United Nations: The Universal Declaration of Human Rights. General Assembly Resolution 217(III). Paris. http://www.un.org/en/universal-declaration-human-rights/index.html (1948)
United Nations: International Covenant on Civil and Political Rights. General Assembly Resolution 2200A (XXI). http://www.un.org/documents/ga/res/21/ares21.htm (1966)
United Nations Economic Commission for Europe: Making Data Meaningful. Part 1: A guide to writing stories about numbers. United Nations, New York and Geneva (2009)
Ward, S.J.A.: Digital journalism ethics. In: Witschge, T., Anderson, C.W., Domingo, D., Hermida, A. (eds.) The SAGE Handbook of Digital Journalism, pp. 164–178. SAGE Publications Ltd, London (2016)

Carlos Pío Del Oro-Sáez He holds a BA and a doctorate in Mathematics from the Universidade de Santiago de Compostela where he is Professor of Applied Economics. He participated in projects in the areas of Contemporary History, Communication and Economics. His main research lines are Economic Statistics and Data Analysis applied to the Social Sciences. Santiago de Compostela, Spain.

Berta García Orosa Journalism Professor at the Universidade de Santiago de Compostela (USC) for two six-year periods. She holds a BA in Information and Communication Sciences and Political and Administration Sciences, as well as a Ph.D. in Journalism from the USC in Spain. She has been part of the Novos Medios research group since it was created. She has published several articles and books on organizational communication, journalism and cyber-journalism. Santiago de Compostela, Spain.

Nuria García Guillín BA in Journalism from the Universidade de Santiago de Compostela. Santiago de Compostela, Spain.

Neologisms, Metaphors and New Concepts for Writing Media News

Áurea Fernández-Rodríguez and Iolanda Galanes-Santos

Abstract This chapter looks at the terminological difficulties faced by news writers when reporting on topics of current interest. It proposes the linguistic strategies needed to restate the conceptual innovations, new names, foreign words, metaphorical expressions and semantic and conceptual changes that characterise specialised terminology under consolidation in journalistic discourse. An explanation is provided on the specialised neologism concept and how to identify it in a specialised press corpus on the world economic crisis. Terminology management strategies are likewise put forward, paying special attention to the cultural peculiarities reflected in some specialised terms and in terminological metaphors.

Keywords Specialised terminology · Metaphor · Neologism
World economic crisis · Translation

1 Introduction

The economic crisis has had a notable presence in all media from 2007 onwards, and particularly since September 2008, coinciding with the fall of Lehman Brothers, one of the most prominent investment banks in the world. The communicative topics of interest related to the crisis range from the debate about its causes to its effects on productive sectors and on society, as well as the international contagion and its many consequences. All of them have consolidated a journalistic discourse on the crisis in the last decade. The data offered by Chavero (2015: 92 and ff.) are just one example. The authors' analysis of the media discourse during the 2008–2011 Spanish government's term of office indicates that economic issues accounted for 36% of the media agenda between September 2008 and December 2009. Moreover, the labour reform derived from the same covered 37.4% of the agenda between April and June

Á. Fernández-Rodríguez (✉) · I. Galanes-Santos
Universidade de Vigo, Vigo, Spain
e-mail: aurea@uvigo.es

I. Galanes-Santos
e-mail: iolag@uvigo.es

M. Túñez-López et al. (eds.), *Communication: Innovation & Quality*, Studies in Systems, Decision and Control 154, https://doi.org/10.1007/978-3-319-91860-0_20

2010, while, the breakdown of social dialogue resulted in the crisis occupying 43.1% of the agenda in the first quarter of 2011.

The profusion of the world economic crisis in newspapers and its permanence in the informative agenda over a prolonged time period have converted this crisis into a discursive event (Moirand 2008: 4). Its most noteworthy features, in addition to those already mentioned, are firstly its presence in several of the genres and sections of the written press, right from opinion articles to the news on current events, including of course the section on economy. Secondly, it is an event that has left its medium to long term mark on discourse produced on other topics. For example, any current news on the U.S. financial sector is likely to take into account the situation before and after the subprime mortgage crisis, in the same way as one would essentially refer to the real estate bubble and its bursting, when informing about the Spanish real estate sector. These singularities distinguish the 'discursive event' from the 'discursive moment', conceived as an event that disappears fleetingly from the press, i.e., just as soon as it appears on the scenario.

It is likewise a complex, international phenomenon that has already been coined by specialists as a 'crisis of crisis'. Even though it was rooted in the US real estate sector, the peculiarities of the US financial system led to a rapid contagion and, therefore, to an international financial and economic crisis. This crisis took specific forms and characteristics in each country. In the words of Pernias (2014: 26) "más que de una crisis, en singular, podríamos referirnos a ella, como una sucesión de crisis, en la que se han ido encadenando distintas situaciones en una relación causa-efecto en unos casos y, en otros, como crisis paralelas que se autoalimentan en un círculo vicioso".[1] Therefore, we can identify different types according to the latitude and the sector most affected: mortgage crisis, financial crisis, economic crisis, employment crisis, real estate crisis, debt crisis, austerity crisis, etc.

The vision that newspaper journalists from every culture have about the world economic crisis is conditioned both by its evolution in that particular country, as well as by the characteristics of its economic system and its crisis history, to which we should always add the intentionality of the transmitter. All this may imply that the same concept derived from the crisis is explained differently by the respective media of each country and even differently from one media to another (Arrese 2015).

The incursion of the U.S. 'subprime loan mortgage' into the international media due to the collapse of Lehman Brothers (September 2008) meant that newspaper journalists from non-English speaking countries had to seek new names for concepts such as this, which in the past were exclusively linked to the legal and economic system of that country. In order to disseminate this information, each linguistic community transferred these concepts to their (specialised and general) press using different strategies ranging from an initial use of foreign words (from American English, for example 'subprime credit') to their own national terminology later (for example in Spanish, *hipotecas basura*-trash mortgages, *crédito de riesgo*-risky credit, *crédito*

[1] "Rather than a crisis in the singular, we could refer to it as a succession of crises, which in some cases led to a cause-effect chain reaction relationship, while in others, it led to a parallel crisis which fed into a vicious circle".

de alto riesgo-high risk credit, *hipotecas de mala calidad*-poor quality mortgages, *hipotecas tóxicas*-toxic mortgages, etc.) which not only contain the conceptual characteristics of the culture of origin but also the cultural values of the receiving society, which are incorporated into the conceptualisation of this element in that place. The manner in which this naturalisation was done, the way in which such foreign crisis concepts are explained in each culture and moreover which concepts are self created by each one of them, are issues that our crisis terminology research project addresses.[2]

2 The Role of the News Writer in the Introduction of Neologisms

The dictionary from the Spanish online daily newspaper *El País* defines a news writer as: "Persona que redacta, especialmente el que lo hace como trabajo en un periódico editorial, emisora de radio, etc".[3] Despite the vagueness of the definition, we understand that the job consists of ordering a compilation of several ideas in order to transmit them through different media: written, digital, audiovisual, etc.

The news reporter and/or writer perform a number of functions: research, analysis of sources, drafting, editing and correction of texts. In their role as mediators that inform and sometimes opine, news writers, whether columnists, writers or commentators, must be subject to accuracy, independence and immediacy requirements, and furthermore find ways to perform a precise and quality job. However, despite swift and rigorous reporting becoming one of the maxims of journalism, these two qualities do not always go hand in hand. To the above difficulties, we can add that communications professionals are mediators who, on many occasions, just like translators and interpreters, handle specialised concepts without necessarily being specialists in the field, which increases the risks associated with their job that furthermore must be delivered in record time.

One of the most studied and problematic aspects for any mediator is the linguistic aspect, especially, accuracy of the specialised terms, i.e., those that convey scientific or technical content from a specialised knowledge area. From the Communicative Theory of Terminology we can assume that there is not always a single term for each specialised concept and that each term can in turn introduce new conceptual or semantic nuances. Moreover, and beyond the ways recommended institutionally (by the corresponding academy or terminology centre), the purpose of effective

[2]Research project BPH-2012-0121-PC: *Educational and cultural values in specialisation metaphor: the multiple images of the world economic crisis in the written press* (CAPES-DGU, Hispanic-Brazilian Programme for inter-university cooperation) in which two teams of researchers from the University of Vigo and the Universidade de São Paulo participate. The project also has collaborators from the Universidade Estadual de Maringá and the Universidade Estadual de São Paulo in Araraquara. Its work methodology and tools are described in Galanes and Alves (2015).

[3]"A person who writes, especially one who works in a newspaper, publishing house, radio, etc." Available at: http://servicios.elpais.com/diccionarios/castellano/redactor.

mediation requires us to choose the term that is most appropriate to the final text and to the target audience.

Reporting on a discursive event additionally implies acting as a mediator in a developing knowledge area that introduces innovations, through new concepts or new terms. The mediator is, at times, impelled to formulate terminological solutions or to provide explanations for concepts unknown to the reader. One needs to be extra careful in this classificatory and didactic task because the results may not always be neutral.

The language used by economists is framed within specialised languages with its own formal characteristics both at microlinguistic and macrolinguistic levels, to the extent that it can be considered as jargon (Menéndez Alonso 2016). Specialised languages establish a very close relationship between the social norm, the cultural system and the textual structure via the drafting standards and conventions imposed by the communication situation and these, in turn, influence the linguistic options. Despite the fact that the purpose of specalised language is primarily informative, the manner of presenting knowledge and ideas differs from the common language and varies depending on the genre and the discipline.

When transmitting new information or that perceived as new, we need to ask questions like: what does neologism mean and what levels does it affect? And also others like: what are the linguistic strategies used by the writers to the neologism and how do they indicate them in the text?

3 Specialised Conceptual and Terminological Innovations

Language is a living entity that evolves with more or less dynamism, in which a few unusual or forgotten words are left behind and new ones appear, either to replace them or to respond to recently formalised benchmarks or concepts. Ferdinand de Saussure, the pioneer of modern Linguistics, considers linguistic sign as a dynamic unit because it evolves over time and is unchangeable, i.e., we cannot modify it at whim. It evolves within a particular context, under specific conditions and as a result of an implicit agreement by the user community which accepts such linguistic change. In a specialised field, these innovations are mainly reflected in the specific terminology. The term coined for the "new words" is neologisms, and these have also been extended to terminology.[4]

The crisis has led to a constant renewal of terminology, which has been addressed in the respective cultures from their own peculiarities (Fernández and Galanes 2015). The concept of neologism is linked to novelty and this, in turn, can manifest itself at different levels.

We therefore see the emergence of new concepts and terms that had not been known earlier such as recasting capitalism (*refundación del capitalismo* in Spanish) (Luna 2016). In other cases, the novelty is purely linguistic and manifests itself only

[4]Despite the attempt to disseminate a specific term, *neonyme* by Rondeau (1983).

at a formal level. This implies coining new names for existing concepts in other cultures such as subprimes mortgages (*hipotecas de alto riesgo*) or in other registers, such as risk premium (*prima de riesgo*). In some so-called "novel" variants there is an underlying metaphor like in *hipoteca basura*, *hipoteca de baja calidad*, *hipoteca loca*, ('trash mortgage, low-quality mortgage, crazy mortgage', etc.) wherein the same concept is presented under different perspectives, sometimes with impact on its connotation also at the conceptual level. The metaphorical terminological variants are of great interest for the observation of cultural aspects as we will see later.

A third type of novelty is the one concerning meaning, where new meanings are attributed to pre-existing terms. There is a difference between (i) those terms that have just added a new connotation, for example, the perception of what it means to be a *mileurista* (a person earning around €1000/month) (has changed after the crisis), (ii) those that add a new meaning such as *austerity* and (iii) those that completely change meaning (such as *bailout*). Once again the figurative uses of the metaphorical type gain prominence, these being a vehicle of subjective aspects attached to specialised terms. All these metaphorical neologisms are initially recorded in the written press and, given their novelty, have had to be explained by writers whose language also includes scientific-technical discourse as well as the so-called sectoral discourse (Medina Montero, J. Francisco 2016: 156).

Neologism understood as "vocablo, acepción o giro nuevo en una lengua"[5] (DRAE) can be created within our own language or can come from another language-culture. Its classification has long been debated, but due to its complexity, we find approaches that are based on semantic, functional or origin criteria. On the other hand, Cabré applies formation criteria to differentiate between the several neologisms types: formal neologisms, semantic neologisms, syntactic neologisms, loans and adapted loans (2006: 231). Despite the fact that this classification helps us to sort out neologisms, it does not however guarantee a single objective and systematic approach to identify them all because some are more difficult to identify than others. Thus, for example, semantic neologisms are not easy to detect because they do not show any "formal evidence of a change of meaning" (Cabré 2006: 240). How then can we know when a neologism is being used? In this context, Mª Teresa Cabré (1993: 441) refers to neologisms as terms that meet most of the following criteria: diachronic (if it appears on a recent date), lexicographic (if it is not recorded in dictionaries), formal and/or semantic instability (if signs of instability exist, whether phonetic, typographic, morphological or semantic), and psychological (if the word is considered to be new by the language community). This set of criteria is complemented by another (Freixa 2002a, b) set of identification sub-criteria: the reader does not understand the first time he/she sees the term (semantic opacity), the referent is new, new name, a unit from another linguistic system, the text marks the neologicity[6] as we will see in our analysis.

[5]"A word, meaning, or a new twist in a language".

[6]The analysis has become more comprehensive ever since the reflections made by Rey (1976) who defines a neologism as a word used recently in exchanges and one that is not registered in the metalinguistic corpus, to the studies carried out by Cabré (1993 and later ones).

4 Denominative Strategies and Didactic Lexical Innovation Strategies

The preceding lines have described the conditions in which the press writer confronts, understands, interprets and expresses the innovations generated by the communications agenda. It is now time to ascertain what restatement strategies are used and identify which restatement marks are offered by the discourse itself.

We refer to this phenomenon as *translation neology* meaning that the writer performs a transfer operation, although this does not necessarily always occur between words from different languages. To illustrate this section, we include examples from the extensive written press corpus that we have complied for preparing the *Dictionary of the World Economic Crisis* (DiCEM) (in progress), and in particular from the Spanish press[7] subcorpus.

It is important to highlight that Jakobson (1959/2000) and other well known intellectuals like Peirce or Umberto Eco identify three types of translations: interlinguistic, intralinguistic and intersemiotic (Pym 2016: 170), of which only the first type involves two different linguistic systems. The task carried out by the writer as mediator on many occasions covers the three types of transfer processes when faced with innovation.

Thus, interlinguistic translation occurs when the writer must find an equivalence for a foreign term, as was the case in the recent economic crisis with English expressions such as: subprime, stress test, credit default swap, core tier 1, quantitative easing, sale&lease-back, midswap, minijob, credit event, haircut, monoline, saving and loans, credit boom, script dividend, covered bonds, crowding out, Grexit, stand still, real estate stock, conduits, swap lines, company cost, cross selling, seniority, short sellers, steering committee, twin peaks. The neological nature of these names is related to a new concept or also with its innovative nature in Spanish journalistic discourse. We have thoroughly checked (Galanes 2017) that none of the terms cited was documented in these Spanish newspapers before 2006, that is, from pre-crisis times. The corpus contains many other foreign words but these have not been considered as neologisms because they appeared in such press earlier.

We can speak of intralinguistic transfer whenever we document certain very specialised technical terms in our corpus taken from the general press. That is to say, transfer takes place from the different language and research or technical sector registers to the general press. The use of technicalities in consumers goods may confer greater prestige to the product and encourage purchase, albeit misleadingly. This latter vulgarised group of terms from the world economic crisis can include: *Participaciones preferentes*[8]- Preference shares, *obligaciones subordinadas*-subordinated debentures, *prima de riesgo*-risk premium, *negative equity*, etc.

[7]This corpus contains all media news on the economic crisis and its many manifestations published every Friday in the newspapers *El País* (EP), *El Mundo*(EM) and *Expansion* (EXP) from 2007 to 2013, as well as the same items published in *Expansion* every Friday from 2014 to 2016 (with 6,821,213 tokens and 78,951 types).

[8]Also known as "*preferentes*" in Spanish.

Finally, newpaper writers do an intersemiotic translation when they present data through graphs in the written press or when they illustrate testimonies or news using audio-visual resources.

4.1 Denominative Strategies

When writing about an event that involves a discourse innovation, news writers are forced to adopt a strategy that permits delivering "new" specialised content to the target recipient: the reader. Moreover, such action may also be limited by the entity's drafting rules (contained in the corresponding style guide), the more or less purist policy applied to linguistic loans by the editorial staff of the pertinent media, etc.

Communications media also play a relevant role in disseminating innovation, while ensuring intelligibility and connection with the reader. The naming and conceptual instability implied by innovation triggers a range of names that Guilbert calls "*foisonnement neologique transitoire*", considered as an index of neologism (Dury and Picton 2009 *apud* Humbley 2015: 55) since they are examples of naming and/or conceptual instability.

Faulsticht (2002: 88) presents the possible scenarios that can lead to the introduction of a loan. These in turn arise from a particular terminology praxis shown below. Sometimes, several of these terminology strategies appear concatenated and are stages of the same naturalisation process of a new concept. To illustrate these strategies we include examples in Spanish from the world economic crisis.

- The first possibility is that we maintain the original English term (bad bank or subprime mortgage loan) in Spanish.
- The second possibility is that a hybrid form be created to designate the new form (*hipoteca subprime*-subprime mortgage loan) and a complementary form in the national language (*hipoteca convencional*-conventional mortgage), such that both reflect the total number of mortgages. These two terms are listed as cohyponyms of the hyperonym *hipoteca*-mortgage. The systemic nature of the specialised terms becomes evident when we have to rename an existing concept using a complementary term (Roldán 2014: 17) to facilitate the introduction of innovation.
- There may also be the case that a hybrid form is generated or that several alternative vernacular forms exist (*hipoteca subprime*-subprime mortgage/*hipoteca de riesgo*-risk mortgage, *hipoteca de alto riesgo*-high-risk mortgage, *hipoteca de mala calidad*-bad quality mortgage, *hipoteca basura*-trash mortgage). Each of the vernacular forms is generated by highlighting one or more semantic features of the original, and sometimes through metaphorical procedures.
- It is also possible to abandon the original form 'bad bank' in favour of a translation (*banco malo*) that respects the initial English structure with word for word transfer, as well as the syntactic relationships within the syntagm and not in the entire unit of sense, as would be the case in a more elaborate terminology practice.

- Yet another possibility is that a new adapted form (*suprima*) be created together with the English one (subprime). Normally these adapted forms suffer greater instability since they are not motivated lexical units that are semantically opaque to the speaker.

The result of this terminology profusion will be a set of denominations. This breaks the rule that each concept must have one (and only one) term, and vice versa. Of them all, just one will function as the reference form. This is usually the one that belongs to the standard variety and is a strong candidate for lexicalisation, that is to say, inclusion in dictionaries. The selection of that word as against another is done through formal suitability criteria, conceptual precision (semantic) and usually also of use, although it may not always be the most used one. Becoming a reference does not imply immediate introduction of the innovation, but requires a period of time for the concept to remain in the current panorama and to reaffirm acceptance of the proposal. In the present case, the name hipoteca de alto riesgo-high-risk mortgage is the one that has gained such a condition.

The search for the reference name is preceded by many others that had more or less lasted in time. Such name subsets generated from the time innovation is introduced to its lexicalisation (i.e., appearance in dictionaries) are not stable in terms of composition and forms. Within this subset, one can distinguish at least one reference name prescribed by academic authorities or by the linguistic services of that language (*hipoteca de alto riesgo*-high-risk mortgage) and other alternative names (*hipoteca subprime*-subprime mortgage, *hipoteca burbuja*-bubble mortgage, *crédito subprime*-subprime credit, *crédito subestándar*-substandard credit) of which one part has a connotative load, and hence is reserved for certain uses, as is the case with *hipoteca basura*-trash mortgage, *hipoteca tóxica*-toxic mortgage, *hipoteca problemática*-problem mortgage, *credito dudoso*-doubtful debts, which are reserved for less formal uses and involve a certain approach (negative in this case) to the concept.

Terminological variation is an indication of an emerging neology (Humbley 2015: 55) and rather than reduce it, our objective is to study it from the communication perspective of Terminology with the pertinent contributions from Discourse Analysis, in order to know the forms, study the associated nuances, and "sort out" the naming and conceptual instability. Awareness of the semantic and pragmatic aspects of the term are of utmost importance for the practice of linguistic mediation (and for traditional Translation and Interpretation) as communicative activity. The formal and conceptual aspects are respectively catered to by Linguistics studies or by studies of the respective specialisation or scientific areas.

4.2 Marks of Translation Neology

The Communicative Theory of Terminology (Cabré 1999) states that the term performs a double function of representing and communicating specialised knowledge.

This dual role is also present in specialisation neologisms. Our studies (Galanes 2016; Fernández 2016) focus on how specialised concepts from other cultures are communicated in journalistic discourse. That is to say, we work with secondary neology, the one responsible for national adaptation of foreign neologisms, and which is developed in the field of translation (Hermans and Vansteelandt 1999: 37).

One of the tasks proposed is to identify the strategies used by press writers. When introducing a new concept, which may not be understood by the target audience, these writers prefer to offer an explanation, or add the original term, develop a definition, etc. The word marks used in French have been described in Hermans and Vansteelandt (1999) and the Portuguese ones in Alves (2011: 98). For Spanish we have made a first approximation in Galanes (2017), which is being completed in this chapter.

It is about the word marks that the writer includes when presenting the required information so that readers are able to understand the term. All word marks have an underlying transfer operation, either involving extra or intralinguistic information, which beyond doubt, is a neological mark, and therefore can be added to the lexicographic, chronological, psychological and instability criteria that characterise neologisms.

- Graphical marks include different types of letters such as italics, single or double quotation marks, parentheses or brackets, etc.
- Metalinguistic labels that introduce a name through formulae such as: *conocida*-known, *lo que se conoce*-what is known, *denominada*-termed,*llamada*-called, *término usado*-term used, *en la jerga económica*-economic jargon, *técnicamente*-technically *speaking*, *literalmente*-literally, *en ingles-in English*, etc.
- Reference or designation marks that introduce an object or a concept to which the name can be applied: *como*-such as, *similares*-similar, *en términos de*-when referring to, etc.
- Translation or adaptation marks of the neological term which in addition to graphical symbols (dashes, parentheses) can also be used with the conjunction "or".
- Explanatory marks that introduce typological or conceptual information together with the neological term, using formulae such as: *se trata*-it is about, *consiste en*-regarding *(something)*, *es decir*-i.e., or with the introduction of an apposition preceded by comma, etc.
- Defining marks which introduce written information using definition techniques. Expressions such as *que es*-which is, etc. can also be used in addition to the graphical marks.

The above marks can appear together in texts, as shown in the following examples. Although we had initially identified them through the neologisms of foreign origin, this classification is equally valid for other "novelties" described below.

Example of marks for neologisms and new objects and loan neologisms:

La quiebra de Lehman Brothers, que fue el momento más agudo de una crisis financiera iniciada en verano de 2007 como consecuencia de la infravaloración de los riesgos de las hipotecas de alto riesgo ("subprime") y productos financieros creados en torno a ellas, hizo tambalear el sistema financiero internacional y generó la peor recesión económica mundial desde 1929. <EXP_13/09/2013_POR_EFE>

Example of marks for naming neologisms based on metaphors:

El desplome adicional del 10% que sufren las acciones de la automovilística en la preapertura de Wall Street refleja el grado de incertidumbre que dejan sus cuentas, más aún cuando ayer mismo Standard & Poor's volvió a rebajar surático crediticio, en terreno de 'bonos basura', a 'B-', seis escalones pordebajo del 'grado de inversión'. <EXP_01/08/2008_EMP_EXPANSIÓN.COM>

Example of marks for conceptual/or semantic neologism (connotation, meaning, concept):

Durante los años de bonanza, el gasto público se aumentó más allá de lo razonable. Luego, la crisis se enfrentó aumentándolo aún más. Lo que vino después, que equivocadamente algunos llaman "austeridad", es apenas un proceso de aceptación de la realidad. Es imposible gastar permanentemente más de lo que se ingresa. La alternativa a reducir el déficit público es cometer un "deudicidio": gastar hasta que la deuda pública se haga insostenible y resolver el problema con una nueva crisis. Estamos a tiempo de evitarlo. Hagámoslo. <EXP_30/01/2015_OPI_EXPANSIÓN>

5 Terminology Metaphors: The Cultural Contrast

Metaphors are a very effective way of transmitting economic information and introducing new concepts. They are used to explain a specialised reality through terms from other knowledge areas or words from the common language.

The vast literature developed in recent years around their definition and classification reveals the complexity of the subject matter. Numerous studies have been carried out on metaphors within specialisation languages in different thematic fields from politics to economy (cf. Arrese 2015) and including informatics (L'Homme 1997, 2004) and medicine (Vandaele 2005, 2006, etc.). They conclude that cultures generally share the same conceptual fields while languages systematically import metaphorical concepts from English, thus showing little creativity.

Most of these studies lie within the framework of cognitive semantics (cf. Lakoff and Johnson 1980; Taylor 1989) whose vision defends a high degree of universality of the metaphorical structures. From this position it is argued that metaphors allow us to structure complex concepts from other more specific ones because "metaphor is primarily a matter of thought and action and only derivatively a matter of language" (Lakoff and Johnson 1980: 154). Therefore, they consider that the concept existed prior to the name.

On the other hand, from a linguistic approach which could be called structuralism or more specifically an example of stylistic grammar, Tamine (2011: 50–51) regrets that American linguists obviate the differences among the different forms adopted by the same metaphorical concept by laying emphasis only on fields of experience in its entirety and not on isolated concepts, and therefore do not take into account the name varieties or even the semantic values acquired according to the moment and users that reactivate them (Gardes-Tamine 2011: 61). This can be exemplified in the crisis terminology through the terms *ajuste*-adjustment or *austeridad*-austerity.

Specialisation metaphors have also been the object of translation studies, which should make us doubt about the universality of the conceptual metaphors of Lakoff and Johnson. One point of interest lies in the many strategies used for translating metaphors. If all metaphors were of the conceptual type and were hence present in the same way in all cultures, then it would be senseless to analyse several transfer strategies.

Therefore, one of our objectives is to perform a multicultural contrastive analysis of the terminology metaphors in economics, in order to highlight the existence of culturally marked metaphors. As an example we have analysed the translation of the neologism 'Bad Bank' in a case study.

6 Case Study: *Banco Malo*-Bad Bank

A bad bank is a financial institution whose balance sheet concentrates toxic assets or doubtful profitability investments from other banks. The term 'bad bank' has had a strong presence in the British press from the beginning of the crisis where verbal variants such as: 'bad bank', 'bad' bank, 'bad-bank', bad-bank, 'bad bank' or 'aggregator bank' have been used. This phenomenon burst into the Spanish newspapers scenario only after 2009, where the anglicism was initially accompanied by its translation *banco malo*, with different variants. From 2011 onwards, the translated form is preferred over the English one. Moreover, the alleged Spanish public bad bank (Sareb) is a financial institution that accumulates toxic assets from other Spanish institutions, usually from the real estate market. This semi-public company, as a financial instrument (which is not really a bank), has a term of 15 years from its inception to manage the real estate assets and sell them profitably.

The French press uses different words as synonyms: *structure de défaisance*, *structure de cantonnement*, « banque-poubelle » « banque-hôpital », en anglais *bad bank.*

Structure de défaisance and *structure de cantonnement* were used before the crisis, and as we can see, there is no indication that this is a bank, and hence only the context might indicate that reference is being made to the banking sector and not to any other company as was the case with Peugeot in the 1980s. These two names refer to two different concepts because *structure de defaisance* implies the creation of a new entity for transfer of the problem or bad debt assets; while *structure de cantonnement* implies that the assets do not leave the entity. In some cases, and as a clarifying strategy, reference is made to the entity using the Latin loan *Consortium de Réalisation* (*CDR*) used in the 1990s to purge Crédit Lyonnais and well known to the public. Furthermore, whenever reference is made to the Belgium bad bank, the term used is *banque résiduelle*. This shows that the translation of the English term *bad bank* includes different models of companies to manage doubtful assets, but each culture has opted for names that reflect specific realities, which may include the types of assets and the nature of the debtors and creditors.

7 Final Considerations

This chapter has therefore delivered the objectives that had been set out. After describing the conditions that the writer has to face in order to understand, interpret and express new concepts created within the economic context, we have been able to show with concrete examples related to the crisis phenomenon that news writers use different strategies to clarify and popularise these new concepts. It has been verified that this process of incorporating neologisms in the press not only imports traits typical to the culture of origin (mainly Anglo-American culture) but also the values of the receiving culture. The analyses of the word variants in context, as well as the marks that permeate journalistic language, have been crucial for this purpose.

References

Alves, I.M.: Neologia tradutiva em textos de Economia. ReCIT **2**, 97–109 (2011)

Arrese, A.: Las metáforas de la crisis del euro en la prensa española. Commun. Soc. **28**(2), 19–38 (2015)

Cabré, M.T.: La clasificación de neologismos: una tarea compleja. Alfa, São Paulo. **50**(2), 229–250. http://seer.fclar.unesp.br/alfa/article/view/1421/1122 (2006)

Cabré, M.T.: La terminología. Teoría, metodología y aplicaciones. Antártida/Empuries, Barcelona (1993)

Cabré, M.T.: La terminología. Representación y comunicación. Una teoría de base comunicativa y otros artículos. Institut Universitari de Lingüística Aplicada, Universitat Pompeu Fabra, Barcelona (1999)

Chavero, P.: Prensa y política en tiempos de crisis: estudio de la legislatura 2008–2011. Centro de Investigaciones Sociológicas, Madrid (2015)

Faulstich, E.: Variação em terminología. Aspectos de Socioterminologia. In: Guerrero Ramos, G., Pérez Lagos, M.F. (eds.) Panorama actual de la terminología, pp. 65–91. Comares, Granada (2002)

Fernández Rodríguez, Á., Galanes Santos, I.: La crise hypothécaire et ses dénominations. Babel. Revue de la Fédération Internationale des Traducteurs FIT. **61**(2), 265–282 (2015)

Fernández Rodríguez, Á.: Del concepto a la denominación en traducción económica. El caso de activo. Revista de Filología e Lingüística Portuguesa. **18**(1), 116–144. http://dx.doi.org/10.11606/issn.2176-9419.v18i1p115-144 (2016)

Freixa, J.: Reflexiones acerca de las causas de la variación denominativa en terminología. In: Guerrero Ramos, G., Pérez Lagos, M.F. (eds.) Panorama actual de la terminología, pp. 107–116. Granada, Comares (2002a)

Freixa, J.: La variació terminológica. Anàlisi de la variació denominativa en textos de diferent grau d'especialització de l'àrea de medi ambient. Ph.D. dissertation, Universitat Pompeu Fabra, Barcelona (2002b)
Galanes, Santos I., Alves, I.M.: Metodología de trabajo para el estudio de las múltiples imágenes de la crisis económica en la prensa escrita. In: Gallego-Hernández, D. (ed.) Current Approaches to Business and Institutional Translation, pp. 77–88. Peter Lang, Bern (2015)
Galanes Santos, I.: Neologismos conceptuales y referenciales: aportaciones desde la neología traductiva de la crisis. In: Congreso Internacional Entretextos [in press] (2017)
Galanes, I.: La crisis económica mundial. Un concepto complejo con múltiples denominaciones. Revista de Filología e Lingüística Portuguesa, 18(1), 5–41. http://dx.doi.org/10.11606/issn.2176-9419.v18i1p5-41 (2016)
Gardes-Tamie, J.: Au coeur du langage. La métaphore. Honoré Champion, Paris (2011)
Hermans, A., Vansteelandt, A.: Néologie traductive. Terminologies nouvelles **19**, 37–43 (1999)
Humbley, J.: La néologie dans les langues de spécialité. In: Alves, I.M., Pereira, E.S. (eds.) Neologia das línguas románica, pp. 39–74. Humanitas, São Paulo (2015)
Lakoff, G., Johnson, M.: Metaphors We Live By. The University Press, Chicago (1980)
L'Homme, M.C.: Méthode d'accès informatisé aux combinaisons lexicales en langue technique. Meta **42**(1), 15–23 (1997)
L'Homme, M.C.: La Terminologie: principes et techniques. Presses de l'Université de Montréal, Montréal (2004)
Luna Alonso, A.: La refundación del capitalismo. Estudio terminológico en la prensa francesa y española. Revista de Filología e Lingüística Portuguesa. **18**(1), 145–174. http://dx.doi.org/10.11606/issn.2176-9419.v18i1p145-174 (2016)
Medina Moreno, F.J.: La metáfora en el léxico futbolístico: el caso de la actividad deportiva en español, y algunas propuestas de traducción al italiano, pp. 155–202. http://www.termcoord.eu/wp-content/uploads/2016/05/J.F_Medina_Montero_La-met%C3%A1fora-en-el-l%C3%A9xico-futbol%C3%ADstico_2.pdf (2016)
Menéndez Alonso, E.J.: El economista que regresó a Babel. Un ensayo sobre la jerga de las finanzas. Pirámide, Barcelona (2016)
Moirand, S.: Les discours de la presse quotidienne. Observer, analyser, comprendre- PUF, Paris (2008)
Pernías Solera, S.: Crónica de la crisis económico-financiera. Revista de Derecho de la Unión Europea **28**, 25–58 (2014)
Pym, A.: Teorías contemporáneas de la Traducción. Materiales para un curso universitario. [2ª ed., rev. trad: Noelia Jiménez, Maia Figueroa, Esther Torres, Marta Quejido, Anna Sedano, Ana Guerberof, Helena Romero López, Melisa Monaco]. Intercultural Studies Group, Tarragona (2016)
Rey, A.: Néologisme, un pseudo concept? Cahiers de lexicologie **28**, 3–17 (1976)
Roldán Vendrell, M.: Representación del conocimiento especializado y terminología multilingüe. In: Roldán Vendrell, M. (ed.) Terminología y comunicación científica y social, pp. 1–23. Granada, Comares (2014)
Rondeau, G.: Introduction à la terminologie. Gaëtan Morin, Chicoutimi (1983)
Taylor, J.R.: Linguistic Categorization: Prototypes in Linguistic Theory. Clarendon Press, Oxford (1989/1995)
Vandaele, S.: Métaphores conceptuelles et fonctions lexicales: des outils pour la traduction médicale et scientifique. In: Actes du IIIe congrès international de traduction spécialisée, Univ. Pompeu Fabra, Barcelona, 4–6 March 2004, pp. 275–286 (2005)
Vandaele, S.: Conceptualisation métaphorique en biomédecine: importance pour le processus traductionnel. In: Congrès de l'ABECAN (Associação Brasileira de Estudos Canadenses), nov. 2005, Gramado. In: Nubia Hanciau (org.), Brasil/ Canadá: visões, paisagens e perspectivas, do Ártico ao Antártico. Ed. da FURG, Rio Grande, pp. 281–294. https://papyrus.bib.umontreal.ca/jspui/bitstream/1866/1443/1/van06gram.pdf (2006)

Áurea Fernández-Rodríguez Degree in Romance Philology and Doctor in French Philology from the Universidade de Santiago de Compostela. She is currently senior lecturer at the Department of Translation and Linguistics at the Universidade de Vigo. Her research areas devoted to Translation Studies include literary translation research from and into different cultures, the history of translation and translators; the study of translation flows and terminology. Vigo, Spain.

Iolanda Galanes-Santos Degree in Romance Philology and Doctor Galician-Portuguese Philology by the Universidade de Santiago de Compostela. She is currently senior lecturer at the Department of Translation and Linguistics at the Universidade de Vigo. Her research areas devoted to Translation Studies and Terminology. She was leader in a university cooperation project with the University of São Paulo in order to study the imagery of the crisis through the terms used in the press. Vigo, Spain.

Part III
Corporate and Institutional Communication

Generation *Co*. Tendencies of XXI Century's Communication Management

Miguel Túñez-López, Karina Paola Valarezo-González and María Isabel Punín-Larrea

Abstract Artificial intelligence, big data, geolocation, mobility, augmented reality, Internet of Things, machine learning, augmented reality or business intelligence have become fundamental in the way of understanding and approaching communication management in the XXI century, oriented into a new way of doing and understanding relationships to encourage co-creation and co-protagonism with audiences. Through *liquid contents*, the modes of communication are adjusted to the new scenarios in transmedia narratives. Communication has been installed in a systemic framework in which everything is interrelated. You can no longer act thinking about the public, but being the public.

Keywords Liquid communication · Internet of things · Geolocation Organizational communication · Transmedia

The way we socially interact has been transformed in such a way in recent years that we have derived the scenario of our activity of personal and social communication to platforms that hybridize online and offline. The Internet has opened an infinite universe of possibilities for intercommunication but, at the same time, it has become a spider web in which we fall, trapped, and with limitation of movements. It is a constant: every day we open more to online environments to exercise offline relationships from mobility terminals. Each time we think we are freer, more autonomous but, as that feeling grows, our dependence on support and connection increases.

M. Túñez-López (✉)
Universidade de Santiago de Compostela, Santiago de Compostela, Spain
e-mail: miguel.tunez@usc.es

K. P. Valarezo-González · M. I. Punín-Larrea
Universidad Técnica Particular de Loja, Loja, Ecuador
e-mail: kpvalarezo@utpl.edu.ec

M. I. Punín-Larrea
e-mail: mipunin@utpl.edu.ec

M. Túñez-López et al. (eds.), *Communication: Innovation & Quality*, Studies in Systems, Decision and Control 154, https://doi.org/10.1007/978-3-319-91860-0_21

It is logical to think that if the communication between individuals is transformed also the organizations communication is substantially modified. Organizations are the result of the activity between the individuals who compose them and of those who direct them. Moreover, in a systemic environment such as the one that characterizes the 21st century, the result no longer depends only on the activity of the members and leaders, but also on the behavior (proactive, reactive or inactive) of each entity with the systemic environment in which each organization is immersed.

We are in an interesting moment in which the organizational relations are increasingly open and more interdisciplinary. They are also more complex because the activity of individuals and organizations begins to add the capacity of things to join autonomously to that relational network. In fact, social and technological advances generate a new scenario and, by extension, a new way of understanding and approaching communicational management, always based on promoting mimicry with the action modes of those with whom it relates or to which it is directed.

With the profound transformations of the last two decades, the axis of gravity of communicative management has shifted: it has ceased to rely on the needs of organizations and has moved to pivot on the behavior and attitudes of individuals. As it has been done up to now, this is not a question of managing for the public, but of communication managing and of how the public communicates. The important thing is not to manage the communication thinking in the public but to manage it being public.

Almost everything is already connected with everything. Today, there is a new scenario for human relations caused by the possibility of connecting to the Internet through wireless terminals and low-cost tariffs. This dynamic has not only rapidly transformed the universalization of access capacity and the possibility it confers for individuals to be the proactive in order to act as issuers of all type of messages. Moreover, being socially interconnected, this capacity of broadcast is understood to have a potential audience that can become massive. We have incorporated these changes in a progressive way into our lives, but seen from the perspective of two tiny decades, if we talk about accelerated transformation, we are likely to be short in the way we relate and communicate and we should probably admit that this is a whole revolution in progress because it implies a radical and unstoppable change that has, already, led us to a different way of doing and tackling things in a world that is increasingly overlapping which is offline and what is done online.

In these two decades we have seen how we go about socializing from one to one or in small groups to interacting all with everyone, deriving those relationships in online environments towards changes in the offline world, that is, from the technological terminal to the personal social life. If communication modes change, it is logical to also vary the message's flows, which, for the construction of the story or of the event, have evolved from a discursive and unidirectional model (from one to several) to another interactive model between the participants and in several supports, simultaneously. That is to say, now the narrative is produced with multiple actors who become emitters with bidirectional participation (back-feeding the story) in more than one channel, simultaneously.

Communicating socially in this interactive and transmedia way implies a change in global functioning structures that not only affects individuals, but also forces organizations to redefine their communication behavior in several aspects (Túñez López 2015). It is necessary that entities of all kinds adapt to it because individuals have conquered the ability to collectivize their prosecution of the products, services, practices and attitudes of the organizations with which they relate. Through channels of return, publics have also gained the possibility of responding to the messages that organizations send them, that is, those used by the entities themselves, or through media that reach the public, but exclude the Organizations (Túñez López 2015).

So, publics are no longer the final receivers in the communication process: they become senders. This means that the universalization of the possibility of being issuers that can reach mass audiences gathered with the ability to articulate bidirectional relations means that we can say and be heard and this breaks the discursive monopoly that until now, Organizations had about themselves or, what is the same, about their activities and about their products/services. Moreover, although it has surely been in a non-conscious way, we have even changed our usual process of behavior as buyers by moving the acquisition in offline scenarios to the purchase of online environments.

Before the digital era, we first experienced the product and then we bought it. Now, we first buy and then experience it. Before, we said satisfaction with the purchase or with the product in reduced circles. Now, the comment on the Internet is encouraged because this evaluation has become a guarantee of credibility for new buyers who, in exchange for buying online before experimenting offline, go to the valuation of others who have already known the product and have left their impressions in the network. These assessments are made either in spaces of the organization that offers the product/service either in other spaces or supports outside of it and, what is more important for the communication management or relations, without the effective possibility of inferring about it.

The proactivity of users forces organizations to orient their way of relating and communicate internally and externally towards a revalorization of their online and offline reputation, generating credibility in their services or their products and practices. For organizations, the ultimate goal is to transfer trust. The actions that are developed can not be limited to the communicative action. It is not enough to say, it must be done. The commitment to take care of the external discourses or messages and the way to approach the relations with the publics has to be extended to the organizational culture.

The way of doing things has to be in line with the contents and the ways of saying things because it is increasingly essential to unify what is said and what is done and at the same time, the expectations that are generated.

This new way of public behavior incorporates the integrity (Kotler 2011) of the organizations as the central component of the way to manage communication, but also of the way of doing things because without credibility and without trust it is unthinkable to obtain a loyal and prolonged relationship, oriented and sustained in the construction of an organization good image and a favorable reputation.

As netizens we take or consolidate the decisions to purchase products or services motivated by the information we find on the Internet (Celaya 2009: 24) and we are the protagonists and promote this collaborative organizational communication through a participatory attitude that encourages dialogue in social networks or forums or, in a generic way, in any space that allows or promotes the evaluation of products or services.

This leadership of users and their independence to interact, or not, in the network outside the organizations or in response to initiatives promoted by those organizations, increases the need for organizations to combine actions in formats or media aimed at mass audiences and, in parallel, to create channels of direct contact with their own public. The goal is for organizations to be able to take the initiative to communicate when they need it and, more importantly and less damaging, economically, to be able to do so without having to resort to intermediaries or external means.

1 The Interactive Symmetry

The organizations of the 21st century are installed in a systemic conglomerate. As so, most naturally, everything is interrelated and, because of that, it becomes necessary to adopt a holistic attitude as it makes sense the synergy between all the parts, even between those that are not directly incorporated to the dynamics of the organization.

Therefore, it is not possible to plan, execute and analyze things in a single mode. Instead, it is necessary to assume that a change must go beyond the mere intentions of making a messages renewal or to undertake an apparent modernization of the ways of saying or channels to transmit. In fact, when Kotler (2011) talks about integrity, he is also referring to the need for transparency in the way things are done. Therefore, it is understood that the way to proceed is to seek an unification of three fundamental elements: the way of doing things, the message that is transmitted—the way of saying things—and the expectation that is generated on the things that are offered, not vice versa.

From the point of view of communication management, adapting to these new social dynamics means conforming to the conventions and languages used by all. It is almost a neophyte task that accedes to a new culture: to observe to know, to assume and to integrate.

It is, in short, learning to be one more in the group. For organizations, this integration is equivalent to assuming that they must humanize, approach and mimic the group or community they want to be part of and be seen as one more member, in a group that interacts. In online environments, it means to do so by ratifying what has been conceptualized as a commitment of real bidirectional interaction 2.0 (Túñez López and Sixto García 2011): it is no longer enough to be in a group, but rather to participate actively by taking on the conventions and rules of the collective and of the support in which its members are interrelated.

The objective must be to plan in a position enabling to adjust communicative processes to the needs of the organization and, at the same time, to the expectations of individuals in, as we said before, a systemic social environment. It influences everything: from the orientation of the internal labor climate due to its impact on productivity, to the fact that the strategic planning of the 21st century inevitably involves incorporating into the action plans of other disciplines, that until now, were not usual in strategies of communication design such as the contributions of artificial intelligence or neurology. This is because this new way of understanding communication management starts from the inevitable need to know better how we relate.

It seems obvious, but not all organizations have assumed and incorporated that change of role and remain in a discursive linear communication attitude. The future seems to be oriented to ensure that the relations between the organization and its publics are produced in such a way that the organizations communication with the society fits and aligns with the new communication guidelines in which society has vertiginously settled.

This proposal also includes the communication models overwhelmed by the changes resulting from the social transformations imposed by the democratization of access and the ability to broadcast on the Internet. A receptive attitude such as that described in the bidirectional symmetric model (Grunig and Hunt 1984) is no longer enough, but in what, projecting the typology contributed by Grunig, it was proposed to label as the "*model of interactive symmetry:* relationships that are develop between actors with independent behaviors but located in levels of equality and mutual influence, in which both have mass emission capacity, initiative in the management of their relationships, and individual or group response to their communications" (Túñez López and Altamirano 2016).

Although it is associated with critical positions with systemic theory, Grunig himself refers that the interdependence of the organization with the environment in dealing with relations with audiences must be "a process of symmetrical and bidirectional communication, a dialogue in Place of monologue" (2003, 413) and for this it is advisable to understand "relations with the community as a mechanism to help the organization to solve important problems" (2003, 414).

Individuals, groups, organizations or, globally, actors who belong, relate, who are affected by the organization or who can influence their behavior could be considered their target audience. There is no single public but a sum or a confluence of audiences, according to the activity that the organization carries out at any moment and the action of communication that it wants to implement or of the supports/channels that it uses in doing so. One should not think of a single classification of the public with social, anthropological or territorial variables because the same individual can belong to different public categories if they are different organizations or activities.

In many cases, the concept of public is assimilated to the stakeholders, probably by a translation of concepts from the scope of the company. Stakeholders are groups or individuals that may have effects on the organization or that may be affected by it (Freeman 1984) and are interested (Coombs 2000) or affect or can be affected by it (Grunig 1989; Daugherty 2001).

Some authors limit the concept of stakeholders to actors who bet on the company because of the trust it deserves, such as employees, suppliers, customers or community (Mazo 1994: 339) and differentiate it from shareholders, which would encompass those agents directly interested in the ownership mechanisms of the company, such as financial institutions, shareholders or, among others, senior managers.

Other experts refer to the public as a collectivity of constituents or individuals to whom an organization is directed and who can be influenced by it. In studying the concept, Míguez clarifies that "although its first uses in the discipline are very early, it is less used than the term stakeholder and has aroused less interest among theorists (…) most authors in the field of public relations who speak of constituents in their texts often use this concept simply as a synonym for public or stakeholder" (2010, 59).

The first definitions of public, however, are linked to a set of people who do not have to be in contact, who hold different views on the same subject and who organize their discussion in an organized way (Dewey 1927; Blumer 1953; Park 1972) with the ability to think and reason until creating a discourse (Price 1994).

From the communication management of an organization and from public relations, the need to identify audiences as the target audience for actions requires that the concept of public be defined in order to substantiate the definition and define its objectives. In each entity, it is necessary to provide the public with identity to move from a concept to a recipient, to segment them to determine actions, channels and contents, and singling them out because there is never a public but infinite potential audiences.

Thus, public is a polysemic concept and, as such, public and audience have always been a permanent controversial reason for delimit that they both represent what is their nature or what are their functions.

From the Middle Ages, the identification of public is overcome with the interests or affairs of the feudal elite. So, public it's referred to as the collective interest and it becomes analogous to the management and patrimony of the Administrations and the different governments, according to the territorial structuring of each State. Public is, therefore, a concept that characterizes modern states, although we also use it to refer to a group of individuals as a social group, to take part in common issues and, among other meanings, to a social space.

> Modern societies have just coined the meaning of the public in the wake of the printing boom in the 16th century, which involved the collective identification of all readers as a public of literary creation. It was a bourgeois and aristocratic public. (Túñez López 2012: 170).

Years later, from the 17th and 18th centuries, it is also beginning to use public to refer to people who have read a book or who have attended some kind of cultural event, such as a concert or a theatrical performance. "It begins to be applied to refer to those who attend acts of political content, especially with the rise of debates, criticisms and gatherings in cafes and salons" (Túñez López 2012: 170).

There are several disciplines that have tried to narrow or differentiate the different meanings of public and audience in the contributions made from social psychology, sociology, anthropology, economic sciences or political sciences are a good part of the definitions from which we later try to delimit the concept from the communication sciences in order not to fall into the simplicity of making a mere identification of the public with the recipient of the message. However, public and audience are sometimes exchanged as synonyms, although this is limited to those who consume or view a message of the global audience to which that action or product is directed.

2 Co-creators and Re-Emitters

The need to identify in detail the audiences and to know their particularities in order to be able to mimic with them, shows the importance of orienting the analysis of this relationship to be able to know, in depth, their habits as a group, but also to have a detailed knowledge of the individuals who form it, considering to try to manage personalized relationships.

In online environments, without neglecting the need to promote and group own public through the creation of communities, the main goal is to profile of the global audience since this is the only way to profile and segment the different publics, individuals that can be reached through different media and products (social media, blogs, influencers, etc.).

In fact, in the online world, the community has a symbolic dimension: it is composed by members who have common traits or preferences, who decide that they have a common goal or purpose and share it, interacting with the organization or with each other in virtual environments. As Bernays anticipated and, later, completed other researchers, from relationship management we could validate the consideration of public as any group of individuals on whom it is intended to influence a public relations program (Lesly 1969) or as groups with which communicates or is wanted or should be communicated (Harrison 2002).

The concept of community also varies or is readjusted with the rise of social relations and with the revitalization of the community transferred to online environments. The community, incorporated into public relations management, since 1988, before the takeoff of the internet, was considered as a territorial or geographical grouping but also as a group of people. Kruckeberg and Stark realized at the time that the impact of new technologies on society could act as an agent that would dilute the sense of community as a group of people who organize their activity around common ends. The proposal of Kruckeberg and Stark was that public relations should be reoriented to maintain the sense of community through the change of the purely persuasive strategies by new formulas of social participation.

But the evolution of events has gone in the opposite direction. The initial fear was that there would be social isolation from the individual because one would end up locked in the use of technological tools. What happened was that there was a revitalization of virtual group actions that were collectivized through the Internet.

The virtual multiple relationship that allows the internet has favored the emergence of online groupings integrated by members who share the same lifestyle, or similar hobbies/concerns, that are geographically dispersed, that promote links between them and that generate ways of influencing or being influenced by organizations.

This evolutionary extension of the concept of community allows us to speak today of communities as groups of public that can be promoted by the organization, and, when promoting, the entity is guaranteed the possibility of direct contact with them. In the dynamics of the community, however, the organization doesn't need to always adopt a role of visible protection of the collective process. Fundamentally, the interest in the community allows to identify a specific public, similar to the profile that the organization seeks, and that is be able to reach without mediators.

Publics have stopped being consumers to become prosumers (producers and consumers of content, at the same time), but we may have to admit that they are no longer prosumers because they have become adprosumers, that is, they have added the possibility of advertise, publicize and evaluate (*ad*) their roles of producing and consuming content.

Current audiences perform buyer, consumer and product *recommendations* "before, because they are looking for buying opinions; during, because they communicate their impressions through social networks, blogs and message systems; and then, because they generate opinions and reputation" (Caro et al. 2014: 942). This change of roles shows that the management of communication is an activity in constant change because the relations of organizations with their public, or with the public in general, must be developed taking into account that the barrier between online and offline environments is each day more blurred. The variation not only affects the environment or the support. In reality, there is a change in the behavior of users that can not go unnoticed: their level of involvement.

The new attitude of the public in online environments is of active participation. A new communication based on the dialogue between equals means that both organizations and public participate by taking the initiative to manage the relationship and to decide the content of information that is exchanged and diffused.

Looking for tags, the new user profile would allow us to talk about the Co Generation: co-creators and content co-creators, as well as re-senders. The selfie, the photo of a work celebration or the meeting of friends that goes up to the social network, the infographics that explains, in detail, a fact, the complaint recorded on video, a live concert through social networks, the meme… users are not recipients but co-generators. The interactive symmetry is that: the emitter and receiver are promoters of ideas and content developers, regardless of who is the issuing actor and who the receiver.

3 Net Contents and Structures

For organizations, changes in the patterns of behavior of the individuals to which they are interested increase the awareness in planning and implementing good relations management at all levels: with the target on which visibility is sought, on which they want to capture as leads, with customers or with any user interested or involved in their proposals.

In order to manage credibility, it is necessary to promote the public evaluation of the actions and products/services that are offered, to promote transparency and to monitor the organizational culture and to maintain honest actions. It is clear that this list of actors with whom to manage relations must incorporate internal audiences and add *employer branding* in the referents indispensable for a correct management of the current organizational communication. The workers are external ambassadors of the brand and, for external audiences, they are a guarantee of credibility in their contributions on the effectiveness of the entity to which they belong and the quality of the proposals that organization makes.

The communication and the management of internal relations are worked thinking about the external projections, but also to attend to a greater degree of motivation in the internal development. It seeks to increase the sense of belonging because it is known that if the work climate is improved, the implication in the work increases and, by extension, the profitability. Managing the internal public to reinforce the brand of the organization is therefore to achieve higher levels of empathy with the organization so that through engagement increase the indices of identification with the organizational mission and culture.

We must not forget that the overlap of online and offline means that the scenarios, media and communication channels are also in the process of change. If the sender and receiver reorganize their roles, so does the channel and how to manage access to it. New ways of managing information relationships are given because the management of information dissemination to be transmitted to society has to adjust to the new media. Mass media and non-organization broadcast channels continue to be a fully effective tool. Own means and communities complement them, do not substitute them. Therefore, the new media priorities and the new behaviors of the means to satisfy the changing behaviors of its publics must be guarded.

Like the other organizations, the media promote presence in the new scenarios and adapt their offer and role as online actors that globally seek their space with participatory proposals in the network and, particularly, in specific areas of actions such as the blogosphere, general social networks, professional networks, mass-use applications, etc. The mass media that act in the network know that they have to adapt their language and their material, that the narrative weight is in audiovisual pieces and that the balance in the construction of the agenda has to add the coexistence of the contents that the public must know, because they are interested or affected, and the contents that knows that, even if they escape the criteria of journalistic selection, the public will consume.

The media and, in general, the organizations, have understood that, at the same time that communication ceased to be unidirectional and linear from sender to receiver, users broke loyalty in information consumption and abandoned the single platforms as a stage of collective participation. Until then, participation in the mass communication process was limited to forms of simulated or symbolic feedback (the letters to the director in the newspapers or the opinion of the audience by telephone, in times and spaces delimited) to appear a public protagonism in the communicative process.

The Internet came to upset this idea. Today the stories are generated as we participate in them and when we do, we often change the platform of diffusion. It is the transmedia communication in which the story is constructed with participation and actions simultaneously in different supports. The contents are transferred from the channel and the rupture with discursive communication models is reinforced, since the actors are now proactive in all phases of the process: elaboration, transmission, and reception. Now the model is choral and circular or arboreal, with flows in several senses.

The stage in which the traditional media participated in the web in a symbolic way with the dump of the informational content offline turned to the network is of little more than a decade, but already it is in the Jurassic of the communication management. Today is an activity that no one can think of activating if it intends to gain an online space because, as in the personal life of each of the users, the usual thing is that the stories run simultaneously in different scenarios (from the social network to the web, from the web to the personal network, from the personal network to the social network) and, in all cases, in both directions and adjusting to the particularities of the support. Paraphrasing Bauman (2003), we work with liquid contents so that they fit easily to the languages and the needs of each support as the transfer takes place and a transmedia story is generated.

Liquid structures go beyond the metaphor of adaptation to support. They indicate that, just as a liquid adjusts easily to the container containing it, messages in any language (sound, visual, written) must be easily adaptable to the specific conditions of the carrier or channel. But liquid also means that it is not only proposals fully elaborated by the organization, but it is assumed that on average that are being transmitted is being readjusted and has just taken shape through the contributions of the users who, as already was presented, they should not be deprived of the possibility of being co-processors and co-transmitters of the story. And even, if it is possible, co-protagonists.

The Internet is a scenario of information exchange (search and contribution) and "companies also take advantage of the possibilities of these platforms of interaction and multiple creation to know the profile of their consumer" and, at the same time, give it prominence in campaigns in which one feels involved (Marta et al. 2013: 43).

4 Internet of Things and Data Mining

The need to make adjustments goes beyond making changes in communicative content or techniques. It is a complex task, as are social relations in the context in which the organization unfolds, but the systemic framework in which we move forces us to remember that nothing is independent and everything is connected with everything, what, at times of change, produces cross-transformations, interdependent in the media, attitudes, audiences and products. But this systemic framework also causes changes in the emitters that increasingly interact in scenarios of augmented reality, with which the real physical world is transformed into our devices because they overlap real and virtual environments in which it is possible to interact virtually with other individuals, or with machines, which become actors of the process because they have been endowed with autonomy so that they can initiate themselves a relationship with other machines or with people in online and offline environments.

It is the Internet of Things (IoT) or the automation of objects that overflows the communicative management because it supposes a change of expectation in the global management, with emerging business models among which IoT is not ruled out as a service. Internet of Things goes beyond remote communications between machines (M2 M) as it is the connection through the network of objects with the ability to be managed by other objects, by people or programmed to have autonomy capacity. For organizational communication is a world of expectations because it modifies the way of relating in the network, opens new channels and scenarios, promotes new issuers and whether it is sensor-based IoT—which collect data—as if it refers to actuators—which give autonomy to the thing or object—are drawn unlimited stages of action.

The impact of IoT grows at an exponential rate. By 2016, according to the Gartner[1] report, it was estimated that 6.4 million were connected. The forecast is that this figure multiplies by three and that in 2020 they will be already 20.8 million. The uncertainty lies in how to solve possible security problems, such as meeting the growing need for connection points or how to resolve the storage and processing of all the data that originate.

In any case, to speak of more than 20 million objects connected to the Internet is a sufficiently descriptive figure of the latitude because that means more than 20 million objects with capacity to provide data and to obtain information achievable by itself through other objects connected to the network (Balani 2016). In addition, that object can be anything from which data or information is transmitted from the object or its functions and users, to programmable things, that is to say, on which it is possible to intervene to optimize its way of acting adjusting it to the precise needs or expectations, which connects the IoT with the management of relations and with the obtaining of data for a public hyper-segmentation, of which the organization will have a thorough knowledge.

The IoT allows to act on the reality through programmed actions that are derived from the collection and the management of all type of data. It's the Big Data, the universe of data. The stored information, however, is useless until it is processed. The goal is to transform the data into useful information, the Intelligent Business or the task of managing information so that it can be used in decision making in organizations through the processing of data that are grouped, selected, purified, analyzed, interpreted and reread.

This data mining is oriented to make predictive analyzes of behaviors processing large amounts of data that allow to formulate algorithms. In sum, we look for correlations of data that allow the identification of a pattern through which it is possible to predict results, behaviors, behaviors. It is a matter of anticipating, of being able to know in advance what will be the action of the user to facilitate it or even to guide it in a concrete sense.

All this process is the application of artificial intelligence to things, which characterizes Internet 3.0: the first stage, the Internet 1.0, represented the possibility of interconnecting and exchanging documents and the second one, Internet 2.0, was the

[1] https://www.gartner.com/newsroom/id/3165317.

most visible because, in addition to documents also users could be interconnected, a dynamic that had a more visible impact on the daily lives of people.

The semantic web (3.0) is, instead, a silent change. The performances are still about people, but do not visibly affect the lives of individuals, as occurred in 2.0, especially with social networks. 3.0 is computational intelligence and is oriented to generate languages and tools so that the computers do actions or tasks with which they obtain similar results that would be if it were a person who had executed them.

5 Mobility and Geolocation

IoT is the connection of things in an Internet that does not stop growing in number of users, mainly from devices of connection in mobility. The year 2015 was dismissed leaving in the world the same amount of inhabitants as of subscribed mobile lines. Despite this apparent balance, significant inequalities in infrastructures, access points, connection speed and percentage of population that can be connected continue.

Hootsuite's *Digital in 2017 Global Overview*[2] report speaks of 29% penetration in Africa, 46% in Asia Pacific, 60% in the Midwest, 71% in America and 76% in Europe. The extremes of inequality are abysmal: from 99% coverage in the Arab Emirates or 98% in Iceland to 0.1% in North Korea.

Hootsuite data indicate that one in three individuals does not have a mobile phone. The reverse reading of this data is that two thirds of the world's population (66%) have a terminal, which means 4917 billion users. The number of devices is even greater and projected on the total population is equivalent to 131% in Europe, 127% in the Midwest and 106% in America. That is, more than one phone per capita. In Asia-Pacific the figure drops to 96% and in Africa to 81%.

In any case, the connection from terminals in mobility defines the mode of future connection and determines the management of the communication because it conditions the way of relating the products and the presentation of the contents but it also requires rethinking the creation of web products in responsive formats to ensure that they automatically adapt to the screen size of the device from which each user connects.

The connection in mobility is allied with geolocation to access proximity services, for which it is necessary that we allow ourselves to be located spatially and geographically. This is service information: the combination of mobility and geolocation that facilitates, for example, the use of iBeacon or remitters that organize the communications with geolocalizable users that approach the contact point predetermined by the organization.

[2]https://www.slideshare.net/wearesocialsg/digital-in-2017-global-overview?ref=.

We could consider them a strategic ally of mobility because with the identification of the user, their hyper-segmentation and the use of iBeacon can be made proposals to be adjusted to the user in real time, to be executed imminently.

All individuals are sources of data that help to have a hyper-knowledge of the user to determine predictive models and to single out the relationships through personalized marketing. Until recently, mass media audiences only made it possible to establish conglomerates of audiences with general particulars because classical segmentation criteria are based on standard geographic, demographic and socio-economic data. At present, the trace of data that we leave on the Internet with our performances and the information that is derived from the themes, environments, contents, applications and people—among other factors—with which we interact in the network are an invaluable source of information.

The relationships that are established in Internet are not limited to the individuals, nor to the users of the network, but they end up being extended to all the actors and social agents. Internet facilitates individual and group communication and specializes in the management of communication in organizations. It is necessary to establish an attitude of testimonial presence in a systemic environment in which modern organizations adopt a participative attitude means to be in the network renouncing its potential benefits. But the internet is more than a showcase to be seen, it is a stage to actively participate. Interactivity means reciprocal participation by both parties and requires a change of mentality or attitude in organizations because it is not enough to issue and enable return channels. A real interactivity has to be promoted because the user is proactive, not a passive receiver.

The Internet user seeks, generates and participates in the network by establishing contacts with other individuals. That is the attitude to be assumed by the organizations: from being static to acting on the network as a net surfer because the Internet allows you to listen, converse and relate directly to the public and that implies a transformation of organizational culture and a revolution in the form to manage their communication.

Time to manage for audiences is over. It is time to manage being the public.

References

Balani, N.: Enterprise IoT: A Definitive Handbook. Hatchi, R. (ed.) (2016)

Bauman, Z.: Modernidad líquida. Fondo de Cultura Económica, Buenos Aires (2003)

Blumer, H.: The mass, the public and public opinión. En Berelson, B. y Janovitz, M. Reader in public opinión and communication. Glencoe: The Free Press (1953)

Caro, J., Luque, A., Zayas, B.: Aplicaciones tecnológicas para la promoción de los recursos turísticos culturales. In: XVI Congreso Nacional de Tecnologías de la Información Geográfica, Alicante. In: http://goo.gl/2Svq6l (2014)

Celaya, J.: La Empresa en la web 2.0. Gestión 2000, Barcelona (2009)

Coombs, W.T.: Crisis management: advantages of a relational perspective. In: Ledingham, J.A., Bruning S.D. (eds.) Public Relations as Relationship in Management: A Relational Approach to the Study and Practice of Public Relations. Mahwah Lawrence Erlbaum (2000)

Daugherty, E.L.: Public relations and social responsability. En Hearth, R.L. Handbook of public relations. California: Sage (2001)
Dewey, J.: The public and its problems. In: A. y Walsh, B.A. (eds.) En Boudston. The later works of John Dewey. Illinois: University Press, 1984 (1927)
Freeman, R.E: Strategic Management: A Stakeholder Approach. Pitman, Maryland (1984)
Grunig, J.E.: Publics, audiences and market segments: segmentation principles for campaigns. In: Salmon, C.T. (ed.) Information Campaigns: Balancing Social Values and Social Change, pp. 199–228. Sage, California (1989)
Grunig, J.E., Hunt, T.: Managing public relation. Orlando-Florida: Harcourt Brace Jovanovich (1984)
Grunig, J.E., Hunt, T.: Dirección de relaciones públicas. Gestión 2000, Barcelona (2003)
Harrison, S.: Relaciones públicas, una introducción. Thompson, Madrid (2002)
Kotler, P.: Marketing 3.0. Lid editorial, Madrid (2011)
Lesly, P.: Manual de relaciones públicas. Martínez Roca, Barcelona (1969)
López Sánchez, J.A., Santos Vijande, M.L., Trespalacios Gutiérrez J.A.: Perspectivas de análisis en la creación de valor para el cliente en las relaciones comprador-vendedor: un estudio empírico. Tribuna de Economía, 854 (2010)
Marta, C., Martínez, E., Sánchez, L.: La «i-Generación» y su interacción en las redes sociales. Análisis de Coca-Cola en Tuenti. Comunicar **40**(XX), 41–48 (2013). http://dx.doi.org/10.3916/C40-2013-02-04
Mazo, J.M.: Estructuras de la comunicación por objetivos. Estructuras publicitarias y de Relaciones Públicas. Ariel, Barcelona (1994)
Míguez, M.I.: Los públicos en las relaciones públicas. UOC, Barcelona (2010)
Park, R.E.: The crowd and the public and other essays. Chicago: University of Chicago Press (1972)
Price, V.: La opinión pública: esfera pública y comunicación. Barcelona: Paidós (1994)
Túñez López, M.: La gestión de la comunicación en las organizaciones. Comunicación Social, Sevilla and Zamora (2012)
Túñez López, M.: Modelo de simetría interactiva en Comunicación Organizacional. Revista Mediterránea de Comunicación **6**(2), 5–7, https://doi.org/10.14198/medcom2015.6.2.14. In: http://mediterraneacomunicacion.org (2015)
Túñez López, M., Altamirano, V.: A simetría interativa na Comunicaçao Organizacional. In: Túñez López, M., Costa-Sánchez, C. (eds.) Interaçao organizacional na sociedade em rede. Os novos caminhos da comunicaçao na gestao das relaçoes com os públicos, pp. 13–20. Cuadernos Artesanos de Comunicación, 102. Latina de Comunicación Social, La Laguna, Tenerife (2016)
Túñez López, M., Sixto García, J.: Redes sociales, política y Compromiso 2.0: La comunicación de los diputados españoles en Facebook. Revista Latina de comunicación social **66**, 210–246 (2011)

Miguel Túñez-López Ph.D. in Journalism from the Universitat Autónoma de Barcelona, Spain. Professor of undergraduate and postgraduate courses of the Universidade de Santiago de Compostela, Spain. Expert in communication management in organisations and news making. He coordinates the Ph.D. program in Information and Contemporary Information from the Universidade de Santiago de Compostela. Santiago de Compostela, Spain.

Karina Paola Valarezo González Director of Communication, Dircom at the Universidad Técnica Particular de Loja, UTPL. She holds a Ph.D. in Communication and Journalism from the Universidade de Santiago de Compostela, Spain. She is Professor of undergraduate and postgraduate courses. As a researcher, she has published articles in journals indexed in latindex, Scopus and Isi web of knowledge and chapters of peer-reviewed academic books. She is the main researcher at various research projects and speaker at different academic and research events. Loja, Ecuador.

María Isabel Punín Larrea Ph.D. in communication and journalism. Universidade Santiago de Compostela (Spain), is professor and researcher at the Department of Communication Science from the Universidad Técnica Particular de Loja. She holds a degree in Management and University Quality. She holds a degree in Social Communication Sciences from the Universidad de Sevilla. Universidad Católica del Azuay- Ecuador. She is former director of the BA in Social Communication of presence-based and online modalities at the Universidad Técnica Particular de Loja. Loja, Ecuador.

Creation of Social Media Content and the Business Dialogic Process

Barbara Mazza and Alessandra Palermo

Abstract This paper examines how companies organize content on their social media pages to build relationships with their followers, and stimulate sharing, participation, dialogue, and involvement. To build a good reputation, the company has to cope with online conversations with its present and potential users, and here we examine how this can be most effectively achieved. A company's history, identity, goals, mission, and projects serve to consolidate relationships with its community members. They can not only express their opinions, but also be co-creators. The main goal of this chapters is to understand how a social media company's content is consistent with its identity and communicative style, in the extent it strengthens its credibility and establishes and maintains relationships with its users. The observations in this study can be a starting point and a useful interpretation for scholars and industry, in assessing the role of content in increasing loyalty.

Keywords Corporate communication · Social media content · Facebook Engagement · Corporate reputation

1 Introduction

This paper intends to understand the ways companies use social media to relate to employees, users, and stakeholders.[1] We know that 84% of businesses around the world use social media to converse with different targets, and the one most widely used is Facebook (93%,[2] Social Media Marketing Industry Report 2016), being

[1]See Constantinides (2014), Kane (2015).

[2]The others, in the order, are: Twitter (76%), Linkedin (67%), Youtube (53%), Google+ (49%), Instagram (44%), Pinterest (40%), followed by other social media.

B. Mazza (✉) · A. Palermo (✉)
Sapienza University of Rome, Rome, Italy
e-mail: barbara.mazza@uniroma1.it

A. Palermo
e-mail: alessandra.palermo@uniroma1.it

M. Túñez-López et al. (eds.), *Communication: Innovation & Quality*, Studies in Systems, Decision and Control 154, https://doi.org/10.1007/978-3-319-91860-0_22

utilized by more than a billion and a half active users in the world (Digital, Social & Mobile 2016). A wealth of literature is available on how marketing strategies are adopted to increase user engagement.[3] Conversely, little existing research deals with the way companies manage their presence on platforms. Some studies explain how social media is used in different business activities (Rauniar et al. 2014). The present study focuses on how businesses can organize content, strengthen their credibility, and cultivate relationships. As stated in the literature, following the logic of being online, especially on a social networking generalist, such as Facebook, the principal aim is not to search out potential consumers to guide them in acquiring a product, or to follow-up customers so as not to lose them, and neither to bombard clients with information and promotions. Instead, the overriding goal is to actively encourage a proactive dialogue, which is open to experimentation and the creation of "creative" communities, whereby innovative solutions can be developed and needs, both manifest and latent, can be expressed (Guinan et al. 2014). Referring back, at least in part, to the view of Gerbner (1976), the cultivation of attitudes and beliefs does not merely entail customer loyalty to brands and products, but the long-term construction of a cultural system the center of which is made up of values identified with, and the tacit knowledge of a particular company. The evolution of this system depends on the participation of actors in the constant review of knowledge and processes, promoting the creation of new products and brands through shared experiences. Consequently, effort has been made to understand how some of the most active social media companies present themselves to their consumers.

2 Pre-conditions to Increase Dialogue

Before developing content to increase user dialogue, companies need to consider two key aspects: first, establishing the main communicative goals of being online, and the need to have a good e-reputation; and second, involving consumers to generate engagement, and how this can be achieved.

Firstly, requiring a good reputation entails creating a positive perception not only of the brand, but also of the corporate identity and cultivating it over time (Strategy and David 2016). In terms of corporate communication, having a good reputation means gaining appreciation around corporate identity and personality, that is, around the essential features that make an organization distinct from others. These features are further enhanced by mission, vision, goals, values, and responsible actions. As a whole, identity, which should cover all the aspects so far indicated, represents the ways in which a company creates identification between itself and all its public. Its communication is effective when the corporate identity relates through its content coincides with how the company is already perceived by users and consumers (Dutot et al. 2016).

[3]See Chui et al. (2013), Lee et al. (2014), Koschate-Fischer et al. (2014), Vivek et al. (2014), Lee et al. (2014).

Secondly, generating engagement implies setting up content which stimulates a dialogue aimed at creating a relational network. The setting therefore determines the meaning and orientation of the interactions. Resuming the dialogic theory of Albert J. Sullivan, revised by Pearson (1989), to stimulate the participation of users, a company needs: the commitment of its staff; a clear explanation of ethically relevant content, aimed at increasing the credibility of the company; the creation of conditions that facilitate the accessibility of conversation and guarantee equal treatment of users. The network is really active when such conditions produce trust between the parties, the will and willingness to confront. Under these conditions, the network is also a social capital generator (Putnam 1995). It is in fact a resource that derives from the relationship itself, a tangible result, potentially providing benefits to those directly or indirectly involved in the process. Rheingold (1993) already theorized that social media are sites of aggregations where people talk long enough to build relationships in cyberspace.

In corporate terms, social capital has a major role to play in strengthening the communities that are formed and which are not aimed at pursuing specific interests, such as profit and brand fidelity, but the development of a sense of belonging, shared values and orientations (Adams and McCorkindale 2013; Saxton and Waters 2014). Hence, when the relational network is also a social capital network, over time it more easily facilitates engagement and consolidation based on trust.

Kent and Taylor (2014) indicate that several studies show how companies tend to form a "monologically and asymmetrically" form of dialogue, rather than one which is real and equal. They consult consumers and stimulate them to interact, to know them, to have feedback, to influence them, but not necessarily to cultivate stable relationships. Engagement is, in this case, limited and not durable.

Effective dialogue, however, enhances interaction, and aims at obtaining mutual understanding of meanings and co-creation, in order to foster continuity in the relationship, and to negotiate relationships between different subjects. Dialogue is more than just an exchange of ideas, as people's personalities form and evolve through interaction with others, with participants being aware of their mutual engagement in the process (Johnston 2014). Organizations should also consider that users are involved with symbolic input, especially since they believe and share the meanings it represents (Hollebeek et al. 2016).

For this reason, the choice of business content is essential for establishing effective dialogue, aimed at stimulating participation and engagement. Dessart et al. (2015) distinguish three different types of engagement, which, though, are not mutually exclusive:

- *Affective*, when engagement is fueled by common emotions, and strengthened if linked to shared ethical and social goals. These feelings tend to be long lasting and recurring;
- *Cognitive*, when involvement comes from direct and positive sharing with the company consolidating long-lasting and active mental states;

- *Behavioral*, when the commitment of the consumer depends on motivational drivers. Even with co-creation, online community engagement is facilitated by the exchange of content experiences and information that stimulate active participation.

3 Business Dialogic Behaviors on Social Media

In order to cultivate stable, continuous and trusting relationships between companies, stakeholders and consumers, the participation of all actors involved needs to be encouraged. This process is described through three metaphors elaborated by Leonardi et al. (2013: 7–15), concerning the value of Social Communication for Organizations. Leonardi et al.'s observations are useful to understand the heterogeneous nature of the varying degrees of social exploitation. The model also considers how an enterprise behaves in social media in minimum terms, when mostly limited to the need to be networked, to meet user expectations, and to encourage participation of all actors.

The Leonardi et al. discuss communicative actions, such as *drainpipe*, *sounding board and social lubricant*. In the first case, businesses exploit the platform's potential to a limited extent, merely managing information, mostly one-way. Organizations communicate contents as if they were a drainpipe "leak". The advantage is that the messages given out signal widespread company presence, enabling a company to obtain both user feedback (useful for filling in structural holes, Lin et al. 2001) and also important elements for the updating of company offer. The disadvantage is that only the company determines what is shared with users, thus greatly limiting the creation of virtuous communication circuits. Hence, this mechanism can be said to be equivalent to "dialogic monologue and low-engagement".

When the use of social structure is similar to a *sounding board*, the corporate goal is to create shared views. This mechanism generates, on the one hand, personalization that facilitates the search for people and content on the basis of common interests and, on the other, balkanization, or attempt to hinder exposure to new ideas and reduce forms of cooperation. The principle known as "homophilia" is encouraged, when companies (as with individuals) only seek people who share the same tastes, orientations and ways of thinking (Aiello et al. 2012). This *modus operandi* can create a pleasant experience for the user, but limits the transfer of potentially valuable information which could actually turn out very useful and stimulating for the company. In this case, as stated in the literature, although the dialogue process is equitable in procedural terms, since it triggers sharing, it does not foster social capital.

The third and last metaphor is the use of social platforms as a *social lubricant*. This approach is to feed relational networks and interpersonal communications in order to strengthen the sense of community and membership of social groups around the organization. This favors forms of cooperation and co-creation, and so, common interests. The only risk, however, is that the informal community dispels its attention, with interactions potentially degenerating into "chattering" and exchange of gossip, thereby becoming fruitless in terms of design.

It is also worthwhile considering the structure of the communication process, with the support of more or less intense relational networks. Huang et al. (2015: 2825–2844) propose a breakdown into interactions, proximity and flows.

- *Interactions* are the most common forms in which the exchange translates into message generation and feedback.
- *Proximity* involves a connection between digital groups on shared arguments and between them and the company through additional tools, such as geolocation or provision of media and services through specific applications;
- *Flows* tend to build links between related individuals and encourage greater information exchange. People can even organize themselves to achieve common goals, such as participating in a live event, responding to a crisis, organizing a subcommunity that will complain or cooperate to provide and discuss proposals and suggestions. In all these cases, strong links (Granovetter 1985) can be generated among community components, which are useful for the transfer of knowledge and creativity. It is therefore effective dialogic process that generates social capital.

To summarize the approaches outlined above, in line with the literature,[4] the behavioral continuum of an organization appears to go from adopting a *passive approach* to the use of social media, to an *active one*. In the first approach, the fundamental goal is to listen rumors and feelings from the network. A company merely provides information and input to stimulate user participation, and extrapolates information based on their comments. In the second, users are stimulated to converse and interact as much as possible, to create a continuous and lasting relationship.

4 Set Content to Cultivate Relationships

To stimulate user interaction, content setting is crucial to a business. For this reason, they should consider the most useful and effective content, depending on the characteristics of their followers, their expectations, and the communicative goals of the company.

In this regard, Coombs and Holladay (2015), resuming the work of McGeee (1980), emphasize the importance of setting contents around *ideographs*, namely visual and text elements, which unambiguously specify easily understood concepts. These graphic elements allow an immediate contact to be made between the actors involved in the conversation, thereby creating sharing, stimulating discussion, and building collaborative relationships.

Through these meaningful elements, rites of behavior, and comparison, the definition of community identity is strengthened and stabilized. Expected roles are clarified, boundaries outlined, and the rules of interaction followed, so strengthening practices

[4]Cfr. tra gli altri: Vaast and Kaganer (2013), Anderson and Li (2014), Simula et al. (2015).

and a sense of belonging (Laroche et al. 2012). However, it should be kept in mind that social media text is an open product, something unfinished, and constantly developing. Both companies and users need to be aware of this, with contributions being part of its continual evolution. In the literature, it is argued that content should stimulate users to update content, provide inputs for upgrading products/services, intertwine personal stories with that of the company, not only for profile purposes, but for cultivating the relationship and the sense of community (Bechmann and Lomborg 2013).

The set of content types published by companies are presented in 4 macro-categories, as identified by the literature (particularly through the functionalist approach of uses and gratification): information,[5] entertainment,[6] incentive[7] and socialization.[8]

- *Information content* is the main resource that feeds online dialogue and the reason why users follow business posts. Consequently, companies need to carefully make this content exhaustive and up-to-date. When an individual is satisfied with the content offered, they express it by clicking the links, staying on the longer pages, reading details and discussions, commenting and using the various multimedia features. It should be noted that although the follower is active, it does not mean they are necessarily proactive, as well;
- *Entertainment content* serves to make the platform a fun and enjoyable environment for the user *as* it is stimulating, promoting evasion, hedonistic pleasure, aesthetic enjoyment, and the venting of emotion. Teaser, slogan, word games, for example, exhort the follower to leave comments and, in general, tend to intensify the level of participation and interaction of users;
- *Incentive content* uses remuneration tools that a company offers to users to entice them to participate and collaborate. They are rewards, such as public recognition, obtaining content, gifts and/or products, the awarding of bonus points, prize draws for competitions and/or surveys (Muntinga et al. 2017). This type of content is mainly used to encourage responses to issues which particularly interest an enterprise, but which would not otherwise induce user response. However, greater user involvement is not guaranteed;
- *Socialization content* serves to motivate community members to develop a sense of belonging to a company's world. The more effective the content, the more users are encouraged to co-create. When this happens, conversations increase and common attitudes, languages, and behaviors are developed. Socialization content also contains information and entertainment content.

[5]See de Vries et al. (2012), Cvijikj and Michahelles (2013), Dolan et al. (2015).

[6]See McQuail (1983), Taylor et al. (2011), Eden et al. (2014), Park et al. (2015).

[7]See Cvijikj and Michahelles (2013).

[8]See Leung (2013), Hollebeek et al. (2014).

5 Improvement of the Dialogue Process Is Still Needed

Through this review of the literature, some main trends emerge:

- Businesses are aware of the strategic value of being present and active on social media to intercept users and cultivate relationships;
- Many companies tend to organize content by focusing on information. Although it is a central resource to stimulate users, it is not the most effective way to encourage engagement. Its use must be associated and combined with other types of content;
- Companies still tend to attach more importance to their goals than to being user-centric. For this reason, the dialogue process is very often monologist and engagement levels are still low.

Obviously, although these considerations are the result of several studies reported in the literature, the evolution of the social media use by businesses still needs to be followed and analyzed.

Since the network is characterized by continuous transformations and evolutions, further changes in the near future are likely, particularly concerning how content used to project corporate identity is created. In terms of social media use, the most mature companies will be able to leverage integrated marketing and corporate communications content to present a more organic vision of the company, to be more credible and thus, to stimulate more confidence and engagement.

References

Adams, A., McCorkindale, T.: Dialogue and transparency: a content analysis of how the 2012 presidential candidates used twitter. Public Relat. Rev. **39**(4), 357–359 (2013). https://doi.org/10.1016/j.pubrev.2013.07.016

Aiello, L.M., Barrat, A., Schifanella, R., Cattuto, C., Markines, B., Menczer, F.: Friendship prediction and homophily in social media. ACM Trans. Web (TWEB) **6**(2), 9 (2012). https://doi.org/10.1145/2180861.2180866

Anderson, A., Li, J.: Entrepreneurship and networked collaboration: synergetic innovation, knowledge and uncertainty. J. Gen. Manag. **40**(1), 7–21 (2014)

Bechmann, A., Lomborg, S.: Mapping actor roles in social media: different perspectives on value creation in theories of user participation. New Media Soc. **15**(5), 765–781 (2013). https://doi.org/10.1177/1461444812462853

Chui, M., Dewhurst, M., Pollak, L.: Building the social enterprise. McKinsey Q. **4**, 8–11 (2013)

Coombs, W.T., Holladay, S.J.: Public relations' "relationship identity" in research: enlightenment or illusion. Public Relat. Rev. **41**(5), 689–695 (2015). https://doi.org/10.1016/j.pubrev.2013.12.008

Constantinides, E.: Foundations of social media marketing. Procedia-Soc. Behav. Sci. **148**, 40–57 (2014). https://doi.org/10.1016/j.sbspro.2014.07.016

Cvijikj, I.P., Michahelles, F.: Online engagement factors on Facebook brand pages. Soc. Netw. Anal. Min. **3**(4), 843–861 (2013). https://doi.org/10.1007/s13278-013-0098-8

De Vries, L., Gensler, S., Leeflang, P.S.: Popularity of brand posts on brand fan pages: an investigation of the effects of social media marketing. J. Interact. Mark. **26**(2), 83–91 (2012). https://doi.org/10.1016/j.intmar.2012.01.003

Dessart, L., Veloutsou, C., Morgan-Thomas, A.: Consumer engagement in online brand communities: a social media perspective. J. Prod. Brand Manag. **24**(1), 28–42 (2015). https://doi.org/10.1108/jpbm-06-2014-0635

Dolan, R., Conduit, J., Fahy, J., Goodman, S.: Big social data and social media analytics: tools for exploring social media engagement behavior. In : 2015 ANZMAC Conference: Innovation and Growth Strategies in Marketing. ANZMAC c/o School of Marketing, UNSW Business School, UNSW Australia (2015)

Dutot, V., Lacalle Galvez, E., Versailles, D.W.: CSR communications strategies through social media and influence on e-reputation: an exploratory study. Manag. Decis. **54**(2), 363–389 (2016). https://doi.org/10.1108/md-01-2015-0015

Eden, A., Tamborini, R., Grizzard, M., Lewis, R., Weber, R., Prabhu, S.: Repeated exposure to narrative entertainment and the salience of moral intuitions. J. Commun. **64**(3), 501–520 (2014). https://doi.org/10.1111/jcom.12098

Gerbner, G., Gross, L.: Living with television: the violence profile. J. Commun. **26**(2), 172–199 (1976). https://doi.org/10.1111/j.1460-2466.1976.tb01397

Granovetter, M.: Economic action and social structure: The problem of embeddedness. Am. J. Sociol. **91**(3), 481–510 (1985). https://doi.org/10.1086/228311

Guinan, P.J., Parise, S., Rollag, K.: Jumpstarting the use of social technologies in your organization. Bus. Horiz. **57**(3), 337–347 (2014). https://doi.org/10.1016/j.bushor.2013.12.005

Hollebeek, L.D., Conduit, J., Brodie, R.J.: Strategic drivers, anticipated and unanticipated outcomes of customer engagement. J. Mark. Manag. **32**(5–6), 393–398 (2016). https://doi.org/10.1080/0267257x.2016.1144360

Hollebeek, L.D., Glynn, M.S., Brodie, R.J.: Consumer brand engagement in social media: conceptualization, scale development and validation. J. Interact. Mark. **28**(2), 149–165 (2014). https://doi.org/10.1016/j.intmar.2013.12.002

Huang, Y., Singh, P.V., Ghose, A.: A structural model of employee behavioral dynamics in enterprise social media. Manage. Sci. **61**(12), 2825–2844 (2015). https://doi.org/10.2139/ssrn.1785724

Johnston, K.A.: Public relations and engagement: theoretical imperatives of a multidimensional concept. J. Public Relat. Res. **26**(5), 381–383 (2014). https://doi.org/10.1080/1062726x.2014.959863

Kane, G.C.: Enterprise social media: current capabilities and future possibilities. MIS Q. Executive **14**(1). ISSN 1540-1960 (2015)

Kemp, S.: Digital, social & mobile in 2016. In: https://wearesocial.com/uk/special-reports/digital-in-2016 (2016)

Kemp, S.: Digital in 2017: global overview. We are social's compendium of global digital statistics. In: https://wearesocial.com/blog/2017/01/digital-in-2017-global-overview (2017)

Koschate-Fischer, N., Cramer, J., Hoyer, W.D.: Moderating effects of the relationship between private label share and store loyalty. J. Mark. **78**(2), 69–82 (2014). https://doi.org/10.1509/jm.13.0075

Laroche, M., Habibi, M.R., Richard, M.O., Sankaranarayanan, R.: The effects of social media based brand communities on brand community markers, value creation practices, brand trust and brand loyalty. Comput. Hum. Behav. **28**(5), 1755–1767 (2012). https://doi.org/10.1016/j.chb.2012.04.016

Lee, D., Hosanagar, K., Nair, H.: The effect of social media marketing content on consumer engagement: evidence from facebook. Available at SSRN, 2290802 (2014)

Leonardi, P.M., Huysman, M., Steinfield, C.: Enterprise social media: definition, history, and prospects for the study of social technologies in organizations. J. Comput. Mediated Commun. **19**(1), 1–19 (2013). https://doi.org/10.1111/jcc4.12029

Leung, L.: Generational differences in content generation in social media: the roles of the gratifications sought and of narcissism. Comput. Hum. Behav. **29**(3), 997–1006 (2013). https://doi.org/10.1016/j.chb.2012.12.028

Lin, N., Cook, K.S., Burt, R.S. (eds.): Social Capital: Theory and Research. Transaction Publishers. ISBN: 978-0-202-30643-8 (2001)

McGeee, M.C.: The "ideograph": a link between rhetoric and ideology. Q. J. Speech **66**(1), 1–16 (1980). https://doi.org/10.1080/00335638009383499
McQuail, D.: McQuail's Mass Communication Theory (1983)
Muntinga, D.G., Moorman, M., Verlegh, P.W., Smit, E.G.: Who creates brand-related content, and why? The interplay of consumer characteristics and motivations. Digital Advertising: Theory and Research. ISBN: 978-1-138-65442-6 (2017)
Park, J., Hill, W.T., Bonds-Raacke, J.: Exploring the relationship between cognitive effort exertion and regret in online vs. offline shopping. Comput. Hum. Behav. **49**, 444–450 (2015). https://doi.org/10.1016/j.chb.2015.03.034
Pearson, R.: Beyond ethical relativism in public relations: coorientation, rules, and the idea of communication symmetry. J. Public Relat. Res. **1**(1–4), 67–86 (1989). https://doi.org/10.1207/s1532754xjprr0101-4_3
Putnam, R.D.: Bowling alone: America's declining social capital. J. Democracy **6**(1), 65–78 (1995). https://doi.org/10.1353/jod.1995.0002
Rauniar, R., Rawski, G., Yang, J., Johnson, B.: Technology acceptance model (TAM) and social media usage: an empirical study on Facebook. J. Enterp. Inf. Manage. **27**(1), 6–30 (2014). https://doi.org/10.1108/jeim-04-2012-0011
Rheingold, H. (1993): The virtual community: Finding Connection in a Computerized World. Addison-Wesley Longman Publishing Co., Inc., Boston
Saxton, G.D., Waters, R.D.: What do stakeholders like on Facebook? Examining public reactions to nonprofit organizations' informational, promotional, and community-building messages. J. Public Relat. Res. **26**(3), 280–299 (2014). https://doi.org/10.1080/1062726x.2014.908721
Simula, H., Töllmen, A., Karjaluoto, H.: Facilitating innovations and value co-creation in industrial B2B firms by combining digital marketing, social media and crowdsourcing. In: Marketing Dynamism & Sustainability: Things Change, Things Stay the Same… (pp. 254–263). Springer International Publishing, Cham. https://doi.org/10.1007/978-3-319-10912-1_84 (2015)
Stelzner M.A.: Social media marketing industry report. social media examiner. In: https://www.socialmediaexaminer.com/wp-content/uploads/2016/05/SocialMediaMarketingIndustryReport2016.pdf (2016)
Strategy, V.D., David, W.: CSR communications strategies through social media and influence on e-reputation. Manag. Decis. **54**(2), 363–389 (2016). https://doi.org/10.1108/md-01-2015-0015
Taylor, D.G., Lewin, J.E., Strutton, D.: Friends, fans, and followers: do ads work on social networks? J. Advertising Res. **51**(1), 258–275 (2011). https://doi.org/10.2501/JAR-51-1-258-275
Taylor, M., Kent, M.L.: Dialogic engagement: clarifying foundational concepts. J. Public Relat. Res. **26**(5), 384–398 (2014). https://doi.org/10.1080/1062726x.2014.956106
Vaast, E., Kaganer, E.: Social media affordances and governance in the workplace: An examination of organizational policies. J. Compu.-Mediated Commun. **19**(1), 78–101 (2013). https://doi.org/10.1111/jcc4.12032
Vivek, S.D., Beatty, S.E., Dalela, V., Morgan, R.M.: A generalized multidimensional scale for measuring customer engagement. J. Mark. Theor. Practice **22**(4), 401–420 (2014). https://doi.org/10.2753/mtp1069-6679220404

Barbara Mazza Associate professor at the Department of Communication and Social Research from La Sapienza, University of Rome. She leads research projects focused on business corporate communication, especially on financial communication, social enterprise, social media engagement and employer branding. She is director of Unimonitor.com, Observatory of the National Conference of Communication Sciences degree. Roma, Italy.

Alessandra Palermo Ph.D. in Communication, Research and Innovation at Sapienza University of Rome. She deals with institutional communication and the communication of Italian and European Universities. She is part of the «Observatory Scienze.com». Among the recent publications:

"The Impact of the Scientific CyberJournalism on Facebook" (VIII International Conference on Cyber Journalism, 2016) and "The Social Responsibility of Journalists in Sporting Narration: Media, Diversity, Sport" (ComunicazionePuntoDoc, 2017). Roma, Italy.

Employer Branding in the Digital Era Attracting and Retaining Millennials Using Digital Media

Paula Arriscado, Helena Quesado and Bianca Sousa

Abstract The 21st century brought the greatest transformation since the industrial revolution. In this moment of change, the first and most important shift is to attract, develop and retain talent, namely digital talent; talent from the Millennial Generation or Y Generation. This paper aims at reflecting on how an employer branding (EB) strategy can be the answer, through the delivery of value proposition based on functional, emotional and symbolic benefits, consistent with the needs of this new generation. To carry out this mission, companies need to have an employer brand with a strong identity matrix and, at the same time, need to know how to communicate, using digital codes and channels. There are several digital tools that support EB for Millennials. However, the implementation of the company's values and vision in the processes and methodologies, transporting the organizational personality from the inside out, will always be a complex challenge, mostly in this era of volatility and complexity.

Keywords Employer branding · Digital media · Millennials · Digitalisation

1 The Digital Era: Transforming Work Relationships

The 21st century is characterized by the digital era, the greatest transformation since the industrial revolution. We are talking about one of the most revolutionary periods, where the knowledge of the world multiplies every minute, consolidating a way of being, doing, living and working in a digital approach. In fact, technology around

P. Arriscado (✉)
IPAM—The Marketing School, Laureate International Universities, Oporto, Portugal
e-mail: paula.arriscado@salvadorcaetano.pt

P. Arriscado · H. Quesado · B. Sousa
Grupo Salvador Caetano, Vila Nova de Gaia, Portugal
e-mail: helena.quesado@salvadorcaetano.pt

B. Sousa
e-mail: bianca.sousa@salvadorcaetano.pt

M. Túñez-López et al. (eds.), *Communication: Innovation & Quality*, Studies in Systems, Decision and Control 154, https://doi.org/10.1007/978-3-319-91860-0_23

us is transforming the way we live, teach and learn, but also our social interactions. Digitalisation, mobility, artificial intelligence, automation, data analytics, business intelligence are a reality of this new world (Leonhard 2016). And it is not just our daily lives that are being affected. Digitalisation leads to new markets, new forms of business and new ways of working. In the digital age, the examples are endless. Currently, the largest taxi company in the world has no taxis (Uber) and the largest film distributor does not have a movie theatre (Netflix). In 2012, The Wall Street Journal reported that Kodak was preparing to start bankruptcy, despite 14 years before employing 170,000 employees and being the world leader in the sales of photo paper. In 2007, Nokia led the manufacture of mobile phones worldwide, but in a few years it almost disappeared due to its lack of adaptation to smartphones. With these examples, it is noteworthy that, in this new era, past success does not guarantee even the existence in the future. And as D'Souza (2015) said: "In today's era of volatility, there is no other way but to re-invent." It is in this paradigm that today's organisations have to operate. To this end, they are increasingly developing workspaces that allow communication, collaboration and connectivity between the different employees.

In fact, vertical companies that hire full-time employees to work eight hours a day always in the same place are becoming outdated. Nowadays, work can be done anytime and the office is anywhere, as employees must stay connected through smart phones, tablets and other technologies. At the same time, companies become more and more horizontal in terms of hierarchy. Workplaces need to be open to innovation, to take leadership online, to adopt new digital methods, to support virtual and global collaboration and to integrate the benefits of the smart machines with their talented people.

Despite the importance of the technologies, the attention must be on the employees, fostering a human-centred workplace and the employee experience. The introduction of this new work models can deliver huge advantages such as: increased productivity, reduced costs, mobility, crowd sourcing, flexibility, increased capacity to adapt to the complex market. The main goal is to break down communication barriers, transforming the employee experience by fostering efficiency, innovation, collaboration, communication and connectivity. Employees need also to be able to follow the fast pace that characterizes technological progress, guided by transversal competences of adaptability, critical thinking, creativity, flexibility, emotional, intercultural and virtual intelligence (Davies et al. 2011). We are especially referring to the Millennials; a generation that in 2020 will represent more than 1/3 of the world's workforce. A finding that reminds us of what John F. Kennedy once said: "the future promise of any nation may be directly measured by the present prospects of its youth".

Organizations are not exempt from this tsunami of innovation. The question that arises is whether they will drive the change or be driven by it. And in this journey, the first and most important keystone is to attract, develop and retain talent, namely digital talent; talent arising from the Millennial Generation or Y Generation (born between the years 1980 and 2000).

2 Millennial Generation: Always Connected

The Millennial Generation is known to be always connected. But what describes it? First of all, we are talking about the generation that will be able to follow the technological revolution, reducing the gap between the new reality and the obsolete procedures of traditional companies. Since technology is an extension of this generation, it is considered that they have the digital influence necessary for creative information analysis, for problem solving, for the development of new products and services, as well as for new ways of working (Colbert et al. 2016). However, for organisations to create environments that enhance these benefits, they need to know the goals, aspirations and desires of this generation. That's because, at a time when the market is becoming more competitive and technological progression is faster, power shifts from the company to the customer and from the company to the employee. Therefore, one of the main missions of companies is the creation of cultures and environments that enhance the personal and professional development of their employees, integrating them into a team that strives toward a common goal (Ahmad and Daud 2016).

Millennials were born between 1980 and 2000, following all the technological and globalised transformations developed since then. If by 1980 users of mobile phones and computers were rare, today average smartphone users now check their equipment every minute, hundreds times a day (Colbert et al. 2016). In this scenario, Millennials developed a different view of the world, as they were forced to make rapid changes, getting used to it and learning to react. As technology is considered a vital component in their daily lives, they want information that is quick and easy, accessible at any time and place. We do not speak merely of informative content; we refer to entertainment that engages them in unique and unforgettable experiences. Fans of videos and multimedia content, the use of multiple digital devices and multitasking are now a reality. They spend more time watching videos on their smartphones than watching TV, spending more than five hours online (Telefónica 2016; ESA 2016).

The social relations of this generation have been digitalised, but in harmony with a high sense of community and a strong social, economic, environmental and health concern. They increasingly value the practice of physical exercise, healthy eating and environments that allow these variables to be conjugated through local, organic fair trade (Barton et al. 2012). We also speak of a nomadic and liberal generation, guided by the adventurous spirit, which makes them travel the world, discovering new cultures, tastes, smells, that is, new experiences that lead to new social, cultural, political and economic values.

In the workplace, Millennials are also looking for new experiences, different sensations and ethics in companies. In this respect, they are ambitious, enterprising, and result-oriented. They aim for rapid growth and enjoy teamwork. However, they are not so committed to companies (they spend an average of two years in the same job), and if a company does not reflect their values and sense of ethics, they rapidly abandon it (Ahmad and Daud 2016). This generation values flexible and emotional benefits, challenges, new ways of working, a clear career path, work-life balance,

Table 1 Millennials lifestyle, skills and motivations

Lifestyle	Skills	Work motivations
Fans of videos and multimedia	Digital influence	Flexible and emotional benefits
Spend more than 5 h online	Ambitious	Challenges
High sense of community	Entrepreneurs	Clear career path
Health concern	Results oriented	Work life balance
Nomadic and liberal	Adaptation to change	Recognition
Adventurous spirit	Team workers	Feedback
Search for experiences and entertainment		Mobility collaborative workspaces

Adapted from: Manpower Group (2016), Deloitte (2017), Miguel (2015), Telefónica (2016), Freitas (2014)

recognition, feedback, mobility and collaborative workspaces. Flexibility is now the key word in an organizational culture that needs to support collaboration, communication and connectivity, based on passion and purpose. We are talking about flexible time, flexible role and flexible location. Those work environment characteristics are strongly linked to improved performance and Millennials retention. They also support higher levels of productivity, engagement, well-being, happiness and health of this generation. We are facing a generation focused on career and evolution. More than the ability to lead, the money and the stability; purpose, positive social impact and passion are the differential factors in a job proposal. It is now time to rethink HR practices using marketing behaviours, and promoting employee experience to attract, retain and develop this generation. Table 1 summarizes Millennials lifestyle, skills and motivations.

3 Employer Branding Strategy: Promoting Experiences and Value

As previously described, in an increasingly volatile world, the attraction, development and retention of Millennials are critical success factors for organisations seeking prosperity and success. In this regard, the implementation of an EB strategy coherent with the needs and aspirations of this generation may arise in response.

Employer branding is a specific term created in the 1990s by Ambler and Barrow (1996, p. 185), defined as "the package of functional, economic and psychological benefits provided by employment and identified with the employee company, with the primary role of provide a coherent framework for management to simplify and focus priorities, increase productivity and improve recruitment, retention and commitment". Since its origin, EB has gained increasing importance in the area of people management, mainly due to its potential to retain and attract talent, which consequently promotes the capacities and competitiveness of organizations. Many authors point out that this strategy is one of the few long-term solutions to the problem of "talent shortage", as well as a solution for the development of aligned, engaged and engaged work teams; a fundamental condition to feed the desires of this generation (López et al. 2014; Stariņeca and Voronchuk 2014). "Employment brand is a longer-term proactive solution designed to provide a steady flow of applicants. Employment branding is the process of developing an image of being a «great place to work» in the minds of the targeted candidate pool" (Kapoor 2010, p. 53). However, for organizations to enjoy these benefits, they need well-designed and developed EB programs that reflect the needs and desires of employees (Thomas and Jenifer 2016). "If an organization wants its employees to live up to the company brand promise, it clearly needs to understand what drives their sense of engagement or commitment" (Kapoor 2010, p. 51).

EB can then be seen as a holistic process, and at its centre is an attractive job proposal, that is, the Employee Value Proposition (EVP). EVP captures the essence of what the organisation wants it to be in the minds of potential and current contributors (Benz 2014). Through it, each worker is unique and feels part of a promise that grows from their experience. This can be seen as a synthesis of the unique attributes and benefits that will lead individuals to want to be part of an organization's everyday life (Vaijayanthi et al. 2011). And as in a traditional branding strategy, we are faced with a promise of value based on functional, emotional, and symbolic benefits (Neil and Strauss 2008; Balmer 2012; Aaker 2010). They seek relevance to the brand (Aaker 2011) and create employer brand attractiveness (Verma and Ahmad 2016). EPV conveys imagination, personality, emotions, resulting in an inspiring idea, a purpose that unites business, ideas and people (Roper and Parker 2006; Goldfarb et al. 2009).

Culture and organisational environment, employee experience (tangible benefits such as salary as well as intangibles such as recognition and career advancement opportunities), organisational integrity and reputation, internal communication, performance appraisal systems, social responsibility, external marketing, performance of the organisation, leadership, security, stability, among many others, are variables to be developed in an integrated Employer Branding strategy (Jain and Pal 2012; Dahl and Peltier 2013). Next-year success will depend, therefore, on the organizations' ability to extend their traditional marketing mix to include corporate reputation and align external and internal brands with integrated marketing that recognizes the differences in all stakeholders and acts accordingly (Aydon Simmons 2009; Luxton et al. 2014).

We speak of an era driven by emotions and essentially values, where proposals must involve the functional, emotional and spiritual benefits and be conceived with the "many to many" collaboration. The storytelling and the use of metaphors that identify with the body, heart and human spirit is an aspect to be used with all stakeholders and, in particular, with the workers, in a fuller and integrated perspective. We talk about value marketing with consumers, channel partners and employees (Kotler et al. 2010).

3.1 The EB Strategy in Practice: The 10 Steps

The implementation of the company's values and vision in the processes and methodologies, transporting organizational personality from the inside out, will always be a complex challenge. But in practical everyday life, what stages and steps are necessary to implement an EB strategy? We have thus selected 10 essential steps for an organisation that intends to implement an EB strategy.

1. The diagnostic phase is crucial for any organisation wishing to walk towards an EB strategy (internal and external diagnosis). The diagnosis answers two questions: Where are we? and How did we get here? This establishes a framework of the internal and external context, as well as the critical analysis of the existing processes, the feeling of interventions and practices of other organizations that can be their competitors and/or sources of inspiration;
2. After the diagnosis, it is essential to develop a strategy based on HR goals, but, above all, integrating marketing concepts whose mission is to deliver value to all the stakeholders, respecting the corporate brand and enhancing the values of the organization from the inside out;
3. This strategy should be aligned with the business strategy, through the achievement of results, efficiency gains and by taking care of future sustainability, so the commitment of each worker with the aspirations of the organization, business success and financial health is important;
4. The top management and leadership of the company must be committed to the employer branding strategy, to ensure the company delivers on its promises, and they should lead by example, transforming each value in daily behaviour;
5. The definition of goals, intervention areas and structure of a multidisciplinary team is also fundamental. Only this way the specificities of each one are safeguarded, but at the same time, they work in alignment with the whole, allied to the digitalisation of processes, with professional flexibility and the reconciliation of personal and professional life also being a requirement;
6. After the strategy formulation, it is time to develop an amount of policies, processes, activities and tools that allow a people integrated management system that transforms employees in true ambassadors of the company brand. However, it is essential to communicate and insistently train all the members of the team, preferentially using online tools, without ever disregarding personal contact;
7. Communication, as we already demonstrated, becomes an essential dimension of an EB strategy. As so, it is important to introduce marketing skills on human resource management, defining targets, touchpoints and inspirational messages;
8. In the development of the policies, processes and activities, it is important to attend on symbolic, functional and emotional benefits that can be delivered to all stakeholders, because we are in an era driven by emotions, experiences and essentially by values;
9. Communication, commitment and engagement are critical success factors for the sustainability of the EB strategy, and consequently for the preservation of business and consolidation of organizational culture;

10. When the EB strategy reaches a mature state, it is time to measure its implication on customer (internal and external) orientation and satisfaction, applying contingency measures that respond to the challenges of the organization and its stakeholders.

3.2 EB Strategy: Connecting with Millennials Using Digital Media

After analysing the environment, reflecting on the characteristics of the Millennials and considering the practical development of a strategy of EB, the question arises: how can companies adapt their Employer Branding strategy in order to respond to the characteristics of this generation and take advantage of the potential of digitalisation?

In effect, the 10 steps described above also meet a set of five Millennials expectations and other challenges of digitalisation. First, they are in tune with this generation because: (1) they result from the collection of their opinions (the internal customer voice); (2) they translate into a mission for the company that will only work if it responds to the challenge of each worker; (3) they require communication, transparency and recognition; (4) they reflect a leadership style that gives autonomy, responsibility and example; and, last but not least, (5) they treat each worker in their individuality, providing challenge, experience and knowledge on an online and offline basis. In terms of digitalisation, a EB strategy (1) requires global transformation of the organization; (2) promotes collaborative work and spirit; (3) allows efficiency gains with savings of resources, time and space; (4) promotes critical thinking, flexibility and adaptability to change; and (5) boosts the most human side of the individual, characterized by creativity, unpredictability and genius.

Undoubtedly, in an era driven by values and emotions, the development of a value promise with purpose and passion, where technology must be an ally rather than a threat, is paramount. It is noted that when Millennials feel their work has meaning and can make a difference in today's world, loyalty and motivation rates increase. Alongside this, as in a traditional marketing and branding strategy, it is important to adapt the package of functional, emotional and symbolic benefits, while developing an integrated communication strategy, which should encompass the digital codes and channels used by the target audience. There are several digital tools at EB's service for Millennials and the use of digital media has the potential of establishing strong relationships between employees and the organization, fostering a digitalisation culture. It is time for companies to re-imagine their policies and ways of acting, to prepare for the future, responding to the challenges and harnessing the potential of this technological revolution.

Finding the right way to communicate is one of the key points of this strategy. This communication has to be done anytime and anywhere, and must be customized according to targets and their messages. This generation is always connected and requires rapid and easy information. It is worth noting that business monologues

Table 2 Millennials and digital communication

How to communicate	Communication tools	Communication purposes
Anytime and anywhere	Instant messaging	Collaboration
Rapid and easy information	Blogs	Connectivity
Bidirectional communication	Enterprise social media	Recognition
Visual communication	E-learning	Feedback
Storytelling	Podcasts	Sharing
	Multimedia	Update
	Gamification	Experience
	Ideation platforms	
	Web conferencing	

Adapted from: Capgemini Consulting (2012), Deloitte (2016), PWC (2011)

are not worthwhile. Bidirectional communication based on mobile technology is the premise of this new century, resulting in a core need for unification of online and offline communication (e.g. instant messaging, micro blogging enterprise social media, e-learning, e-mail, podcasts, ideation platforms, web conferencing, multimedia and gamification). Companies need to select the internal and external communication tools that meet their specifics and that lead to collaborative environments and the involvement of Millennials. As an example, Cisco has launched the Cisco Video communication platform to communicate more effectively and efficiently. In addition, it uses enterprise social software to facilitate collaboration, customization and the delivery of relevant content to each of its employees. And at Kraft Foods, employees can access podcasts from the CEO and other managers to keep up-to-date on the latest corporate strategies and initiatives. Another example is Clube Ser, an internal communication program, supported by a digital platform of Salvador Caetano Goup (a Portuguese leading group in the automotive sector). Based on Group culture and values "Be Caetano", this is an online and offline channel of the employees to employees, where they can interact, know the policies and best practices of the organization, as well as the benefits that the company offers, mostly emotional and symbolic EVP.

Communication with this Generation should thus take on a much more visual and storytelling form, to the detriment of traditional internal communication, in which employees needed to sit down and read long texts. Blogs, internal social networks and streaming videos are more than modern communication tools. They allow bi-directional communication, feedback, and recognition, and encourage the exchange of ideas between all. These characteristics are highly valued by Millennials, because they do not wait. For them, time for change is now and in their own way. Table 2 summarizes the main aspects to be taken into account in the development of a communication strategy aimed at this generation.

In a fast-paced world, in order to deliver a promise of coherent value, beyond communication, it is necessary to adapt the package of functional, emotional, and

Table 3 Functional, emotional and symbolic benefits for Millennials

Functional benefits	Emotional benefits	Symbolic benefits
Payroll	Face-to-face feedback	Work environment
Flexible benefits	Recognition	Collaborative culture
Work-life balance	Career path	Reputation
Technology	Training and developing—upskill regularly	Employee experience
Digital toolbox	Mobility	Employee empowerment
Stability—career enhancement	Career variety	Purpose
	Flexibility—role, location, time	Passion
		Values
		Social responsibility

Adapted from: Manpower Group (2016), Deloitte (2016), PWC (2011)

symbolic benefits. To do so, employers have to start listening, have to be creative and realize that what works with Millennials will most likely work with the rest of the workforce, aligning talent strategy with business strategy. In building this package, it is necessary to keep in mind that this generation seeks more than a job; looking for more than a pay cheque.

This new generation not only wants to be part of the organization, but also wants to actively intervene in its construction, in a more lasting relationship. In this new relationship, payroll and flexible benefits commune with Work-Life Balance, technology and career plan. But also with positive and negative feedback, knowledge, progression and flexibility (Bass 2000; Trevor and Hill 2012). And all this is based on an organizational culture, with values that are practiced, with promises that are fulfilled and with a reputation that protects itself. These are the demands of this generation, which are built, implemented and intensified by an EB strategy. Millennials want to be able to work according to what suits them best. Table 3 summarizes the functional, emotional, and symbolic benefits valued by this generation.

4 EB, Digital Media and the Millennial Generation—Virtuous Circle

The future here presented is already past and present, given the pace at which society moves and, by default, companies and especially those who work there. In this bipolar world, today, speed is the main drive of the humanization of both relationships and spaces. We are facing a new personal, family and business paradigm that demands strong symbols, true values and identities with meaning and essence. Finally, brands with personality and with permanent reinvention capacity, whether they are product

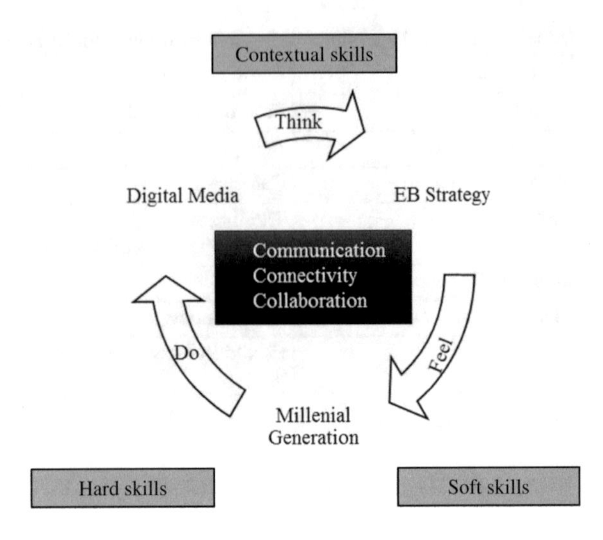

Fig. 1 EB, digital media and the Millennial generation—virtuous circle

brands, employer brands or personal brands (Aaker 2011). We especially refer to the generation that is always online and will have the ability to adapt to the new world, where science fiction will become reality at home, on the street and in companies.

Organizations are now more exposed to this era characterized by volatility, technological progress and competitiveness. We are facing a VUCA (volatility, uncertainty, complexity, and ambiguity) environment that requires for a new generation of leaders that inspire and spread the culture of the company (Lawrence 2013). Therefore, they need to collect the best talent, which is revealed by technical skills (hard skills), attitude (soft skills) and adaptation to the environment (contextual skills). EB, supported by digitalisation in general and digital media in particular, responds to this trilogy by making more prosperous organizations, happier workers, and more sustainable societies. Adaptive skills, flexibility, critical thinking, and creativity are increasingly valued. To foster them, businesses need to create cultures and environments that trigger connectivity, collaboration, and communication among all.

Communication is thus one of the focal points of a Millennial-oriented EB strategy. This must assume the lifestyle of this generation, and should be anytime, anywhere, fast, easy and bidirectional. For this, companies must use a vastness of touchpoints, aligning their communication online and offline. Social networks, due to their multiplicity, ease of use and customization, seem to be the medium with the most potential to achieve the effects of the EB strategy. By using them, preferentially, for the reasons described above, it is possible, in the first stage, to disclose the company's promises (think), then to boost the involvement, heart and soul, in the purpose of the organization (feel). And finally, we must make it happen, translating into practice every corporate value (do). Only then will we have an employer brand that passionately and consistently communicates its functional, emotional, and symbolic benefits.

Concluding, in this journey of attraction, development and retention of talent, communication is the watchword, because it conveys the reconciliation of individual and business expectations, unites managers and workers and responds to business challenges and social stability. EB, digital media and the Millennial Generation are therefore the three cornerstones of a virtuous circle supported by communication, connectivity and collaboration (Fig. 1).

References

Aaker, D.: Building Strong Brands. A Masterpiece. Sophisticated Practical, and Readable Tom Peters. Pocket Books Business, London (2010)

Aaker, D.: Brand Relevance: Making Competitors Irrelevant. Jossey-Bass. A Wiley Imprint, San Francisco (2011)

Ahmad, N.A., Daud, S.: Engaging people with employer branding. Procedia Econ. Finance **35**, 690–697 (2016)

Ambler, T., Barrow, S.: The employer brand. J. Brand Manag. **4**(3), 185–206 (1996)

Aydon Simmons, J.: “Both sides now”: aligning external and internal branding for a socially responsible era. Mark. Intell. Plann. **27**(5), 681–697 (2009)

Balmer, J.M.T.: Corporate Brand Management Imperatives: custodianship, credibility and calibration. Calif. Manag. Rev. **54**(3), 6–34 (2012)

Barton, C., Fromm, J., Egan, C.: The Millennial consumer: debunking stereotypes. Resource document. BCG. https://www.bcg.com/documents/file103894.pdf (2012)

Bass, B.: The future of leadership in learning organizations. J. Leadersh. Organ. Stud. **7**, 18–40 (2000)

Benz, J.: Will purpose rebalance the employee value proposition? People Strategy **37**(2), 10–11 (2014)

Buchanan, J., Kelley, B., Hatch, A.: Digital workplace and culture. Resource document. Deloitte. https://www2.deloitte.com/us/en/pages/human-capital/articles/digital-workplace-and-culture.html (2016)

Colbert, A., Yee, N., George, G.: The digital workforce and workplace of the future. Acad. Manag. J. **59**(3), 731–739 (2016)

Dahl, A.J., Peltier, J.W.: Internal marketing and employee satisfaction and loyalty: cross-cultural scale validation in context of U.S. and German nurses. J. Mark. Manag. **29**, 1030–1055 (2013)

Davies, A., Fidler, D., Gorbis, M.: Future work skills 2020. Resource document. Institute for the Future for the University of Phoenix Research Institute. http://www.iftf.org/futureworkskills/ (2011)

Deloitte: The 2017 Deloitte Millennial Survey [Resource document]. https://www2.deloitte.com/global/en/pages/about-deloitte/articles/millennialsurvey.html (2017)

D’Souza, F.: Our CEO Frank D’Souza Quoted Among the World’s Most Famous Digital Leaders [Web log post]. https://twitter.com/cognizant/status/675304057901215744 (2015)

ESA: Essential facts about the computer and video game industry [Resource document]. http://essentialfacts.theesa.com/Essential-Facts-2016.pdf (2016)

Freitas, I.: Os millennials em Portugal: estudo exploratório a partir de um coorte etário sobre engagement consumidor-marca. Escola Superior de Comunicação Social. http://hdl.handle.net/10400.21/4444 (2014)

Goldfarb, A., Lu, Q., Moorthy, S.: Measuring brand value in an equilibrium framework. Mark. Sci. **28**(1), 69 (2009)

Jain, V., Pal, R.: Importance of employer branding in business up-gradation. Int. J. Res. IT Manag. **2**(11), 6 (2012)

Kapoor, V.: Employer branding: a study of its relevance in India. J. Brand Manag. **7**(1–2), 51–75 (2010)

Kotler, P., Kartajaya, H., Iwan, S.: Marketing 3.0: From Products to Customers to the Human Spirit. Wiley, New Jersey (2010)
Lawrence, K.: Developing Leaders in a VUCA Environment. UNC Executive Development, North Carolina (2013)
Leonhard, G.: Technology vs. Humanity. Fast Future Publishing, United Kingdom (2016)
López, F., Tarodo, A., Lores, S.: Branding: estudio multinacional sobre la construcción de la marca del empleador. Universia Business Review **20**(44), 34–53 (2014)
Luxton, S., Reid, M., Mavondo, F.: Integrated marketing communication capability and brand performance. J. Advertising **44**(1), 37–46 (2014)
Manpower Group: Millennial Careers: 2020 Vision [Resource document]. http://www.manpowergroup.com/wps/wcm/connect/660ebf65-144c-489e-975c-9f838294c237/MillennialsPaper1_2020Vision_lo.pdf?MOD=AJPERES (2016)
Miguel, A.: As roupas que falam: dos e para os millennials portugueses. IADE-U Instituto de Arte, Design e Empresa – Universitário. http://hdl.handle.net/10400.26/11177 (2015)
Neil, W., Strauss, R.: Value Creation: the Power of Brand Equity. Cengage Learning, Mason (2008)
PWC: Millennials at work Reshaping the workplace [Resource document]. https://www.pwc.com/m1/en/services/consulting/documents/millennials-at-work.pdf (2011)
Roper, S., Parker, C.: Evolution of branding theory and its relevance to the independent retail sector. Mark. Rev. **6**(1), 55–71 (2006)
Stariņeca, O., Voronchuk, I.: Employer branding training development for public organisations. Reg. Formation Dev. Stud. **3**(14), 207–220 (2014)
Telefónica: Más allá de los millennials. La generación post-millennial ya está aquí [Resource Document]. https://lacofa.fundaciontelefonica.com/2016/03/28/mas-alla-de-los-millenials-la-generacion-post-millennial-ya-esta-aqui/ (2016)
Thomas, B.J., Jenifer, S.C.: Measurement model of employer brand personality a scale construction. J. Contemp. Manage. Res. **10**(1), 58 (2016)
Trevor, J., Hill, R.: Developing transformational leadership capability. Developing Leaders. 42–47 (2012)
Vaijayanthi, P., Roy, R., Shreenivasan, A., Srivathsan, J.: Employer branding as an antecedent to organisation commitment: an empirical study. Int. J. Global Bus. **4**(2), 91–106 (2011)
Verma, D., Ahmad, A.: Employer branding: the solution to create talented workforce. IUP J. Brand Manag. **13**, 42–56 (2016)
Welch, M., McAfee, A.: Being digital: engaging the organization to accelerate digital transformation [Resource document]. Capgemini Consulting. https://www.capgemini.com/resource-file-access/resource/pdf/being_digital_engaging_the_organization_to_accelerate_digital_transformation.pdf (2012)

Paula Arriscado People, Brand & Communication General Manager at Salvador Caetano Holding (since 2010) and Professor at IPAM—The Marketing School—Laureate International Universities (since 2007). Ph.D. on Brand Management and integrated Communication at the Universidade de Santiago de Compostela (2008), DEA in Public Relations (2001) and BA in Journalism (1995). Between 1991 and 2000, she worked as freelancer in Radio and Press and Business Communication Consultant. She was the Marketing professional of the year 2005 in the automotive sector, Portuguese Marketing Association and Professor of the year 2013 by IPAM. Gaia, Portugal.

Helena Quesado Brand & Communication Coordinator at Salvador Caetano Holding, putting in practice the employer branding strategy of the People, Brand & Communication Division. Has a postgraduate degree in Business Communication (2012) at EGP—Universidade do Porto Business School and a degree in Languages and Modern Literature (English/German) (2005) by Universidade do Porto. Gaia, Portugal.

Bianca Sousa Works at Salvador Caetano Holding (since 2015) has an internal marketing technician, putting in practice the employer branding strategy of the People, Brand & Communication Division. Has a degree in marketing management (2015) by IPAM—The Marketing School—Laureate International Universities and was Student of the Year 2015 by IPAM. Gaia, Portugal.

Copyright and PSM in the Digital Market

Ana María López-Cepeda and Andrea Valencia-Bermúdez

Abstract Broadcasting services in Europe are facing changes that, in the case of Public Service Media (PSM), are complemented by a serious economic and trust crisis that force them to rethink their model. One of the core pillars of this change is the adaptation to the digital market. To do so, a modernization in the regulation of copyrights in PSM is needed. In this regard, European regulation for broadcasting services has become obsolete, so there is a need to move towards a system that benefits online transmissions, digital transmissions and the European Single Market. In the case of public broadcasting services, more policies of change are requested related to the concept and mission of public service. The aim of this research is to assess the situation, problems and European policies on copyrights and public service media in a moment when the EU has proposed a legislative package for its modernization in the digital market.

Keywords Copyright · Public broadcasting · Digital market · European policies

1 Introduction

The main copyright policies in Europe have become obsolete and need an immediate change. "The framework for copyright licensing should thus be modernised on a technologically-neutral basis in order to cater for the wide range of devices and platforms available to viewers and listeners" (European Broadcasting Union 2016a: 2). This situation affects a large number of stakeholders—such as authors, creators, producers and cultural industries-, and audiovisual media are among them, especially PSM. The latter are in the middle of an economic and model crisis that force them to adapt to the new digital market while maintaining their public service mission.

A. M. López-Cepeda (✉)
Universidad de Castilla-La Mancha, Cuenca, Spain
e-mail: ana.lopezcepeda@uclm.es

A. Valencia-Bermúdez
Universidad de Santiago de Compostela, Santiago, Spain
e-mail: andrea.v.bermudez@gmail.com

M. Túñez-López et al. (eds.), *Communication: Innovation & Quality*, Studies in Systems, Decision and Control 154, https://doi.org/10.1007/978-3-319-91860-0_24

The EU has started to work for the modernisation of copyright policies in order to achieve a digital single market. That process has its origin in 2013, but it was last year when it became more active in this regard (López-Tarruella 2016). The situation is not simple "taking into account that the regulation of copyright is still governed by the principle of territoriality" (López-Tarruella 2016: 105).

The main proposals suggest changes on the regulation of copyright that will particularly affect audiovisual media services in four areas: a. online transmissions; b. digital retransmissions of radio and TV programmes; c. increase in European works in VoD services; and d. geoblocking. As regards public broadcasting services, these ideas are complemented by an imminent need to reformulate the concept of public service within the new digital market in order to jointly manage copyrights in pursuing its public mission. Nevertheless, and in the absence of an entire process of change, the EU has therefore not yet included proposals on other elements that raise doubts among broadcasting operators, such as copyrights, cloud-based content and other new technologies (European Broadcasting Union 2014).

Against this background, the article provides an analysis of the main European policies on copyright in the digital market, undertaking an overview of the current situation, main problems, proposals and their potential impact on public broadcasters. To this end, an analysis of the main legislative proposals is carried out together with a study of the contributions from scientific literature on the subject. First, the research analyses the key measures on copyright that affect all broadcasters as a whole, including public ones, and finally it focuses on those policies that affect exclusively to public service broadcasters when managing copyrights and related rights.

The novelty of this research is that, although there are studies on copyright in the digital market (López-Tarruella 2016; Castets-Renard 2016), any of them has focused solely and exclusively on audiovisual media services in general and public service media in particular. In this regard, these services should face a redesign in the new digital market. The recommendation of the EU is to "encourage all TV operators to have a presence in all platforms. There already exists the Internet Protocol Television (IPTV), developed from the video-streaming, to converge with the connection of the network of networks, TV on demand, and the metamorphosis of the five screens: TV, computer, console, smartphone and home cinema" (Campos 2013). The relevance of the article lies in presenting and explaining a wide compendium of rules whose effects would result in reforms of public service media, which will affect their needed remodelling in the new market structure.

2 Main Copyright Policies that Affect Audiovisual Media Services

In 2016 the EU published a package of copyrights where it proposed "two directives and two regulations to adapt the EU copyright rules to the realities of the Digital Single Market" together with "an explanatory Communication, as well as an extensive Impact Assessment on the modernisation of EU copyright rules[1]" (IRIS Merlin database 2016). Among all these measures, those that concern public service media are: (a) the proposal for a regulation laying down rules on the exercise of copyright and related rights applicable to certain online transmissions of broadcasting organisations and retransmissions, COM (2016) 594 final, 14 September 2016; and (b) the proposal for a directive of the European Parliament and of the Council on copyright in the digital single market, COM (2016) 593 final, 14 September 2016, the latter with regard to the promotion of European audiovisual works via on demand services. It should be also considered within this package—although above mentioned—the proposal for a regulation of the European Parliament and of the Council on ensuring the cross border portability of online content services in the internal market COM (2015) 627 final, 09 December 2015, on geoblocking.

2.1 *Policies for Online Transmissions*

The access to online contents still represents a small portion of the total time of video watching. Nevertheless, online video watching has increased since the emergence of broadband networks in the European Union (Fontaine and Grece 2015: 32).

[1]Communication from the Commission to the European Parliament, the Council, the European Economic and Social Committee and the Committee of the Regions: Promoting a fair, efficient and competitive European copyright-based economy in the Digital Single Market: COM(2016)592; Proposal for a Directive of the European Parliament and of the Council on copyright in the Digital Single Market: COM (2016) 593 final, 14 September 2016; Proposal for a Regulation laying down rules on the exercise of copyright and related rights applicable to certain online transmissions of broadcasting organizations and retransmissions: COM (2016) 594 final, 14 September 2016; Proposal for a Directive of the European Parliament and of the Council on certain permitted uses of works and other subject-matter protected by copyright and related rights for the benefit of persons who are blind, visually impaired or otherwise print disabled and amending Directive 2001/29/EC on the harmonization of certain aspects of copyright and related rights in the information society: COM (2016) 596; Proposal for a Regulation of the European Parliament and of the Council on the cross-border exchange between the Union and third countries of accessible format copies of certain works and other subject-matter protected by copyright and related rights for the benefit of persons who are blind, visually impaired or otherwise print disabled: COM (2016) 595; Commission Staff Working Document—Impact Assessment on the modernization of EU copyright rules—Accompanying the document "Proposal for a Directive of the European Parliament and of the Council on copyright in the Digital Single Market" and "Proposal for a Regulation of the European Parliament and of the Council laying down rules on the exercise of copyright and related rights applicable to certain online transmissions of broadcasting organizations and retransmissions of television and radio programs": SWD (2016) 301 (IRIS Merlin database 2016).

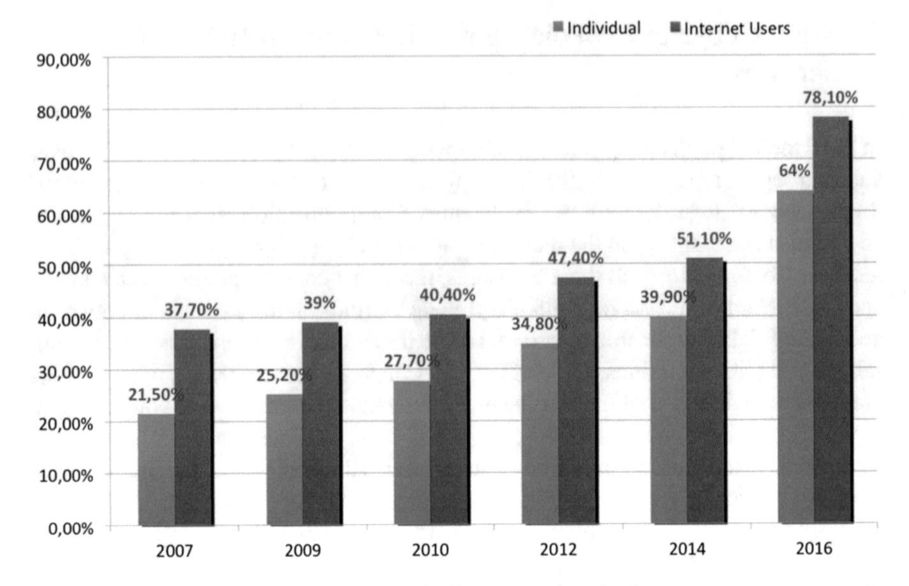

Fig. 1 Percentage of EU citizens that use the Internet to download or access to games, music, movies or pictures (2007–2016). *Source* European Commission (2016)

The figure has risen in recent years. While in 2010 only 27.7% of the EU citizens used the Internet to download or access games, music, movies and pictures, in 2016 the figure had increased to 64%. The figure is higher when surveying Internet users in the last three months. Data evidence an important gap between ages, which shows a growing trend in the short-medium term (16–24 years: 91.3%; 25–54 years: 72.6%; 55–74 years: 36.8%) (European Commission 2016) (Fig. 1).

This increase in demand of online services has led to new concerns for all audiovisual media services, in general, and for public media in particular. Not only audiovisual media services should adapt themselves to the new digital market, but they are also facing new challenges linked to the traditional system of copyright management. This is because they find themselves in the need to retransmit a large number of products and services that include very diverse contents protected by copyright. "This requires a complex clearance of rights with a multitude of right holders (…) In order to make their services available across borders, broadcasting organizations need to have the required rights for the relevant territories and this increases the complexity of the rights clearance" (COM/2016/594 final: 1).

The solution envisaged by the EU consists in the application of the country of origin principle to broadcast-related online services. Lopez-Tarruella (2016: 113–114) explains in detail this measure:

> The proposal indicates that its lawfulness is regulated by the law of the State of origin of the service. The authorities of the Member States from which the service is received cannot apply its law to verify its legality. Only the law of the state of origin could be applied, since it is where the alleged infringement takes place. That means that they should confine

themselves to recognize the legality of the service it complies with the requirements of the state of origin.

This principle will be applied for using, accessing or offering a broadcast-related online service. That is the option under the proposal for a regulation of the European Parliament and of the Council laying down rules on the exercise of copyright and related rights applicable to certain online transmissions of broadcasting organisations and retransmissions of television and radio programmes, which is already applied in satellite broadcasting (Council Directive 93/83/EEC of 27 September 1993 on the coordination of certain rules concerning copyright and rights related to copyright applicable to satellite broadcasting and cable retransmission).

However, the proposal for a regulation (COM/2016/594 final: art. 1) applies the country of origin principle only to broadcast-related online services. These are understood as those services "consisting in the provision to the public, by or under the control and responsibility of a broadcasting organisation, of radio or television programmes simultaneously with or for a defined period of time after their broadcast by the broadcasting organisation as well as of any material produced by or for the broadcasting organisation which is ancillary to such broadcast" (COM 2016, 594 final: art. 1 a). The EU states that the fact of including online transmissions not connected to a retransmission (web broadcasting services) would generate legal uncertainty for right holders and could lead to a lowering of the level of protection, as online operators can easily move their business establishment within the EU (COM 2016, 594 final: 6).

This has been the subject of much controversy. According to the European Parliament's Committee on Internal Market and Consumer Protection—IMCO—(2017: 3) "all online services provided by broadcasters should be included in the scope of the COO provision" and places particular emphasis on the fact that public broadcasters should use all possible channels to reach all their audiences.

Even so, the application of the country of origin principle is expected to imply economic and time savings in the acquisition of rights endured by broadcasters when carrying out cross-border online transmissions (SWD/2016/302 final). This is an important measure for public service media in order to adapt themselves to the new digital scenario.

2.2 *Policies on Digital Retransmissions of Radio and TV Programmes*

The access to a digital single market requires to facilitate the management of copyrights and other related rights in digital retransmissions of radio and TV programmes. In this case, the operators that package a large number of broadcasting channels may have problems in acquiring all the needed rights to retransmit programmes of broadcasting organizations (COM/2016/594 final: 2), especially when they belong to other Member States. The EU has opted to facilitate "the clearance of rights for retransmission services provided over closed networks (other than cable), by introducing rules on mandatory collective management" (Proposal for a regulation of the European Parliament and of the Council laying down rules on the exercise of copyright and related rights applicable to certain online transmissions of broadcasting organisations and retransmissions of television and radio programmes) (COM/2016/594 final: 4). In the same vein this is applied in cable retransmissions, as envisaged by the Satellite and Cable Directive (Directive 93/83/CEE), whose scope is now also extended to IPTV services (Internet Protocol Television), such as Imagenio, Fastweb, Maligne TV and Free (Delgado and Fernández-Quijada 2007), and other services provided over closed Electronic Communication Networks (ECNs). Therefore, these will benefit from the mandatory collective management of copyright and related rights.

This measure has been also revised and criticised, as it did not include standalone broadcasting services or OTT (Over The Top). "Examples of OTT services may include chat applications, e.g. Whatsapp, Wechat and Facebook Messenger); streaming video services, e.g. blip, Netflix, Amazon Prime, vevo, hulu); voice calling and video chatting services, e.g. Skype, Google Hangout, Facetime; and new services such as videogame streaming, e.g. Twitch)". (Jayakar and Park 2014: 3). These are new business models financed by advertising (YouTube, Google); by revenue-sharing (VeVo, Blonkx, BlipTV); by subscription or pay-per-content (Netflix, Apple, Hulu Plus, Sky, Amazon); and mixed models, announced by the BBC, FreeSat or ITV Player (Huerta et al. 2011: 94).[2] The distinction between IPTV and OTT models is essential, as unlike the former, "in the distribution of OTT content, the network administrator does not interfere" (Villagra and Cavalli 2016: 2.774).

The European Broadcasting Union (2016b) deems it necessary to additionally include IPTV services, "similar closed OTT services and broadcasters' catch-up services offered by third party platforms to fully reflect market evolutions". The Internal Market and Consumer Protection (2017: 4) also points out that the inclusion of OTT services "is necessary to enable cable television providers to offer their clients a service portability through the Internet".

This possibility was already foreseen in the study for the proposal for a regulation of the European Parliament and of the Council laying down rules on the exercise of copyright and related rights applicable to certain online transmissions of broadcasting organisations and retransmissions of television and radio programmes. Nevertheless, it was dismissed as it would entail risks of undermining right holders' exclusive online rights, leading to a reduction of licensing revenues (COM/2016/594 final). These are improvements that, if taken into account, would make it easy the management of rights in both broadcasting services.

[2]Huerta et al. (2011). "Modelos over the top (OTT): regulación y competencia en los nuevos mercados de Internet", en Política Económica y Regulatoria en Telecomunicaciones. GEER y Telefónica, vol. 6, pp. 84–98. Mentioned in Del Pino and Aguado (2012: 63).

2.3 *Policies for the Promotion of European Works on Video on Demand (VoD)*

The growth of VoD services is steady throughout the world. In 2016, 16.9% of the EU citizens surveyed affirmed that they consumed video on demand from commercial services, a percentage that will gradually increase if only Internet users (20.7%) or the target age 16–24 (31.9%) are taken into account (European Commission 2016).

Subscription Video on Demand (SVoD) in the EU has grown significantly in last years, namely a 56% from 2014 to 2015 and it is expected to reach 50 millions of households in 2020. This report highlights that "competition in the SVoD market is also increasing rapidly. Netflix is the current undisputed leader with a 52% share of the market in the European Union, however Amazon is mounting a strong challenge. There are now also many other European groups entering the market including Vivendi (CanalPlay), Sky Plc (Now Tv) and ProSiebenSat.1 (Maxdome)" (European Broadcasting Union 2016, June, 29).

However, despite this rise over the past few years, "EU audiovisual works only constitute one third of works available to consumers on those platforms" (COM/2016/593 final: 2–3). A study of European Audiovisual Observatory carried out by researchers Ene and Grece (2016) show that, for the 75 VOD services analysed throughout Europe, the origin of most of the films wad not European, both if it was taken into account the results for unique films titles and the aggregated results of cumulated film titles (Fig. 2).

In this case as in previous sections, the lack of availability is partly due to the complexity of the process of copyright acquisition. The problem is further complicated in some Public Service Media, as they are limited when offering contents through online services because of their national public service remit. For instance, German public service broadcasters "are not allowed to place TV-feature films and series purchased from third parties online" (Capello 2015: 61).

The solution suggested by the European Union is to facilitate the dialogue together with the obligation of the Member States to establish a negotiation mechanism, a

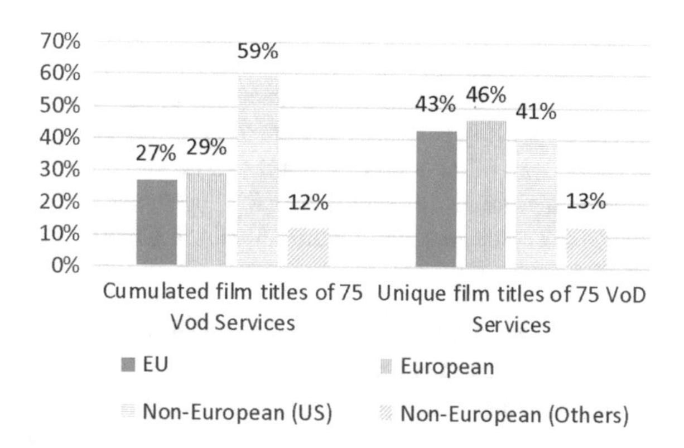

Fig. 2 Cumulated and unique film titles of 75 VoD services (2015). *Source* Ene and Grece (2016)

measure provided for in the proposal for a Directive of the European Parliament and of the Council on copyright in the Digital Single Market (COM/2016/593 final). This body shall meet with stakeholders with the aim of providing professional and external expertise and thus achieving optimal negotiation (COM/2016/593 final: art. 10). It is expected to generate limited costs, as Member States can use existing structures with the technical expertise required (SWD/2016/302 final). In all events, they should report the Commission on the mechanism within 12 months after the directive comes into force (COM/2016/593 final: art. 21.1).

The measure is part of a wider policy to face the multiple factors that explain the limited availability of European audiovisual works, especially films, throughout the European Union. In this sense, public service broadcasters should play an essential role: if they do not promote national and European audiovisual works, American films and formats would dominate the digital market. The aim is to promote culture and audiovisual industries from different Member States, as it has been trying since the adoption of the Television Without Frontiers Directive and its subsequent amendments, as well as to establish new measures to foster its outreach, especially through Public Service Media. The European Broadcasting Union (2017, May 03) points out that PSM are the biggest investors in European audiovisual content, "spending EUR 18.2 billion per year". Nevertheless, changes in the digital market requires new policies towards this direction.

2.4 Policies Against Geo-Blocking

Geo-blocking is the practice through which the "access to Internet sites based on other EU countries or with different prices depending on the location of the customer" (Urquilla 2015: 46) is denied. This is one of the main problems facing the European Union in creating a digital single market. It is a settled practice among broadcasters. A Commission research on electronic commerce published in March 2016 showed that "82% of the public service TV broadcasters and 62% of commercial TV broadcasters who responded to the inquiry implemented at least one type of geo-blocking to their online services" (SWD/2016/301 final: 24).

However, social demands call for the removal of the project. In an interview conducted to 26,586 European citizens by Flash Eurobarometer 411 (2015), they would be willing to access to almost 30 per cent of audiovisual contents addressed to users from other Member States (Table 1).

Despite these results, cross-border "portability" of services it not always possible due to different reasons. Among them, the Commission points out (COM/2015/626 final: 4–5):

- Copyrights and their territorial application, as in some instances it is difficult to obtain multi-territorial licences.
- "Right holders may decide to limit the territorial scope of licences granted to service providers and/or to confine a service to a particular territory".

Table 1 Percentage of contents in which EU citizens surveyed would be interested in accessing at through Internet services addressed to users from other member states

Kind of contents	Percentage (%)
Audiovisual contents	29
Music	23
Sports	15
Electronic and digital books	10
Online or downloaded games	8
None	47
Do not know	3

Source Flash Eurobarometer 411 (2015)

- "Market-readiness" of works, i.e. "how visible they are for potential licenses, and how easy is to license them".
- The gap between content offers and their uptake.

A large part of the Internet industry prefers to be subjected to a single jurisdiction, as it is "much more profitable to license rights State by State" (López-Tarruella 2016: 117). That is the option initially adopted by governments to ensure market safety. Against geo-blocking, there are those who consider that the original Internet architecture should be maintained, or that the reasons of geo-blocking respond to interests of copyright defenders (Trimble 2016: 51).

The European Commission (2015, May 06) considers that the obstacles of geo-blocking lead to loses in goods and services for citizens: "Only 15% shop online from another EU country; Internet companies and start-ups cannot take full advantage of growth opportunities online: only 7% sell cross-border". Therefore, the European Union has opted in the last decade for a policy against geo-blocking, which is visible in various measures:

The first measure is the Directive 2014/26 related to the collective management of copyrights and relative rights and to the granting of multi-territorial license rights on musical works for its use online in the internal market. "In recent proposals, the European Commission suggests extending this system to other areas" (López-Tarruella 2016: 119).

The most recent anti-geo-blocking policies mainly focus on two proposals for regulation of the Commission and the European Parliament. The first of the proposals, from December 2015, is for ensuring the cross-border portability of online content services in the internal market (COM 2015, 627 final); the second one, from May 2016, is for addressing geo-blocking and other forms of discrimination based on customers' nationality, place of residence or place of establishment within the internal market (COM 2016, 289 final) (Trimble 2016).

In the first proposal, the EU aims to ensure "that subscribers to online content services in the Union, when temporarily present in a Member State, can access and use these services" (COM 2015, 627 final: art. 1), being paid or not, whenever there is a register allowing to check consumers' Member State of residence.

This option may create certain problems, such as "unless providers are expected to rely on users' own declarations of their status, providers will have to implement geolocation, location tracking, and authentication mechanisms to verify that particular users are eligible to enjoy portability at a given moment" (Trimble 2016: 54). At the moment, in the agreement of EU negotiators on new rules allowing European citizens to travel and enjoy online contents across borders, it was found that providers of online content services will verify the country of residence of the subscriber using means such as payment information, the existence of an Internet contract or checking the IP address (European Commission 2017, February, 07). Nevertheless, this requirement is not applied to free services, so "we can also ask if this exclusion of free services encourages the evolution of economic models towards external funding of content, for instance by advertising, which permits services providers to escape the obligation to ensure portability" (Castets-Renard 2016: 393).

Criticism is also focused on the inaccuracy of the temporal concept, as the proposal of regulation or the one of the State of residence do not indicate the time limit that one can be in a different State other than the one of residence to avoid geo-blocking (López-Tarruella 2016: 123).

The ambiguity is also perceived in other dimensions, such as the guarantee of a minimum quality of access (European Economic and Social Committee 2017: 87). Due to the choice of this regulatory mechanism (regulation) and not a directive, Member States may either clarify these concepts.

The second proposal (COM 2016, 289 final) is applied to traders in the sale of goods and services, who are forbidden to "block or limit customers' access to their online interface for reasons related to the nationality, place of residence or place of establishment of the customer", although the proposal establishes as a limitation "to ensure compliance with a legal requirement in Union law or in the laws of Member States in accordance with Union law" (COM 2016, 289 final: art. 3) (Trimble 2016: 53).

3 Definition of Public Service Media and Copyright

European public service broadcasting is facing an economic and model crisis with a drop in budgets, while it has tended to recovery in recent years:

> The nominal budget increase (of European public service broadcasters) was 2.7%, 33.050 million euros in 2006 to 33.960 million in 2011, but the real evolution –excluding inflation- represents a negative margin of 7.4%. The declines are produced in 2008 and 2009, while the slight recovery is observed in 2010 (Campos 2015: 197).

Against this background, the challenges of public service broadcasters lie on their adaptation to social demands, changes in their governance and funding models, and their positioning in new media (Campos et al. 2015: 9) and digital scenarios. In this regard, and in the field of copyright and related rights in the digital market, besides the measures above-mentioned, there is a clear need to rethink its definition and mission of public service, as well as to solve accountability systems.

The options to negotiate copyrights and related rights will depend on the concretion of these concepts. It has already been indicated the possibility to find a link between State aid within the meaning of the article 107.1 of the TFEU[3] and the establishment of attractive copyright rates for certain operators, if only there were a greater rates regulation, remaining under public control and being available to competent national authorities (Comisión Nacional de los Mercados y la Competencia 2015: 25) (CJEU Vent de Colére. Judgement, 19 December 2013, case 262/12).

For State aid in broadcasting to be considered included in the article 106 TFEU[4] and compatible with it they have to meet the following criteria, according to the TJEU (Casado 2008: 72):

> The service in question should be a service of general economic interest and be clearly defined as such by the Member State (definition). The company concerned should have been explicitly tasked by the Member State for the provision of such service (mission). And the application of competition rules of the Treaty (in this particular case, the ban on State aid), should prevent the compliance with the specific tasks assigned to the company, and the exception to such rules should not affect the development of exchanges to an extent contrary to the interest of the community (proportionality principle).

Therefore, if the establishment of special copyright rates for certain broadcasters were to be understood as a State aid in the terms above-mentioned, it is necessary to specify the mission and definition of public service, as well as to comply with the proportionality principle. Nowadays, however, it appears to be necessary a excessively forced interpretation of jurisprudence to consider copyright rates as a State aid.

Notwithstanding, different rates can be applied as long as the distinction is objectively justified. Indeed, an extensive case-law[5] states that if there were dissimilar

---

[3]Article 107.1 TFEU: "Save as otherwise provided in the Treaties, any aid granted by a Member State or through State resources in any form whatsoever which distorts or threatens to distort competition by favouring certain undertakings or the production of certain goods shall, in so far as it affects trade between Member States, be incompatible with the internal market".

[4]Article 106 TFEU. In the case of public undertakings and undertakings to which Member States grant special or exclusive rights, Member States shall neither enact nor maintain in force any measure contrary to the rules contained in the Treaties, in particular to those rules provided for in Article 18 and Articles 101–109. Undertakings entrusted with the operation of services of general economic interest or having the character of a revenue-producing monopoly shall be subject to the rules contained in the Treaties, in particular to the rules on competition, in so far as the application of such rules does not obstruct the performance, in law or in fact, of the particular tasks assigned to them. The development of trade must not be affected to such an extent as would be contrary to the interests of the Union.

[5]Case C-52/07 Kanal 5 Ltd. and TV 4 AB v Föreningen Svenska Tonsättares Internationella Musikbyrå (STIM) upa (Reference for a preliminary ruling from the Marknadsdomstolen).

Judgement of the Court of 14 February 1978. United Brands and United Brands Continental/Commission.

Judgement of the Court of 13 July 1989. Tournier.

Judgement of the Court of 15 March 2007. British Airways/Commission

conditions to equivalent services, this could result in a competitive disadvantage, unless such a practice may be objectively justified. The Commission, in its communication, Guidance of the Commission's enforcement priorities in Article 82 of the EC Treaty to abusive exclusionary conduct by dominant undertakings (2009/C 45/02), states that a practice will be objectively justified as long as it "is objectively necessary of by demonstrating that its conduct produces substantial efficiencies which outweigh any anti-competitive effects on consumers. In this context, the Commission will assess whether the conduct in question is indispensable and proportionate to the goal allegedly pursued by the dominant undertaking".

The concept of mission and public service are essential to objectively justify practices. In the new digital scenario, a review must be carried out to incorporate all the instruments, spaces and platforms and thus allow public service broadcasters the performance of their distinguishing function.

For the moment, the proposal for a revision of the Directive 2010/13/EU (COM/2016/287 final) extends the concept of audiovisual media services, in which are included both linear and non-linear services, but also video distribution platforms.

4 Proposals for the Challenges of the Digital Single Market

The attitudes of the media have changed in the last decade both in ways of accessing and consuming and production of contents. Public service media should continue to exercise their functions in this changing scenario by adapting themselves to the new consumption and production needs. These needs include problems derived from the regulation of copyrights and related rights, an obsolete regulation in the EU.

The main European policies aim at remodelling the management system of copyrights to reach a digital single market, a change that will affect public service media. The proposals linked to online transmissions and digital retransmissions of radio and TV programmes point to an economy in the negotiation of copyrights by operators. Furthermore, policies for the inclusion of audiovisual works in VoD services and a trend to reduce geo-blocking are extended. However, and despite the upsides of these proposals, there are also restraints that are to be unavoidably approached. On the one hand, all possible services offered through the digital market should have a presence in the measures. The EU proposals enable progress in the single rights management through the country of origin principle for online ancillary transmissions and in the enforceability of collective negotiation for digital retransmissions of radio and television programmes through IPTV. However, non-ancillary transmissions are not included in the former nor in the latter are the retransmissions through OTT platforms. Neither this package of measures includes issues of relevance as copyright, social networks and other technologies.

On the other hand, the creation of a digital single market also generates the problem of implementing the same criteria in different Member States with different national regulations. Maybe the solution does not lie in creating a single European copyright title, as it is matter that "has a direct impact on vested national interests"

(López-Tarruella 2016: 108), but its is now when we have to address the complete remodelling.

As regards public service media, it is time to take advantage of all these possibilities offered by technologies, together with the changes proposed by the EU and a rethinking of the concept of public service. The shift from Public Service Broadcasting to Public Service Media has already finished and it is now necessary to expand the concept including convergence between traditional television and all those services offered through the Internet.

Acknowledgements Results of the research are part of the project *Indicators related to broadcasters' governance, funding, accountability, innovation, quality and public service applicable to Spain in the digital context* (CSO2015-66543-P), belonging to the Spanish National Plan for Scientific and Technical Research and Innovation, a national subprogram of Knowledge Creation from the Spanish Ministry of Economy and Competitiveness.
The translation of the chapter is financed by the research group "Comunicación Pública, Poder, Derecho y Mensaje" (Public Communication, Power, Law and Message) from the Universidad de Castilla-La Mancha.

References

Campos, F.: El futuro de la television europea es híbrido, convergente y cada vez menos público. Revista Latina de Comunicación Social, 68. In: http://www.revistalatinacs.org/068/paper/970_Santiago/04_Campos.html. Accessed 09 May 2017 (2013)

Campos, F.: Financiación e indicadores de gobernanza de la radiotelevisión pública en Europa. In: Marzal, J.J., Izquierdo, J., Casero, A. (eds.). La crisis de la televisión pública. Universitat Autónoma de Barcelona, Universitat Jaume I, Universitat Pompeu Fabra, Universitat de València (2015)

Campos, F., Fernández, T., Valencia, A.: Las políticas editoriales de las radiotelevisiones públicas europeas para el uso de los nuevos medios sociales. Revista Brasileira de Políticas de Comunicaçao **6**, 7–23 (2015)

Capello, M.: Online activities of public service media: remit and financing. IRIS Special 2105–1, European Audiovisual Observatory, Strasbourg. Mentioned in the Commission Staff working document impact assessment on the modernization of EU copyright rules (SWD/2016/0301final) (2015)

Casado, L.: La configuración de la televisión de titularidad del Estado como servicio público en la Ley 17/2006, de 5 de junio, de la radio y la televisión de titularidad estatal. R.V.A.P., 80, 55–109 (2008)

Castets-Renard, C.: Digital single market: the European Commission presents its first measures in Copyright Law. French Review **7**, 388–401 (2016)

Comisión Nacional de los Mercados y la Competencia: Informe sobre el borrador de orden por la que se aprueba la metodología para la determinación de las tarifas generals en relación con la remuneración exigible por la utilización del repertorio de las entidades de gestión de derechos de propiedad intelectual. IPN/CNMC/0020/15. In: https://www.cnmc.es/expedientes/ipncnmc02015 (2015)

Del Pino, C., Aguado, E.: Internet, Televisión y Convergencia: nuevas pantallas y plataformas de contenido audiovisual en la era digital. El caso del mercado audiovisual en España. Observatorio Journal, 6(4), 057–075 (2012)

Delgado, M.: Fernández-Quijada, D: IPTV: Estructura de mercado y tipología de la oferta en España. Zer **22**, 413–428 (2007)

Ene, L., Grece, C.: Origin of films in VoD catalogues in the EU. European Audiovisual Observatory (2016). http://www.obs.coe.int/documents/205595/264625/DG+CNECT+-+Note+4-2015+-+Origin+Of+Films+In+VOD+Catalogues+In+The+EU28.pdf/

European Broadcasting Union, EBU: Copyright Guide. Practical Information for Broadcasters (2014). https://www.ebu.ch/files/live/sites/ebu/files/Publications/EBU-Legal-Copyright-Guide.pdf

European Broadcasting Union, EBU: Policy Sheet. Copyright Licensing (2016a). https://www.ebu.ch/files/live/sites/ebu/files/Publications/Policy%20sheets/Copyright%20Licensing_EN.pdf

European Broadcasting Union, EBU: More online TV programmes across borders, safeguards for contractual freedom and territoriality (2016b). https://www.ebu.ch/files/live/sites/ebu/files/Publications/Policy%20sheets/Flyer_EU%20Regulation%20on%20Boadcasters%20Online%20Transmissions.pdf

European Broadcasting Union: European Subscription Video-On-Demand subscriber to reach 50 million by 2020. Press release (2016). https://www.ebu.ch/news/2016/06/european-svod-reach-50-mil-2020

European Broadcasting Union: PSM are leading investors in European Content. Press release (2017, May 03). https://www.ebu.ch/news/2017/05/psm-are-leading-investors-in-original-european-content

European Commission: Communication from the Commission—Guidance on the Commission's enforcement priorities in applying Article 82 of the EC Treaty to abusive exclusionary conduct by dominant undertakings (2009/C 45/02) (2009)

European Commission: A Digital Single Market for Europe: Commission sets out 16 initiatives to make it happen. Press release (2015, May 06). http://europa.eu/rapid/press-release_IP-15-4919_en.htm

European Commission: Digital Single Market (2016). http://digital-agenda-data.eu/charts/analyse-one-indicator-and-compare-countries#char

European Commission: Digital Single Market: EU negotiators agree on new rules allowing Europeans to travel and enjoy online content services across borders. Press release (2017, February 07). http://europa.eu/rapid/press-release_IP-17-225_en.htm

European Commission: Communication from the Commission to the European Parliament, the Council, the European Economic and Social Committee and the Committee of the regions: Towards a modern, more European copyright framework. (COM/2015/626 final)

European Commission: Proposal for a regulation of the European Parliament and of the Council on ensuring the cross border portability of online content services in the internal market. (COM/2015/627 final)

European Commission: Proposal for a Directive of the European Parliament and of the Council amending Directive 2010/13/EU on the coordination of certain provisions laid down by law, regulation or administrative action in Member States concerning the provision of audiovisual media services in view of changing market realities. (COM/2016/287 final)

European Commission: Proposal for a regulation of the European Parliament and of the Council on addressing geo-blocking and other forms of discrimination based on customers' nationality, place of residence or place of establishment within the internal market and amending Regulation (EC) No 2006/2004 and Directive 2009/22/EC. (COM/2016/289 final)

European Commission: Proposal for a Directive of the European Parliament and of the Council on copyright in the Digital Single Market. (COM/2016/593 final)

European Commission: Proposal for a regulation of the European Parliament and of the Council laying down rules on the exercise of copyright and related rights applicable to certain online transmissions of broadcasting organisations and retransmissions of television and radio programmes. (COM/2016/594 final)

European Commission: Commission Staff working document impact assessment on the modernisation of EU copyright rules. (SWD/2016/301 final)

European Commission: Commission Staff Working. Document Executive Summary of the impact assessment on the modernization of EU Copyright rules. (SWD/2016/302 final)

European Economic and Social Committee: Proposal for a Regulation of the European Parliament and of the Council on ensuring the cross-border portability of online content services in the internal market (2017)
Flash Eurobarometer 411: Cross border access to online content (2015). http://ec.europa.eu/COMMFrontOffice/publicopinion/index.cfm/Survey/getSurveyDetail/instruments/FLASH/surveyKy/2059
Fontaine, G., Grece, C.: Measuremented of fragmented audiovisual audiences. European Audiovisual Observatory (2015). http://www.obs.coe.int/documents/205595/264625/DG+CNECT+-+Note+1-2015+-+Measurement+of+fragmented+audiovisual+audiences.pdf/4222c549-9133-4f6e-bdbb-e3bdb0d7272b
Internal Market and Consumer Protection, European Parliament: DRAFT OPINION on the proposal for a regulation of the European Parliament and of the Council laying down rules on the exercise of copyright and related rights applicable to certain online transmissions of broadcasting organizations and retransmissions of television and radio programs (2017)
IRIS Merlin database. (2016). http://merlin.obs.coe.int/cgi-bin/search.php?action=show&s[result]=short&st[]=1100
Jayakar, K., Park, E.: Emerging Frameworks for Regulation of Over-The-Top Service on Mobile Networks: An International Comparison. TPRC Conference Paper (2014). https://ssrn.com/abstract=2418792
López-Tarruella, A.: La reforma del sistema de los derechos de autor en la Unión Europea. Estado de la cuestión. Revista La Propiedad Inmaterial **22**, 101–139 (2016)
Trimble, M.: The role of geoblocking in the Internet legal landscape. IDP. Revista de Internet, Derecho y Política **23**, 45–58 (2016). https://doi.org/10.7238/idp.v0i22.3076
Urquilla, A.: El Mercado único digital será la clave del éxito para el comercio internacional en la Unión Europea. Realidad y Reflexión **42**, 39–54 (2015)
Villagra, D.R., Cavalli, A.R.: Analysis and influence of economical decisions on the quality of experience of OTT service. IEEE Latin America Transactions **14**(6), 2773–2776 (2016)

Ana María López-Cepeda Degree in Journalism and Ph.D. in Communication and Journalism from the Universidade de Santiago de Compostela (USC). Also, she has a Law degree from the Universidad Nacional de Educación a Dsitancia (UNED). She is Professor at the Faculty of Journalism from the University of Castilla-La Mancha. She is specialised in Media Structure, Public Communication Policies and Audiovisual Policies. Cuenca, Spain.

Andrea Valencia-Bermúdez BA in Journalism and Ph.D. candidate in Communication and Contemporary Information from the Universidade de Santiago de Compostela. Also, she is finishing her studies in Applied Languages and Translation at the University of Vic. She is entrepreneur and collaborates as external researcher with the European Broadcasting Union (EBU). Palma, Spain.

Corporate Communication, Marketing, and Video Games

Carmen Costa-Sánchez and Barbara Fontela Baró

Abstract In 2009, the video game industry received the consideration of cultural industry in Spain. From that moment until now, numerous video game development companies have emerged and aimed to consolidate the business sector as well as distribute and position their products. Marketing and corporate communication functions can help to make the creative proposals of the businesses visible, and the sector is demanding professional profiles with these specializations. However, it is a field that has been little approached from research and teaching perspectives. This work seeks to know the characteristics of this cultural industry in Spain and focus the attention in the new professional field that is emerging for communication, marketing and video game experts.

Keywords Corporate communication · Marketing · Video games · Public relations · Business communication

1 The Video Game Industry in Spain. Current Situation

On March 25th 2009, the Culture Commission of the Spanish Congress unanimously voted a PSOE initiative that recognized Video games as a Cultural Industry. This action positioned the gaming sector at the same level of other cultural industries like cinema that traditionally enjoyed the protection, promotion and public subsidies. Despite the importance of this recognition, other determinants have slowed down the industry's takeoff, such as the lack of tradition, the absence of any specific training, and the need of a big amount of funds to start large own projects.

C. Costa-Sánchez (✉) · B. Fontela Baró
Universidade da Coruña, A Coruña, Spain
e-mail: carmen.costa@udc.es

B. Fontela Baró
IE Business School, Madrid, Spain
e-mail: barbara_fb94@hotmail.com

M. Túñez-López et al. (eds.), *Communication: Innovation & Quality*, Studies in Systems, Decision and Control 154, https://doi.org/10.1007/978-3-319-91860-0_25

An imbalance between the role of Spain as consumer market and as content producer is detected. In fact, Spain is the fourth/fifth European market in terms of consumption (following UK, Germany, France and Italy). However, only 2% of the products that Spain consumes are produced in the country. In a year with big revenues, as 2007, only 150 games were developed in Spain (Rodríguez and Pestano 2012).

Additionally, the economic crisis has also had consequences for this cultural industry. The revenue evolution of the sector from the year 2000 shows how in 2007 revenues reach a peak (1454 million euros) and start its decline, achieving only a 1083 million euros consumption in Spain in 2015 (AEVI 2015a, b, 2016).

As it has been explained in previous papers, the video game industry has been recently recognized as a cultural industry and the first actions to protect it are being hopeless and slow because of the crisis that affects the whole of the cultural industries, harmed by the austerity measures imposed by the Government (Pérez Rufí 2015).

There are currently 480 videogame development companies in Spain. The picture of their main characteristics shows that most of them are microenterprises (they have less than 10 employees) and young (the majority them have less than 5 years of history). Cataluña (24.8%) and Madrid (26.2%) are the hubs that concentrate these organizations, followed by Valencia (11.4%) and Andalucía (8.6%). However, there are companies that have consolidated in the sector, as Pyro Studios (the creator of Comandos, which sold 5 million copies), FX Interactive (Imperivm, Runaway 2, Sparta), Mercury Steam, Virtual Toys, and Novarama (Invizimals). The Spanish gaming industry achieved a 32% increment in its workforce in 2015, reaching the number of 4460 professionals, only 17% of which are women (DEV 2016).

Despite the growth rate is high, the Spanish companies face difficulties originated by the aforementioned causes and the singularities of the sector, which is an oligopoly of American and Japanese companies. Some of the main difficulties that obstruct the business activity of the national companies in this sector are:

In first place, the difficult distribution of the product in an oligopoly, where the *majors* produce and distribute games regardless of being their own product (originated in their studio) or externalized (arranged with a third party).

Secondly, launching a product needs a large volume of funding. The little Spanish studios necessarily and generally go for mobile projects that require more limited investment in human and technical resources, as well as less developing time.

The cost of a big project is higher than the cost of a film. It is not accessible for a small business. In a market captured by multinationals, a high number of copies sold and high expenses for commercialization are required to achieve profitability (Checa Godoy 2009). This element is related to one of the singularities that is common to all cultural industries: the big uncertainty about the success of the product. Nothing guarantees that the game will achieve enough success to recover the costs. If we also add high production costs, the equation becomes more complicated for small companies that can't afford to balance out successes and failures. From here on, the need of larger investment—to acquire intellectual property rights, to produce quality products, to promote the products, etc.—could lead to the creation of products that can be easily amortized among thousands of users of diverse characteristics; that is, products thought for broad and large global markets, which could lead to a harmful effect of homogenization of culture and production (De Aguilera 2004).

In third place is the lack of a planned and strategic institutional support. In an international comparison, Canada has become a globally relevant industrial player, besides United States and Japan. In the European level, United Kingdom and France have transformed the video game industry into a strategic sector. In Spain, besides the fact that the cultural industry recognition is recent, the institutional support belongs to a bigger framework of support to the cultural and creative industries that are related to technology. This support has a strategic potential that still should be developed. There are subsidies at the regional level (each Autonomous Community allocates a different budget to them), and at the national level (which are divided between the Culture and Industry Ministries). The association of video game development companies focus its attention in the fiscal policies that are already applied to the film industry and that would boost the sector (DEV 2016)

In fourth place is the training factor. Despite Spanish universities have recently added courses related to the industry to their curriculum, and specific professional and postgraduate training is rising, this is a recent and novel phenomenon that needs time to explore its way. “The problem is the lack of consistency between the current offer and the needs of the companies—despite in the last years some professional schools have launched some professional game developer training offer—which is added to the lack of sensibility in the government administration to accelerate the approval of new degrees that are being created after the Bologna Plan” (Iglesias Redondo 2015: 199). In other side, if the market can’t absorb specialized professionals, they will leave the country and integrate as human resources in the industries of other nations.

2 Effective and Potential Audiences. The Video Games Users

Video games have experimented an interesting evolution regarding its relationship with the audience and its social legitimization. At first, the gamer was stereotyped as a lonely and nocturnal teenager that would spend the free time playing in the computer, reaching the point to associate gamers with unhealthy and violent behaviors.

Since then, the concept has been normalized. The range of the audiences has expanded (women or elderly, for example, that had been apart from the new cultural product until now) and its potential for training and simulation is being studied.

In relation to this normalization, the gamification phenomenon has spread and it is now applied to a big variety of fields, such as business, education or public relations (Costa-Sánchez 2014). The evolution of the social understanding of video games is a good example of the improvement of the reputation of a sector that was demonized and has become a nice entertainment option for children and adults.

Regarding users and according to the data from ISFE (the Interactive Software Federation of Europe), there are 15 million video game users in Spain, which is

Table 1 Best-selling videogames and their platform as December 2016

Ranking	Game	Platform
1	FIFA 17	PS4
2	POKEMON SUN	NINTENDO 3DS
3	POKEMON MOON	NINTENDO 3DS
4	JUST DANCE 2017	WII
5	SUPER MARIO MAKER	NINTENDO 3DS
6	GRAND THEFT AUTO V	PS4
7	FINAL FANTASY XV	PS4
8	WATCH DOGS 2	PS4
9	BATTLEFIELD 1	PS4
10	CALL OF DUTY: INFINITE WARFARE	PS4

42% of the total population of the country. This number positions our country in the top four of the European nations with more gaming users along with France (62%), Germany (52%) and United Kingdom (40%). Spanish gamers devote 6.2 weekly hours in average to this activity.

In addition, playing video games is not only for children or teenagers anymore. According to a recent study published by AEVI, "Videogames and adults", almost 40% (38.9%) of the adult Spanish population plays video games: 45.3% of men and 32.8% of women. From the total, 26.2% of them declared being frequent players. Regarding the preferred device, Spanish users prefer the desktop game console (61.1%). Men use more the desktop console (66.2%) and women use more the Smartphone (57.6%). Among the conclusions of the study stands out that the adult gaming population will grow because of a simple reason: young parents are familiarized with technology.

The industry differentiates two main gamer profiles: the hardcore or traditional gamer and the casual gamer. The first one plays many hours and has years of experience. He or she is demanding with the games and the brands, is well informed, knows the market, and shares information in the Internet. He or she also has a double consumption behavior: buys more games and more expensive ones, but also pirates more. The second one plays only a few hours, is novice, and has probably started playing with the Wii or Nintendo DS. Regarding game preferences, the casual player is more malleable because he or she follows superficial information (generalist, packaging…). Concerning buying behavior, the casual player acquires less and cheaper games, however, he or she doesn't pirate and has bought a lot of consoles.

The ranking of best-selling games in the past years in Spain points to the success of football franchises—specifically FIFA—action franchises—in particular Call of Duty or GTAV—as well as some casual games such as Just Dance. Sony platforms are the favorite ones of Spanish gamers, followed by Nintendo and Wii (Table 1).

The best-selling list in the past years doesn't leave room for Spanish companies. Top places in the rankings are taken by successful franchises with loyal users that assure a good reception. The challenge for the Spanish companies is to make the product reach its potential users at a global level. Given than companies are facing a rich and varied offer, sometimes even free; the product positioning has become one step more in the process of developing a game, causing an important demand of specialized professionals in this area.

3 Corporate Communication and Marketing Applied to the Video Game Industry. Foundations and Tools

Corporate Communication may be a convenient ally for the gaming sector. As other sectors have confirmed, it is a strategic management tool that allows to reach business goals in a highly saturated and competitive market. The goal of corporate communication is to "achieve sufficient reputation for the company and stablish quality and continuous relationships with stakeholders. In short, the goal is to achieve everyone's trust" (Dircom 2013: 36).

Given that gaming is a young industry in Spain, new studios can find the solution to different needs in Corporate Communication.

In first place is the need of visibility. The digital market is saturated by initiatives developed by majors, indies or fans, so catching customers' attention is the main business goal. The AEVI secretary acknowledged this need when he explained that the industry "needs support that gives access to promotion channels, media, funding for projects—both private and public—sponsorships of editors that have a relationship with the game developer companies" (Iglesias Redondo 2015: 200). A good product can be unnoticed in a context of overproduction. Communication may contribute to create awareness for brands, professionals and products.

In second place is the need for differentiation. Uncertainty is always threatening the cultural industries. There isn't any guarantee of significant success. Communication can translate into added value that differentiate the product from competitors. Exploiting the differentiation as a uniqueness factor that helps engage users or potential customers could be useful to minimize the uncertainty risk.

Third is the need for positioning. The digital product needs to be positioned in the online stores for applications and games, or in the online gaming platforms. Marketing and communication techniques need to be combined to achieve this objective.

The need to manage the institutional and business relationships is forth. The relationships and synergies between small companies can contribute to the benefit of all. Besides, given that gaming is a sector with no tradition in Spain, relationships management is especially important to create awareness and networks. The relationships with institutional organizations are also important because they can become customers at certain point, as well as because they can contribute to the improvement of the sector (by means of subsidies, debate forums, etc.)

All these needs are integrated in traditional corporate communication goals. However, the characteristics and specifications of the sector are changing. Next, some traditional Corporate Communication tools applied to the gaming sector in Spain are reviewed.

3.1 Media Relations

External Communication is formed by a group of messages sent by any organization to its external audiences to maintain or enhance its relationships with them, project a positive image or promote its activities, products or services (Andrade 2005).

Media relations are one of the most outlined external communication activities. They are relationships that allow to create awareness of an informative speech in society through media. This is known as publicity: information given by an external source that media employs because it has informative value. Contrary to advertisement, this channel is not controlled by the company because media is not paid for diffusing the message (Túñez-López 2012). The advantages of media relations,—opposite to traditional advertisement—are the credibility, the ease of persuasion, and the broader information that it offers.

Press releases and press conferences are the two most used publicity techniques, and nowadays they are the two main ways to access the media agenda. The use of press releases by communication departments is now consolidated and they are integrated in the routine of journalists.

The press release is equivalent to a new sent by the organization. However, it is written from a journalistic point of view. Nowadays, the possibility of a press release 2.0 is being discussed. In comparison to the traditional document (1.0), it would adopt the characteristics of the multimedia and hyper textual language, and it would be shareable in social media (Costa-Sánchez and Piñeiro-Otero 2013).

The gaming industry is interested in one hand in the relationship with the generalist media, that usually have a section related to technology or new trends in their publications or shows. On the other hand, it is also interested in the relationships with the journalists, media and channels specialized in video games as primary resource. The characteristic of the target audience of the product should also be considered, since media content is basically information about it. For example, when talking about a game designed for women, specialized publications targeted to women can be an appropriate option.

As Muñoz and Sebastián (2010) explain, the daily task of a communication agency in the video game industry is to keep the media informed about the news that can arise in relation to the studio, developer, distributor, and product in progress. According to the specialized journalist Maeso (2015), the marketing and communication media relations are essential tasks in the companies in this sector. Designing specific opportunities, materials and content for specialized press and different ones for generalist press is a good communication strategy for the PR teams.

Íñigo Vinós (2013), from the agency Logic Activity, states that when talking about a specific game, three or more events that can be communicated can take place: the new, the preview and the review. The new reveals that the product is under development; the preview shows pictures of the game along with information prior to the launching; and the review includes all the available material (pictures, videos, product sheets, etc.) and happens when the game is ready to be analyzed by specialists. The main obstacles of this job are two. First is the lack of differentiation between advertisement and communication in some media, which causes the loss of the trust of the users, that end up being suspicious of the messages. Second are the difficulties to obtain visibility in a globalized market with no barriers of entry, in which the Spanish talent design, coding, and graphics talent—is not in its rightful place.

One of the baseline specialized media for the Spanish industry is Hobby Consolas magazine, which is edited in printed and online formats. The magazine was founded in 1991 and is currently edited by Axel Springer Spain. The publication is specialized in current news about the gaming industry. Besides news, it includes analysis of the launchings, in depth reviews, and ratings of each game. It aims to provide all the necessary information to make the purchase decision to the reader. It is published monthly and in Spain it has 234,000 readers (EGM).

On the other hand, nowadays there is a big variety of online websites that are specialized in this topic. It is possible to outline the case of 3djuegos, a website launched in 2005 which is self-defined as "a magazine about video games completely open to new ideas and to all types of players. A community that is being built as a team and is focused in video games, the culture of gaming". The big number of blogs and websites focused on the topic may help to spread information in the press release format, although they may be overlooked in comparison with other journalistic publications.

Media relations must be continuous and be watched. As a game is developed, specialized media are informed. They are taken to the studio to see the evolution of the game and to talk with the team that is creating it, they try beta versions before it is released to the market, they are invited to launching events… specialized press tracks in detail the whole process.

The challenge is to jump to the generalist media, especially not specialized shows in radio and television. In this case, it is required to explore creative options, organize an event that attracts attention, or look for an unusual picture to communicate to them.

On the other hand, local media are still an interesting platform to reach society. Small companies may have better access to local offices and media in their local network. Turning local media into collaborators could become an appealing *publicity* tool to position the company in society. Even gaming products are launched globally, and the impact of the local media is smaller, their informational role shouldn't be underestimated, and either their role as links to catch the attention of bigger generalist media.

Regarding press conferences, in the videogame industry these are related to the launching of the products, and they are usually specialized events or fairs. This topic will be covered in the corresponding section.

Table 2 Most subscribed YouTube channels in Spain

Ranking	User	Subscribers
1	elrubiusOMG	24,435,833
2	Vegeta777	18,117,863
3	TheWillyrex	11,932,017
4	iTownGamePlay	8,376,869
5	Willyrex	8,242,378
6	aLexBY11	6,806,572
7	luzugames	6,688540
8	El Rincón de Giorgio	6,534,402
9	Toys and Funny Kids Surprise	6,499,378
10	ExpCaseros	6,434,147

Source Socialblade

3.2 **Influencer** ***Relations***

Halfway between marketing and public relations, the relationships with influencers, essentially youtubers, have become a way to introduce, differentiate, and prescribe products, especially video games. It should be outlined that gaming is the topic of the moment in the content created by the most subscribed youtubers in Spain, such as ElrubiusOMG, Vegeta777, Auronplay, and TheAlvaro845 (Table 2).

When a company wants to spend part of its budget in influencers, the communication agency or the brand contact the specialized agencies which perform a strategic search for the profiles that best match the brand (Benítez and Navas 2016).

Even the recommendation of an *influencer* can be spontaneous, the video game companies usually send exclusive material for them to know the game and recommend it. The videos made by *influencers* show how they play, explain tips to complete levels and comment on interesting facts related to the game. Their position as *influencers* causes that the followers that emotionally and rationally connect with them also connect with their recommendations.

On the other hand, brands should consider that gaming content is successful in YouTube because it raises interest among the gaming fans but also among non-gamers. A 2014 Google survey among consumers that reported watching video game content in YouTube revealed that only part of this group (37%) saw themselves as "gamers". When they were asked about their motivations to watch videos about games they mentioned "entertainment" and "humor" as well as "strategy learning" and "advice about games". YouTube can also be an appropriate channel to promote the business, brand, product, or relationships with the audience because it creates a favorable environment for the use of the product with potential users.

Moreover, this can also mean that companies and brands in the gaming industry can form alliances with other brands that target young segments to engage those audiences that are less used to traditional media (Burger-Helmchen 2014).

The content about video games in YouTube is rich and varied: news and reviews, instructive videos, users that make comments while they play, and the streaming of competitions. The video guides that help other players defeat their enemies, find treasures and improve their pace are also very popular in YouTube. This genre is called "Let's Play" and proposes an experience in which the viewer attends the journey of the YouTuber along with him or her, allowing the observer to live the gamer's successes and mistakes and apply different strategies.

Small companies find in YouTube an ally with great potential. Given that their audience is in YouTube, they should have their own channel in this network, either supported by influencers or not. On the other hand, as it is usual when lacking a big budget, creativity may help them achieve product awareness, differentiation and affinity goals.

3.3 Planning and Participating in Events

During a year, there are numerous big events, exhibitions, meetings, and fairs at the national and international level that have the video game industry as protagonist. These events are very important for the companies in the sector: editors, developers, majors, and independent ones. From a marketing point of view, the main goals of these events are (Muñiz 2010):

- Launching new products.
- Strengthening the reputation of the company.
- Building or expanding the list of potential customers.
- Reinforcing the sales process.
- Selling.
- Reaffirming the relationships with current and future customers.
- Studying the market and the purchase decision making of the visitors.
- Studying competitors.

The most famous fair in the world is the Electronic Entertainment Expo, known as E3. Yearly organized at the Los Angeles Convention Center, it gathers the main actors of the sector since 1995 during one week in which the fans from all over the world expect launchings of new games.

Professionals from the industry such as producers, distributors, sellers, journalists, and developers have preferred access to the event, although it has recently open some of the presentations to the public. The big announcements of the industry have historically taken place at the E3. For big companies, the stage is an area in which they plan every movement and announcement to reinforce their sales in the next months. Those E3 in which new hardware is launched are especially renowned. Big companies such as Sony, Nintendo, Microsoft and editors such as Activision,

Electronic Arts, Ubisoft, Bandai Namco or Square-Enix show their main proposals for the rest of the year and the next season. It is considered the most important gaming fair in the world.

Tokyo Game Show (Japan) or Gamescom (Germany) are other of the most important international events.

Spain also has baseline events for the companies in the industry such as GameLab (Barcelona), the Video Game and Interactive Entertainment International Congress. The meeting includes talks from experts and industry professionals and the opportunity to launch new products. In 2016, GameLab speakers were industry professionals such as Ron Gilbert, Mokey Island creator; John Romero, author of the first Doom; and Enric Álvarez, director of the Spanish studio MercurySteam. In fact, the first gameplay of Raiders of the Broken Planet, the new game from this Spanish company, was presented in the event. Besides talks, the event includes an Indie Hub, an area in which independent Spanish studios show their games under development [hobbyconsolas]. In addition to the distinguished international speakers attending, GameLab is a meeting point for small and independent studios and professionals.

GameLab in collaboration with the Spanish Interactive Arts and Sciences Academy presents the National Awards of the industry since 2008. These awards are the major recognition in the country for gaming companies and professionals. The jury is formed by professionals that are members of the event organization, and it decides which ones have been the best projects of the year in different categories such as graphic art, technology, game design, sound, and music. Being one of the awarded projects secures the attention of the media and the customers and it is a guarantee and a good cover letter to start relationships at higher levels.

Madrid Games Week, other of the yearly events of the industry organized by AEVI (Spanish Video Game Association) will this year be transferred to Barcelona and renamed as Barcelona Games World. There is little information about its content but it's expected to have a large exhibition in the Montjuic facility with big eSports arenas and an outdoors Party area. The organizers have confirmed that the first day will only welcome professionals and media and the rest of the days will be open to the public.

The number of events celebrated every year in and outside Spain are an opportunity for companies to stablish alliances for future projects and create awareness for their products and specializations. They are also events that call the attention of the media, being a good platform to introduce the company and its professionals.

The talks that they include in their agendas can be organized as debates, panels or the traditional press conference format. However, press conferences that create most interest are the ones from big brands and high executives from the sector. Moreover, it can't be ignored that businesses have now more channels to directly contact journalists and potential or effective customers under the new 2.0 paradigm, making campaigns and trailers more constant and not exclusive for big events in which standing out among big projects is a challenge.

3.4 Business and Institutional Relations

Communication planning in this industry will also require the development of strategic business and institutional relationships. The smallest studios and companies can find potential collaborators in other organizations (businesses, institutions, associations, universities, etc.) to develop novel and creative ideas, recycle efforts and find synergies, potential funding or sponsorship, or even new interests for their customers.

The recent normalization of video games as a cultural product in Spain causes that this process, that has already started, will progress. The companies from the industry should help to continue this path. On the other hand, Universities are driving new studies oriented to train industry professional. This relationship can bring mutual benefits in the medium and long term.

Independently, gamification as a transversal process to involve participants in reaching goals is approved by professionals and scholars (McGonigal 2011). Starting a line of research or work around it could yield successful results with important effects beyond the gaming industry.

4 Corporate Communication as a Part of the Management Strategy

Developing an attractive and interesting product with a good design is not enough. Big studios and big companies in the video game industry know it and it is easy for them to allocate a large part of their budget for marketing and communication functions to accompany strategic activities and each of the new launchings.

For smaller companies, marketing and communication get less attention in comparison with product design and development. However, they need to be considered as functions included in the creative processes since the start of it, just as defining the target is something they consider from the beginning.

In a context that is saturated with new applications, video games, gaming technologies, entertainment, and education, looking for a place to position a product is risky for the company, that should use corporate communication as a strategic tool to grow. Relationships with the customers, awareness and positioning are usually outcomes of a good communication strategy. Therefore, the industry is demanding more professionals that know the foundations of communication in a field of high specialization.

On the other hand, higher education institutions consider the gaming industry an employment sector for the present and the future, and should drive efforts to train new corporate communication and sales and marketing professionals in this area. As two sides of the same coin, their success is linked.

In order to be practical and delimit the scope of the study, the current work has been limited to explore some of the aspects of external communication that could be of interest for the companies in the gaming industry. The potential of internal communication has not been addressed, even it also exists. Small companies challenged by survival might find among these lines the fundamentals for a new mentality that integrates corporate communication as part of their strategic business and product management. The continuation of this line of reflection is a need for professionals and scholars.

References

Andrade, H.: Comunicación organizacional interna: proceso, disciplina y técnica. Netbiblo, A Coruña (2005)

AEVI.: Estudio Videojuegos y Adultos. AEVI, Madrid (2015a)

AEVI.: Anuario del videojuego. AEVI, Madrid (2015b)

AEVI.: Anuario del videojuego. AEVI, Madrid (2016)

Benítez, A., Navas, V.: El fenómeno fan y la figura de influencers en redes sociales. Universidad de Sevilla, Sevilla (2016)

Burger-Helmchen, T.: Communities of players: scale and scope, and the effect of social medias in the video game industry. J. Commun. Res. **6**(2), 153–174 (2014)

Checa Godoy, A.: Hacia una industria española del videojuego. Comunicación **7**(1), 177–188 (2009)

Costa-Sánchez, C.: La Narrativa Transmedia como aliada de la Comunicación Corporativa: estudio del caso# Dropped by Heineken. Commun. Soc. **27**(3), 127–150 (2014)

Costa-Sánchez, C., Piñeiro-Otero, T.: Estrategias de comunicación multimedia. UOC, Barcelona (2013)

De Aguilera, M.: La institucionalización de una industria cultural. Estructura y desafíos de la industria de los videojuegos. Telos **59**, 38–44 (2004)

DEV.: Libro blanco del desarrollo español de videojuegos 2016. DEV, Madrid (2016)

Dircom.: Manual de la comunicación. Dircom, Asociación de Directivos de la Comunicación, Madrid (2013)

Iglesias Redondo, C.: Retos del desarrollo de videojuegos en España. adComunica **9**, 197–201 (2015)

Maeso, G.: El papel de los medios de comunicación. In: DEV, Libro blanco del desarrollo español de videojuegos 2015, pp. 65–67 (2015)

McGonigal, J.: Reality is Broken. Penguin Press HC, London (2011)

Muñiz, R.: Marketing en el siglo XXI. Centro de Estudios Financieros, Madrid (2010)

Muñoz, D., Sebastián, A.: La relación de los videojuegos con los medios de comunicación. In: Carrillo, J., Sebastián, A. (coord.) Marketing Hero. Las herramientas comerciales de los videojuegos, pp. 201–232 (2010)

Pérez Rufí, P.: Modelos de producción del videojuego en España. adComunica. Revista de Estrategias, Tendencias e Innovación en Comunicación **9**, 97–115 (2015)

Rodríguez, V., Pestano, J.M.: Los videojuegos en España: una industria cultural incipiente. Ámbitos **21**, 361–379 (2012)

Túñez-López, J.M.: La gestión de la comunicación en las organizaciones. Comunicación Social, Sevilla (2012)

Vinós, I.: Comunicar en el sector de los videojuegos. Topcomunicación. In: goo.gl/iyJcEA (2013)

Carmen Costa-Sánchez Professor of Corporate Communication in the Graduate studies of Audiovisual Communication of Universidade da Coruña. Doctor in Communication by Universidade de Santiago de Compostela (USC). Extraordinary Doctorate Award by the Faculty of Communication Sciences of USC. A Coruña, Spain.

Bárbara Fontela Baró Student at the IE Business School Master Management. She previously received grants from the Amancio Ortega Foundation and the Pedro Barrié de la Maza Foundation for studying abroad in Seattle, USA. She has been Corporate Communication Department Intern at Gas Natural Fenosa and Research Intern at Universidade de A Coruña. Madrid, Spain.

Virtual Tourist Communities and Online Travel Communities

María-Magdalena Rodríguez-Fernández, Eva Sánchez-Amboage and Oscar Juanatey-Boga

Abstract The implementation of information and communication technologies in the tourist industry has resulted into a change of paradigm, both for companies, when designing their communication strategies and for consumers, when planning their holidays. The Internet has become the major source of information for travellers, as well as a relevant platform to develop tourism businesses. In this regard, the 2.0 web provides mechanisms to contact and encourage consumer's loyalty for tourism companies, through the use and adaptation of software applications and programs, where social media play a relevant role. Behind this context and due to the intangibility of tourist products and services, the eWOM has an impact on consumer's behaviour, so companies should be aware of this and adequate their strategies with a view to satisfy an increasingly informed, demanding and experienced tourist. All these changes in the sector have made online travel communities a reference for tourists when planning and managing their travel reservations.

Keywords Online communication · Tourism · Consumer behaviour · eWom Virtual communities

M.-M. Rodríguez-Fernández (✉) · E. Sánchez-Amboage · O. Juanatey-Boga
Universidad de A Coruña, A Coruña, Spain
e-mail: magdalena.rodriguez@udc.es

E. Sánchez-Amboage
e-mail: eva.amboaxe@gmail.com

O. Juanatey-Boga
e-mail: oscarjb@udc.es

M. Túñez-López et al. (eds.), *Communication: Innovation & Quality*, Studies in Systems, Decision and Control 154, https://doi.org/10.1007/978-3-319-91860-0_26

1 Tourist Communication Online

If we look back and we take stock of all changes within this context, as a result of the influence of Information and Communication Technologies (ICT), we can better understand that business organizations and consumers are not the same. Information and communication technologies have evolved rapidly and revolutionized all aspects of social and economic life.

Companies have had to adapt themselves to new virtual environments and to design different communication strategies, so as to be competitive. On the other hand, consumers have become the key element of organizations. They have a proactive attitude instead of a reactive one, which affects their behaviour and, hence, the decision-making process when purchasing products and services.

Against this background, Sanagustín (2010, as quoted in Celaya 2011), identifies the factors that have transformed the model of business communication, including:

- The increase of smarter consumers, called "*crossumers*", a certain type of consumers that investigate further before purchasing, question what brands say and have previously certain marketing knowledge.
- "Infoxication", a concept that represents the information overload to which we are subjected.
- The economy of attention, where it is necessary to compete for the attention of consumers.
- "Influencers", people of interest in their sphere of influence aided by social media to extend their recommendations.
- Trust, a consequence of the loss of credibility in traditional media. It is preferable to listen to individuals than representatives of companies.
- Network consumers, due to the loss of TV audience against online media.

Extrapolated all these factors to a complex sector as tourism, with different specificities due to the intangibility of their products, we can thus understand the impact they had on tourism, where the ICT have also played a key role.

Abuin (2016) points out that the constant updating as regards the use of Information and communication technologies is essential to achieve a more effective communication on tourism products or services. Also, he argues that it is necessary to take advantage of all available channels to reach the potential tourist.

In this regard, before discussing about tourism communication, it is appropriate to previously mention the web 2.0., which has been somewhat the responsible of many of changes affecting tourism, as it has helped the tourist to become the main protagonist of the network. This is the reason why designing communication strategies is no longer so simple. The web 2.0. shows evidence of a "period in which the web starts to be used for which it was created: sharing, collaborating, providing, editing and, above all, communicating people with people" (Marín 2010).

Tourism communication is complex, so making proper planning of their own strategies becomes a primary task. Current environments somewhat help companies and tourist destinations to develop communication actions to create, maintain or improve their image, and influence the decision-making process of tourists, raise their interest, obtain their loyalty and ensure that they have a rewarding experience to make them good opinion leaders.

In this context, online communication in tourism has become a key element. The Internet and social media has revolutionized the way to reach the consumer in tourism. The multiplicity of contents, which is possible through the Internet, has reduced travellers' threshold of attention towards traditional communication (Ejarque 2016).

When discussing about the Internet, we should think that we are dealing with one of the most widely used ICT in the world, whose capacity enable the immediate contact between people, the knowledge exchange, as well as an enormous amount of advantages both at industry and personal level (Sánchez 2015). Therefore, the Internet is both a communication and an information storage tool (Obercom 2014).

Gonçalves et al. (2013) reveal that the use of the Internet as a strategy for strengthening the brand in tourism has three benefits: (1) economic, via website promotions and the increase of demand of a certain product; (2) informational, by directly contacting consumers and collecting their opinions; and (3) branding, by contributing to position the brand, making it more attractive for the customer.

As regards social media, it must be pointed out that they have become the true protagonists of the beginning of the 21st century. "Media that, in their different formats, facilitate access to users' segments that can be classified according to their interests, concerns, feelings, ideologies, etc." (Gomis 2009). These websites meeting points for Internet-users that maintain their followers updated and informed through emotional states, pictures, queries and games.

According to Hudson and Thal (2013), the consumer decision process has radically changed and, in the last decade, it has emerged a more sophisticated vision of how consumers engage with brands.

It is true that social media in tourism play an increasingly important role as platforms for personal travel reviews and where stories, warnings, advices and recommendations have an impact on purchasing decisions of a tourist product or service, even modifying the perception on those products or services of future travellers (Kang and Schuett 2014). Likewise, Howison et al. (2014) allude to the impact they have to give visibility to tourism products and companies, being thus considered essential complements to websites.

The social media phenomenon has been key to implement a new way to connect and contact with the tourist-customer, a more direct, customized, participative and faster way, where the image of a company or tourist destination becomes increasingly important.

All these facts have had major impact both on tourism companies when designing their online communication strategies and on their consumers when selecting a product, service or tourism destination. This is why the section below deals with the changes in the tourist as a consumer.

2 The Current Tourist Consumer

The characteristics of the current tourist as a consumer bear little resemblance with those that had in the '70s and the '80s of the previous century.

The changes occurred in the tourism market have led to great variations in their behaviour (Mediano 2002), which have provoked the appearance of distinct tourist groups.

Referring that way to the tourist as a consumer implies talking about both the process of acquiring the tourism product and the participation in the tourism experience, the latter influenced by internal and external factors (Valls 1996).

In this way, various changes have an impact on their behaviour (Serra 2011):

- Population ageing
- Time poverty or lack of time
- Greater environmental responsibility
- Direct ratio between the disposable income and the tourism demand
- Greater segmentation or fragmentation of the market
- Significant use of ICTs
- Reduction in the length of journeys
- Reduction in the time between the reservation and the beginning of the process
- Boom in leisure tourism
- Decrease in the purchase of tourist packages.

As established by Mediano (2002), the changes experienced by the tourist as a consumer in this new millennium may be grouped into three largest groups:

- Changes in the main motivation of the trip
- Changes in products and requested destinations
- Changes in the form of tourism consumption.

In this regard, it can be affirmed that the current tourist consumer has evolved, becoming a new user, who is more informed, demanding and experimented. Tourism influential exercise great power when planning and managing the reservation of products and services.

In their decision-making process, tourists are increasingly guided by recommendations of family members, acquaintances, and information they receive through the Internet and social media (Rodríguez-Fernández et al. 2017).

These new tourists are much more accurate and have higher expectations when travelling. They are travellers who seek to live and feel the tourist destination, as well as to exchange cultural experiences and emotions with other travellers (López Carrillo 2011).

The technological revolution has been imposed and tourists rely more on opinions and reviews made by network experimented consumers than information from companies or tourist destinations.

Caro et al. (2014), López Carrillo (2011) and Marín (2010), indicate that this 2.0 tourist seek the information in many ways and use technology before, during and after the trip.

- Before the trip, collecting information, prices and opinions to decide on purchasing.
- During the trip, communicating their perceptions using social networks, blogs and messaging systems.
- After the trip, sharing their experiences, which generate opinions and reputation of the destination.

Internet users take on an especially important role, exchange information on their trips and this way they influence other users' opinions, who participate in an online conversation and who, in the future, could act based on suggestions, not only from friends or acquaintances, but also from crowds of tourists that post daily pictures, videos and comments on the Internet (Di Placido 2010).

To that extent we are faced with a new tourist typology, independent travellers, who refuse the services offered by travel agents and prefer to arrange their own itineraries (Richards and Wilson 2004). However, one should not ignore that, when planning their journey, recommendations outlined by other tourists on influential social media are a reliable and decisive information channel when choosing a product, service and tourist destination.

It can be concluded that we are faced with a new tourist user with different motivations, needs and behaviours that have revolutionized the sector.

3 The e-WOM and Online Travel Communities

At this point and after reflecting on the changes experienced by the tourist consumer, it should be noted that the e-WOM has become the main source of information before, during and after the trip, since opinions from other tourists that have already enjoyed the experience have proved to generate greater trust and credibility.

In this regard, the extension of the WOM—Word Of Mouth—through the Internet and the so-called 2.0 web is known as *eWOM* (Electronic Word Of Mouth) (Gruen et al. 2006). The *eWOM* represents "all informal communications directed at other consumers about the ownership, usage, or characteristics of particular goods and services or their sellers" (Litvin et al. 2008).

According to numerous studies (Hong 2006; Karakaya and Barnes 2010; Lee et al. 2008; Steffes and Burgee 2009), a positive *eWOM* will generate positive behaviours and these greater purchasing opportunities. Otherwise, a negative eWOM will produce far-reaching effects in the tourist sector (Balagué et al. 2016).

Likewise, the information transmitted on reviews of tourist products and services on the Internet, through an organized system of opinions and scores, is becoming increasingly important and opens a new window of opportunities for all participants (Dellarocas 2003; Pan et al. 2007).

Against this background, it is understandable that online travel communities had become a major tourist attraction for tourists when seeking opinions on potential tourist destinations, services to enjoy (hotels, restaurants, etc.), and receiving information on experiences lived by travellers (Casaló et al. 2012).

To understand the functioning of online travel communities and how they influence on the tourism sector, it is necessary to previously indicate the classification of social media presented in Fig. 1.

According to the contribution by Celaya (2008), and as shown in the above figure, there are three categories of social media or virtual social networks:

- Generalist, that allow free and spontaneous participation and interaction between their members (i.e. microblogging as Twitter, video as YouTube, photography as Instagram, social as Facebook, music as Spotify, messaging as WhatsApp). In this type of social media, the category of "Transports and Tourism" represent 21% of user searches (IAB Spain 2016).
- Professional, able to establish different categories of professionals (i.e. LinkedIn). Tourist companies use them to offer information on the company or to recruit, for example, new employees.
- Specialised, which offer concrete information and a specific environment to communicate and develop both professional and personal activities (i.e. TripAdvisor.com, Minube and Booking.com as regards the tourism sector).

Behind this context and in view of the above-mentioned, virtual communities may work in two different ways:

In the case of generalist social media such as Facebook and Instagram, the virtual community participates in the content that has previously posted by the company managing the profile, through comments, likes and/or shares. In some cases, it may happen that the user of the community generates information within that context, although it is less common. On the other hand, we find specialised social media in tourism, such as Booking.com and TripAdvisor.com, whose users provide content to the community. The company that manages these media uses recommendations, opinions and comments of users as a basis for the sale of their tourist products and services, such as hotel reservations and restaurants.

All these elements of socialization and interaction between customers and entrepreneurs, which make contact between different users possible, as well as the option of commenting and giving opinion on a tourist product and service, create what is known as 2.0. tourism.

Nevertheless, as pointed out in Fig. 1, this category of tourism is not only integrated by social media specialised in tourism, but also by generalist media and, to a lesser extent, by professional media, which are also involved in their organization.

In any case, it is true that online travel communities have sparkled in last years. Regarding the number of users, *TripAdvisor.com* and *Booking.com* stand out above the rest (Hernández et al. 2012).

On the one hand, *TriAdvisor.com* assesses, classifies, prioritize and arrange systematically different facilities, products and tourism products. For tourist businesses, this community of travellers is a means to manage their digital reputation. From the consumer's perspective, the online portal enables them to learn the opinion of customers that have used the same service, both information on a tourist destination,

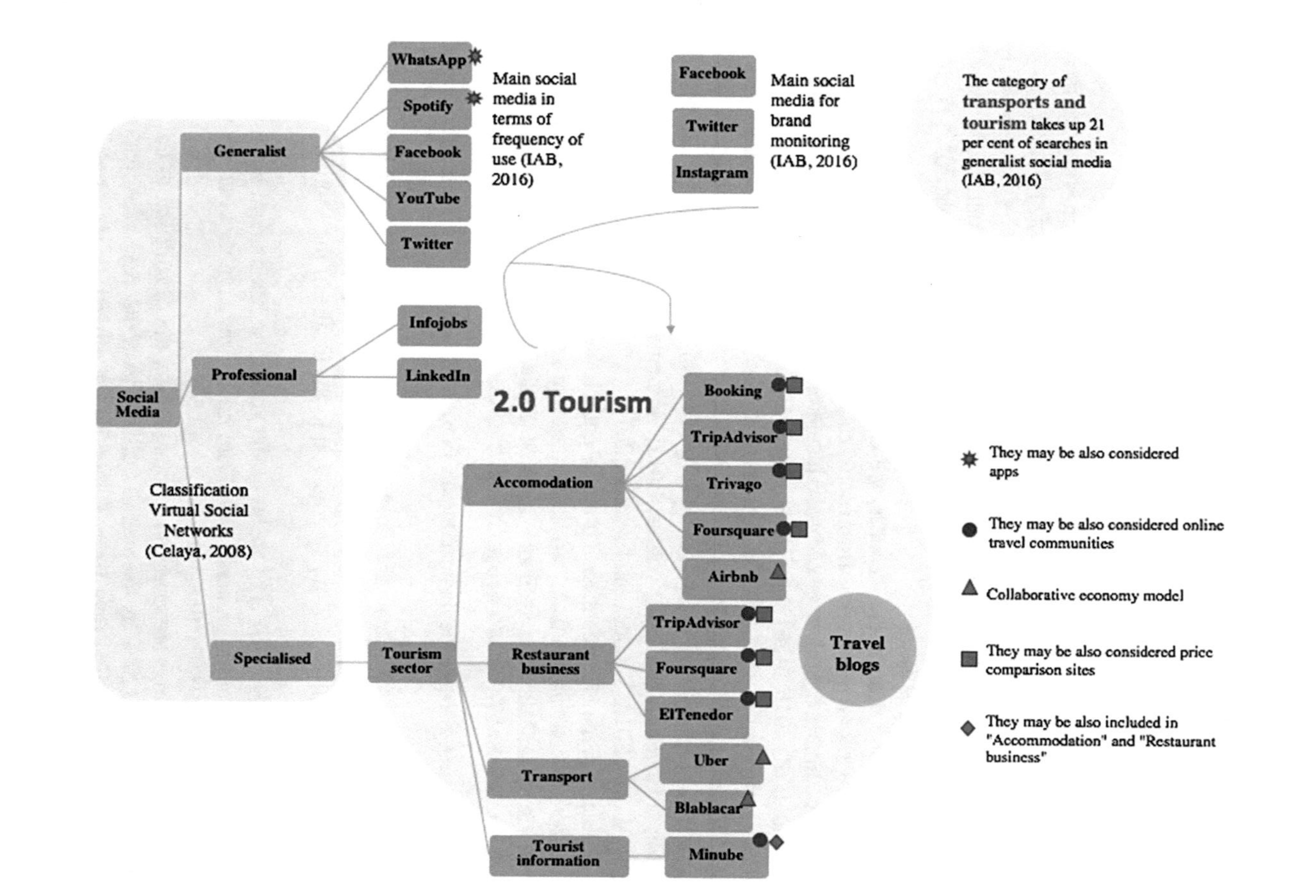

Fig. 1 Classification of social media and online travel communities. *Source* Prepared by authors

accommodation, restaurant and flight (Law 2006; O'Connor 2010; Memarzadeh and Chang 2015; Keates 2007).

On the other hand, the procedure used by *Booking.com* to obtain their users' opinion is very simple (Gutiérrez et al. 2014). Two or three days after checking out, customers receive an email where they are invited to make a review on their stay. Having the user made the comments, the text cannot be modified, whether from the user or *Booking.com*. This way, opinions become a useful source of first-hand information on the experience and satisfaction of the user at that accommodation, which serves a guide for future customers.

Both *Booking.com* and *TripAdvisor.com* offer similar information, although there is actually a significant difference between the two web portals. On *Booking.com*, only hosted consumers can have their say on the hotel, while on *TripAdvisor.com* any logged user can enter their comments, which has given rise to controversy within the scientific and business environment (Balagué et al. 2016; Mellinas 2012; Palmer 2013; Shankman 2013).

However, and despite the discords, we cannot overlook the importance which nowadays represent these online travel communities in the purchase decision process of tourists. In this regard, several research support the assertion (Ayeh et al. 2013; Balagué et al. 2016; Bjørkelund et al. 2012; Bold 2012; Chevalier and Mayzlin 2006; Costantino et al. 2012; De Albornoz et al. 2011; Dickinger and Mazanec 2008; Filieri and McLeay 2014; Gretzel and Yoo 2008; Grinshpoun et al. 2009; Korfiatis and Poulos 2013; Law 2006; Mayzlin et al. 2014; Mellinas et al. 2016a, 2016b; O'Connor 2008, 2010; Vásquez 2011; Vermeulen and Seegers 2009; Wichels 2014; Wilson 2012; Yacouel and Fleischer 2012).

In line with the aforementioned matters, authors such as Limberger et al. (2016) offer the key conclusions from various studies that analyse the quality of hotel services through information available in social media specialised in tourism, such as *Booking.com* and *Tripadvisor.com*, in which it is included the following:

- Tourists using social media specialised in tourism trust the recommendations of other users.
- The most trusted comments are those which provide full and detailed information on the tourist service.
- Users of social media use to pay attention to the profile of the commentator and the posting date.
- The quantity of comments in social media specialised in tourism has a positive relation with hotel recommendations. The greater the number of comments, the more positive the evaluation.
- Although the ratings are made on a hotel service, the satisfaction of the guest is influenced in a 11.38% by the overall satisfaction of the tourist destination.
- A 97% of satisfaction may be explained by the criteria: room, prize, cleaning, check-in and check-out.

As a final thought, it is appropriate to point out that, at present, online travel communities are involved in all stages of the trip (before, during and after) or, as specified by Google (2016), in all different micro-moments of a trip. These states or micro-moments require multiple planning that can be developed in different periods, in which consumption trends are different and affect the purchase decision of a particular tourist experience. Understanding the motivations of a community of users enable planners/organisers to design and operate more efficient and profitable online communities for the business or tourist destination.

Acknowledgements Results of the research are part of the project *Indicators related to broadcasters' governance, funding, accountability, innovation, quality and public service applicable to Spain in the digital context* (Reference CSO2015-66543-P), belonging to the Spanish National Plan for Scientific and Technical Research and Innovation, a national subprogram of Knowledge Creation from the Spanish Ministry of Economy and Competitiveness, co-funded by the European Regional Development Fund (ERDF) from the European Union. It is also part of the International Research Network on Communication Management (REDES 2016 G-1641 XESCOM), supported by the Galician Ministry of Culture and Education Galicia (Ref. ED341D R2016/019).

References

Abuin, L.: Las nuevas tecnologías: herramientas de tangibilización de los destinos turísticos. In: Feijoó, J.L. (ed.) La comunicación en el turismo. Ugerman, Buenos Aires (2016)

Ayeh, J.K., Au, N., Law, R.: Do we believe in TripAdvisor? Examining credibility perceptions and online travelers' attitude toward using user-generated content. J. Travel Res. **52**(4), 437–452 (2013)

Balagué, C., Fuentes, E.M., Gómez, M.J.: Fiabilidad de las críticas hoteleras autenticadas y no autenticadas: El caso de TripAdvisor y Booking. com. Cuadernos de Turismo **38**, 63–82 (2016)

Bjørkelund, E., Burnett, T.H., Nørvåg, K.A.: Study of opinion mining and visualization of hotel reviews. In: Proceedings of the 14th International Conference on Information Integration and Web-based Applications & Services, pp. 229–238. ACM (2012)

Bold, B.: Tripadvisor appoints live Union for Asian expansion. In: Conference & Incentive Travel, p. 6 (2012)

Caro, J., Luque, A., Zayas, B.: Aplicaciones tecnológicas para la promoción de los recursos turísticos culturales. In: XVI Congreso Nacional de Tecnologías de la Información Geográfica, pp. 938–946. AGE, Madrid (2014)

Casaló, L.V., Flavián, C., Guinalíu, M.: Redes sociales virtuales desarrolladas por organizaciones empresariales: antecedentes de la intención de participación del consumidor. Cuadernos de Economía y Dirección de la Empresa **15**(1), 42–51 (2012)

Celaya, J.: La Empresa en La Web 2.0. Gestión 2000, Barcelona (2008)

Celaya, J.: La empresa en la web 2.0: El impacto de las redes sociales y las nuevas formas de comunicación online en la estrategia empresarial. Grupo Planeta, Barcelona (2011)

Chevalier, J.A., Mayzlin, D.: The effect of word of mouth on sales: Online book reviews. J. Mark. Res. **43**(3), 345–354 (2006)

Costantino, G., Martinelli, F., Petrocchi, M.: Priorities-based review computation. In: AAAI Spring symposium: Intelligent web services meet social computing, pp. 2029–2034. ACM (2012)

De Albornoz, J.C., Plaza, L., Gervás, P., Díaz, A.: A joint model of feature mining and sentiment analysis for product review rating. In: Advances in Information Retrieval pp. 55–66. Springer, Berlin Heidelberg (2011)

Dickinger, A., Mazanec, J.: Consumers' preferred criteria for hotel online booking. In: Information and Communication Technologies in Tourism. pp. 244–254. Springer, Vienna, Austria (2008)

Dellarocas, C.: The digitization of word of mouth: Promise and challenges of online feedback mechanisms. Manag. Sci. **49**(10), 1407–1424 (2003)

Di Placido, A.: Interactividad usuario-usuario y redes sociales online en el sector turístico. Análisis de las páginas web turísticas oficiales de las administraciones andaluzas. In: VIII Congreso Nacional Turismo y Tecnologías de la Información y las Comunicaciones, pp. 77–91. TURITEC, Málaga (2010)

Ejarque, J.: Marketing y gestión de destinos turísticos: nuevos modelos y estrategias 2.0. Pirámide, Madrid (2016)

Filieri, R., McLeay, F.: E-WOM and accommodation: an analysis of the factors that influence travelers' adoption of information from online reviews. J. Travel Res. **53**(1), 44–57 (2014)

Gomis, J.: La información turística: del papel a la Red en Manual de Comunicación Turística. De la información a la persuasión, de la promoción a la emoción. Editorial Documenta Universitaria, Girona (2009)

Gonçalves, J., Fraiz, J.A., Manosso, F.C.: Calidad de la experiencia en los hoteles termales de Galicia, España. Un análisis a través de la reputación online. Estudios y Perspectivas en Turismo **22**(3), 492–525 (2013)

Google.: How micro-moments are reshaping the travel customer journey. https://goo.gl/JvxPLb (2016)

Gretzel, U., Yoo, K.H.: Use and impact of online travel reviews. In: O'Connor, P., Höpken, W., Gretzel, U. (eds.) Information and Communication Technologies in Tourism 2008, pp. 35–46. Springer, Vienna (2008)

Grinshpoun, T., Gal-Oz, N., Meisels, A., Gudes, E.: CCR: A model for sharing reputation knowledge across virtual communities. In: Web Intelligence and Intelligent Agent Technologies, pp. 34–41. IEEE Computer Society (2009)

Gruen, T.W., Osmonbekov, T., Czaplewski, A.J.: eWOM: The impact of customer-to-customer online know-how exchange on customer value and loyalty. J. Bus. Res. **59**(4), 449–456 (2006)

Gutiérrez, D., Parra, E., González, Y.: La influencia de la autentificación de la identidad en las valoraciones online de los hoteles, Turismo: liderazgo, innovación y emprendimiento. In: Congreso AECIT, Benidorm (2014)

Hernández, E., Fuentes, M., Morini, S.: Una aproximación a la Reputación en Línea de los Establecimientos Hoteleros Españoles. Papers de Turisme **52**, 63–88 (2012)

Hong, T.: The influence of structural and message features on web site credibility. J. Am. Soc. Inform. Sci. Technol. **57**(1), 114–127 (2006)

Howison, S., Finger, G., Hauschka, C.: Insights into the Web presence, online marketing, and the use of social media by tourism operators in Dunedin, New Zealand. Int. J. Tourism Hospitality Res. **26**(2), 269–283 (2014)

Hudson, S., Thal, K.: The impact of social media on the consumer decision process: implications for tourism marketing. J. Travel Tourism Mark. **30**(1–2), 156–160 (2013)

IAB Spain.: Estudio anual de redes sociales. https://goo.gl/RuHxHD (2016)

Kang, M., Schuett, M.: Determinants of sharing travel experiences in social media. J. Travel Tourism Mark. **30**(1–2), 93–107 (2014)

Karakaya, F., Ganim Barnes, N.: Impact of online reviews of customer care experience on brand or company selection. J. Consum. Mark. **27**(5), 447–457 (2010)

Keates, N.: Deconstructing TripAdvisor. Wall Street J. https://goo.gl/99Uqu5 (2007)

Korfiatis, N., Poulos, M.: Using online consumer reviews as a source for demographic recommendations: a case study using online travel reviews. Expert Syst. Appl. **40**(14), 5507–5515 (2013)

Law, R.: Internet and tourism-part XXI: TripAdvisor. J. Travel Tourism Mark. **20**(1), 75–77 (2006)

Lee, J., Park, D.H., Han, I.: The effect of negative online consumer reviews on product attitude: an information processing view. Electron. Commer. Res. Appl. **7**(3), 341–352 (2008)

Limberger, P.F., de Souza Meira, J.V., da Silva Añaña, E., Sohn, A.P.L.: A qualidade dos serviços na hotelaria: um estudo com base nas online Travel Reviews (OTRS). Turismo-Visão e Ação **18**(3), 690–714 (2016)

Litvin, S.W., Goldsmith, R.E., Pan, B.: Electronic word-of-mouth in hospitality and tourism management. Tour. Manag. **29**(3), 458–468 (2008)

López Carrillo, E.: Las webs experienciales en el sector turístico: el turista como herramienta de comunicación de las organizaciones turísticas. Manual de comunicación turística 305–311 (2011)
Marín, J.: Web2.0, una descripción muy sencilla de los cambios que estamos viviendo. Netbiblo, A Coruña (2010)
Mayzlin, D., Dover, Y., Chevalier, J.: Promotional reviews: an empirical investigation of online review manipulation. Am. Econ. Rev. **104**(8), 2421–2455 (2014)
Mediano, L.: Incidencia del nuevo consumidor turístico en la estrategia de marketing. Revista de Dirección y Administración de Empresas **2**, 99–107 (2002)
Mellinas, J.: ¿Se pueden escribir críticas falsas en TripAdvisor sin ser detec-tado? Personal blog. https://goo.gl/J167Fz (2012)
Mellinas, J.P., María-Dolores, S.M.M., García, J.J.B.: El uso de redes sociales por los hoteles como indicativo de gestión eficiente. Tourism Manag. Stud. **12**(2), 78–83 (2016)
Mellinas, J.P., María Dolores, S.M.M., Garcia, J.J.B.: Evolución de las valoraciones de los hoteles españoles de costa (2011–2014). In: PASOS. Revista de Turismo y Patrimonio Cultural, vol. 14, issue no. 1 (2016b)
Memarzadeh, A., Chang, H.: Online consumer complaints about Southeast Asian luxury hotels. J. Hospitality Mark. Manage. **24**(1), 76–98 (2015)
Obercom.: A Internet em Portugal. Sociedade em Rede 2014. https://goo.gl/nj04cL (2014)
O'Connor, P.: User-generated content and travel: a case study on Tripadvisor.com. Inf. Commun. Technol. Tourism **2008**, 47–58 (2008)
O'Connor, P.: Managing a hotel's image on TripAdvisor. J. Hospitality Mark. Manag. **19**(7), 754–772 (2010)
Palmer, M.: It's the hotel review website that millions use to help choose their holiday. But… Can you trust a single word on TripAdvisor. https://goo.gl/V28LpD (2013)
Pan, B., MacLaurin, T., Crotts, J.C.: Travel blogs and the implications for destination marketing. J. Travel Res. **46**(1), 35–45 (2007)
Richards, G., Wilson, J.: The Global Nomad: Backpacker Travel in Theory and Practice. Channel View Publications, Clevedon (2004)
Rodríguez-Fernández, M.M., Sánchez-Amboage, E., Martínez-Fernández, V.A., Juanatey-Boga, O.: Social media communication as a corporate positioning strategy: the Galician Winemaking Sector in Spain. In: World Conference on Information Systems and Technologies, pp. 141–150. Springer, Cham (2017)
Sanagustín, E.: Marketing 2.0 en una semana. Grupo Planeta, Barcelona (2010)
Serra, A.: Marketing turístico. Pirámide, Madrid (2011)
Sánchez, E.: La Promoción de destinos turísticos termales a través de los medios sociales: Análisis del caso de los balnearios de Galicia y del Norte de Portugal. (Tesis Doctoral) Universidade da Coruña, A Coruña, España (2015)
Shankman, S.: TripAdvisor now removing old negative reviews after hotels renovate. https://goo.gl/XUIxMk (2013)
Steffes, E.M., Burgee, L.E.: Social ties and online word of mouth. Internet Res. **19**(1), 42–59 (2009)
Valls, J.F.: Las claves del mercado turístico. Deusto Ediciones, Bilbao (1996)
Vásquez, C.: Complaints online: the case of TripAdvisor. J. Pragmatics **43**(1), 1707–1717 (2011)
Vermeulen, I.E., Seegers, D.: Tried and tested: the impact of online hotel reviews on consumer consideration. Tourism Manag. **30**(1), 123–127 (2009)
Wichels, S.: Nuevos desafíos en relaciones públicas 2.0: la creciente influencia de las plataformas de online review en turismo. Revista Internacional de Relaciones Públicas **4**(7), 197–216 (2014)
Wilson, T.: Hotels as ready-to-hand recreation: TripAdvisors posting potentiality- for-being in play. Tourist Stud. **12**(1), 70–86 (2012)
Yacouel, N., Fleischer, A.: The role of cybermediaries in reputation building and price premiums in the online hotel market. J. Travel Res. **51**(2), 219–226 (2012)

María-Magdalena Rodríguez-Fernández Ph.D. in Economic and Business Sciences from the Universidade de A Coruña. She is Professor at the Commercialization and Market Research Department of the same university. She is author and co-author of various articles in journals and books. Her lines of research are related to communication, marketing, new technologies and tourism. A Coruña, Spain.

Eva Sánchez-Amboage Ph.D. in Tourism Management and Planning from the Universidade de A Coruña. She was Professor at the Universidad Técnica Particular de Loja. She has a Master in Tourism Management and Planning and a Master in Teaching BAC, FP and foreign languages, both from the same institution. She has a degree in Tourism by the Universidade de A Coruña. A Coruña, Spain.

Oscar Juanatey-Boga Ph.D. in Economics and Business from the Universidade de A Coruña (UDC), master MBA in Management and Business Administration, master in Business Management and Marketing, and master in Business Communication by the UDC. Professor of marketing and market research at the Faculty of Economics and Business at the same university is the author of articles on media and communication strategies. A Coruña, Spain.

Travel Igers: Innovation, Influence and Persuasion Through a Photo Gallery

Mónica Valderrama-Santomé, Ana-Belén Fernández-Souto and Montse Vázquez-Gestal

Abstract Nowadays accessibility science has had enormous advances in technology innovation to enable people with disabilities to participate on equal terms in almost every social area. Among the most effective actions to promote inclusion, the formative aspects stand out from others, and establish a permanent positive and inclusive climate. In this project the design of a multimedia videogame explores technology issues and accessibility and design for all. Main objectives are awareness, mainly for young people, about the needs of those with disabilities.

Keywords Influencers · Advertising · Instagram · Travel

1 Influence Through Recommendations on Social Media. Travel Igers

The ability of advertisers to get the attention of the audience has shifted from conventional to digital media. According to IAB Spain's Annual Social Media Study 2016, 81% of all Internet users between 16 and 55 years old use social media, which represents over 15 million users only in our country. Brands are aware of the importance and power of influence of some opinion leaders and include some influencers in their communication strategy in order to reach not only the youth, but also specific profiles according to the preferences of the audience. Basically, when these influencers show a piece of clothing, a product or a trip they made on their virtual windows, demand soars. Their direct experience with these items becomes a reference and their recommendation goes viral.

M. Valderrama-Santomé (✉) · A.-B. Fernández-Souto (✉) · M. Vázquez-Gestal (✉)
Universidade de Vigo, Vigo, Spain
e-mail: santome@uvigo.es

A.-B. Fernández-Souto
e-mail: abfsouto@uvigo.es

M. Vázquez-Gestal
e-mail: mvgestal@uvigo.es

M. Túñez-López et al. (eds.), *Communication: Innovation & Quality*, Studies in Systems, Decision and Control 154, https://doi.org/10.1007/978-3-319-91860-0_27

The above-mentioned research study also shows that 85% of users follow some influencers on social media and 1 out of 3 follows brands, participates in their give-aways and speaks about shopping often. Fashion and travel (tickets, car rentals and accommodation) are the most prevalent, as they are the most demanded and shared, and their ads in the various digital media have a greater presence.

Regarding the awareness of each social network, almighty Facebook (FB) is known by the entire population surveyed for IAB's study. Instagram is fifth, following WhatsApp, Twitter and YouTube, with an 83% degree of recognition. The prevailing Instagram user profile is a woman between 16 and 30 years old, although more and more users have joined this network in the past year. Moreover, it is the fourth best rated network after WhatsApp. YouTube and Spotify are consumed 2:40 h per week in average, mostly using cell phones and tablets; whereas users mostly check FB on their computers. In addition, Instagram is one of the most common platforms to follow brands: twice as many brands are followed on Instagram than on FB or Twitter, which are undergoing a gradual decrease, although most Internet users use these two networks to inform themselves before a purchase. Finally, we would like to mention that 65% of survey respondents declared that social media had an effect on their purchase decision.

The above-mentioned review of some of the indicators which show the relevance of digital communication and its growing penetration in society has given rise to this study and others, which aim to analyse a new phenomenon by directly observing the supports which allow disseminating purchase recommendations based on personal experience.

And thus begins our description and reflection on a seemingly trivial action, such as *liking* a picture posted by a travel photographer on his gallery or looking at the clothes worn or places visited by a person we know. This creates a communicative flow, highly affected by a positive emotion, which leads us to make a purchase decision without the impact of ordinary advertising. Haven't you seen those photographs of the northern lights in Iceland posted by instagrammers (hereinafter, *igers*) you may follow or know? The amazing colours of this natural phenomenon will have you looking for a plane ticket in seconds, trying to quench your "wanderlust". Otherwise, you will see those images pop up every so often to make sure that idea remains fresh in your mind, because, due to *likes* and positive interactions, more and more ads will appear on our FB wall and on Instagram. We have identified ourselves and we have become a target for that destination we spotted.

This known role of social media in people's lives and their growing number of followers (Fumero and Roca 2007) justifies the interest shown by companies and other types of organizations which target a specific audience and leverage the influence of igers and their personal accounts. Obviously, this informal communication is part of an individual's daily routine nowadays, and it allows people to interact with each other regardless of where they are in the world. Many brands have seen a great business opportunity in this connecting power, and they leverage these tools and influencers to promote their products and services on social media. Thanks to the fact that they allow a more direct relationship with their clients, brands come closer to their clients, time and communication barriers are torn down, and information is massively disseminated while advertising costs drop.

In the last few years, the advantages of virtual media to run communication campaigns have been described (Visón 2010) and social media have been said to bring about many benefits to brands, as their investment is lower when compared to traditional media; they simplify feedback and consumers can express their impressions; and they even contribute to the position of their web site in search engines.

Tourism is also part of this new scene. Current tourism is often referred to as "Tourism 2.0", as tourists are influenced by the opinions of other travellers on many social media when choosing a hotel at their destination or when deciding on a destination. This is known as "word-of-mouth" adapted to Internet times (Dellarocas 2015), and it has been researched since this phenomenon first emerged.

This study will show the power of Instagram and its travel igers, who have been contacted by many brands to have their products mentioned (places, tourist activities, clothing for traveling, etc.). Thanks to this collaboration, brands reach a broader audience and avoid seeming invasive, as they are only mentioned on personal accounts.

2 Tourism Is Committed to Personal Recommendations

The mass media are no longer an attractive showcase for tourism promotion and bet from the sector by personal prescription. To know in a direct way the personalities that do not celebrities think of a small sample to know the work that the influencers develop.

So that, to compose this study, we carried out a qualitative analysis by the direct observation of 4 iger profiles that are listed in several blogs and articles on the recent travel iger phenomenon as the most relevant at the time of this study. We have included some tables with the qualitative data we gathered, specifying the number of *likes*, posts, mentions to brands and other items of interest, such as the application of the engagement formula we have used.

Regarding the goals, the baseline of our research is the analysis of the cases of ordinary people sharing a hobby (traveling) and how they have been able to become relevant in the travel industry thanks to Instagram, thus turning their passion into their job. Therefore, the goal of this study is to analyse the "Travel instagrammer" phenomenon and explore its impact on society, and how society and traveling are changing due to it.

Another goal is to find out how Travel instagrammers obtain their resources (aids, sponsors, own money, work) for their trips and whether, like Youtubers, they can make a living of it.

Products are usually showcased in other sectors. For instance, fashion influencers mostly show clothing and lifestyle items while surrounded by high-end brands, luxury hotels and well-known restaurants.

With regard to this, we intend to analyse another type of iger-influencer, an opinion leader specialized in traveling. This person also shows his or her lifestyle, although very different from what we described above: wonderful landscapes, tips and hacks for traveling for less. For this purpose, we have analysed how many brands were mentioned by these profiles in the first 5 months since their creation, and how many brands were mentioned in the last 5 months prior to this study (December 2016–May 2017).

Our aim is also to prove whether brands take into consideration the engagement created by these influencers' accounts before their collaboration.

3 Reputation in Instagram. The Engagement Formula

3.1 Instagram as a Platform

Instagram is a social network and mobile application of visual nature, designed to take, edit and share photographs and videos.

It was launched in October 2010 for iOS and in April 2012 for Android by Kevin Systrom and Mike Krieger. At the beginning, it was mainly appealing to users because they could apply filters to photos and videos, as well as other effects (frames, temperature filters). Later it also allowed sharing the photos on other social media such as Facebook, Tumblr, Flickr and Twitter.

This network has evolved over time and today, thanks to the introduction of a "Stories" feature, instagrammers can post live videos which disappear after 24 h to have a more direct communication with their audience (at the expense of Snapchat), and know how many users saw their content. This is particularly attractive for younger people, who are very used to using Snapchat.

Three months after its creation, Instagram already had one million registered users. In June 2011 Instagram announced five million users and reached over ten million in September of that same year. In August 2011 the photos uploaded to Instagram amounted to 150 million. This meant great impact and growth in only one year.

Two years after it was launched, in early September 2012, Mark Zuckerberg, Facebook's CEO, announced that Instagram had over 100 million registered users.

In May 2012, every second 58 photographs were uploaded and 1 more user registered. Over 1 billion photographs were uploaded to the network.

The process is simple: first, the user uploads a photo/video from his or her photo gallery or takes a photo directly, then he or she applies a digital filter and when the photo is ready it is shared with the community. It is particularly optimal for sharing places, landscapes and experiences because it is a very visual tool.

Instagram has been defined lately as one of the most prevalent networks in the digital world. One of the keys to its success in advertising may be that it is a platform where ads are not very intrusive, since the user decides whether or not he or she wants to see ads. This freedom to choose may be generating a greater interaction capacity and followers tend to accept advertising better.

3.2 Engagement

Instagram is an application whose main value is the ability of interaction between the brand and the user using a photograph. The app allows a participative dialogue between users and brands, thanks to the high level of interaction, also when the dialogue is participated, authorized or supervised by the brand itself (Caerols et al. 2013).

Herrera (2014) offers a clear and succinct definition of *engagement*: "engagement is the degree of interaction of a consumer and how the consumer commits to a brand".

The formula below to calculate the charisma and influence power on social networks was released on the website of Núñez (2017) as a result of Novaemusik's Social Media Manager Project:

Likes+comments+shares on given day of wall posts made by page on a given day/total fans on a given dayx100.

In 2014 Forrester Research conducted a study revealing that Instagram is the best platform to interact with users. The study compared Instagram with Facebook and Twitter (Fig. 1).

In this chart we can see how brand posts on Instagram generate a 4.21% user engagement, which means that a brand's followers on Instagram are 58 times more committed to the brand than Facebook followers and 120 times more than Twitter followers.

One of the main demonstrations of engagement a brand's updates on social media can get is receiving comments from consumers related to the content published. According to a research study by Newship, there are several theories to explain engagement on Instagram. The first one is related to the type of content published on Instagram. This social network features only photos and videos which, in general, represent the content which best engages users.

Additionally, the fact that Instagram is based on pictures has made its audience global.

User interactions with brands' posts as a percentage of brands' fans or followers

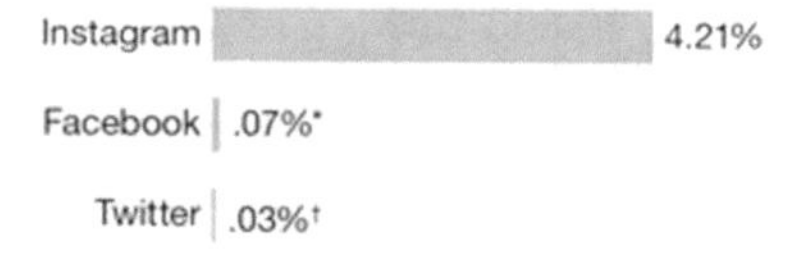

Base: 1,526,388 user interactions on 162 Instagram brand posts
*Base: 1,405,249 user interactions on 329 Facebook brand posts
†Base: 98,298 user interactions on 910 Twitter brand posts

Source: Q1 2014 US Top 50 Brands Social WebTrack

113021 Source: Forrester Research, Inc.

Fig. 1 User-brand interaction rate. *Source* Forrester (2014)

3.3 The Iger as an Influencer

The influencer is a person who makes information from her or his own experiences and opinions with a very high power of recommendation and influence.

The latest report published by Augure (2015) entitled "Status and practices of relationships with influencers in 2015" presents data which shows how the relationship between brands and influencers intensifies.

The relationships with influencers seem to be primarily an efficient manner to increase brand visibility and brand value. 81% of respondents consider that the relationship with influencers is efficient or very efficient.

"Customer loyalty to a brand depends on satisfaction, purchase recurrence and, particularly, on recommendation, which is at the top of emotional bonding" (Best 2007).

"Work with an influencer, they're friends for a day. Help someone become influential and they're friends forever". This quote by Lee Odden perfectly summarizes the motivations which, according to Augure's study respondents, lead influencers to collaborate with brands.

According to this study, 55% of influencers expect brands to help them strengthen their opinion leader status in their community. This is why the strategies aimed at allowing the influencer to express his or her point of view or increase visibility are often a bet.

Improving the quality of their content (45%) is the second motivation. This goal is very much linked to the goal above. Having access to exclusive information, to experts or resources which otherwise he or she would not have accessed are elements which are valued by the influence when collaborating with a brand.

The opportunity to build up the audience's brand image (29%) explains why philanthropic campaigns are preferred.

The economic compensation is the fourth motivation (24% of respondents) and is also linked to the first. In order to become an influencer, the three premises above are necessary and, once the influencer has consolidated as an opinion leader, has quality content and a good brand image, he or she starts receiving economic compensation.

The importance of Instagram influencers is such that, according to a report by eMarketer, Mercedes paid 20,000 euros to an instagrammer to promote the brand or Dom Perignon paid 15,000 euros for a trip to Iceland to an advertiser with many followers.

This document will analyse the phenomenon where travellers, thanks to having an Instagram account which publishes good content, have become travel instagrammers and brands have found in them a person to communicate their brand's philosophy globally or to advertise a specific tourist destination.

3.4 Tourist Promotion Skyrockets with Social Networks

Tourism has undergone a gradual expansion and diversification and has become a thriving economic sector worldwide, even in current downturn times (García García and Garrido Pintado 2013).

Humans are more and more interested in traveling and are less reluctant to stepping into an airplane and traveling thousands of kilometres to reach their destination. This is also so thanks to the possibility to buy cheaper tickets for low-cost flights and to the various accommodation options at a specific destination, such as platforms like Airbnb where tourists may rent a room in an apartment of a full apartment to stay for a few days, or couchsurfing, which is a travellers' network to spend a few nights at someone's home who offers his or her couch for free.

Many use the word "wanderlust", a German term which means "desire to travel" or "passion for traveling", to refer to people who are restlessly looking into new experiences, fantasizing with different places to visit or where they could travel next, new places or adventures they could live along the way.

Many pictures, blogs and movements can be found online under the term "wanderlust".

According to some researchers, there is a wanderlust gene: this "great desire to travel" has a physiological basis linked to the DRD4 gene and to the amount of dopamine released by the brain.

Dawn Maslar, a biologist of the University of Kaplan, claims that this gene may be found in chromosome 11, which may define the likelihood to take risks, curiosity and the interest in exploring new destinations.

In fact, the small sample we analysed included some well-known wanderlusters who were able to make a living out of their passion thanks to their influence power and to the fact that many brands and advertisers found in them a great opportunity to reinvigorate or launch their business.

4 Wanderlusters and Brand Ambassadors

4.1 Molaviajar

A family on the go, made up of Gissi (Poland), Andrian Rodríguez (Galicia, Spain) and little Daniela, their daughter. The family is expecting a baby, as Gissi is pregnant with her second child. Molaviajar first started in 2008 as a way to stay in touch with family and friends. That way, they could keep them up to date on their adventures around the world and, particularly, to follow a trip the couple made in 2009 around Southeast Asia.

Their Instagram feed is full of landscapes, smiles, adventures and giveaways. They are proud of their rootlessness and of having been able to get away and live their family life in an ongoing trip. "We have now managed to catch the attention of companies, which want us to promote their products while traveling around the world", explains Adrian in an interview with Traveler.es.

Their growth rate is spectacular: between November 2016 and May 2017 their number of followers grew by 20,000. They receive many comments, but nothing compared to the celebrity star system in other areas.

They have about 10–50 comments per post and about 1000–4000 *likes* per photo or video. Their YouTube channel is very successful too (Fig. 2).

Among their brand collaborations are AVEXPERIENCE, in 6 videos where they toured Spain by train, and their work as reporters for Hyundai at Brazil's FIFA World Cup. They are also brand ambassadors of Totto backpacks.

Their life philosophy is: "Find challenges in life and go for them. Don't ever settle" (Table 1).

Fig. 2 Screenshot of Molaviajar's Instagram account. *Source* Instagram

Table 1 Summary of the main account details

Instagram handle: MolaViajar
Authors: Adry, Gossi and Daniela
Language: Spanish
Brands: Totto Spain, Disney, Kia
General topics: traveling, life hacks and experiences
Followers: 59,200
Total posts: 1327
Likes on the latest photo: 2781
Comments on the latest photo: 11
Likes on the first photo: 53
Comments on the first photo: 4
Total brands mentioned in the last 5 months: 19
Total brands mentioned in the first 5 months: 4
Engagement May: 6.59

Source Prepared by the author

4.2 *Gary Arndt*

It is incredible that this gallery is not created with photos taken by award-winning authors or Magnum Photos professionals (Fig. 3).

This iger is an American globetrotter and photographer who sold his house in 2017 and has travelled the world ever since. He has already visited 175 different countries. He collaborates with the media. Gary has made the dream of many travellers come true: he left everything behind to become Julius Verne's character, Phileas Fogg—only with a camera on his shoulder.

In an interview with Digital Photography School (2016), Gary explained that he started in this when he didn't know anything about photography. He just bought a Nikon D700 and took a trip. His only goal back then has filling a wall at home with nice photos. From then on, photography gradually became the most prevalent topic in his social media posts, so brands showed an interest in him. At the beginning he paid his trips with money he had saved. However, as his followers on social media grew, brands understood that he had a great potential as an ambassador of their products and services. So several advertisers started to fund his trips and to offer him to travel to different parts of the world to take photos and share them in social media to reach his audience, made up of 197K followers on Instagram.

This wanderluster has been awarded "Travel Photographer of the Year" three times: twice by the North American Travel Journalist Association and once by the Society of American Travel Writers.

Fig. 3 Screenshot of Everythingeverywhere's Instagram account. *Source* Instagram

His Instagram feed includes photos in over 180 countries in all 7 continents, and the quality of his photos has clearly evolved since he first started. Initially, his photos were taken using a cell phone and their quality was very poor. Now more brands are mentioned when photos are posted, as shown in Table 2.

4.3 Kate McCulley

Kate is a 39-year-old instagrammer originally from Boston, Massachusetts. After four years studying Marketing and her whole life dreaming of traveling the world alone, she quit her job in 2011 to fulfil her dream and inspire other women to do the same by providing practical advice and writing about diverse topics, such as working abroad or security in other countries (Fig. 4).

Kate still travels to anywhere in the world, as shown on his Instagram feed. The main appeal of her gallery is that her followers can get to know various countries and enjoy her trips, while she proves one can travel the world alone, without a couple or family. This is a very different profile, as her main target are women. Her

Table 2 Analysis of Instagram account Everythingeverywhere

Instagram handle: Everythingeverywhere
Authors: Gary Arndt
Language: English
General topics: traveling and adventures
Brands: Adove, BBC Travel and his own exhibitions and books
Followers: 197,000
Total posts: 4628
Likes on the latest photo: 3187
Comments on the latest photo: 31
Likes on the first photo: 4
Comments on the first photo: 0
Total brands mentioned in the last 5 months: 4
Total brands mentioned in the first 5 months: 0
Engagement May: 1.72

Source Prepared by the author

Table 3 Analysis of Adventurouskate's Instagram account

Instagram handle: Adventurouskate
Authors: Kate McCulley
Language: English
General topics: traveling
Brands: Kimberleyspirit, Statravel, Mangrovehotel…
Followers: 98,900
Total posts: 2082
Likes on the latest photo: 849
Comments on the latest photo: 18
Likes on the first photo: 3
Comments on the first photo: 0
Total brands mentioned in the last 5 months: 6
Total brands mentioned in the first 5 months: 1
Engagement last month: 0.10

recommendations are basically to show a woman can travel alone easily and safely. In an interview with Ello Giggles, Kate says that life is too short for regrets. At the end of the year, when we first analysed these igers, Kate spent half of the year in London and the other half she travelled around the world. Now, as shown in the screenshot below, she lives in New York (Table 3).

Fig. 4 Instagram account of Adventurouskate. *Source* Instagram

4.4 Alanxelmundo

A young 33-year-old Mexican (Bitacora Viajera) actor, dancer and singer. Alan's life took a 180° turn when he first backpacked India and Southeast Asia. He says that there is nowhere in the world he wouldn't like to visit. He thinks his trips are part of his personal training and growth. His first plan was to travel one or two months per year around different parts of the world, but now he travels for nine months and gets a profit from it. He is no longer an ordinary iger sharing his experiences, but rather the best example of professionalization: he now owns his own online fashion store and has a team managing orders. In addition to collaborations with brands on social media, he has his on website where he sells a variety of products. Although Alan does not reveal how much he makes, he admits that thanks to his Internet site "www.alanxelmundo.com" he makes enough to rent an office, pay five employees and cover the cost of some of his trips (Fig. 5).

Alan asserts that his first intention was not making money off his web site, he was happy with traveling and sharing his experiences with others. Today his web site is profitable and his YouTube channel generates him revenue, as well as the brands

Fig. 5 Screenshot of Alanxelmundo's Instagram account. *Source* Instagram

sponsoring him on Instagram, travel agencies paying for transport, and his online store, which sells sneakers, sweatshirts and caps, among other items. His current life project gives him freedom while his business model generates income (Table 4).

5 Relevance and Charm of Influencers

We had downloadable templates to cover and get to know potential igers in terms of their relevance on the social network in the sector we are interested in. As we mentioned in this text, there are many blogs, web sites and research studies which mention the formulas which, after entering the data we gathered, allow us to accurately calculate the degree of impact or influence an instagrammer may have on the audience. Advertisers' Communication and Marketing departments have realized that they can reach a specific target with a high degree of segmentation of the population in an effective and economical manner. This proven capacity to generate the right stimuli to make the purchase happen makes these agents attractive as opposed

Table 4 Summary of the main account details

Instagram handle: Alanxelmundo
Authors: Alan Estrada Gutiérrez
Language: Spanish
General topics: traveling
Brands: XcaretPark, Dockers, Trips to Canada
Followers: 363,000
Total posts: 2380
Likes on the latest photo: 13,000
Last video views: 38,296
Comments on the latest photo: 78
Likes on the first photo: 333

Source Prepared by the author

to the saturated conventional media and their one-way communication. Therefore, as proved in the analysis of the different accounts representing the target groups willing to find "inspiration" in others' experience to prepare their own, travel instagrammers are a good platform for brands to raise awareness of their products, as they have many followers but, most importantly, a high engagement rate. Something that is also quite obvious is that this influence and the recipient's readiness to become a consumer of the recommended product do not work outside of their commonalities. These wanderlusters will only be a good platform for a brand which has traveling in its symbolic promise and material.

After this study we have also verified that brands do care about user engagement. The evolution from the first to the last months of the process prove how in some cases their influence and the number of brands grow; however, not all posts have the same effect or attract equally their audience. Brand mentions cannot be constant, otherwise authenticity is lost.

References

Augure: Estatus y prácticas de las relaciones con influencers en 2015. http://www.augure.com/es/recursos/whitepapers/informe-influencers-2015 (2015)
Best, R.J. Marketing estratégico. Madrid: Pearson Prentice Hall. (2007)
Dellarocas: Handbook of Research of Education and Technology in Changing Society (2015)
Forrester: Q1 2014 Us Top 50 Brands Social Webtrack. Forrester Research, Inc. (2014)
García García, F., & Garrido Pintado, P.: AGENCIAS DE VIAJE ONLINE EN ESPAÑA: APLICACIÓN DE UN MODELO DE ANÁLISIS DE SEDES WEB. TURyDES **6**(15) (2013)
Herrera, F.: ¿Qué es y Cómo Generar Engagement en Redes Sociales? http://marketingenredesociales.com/que-es-y-como-generar-engagement-en-redes-sociales.html/ (2014)
Instagram: Adventurouskate. https://www.instagram.com/adventurouskate/
Instagram: Alanxelmundo. https://www.instagram.com/alanxelmundo/

Instagram: Everythingeverywhere. https://www.instagram.com/everythingeverywhere/
Instagram: Molaviajar. https://www.instagram.com/molaviajar/
Instagram: What is Instagram? https://instagram.com/about/faq/
Nuñez, V.: Blog de Márketing estratégico. https://vilmanunez.com/formula-engagement-redes-sociales-ctr/ (2017)
Fumero y Roca: Web 2.0. http://www.oei.es/historico/noticias/spip.php?article762 (2007)
Visón, I.: Impacto de la Promoción realizada por las Empresas Dominicanas a través de las Redes Sociales Facebook y Twitter, desde la Perspectiva de los Usuarios. Pre-graduate Thesis, Pontificia Universidad Católica Madre y Maestra, República Dominicana. http://pentui.com/files/PentuiCGI_Tesis_Redes_Sociales.pdf (2010)

Mónica Valderrama-Santomé Coordinator of the Master of Art Direction in Advertising. Faculty of Social Sciences and Communication, Universidade de Vigo. University Lecturer (Ph.D.) and Researcher. She received her Ph.D. in "Advertising and Public Relations" from the Universidade de Vigo, Spain. Award of Ph.D. degree in Social and Legal Sciences (2004). BA in Communication Sciences: Advertising and Public Relations. BA in Information Sciences: Journalist. Universidade de Santiago de Compostela (Spain). Pontevedra, Spain.

Ana-Belén Fernández-Souto Ph.D., coordinator of Ph.D. Program CREA S2i at the Universidade de Vigo (Spain). Professor at Universidade de Vigo since 2000. Guest lecturer at various universities such as UCUDAL (Montevideo, Uruguay), Universidade de Minas Gerais (Brazil), Universidad Autónoma de Querétaro (Mexico); UNAM (Mexico); Universidad de Costa Rica (Costa Rica); Universidad del Norte de Chile (Chile); Universidad de Dubrovnik, (Croatia); Universidade Fernando Pessoa, U do Minho o la Lumsa Universita (Roma, Italy). Pontevedra, Spain.

Montse Vázquez-Gestal Ph.D., Professor at the Universidade de Vigo. Professor of Advertisement Creativity and Advertisement Estrategies. Researcher at the RD group Persuasive Communication from the Universidade de Vigo. Guest lecturer at various universities like Universidad Autónoma de Querétaro and Universidad Nacional Autónoma de México (Mexico, 2001); Universidad de Dubrovnik (Croatia, 2006), Universidade Fernando Pessoa (Portugal, 2008); Universidade do Minho (Portugal, 2009); LUMSA University of Roma (Italy, 2006–2009); La Sapienza, Roma (Italy, 2011). Pontevedra, Spain.

Banking Industry Innovation

Viviana Espinoza-Loayza, Eulalia-Elizabeth Salas-Tenesaca and Aurora Samaniego-Namicela

Abstract Banking industry is a key factor of economic development, although the information and communications technologies have changed in the last years, there are some countries were banking development is still incipient. This chapter show the evolution of the banking industry in the world; including the main innovations, this sector has taken place, and the challenges that it has to face. According with the Millennial disruption index, banking industry is the most risky sector, that is why, banks has to innovate constantly, especially in terms of communication channels and financial products access.

Keywords Banking development · Financial inclusion · Banking innovation

1 Bank History

1.1 Origins

The origin of the bank is not simple, because the banking functions do not have exact historical antecedents with respect to its origins. Guzmán (2011) considers that the banking activity is so important, that it has practically developed together with humanity; Vélez Núñez and Jaramillo (2017) indicate that there is evidence of banking operations from the times of Abraham. The following functions existed in some form: loans, money deposits, and credit letters, these systems were unique and complex (Fig. 1).

V. Espinoza-Loayza (✉) · E.-E. Salas-Tenesaca · A. Samaniego-Namicela
Universidad Técnica Particular de Loja, Loja, Ecuador
e-mail: vdespinoza@utpl.edu.ec

E.-E. Salas-Tenesaca
e-mail: eesalas@utpl.edu.ec

A. Samaniego-Namicela
e-mail: afsamaniego3@utpl.edu.ec

M. Túñez-López et al. (eds.), *Communication: Innovation & Quality*, Studies in Systems, Decision and Control 154, https://doi.org/10.1007/978-3-319-91860-0_28

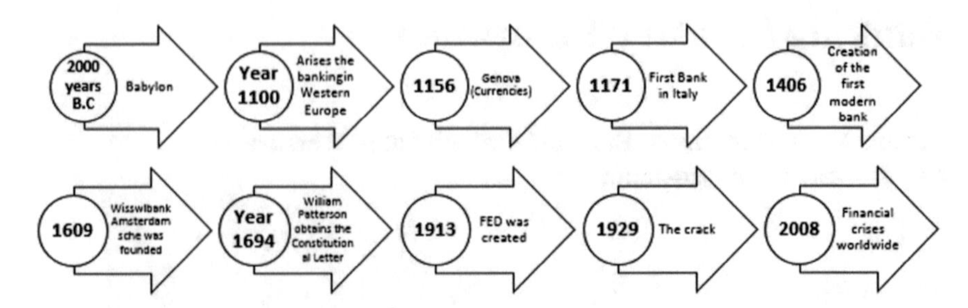

Fig. 1 Banking evolution presents ten important stages. In 1406 the first modern bank was created, through the years this industry has evoluted constantly in order to improve its products and services and help the financial inclusion around the world

The first registers of operations were apparent to banks, which appeared 2000 years B.C. in Fenecia, Asiria and Babylon. These ancient societies used the barter like form of exchange; these operations consisted in loans of grains to the agriculturalists that registered through boards of clay, in which they appeared as a form of register of operations.

Asiria and Babylon stand out for the creation of activities and documents, which used to make banking operations. For example, *trapezita*, specialized in handling currency, the *krematistas* and the *kolobistas*, was developed for coin exchange, and Daneístas, was the notion to commission the completion of deposits.

As mentioned by Orzi (2012) the coining of coins was eventually introduced in the 7th century B.C., the coins of pure gold were primarily symbols of status and reservations of value for the elites, more than system of payment and system of exchange Ingham (2004).

The ancient temples functioned at the beginning as boxes of deposits; this was its main mission. The Temple of Apolo Delfos was a place of custody of treasures and especially the savings box typical for the slaves.

Weber (1923) indicates that the same function of boxes for deposits was integrated in several other temples in Babylon, Egypt and Greece. For this reason, the temples of the antiquity were also considered big lenders.

Vélez Núñez and Jaramillo (2017) mention that during the 1st century the banks in Persia and other territories in the Sasánida empire issued letters of credit known as Sakk. These letters were distributed from califato abasí time, under Harun to the Rashid.

The process of the barter and the use of documents represented the process of buying and selling in the fairs. For the exchange of currencies which indicated a form of banking operations.

Around the years 1100s, the need to transfer big sums of money stimulated the emergence of banking in Western Europe. The first bank in the Middle Ages, were commercial banks, cities such as Siena and Florence became established as rudimentary banking centers. In 1156, the first known currency contracts were produced in Genoa, Italy.

The method of payment through the money extended to all the colonies. With the growth of the commercial activity, the first banks appeared in Italy particularly in Venice in the year 1171.

Vélez Núñez and Jaramillo (2017) present that the first modern bank created in Genoa, Italy was in 1406, Bank of Saint Giorgio. The first banks then appeared during the Renaissance period, in cities such as Venecia, Pisa, Florencia, and Genoa.

The name "bank" derived from the Italian word for bench "banco." Jewish bankers in Florence operated their business by collecting and trading money behind a bench, which reinforce the meaning behind the Italian word. The development of the bank was spread widely to northern Italy, then to all of Europe, and finally around the world.

In 1609, the bank Wisselbank Amsterdamsche was founded in Amsterdam, Netherlands. Banking offices were located by trading centers, the largest of which were during the seventeenth century, in the ports of Amsterdam, London and Hamburg. Some people could participate in the lucrative East Indies trade by buying letters of credit from the banks.

During the 17th century, goldsmiths, workers whose emphasis was working with gold, were responsible for creating gold coins. Adxpansion (2016) goldsmiths became the first bankers, besides those individuals who handled the operations of deposits, loans, and trade. The goldsmiths (in England) and the Lombards (in Italy) devoted their time to buying and selling of foreign currency. Specifically, the trade of precious metals, which helped establish a greater value for their coins.

In 1694, Goldsmith William Patterson obtained the constitutional charter, to create one of the tools that brought Britain to the conquest of nearly three-quarters of the world, for the next two hundred years.

As Ingham (1998) argues, the activity of the banks and other financial institutions has shown lack of consistency. Because it has seen in banking that only half percent of intermediation between saving and investments. Only during the 15th Century, we begin to see the manufacture of money as a social relation that projects to look for the welfare of individuals.

For Mendoza and Mendoza (2009), banks are financial entities, authorized and legally constituted to act as financial intermediaries, receiving money of the surplus agents by means of passive operations (catchments). Additionally, the process to give loans to the deficit agents was a means of active operations (placings). The development of the banking has propagated of considerable way, the development of the telecommunications and computer have carried to that propagate fundamental changes in the banking operations, allowing banks to grow dramatically in size and geographic reach.

The researcher Icke (2012) mentions, banks can loan sometimes more money than the quantity is actually saved. This has been possible thanks to the loan system known as fractional reservation.

For DiverDocus (2016) the Fed is the Federal Reserve System, the central banking system of the United States, which was, establish in 1913. With the enactment of the Federal Reserve Act, created to benefit Americans and the country, by providing stability with how banks did business with money.

Several critics as Lietaer (2005) mentioned in Orzi (2012) stated that banking emerged as a mechanism of debt that comes out from the payment of interest. Today is questionable to affirm that banking is a key piece of development, because of the quantities of money that injects for financing and investment issues.

From ancient times to the present, the functions of the banks have changed over time; banks in their first era were considered rudimentary banking centers that were dedicated to the custody of money and the exchange. Today, they are part of the financial system, which are institutions oriented to credit and payment activities, capital administration, and currency exchange activities too.

Global Finance Magazine (2016) lists the best banks in the world, taking into account categories of corporate banking, consumer banking, emerging markets, border markets, investment banking and trade finance, excellence in management, and many more. Table 1 presents this information.

Table 1 Top ten banks in the world

Bank	Criteria	Stock market capitalisation	Country	Founded
Mitsubishi UFJ Financial (MUFG)	Corporate bank	$89.08 billion	Japan	2005
Bank Bilbao Vizcaya Argentaria, S.A. (BBVA)	Bank of consumption	$56.64 billion	Spain	1857
ING Groep N.V. (ING)	Emergent markets	$57.52 billion	Low countries	1991
Standard Bank	Border markets	$18.09 billion	South Africa	1862
Deutsche Asset & Wealth Management (Part of Deutsche Bank)	Management of active for companies	$39.99 billion	Germany	1870
BNY Mellon	Global custody	$43.18 billion	United States	2007
J.P. Morgan Chase & Co.	Bank of investment	$231.36 billion	United States	2000
Citi	Management of effective	$155.76 billion	United States	1812
HSBC	Commercial finance	$140.87 billion	China	1865
Citi	Market of currencies	$155.76 billion	United States	1812
AlBaraka	Islamic financial institution	$1.21 billion	Baréin	2002
Citi	Subcustody	$155.76 billion	United States	1812

2 Information and Communication Technologies Development

Information and Communication Technologies (ICT) have led to important changes in different business areas, public management, homes, education, among others, which support the development of the activities (Blanco et al. 2008).

The use of ICTs in developing countries makes it possible to transform the delivery of basic services, boost innovation, increase productivity, and improve competitiveness. It has been the concern of government and agencies to strengthen broadband and internet services, to reduce service costs, and increase its use (World Bank 2014) (Fig. 2).

As the number of subscribers to cell phones has grown from 15.5 in 2001 to 99.7 in 2016, individuals using the Internet, since 2001 has grown by approximately 6 times by 2016. More than half of the web traffic comes from cell phones (Kemp 2017). Since 2001 fixed telephony subscriptions have decreased 3 points, active mobile broadband subscriptions has grown from 4.0 in 2001 to 49.4 in 2016, fixed broadband subscribers has grown by 11 from 2001 to 2016, more than 50% of mobile connections worldwide are broadband (Kemp 2017) (Fig. 3).

The number of cell phone subscribers continues growing, cellular subscribers have grown in developed countries.

According to the International Telecommunication Union in 2016 highlights that 95% of the world's population lives in geographical areas covered by a basic mobile cellular network, the 3G mobile network has a lower presence in the rural sector (International Telecommunication Union, ITU 2016) (Fig. 4).

The subscription of mobile phones has grown from 719 million subscribers in 2000 to 2.2 billion in 2005 and to 7.4 million subscribers in 2016. By 2000, 65.2% of subscriptions correspond to developed countries, changing the trend for 2005 where

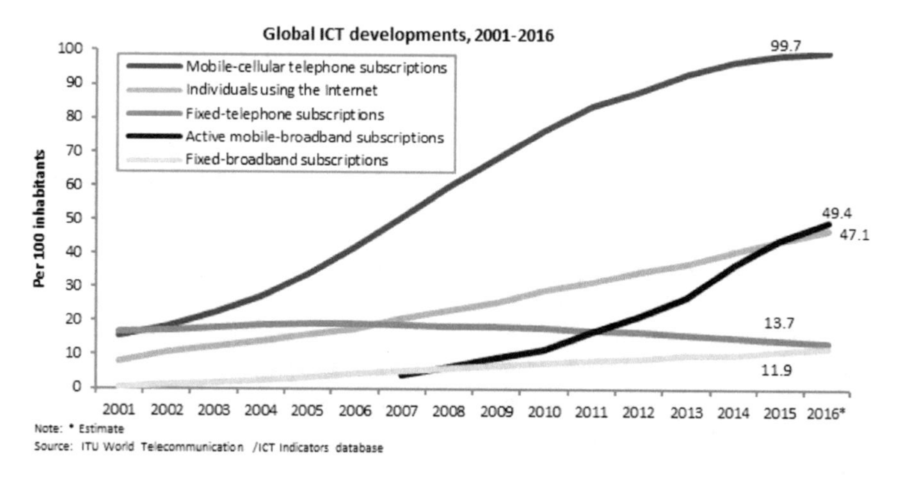

Fig. 2 Shows the analysis of some data on the overall development of ICTs

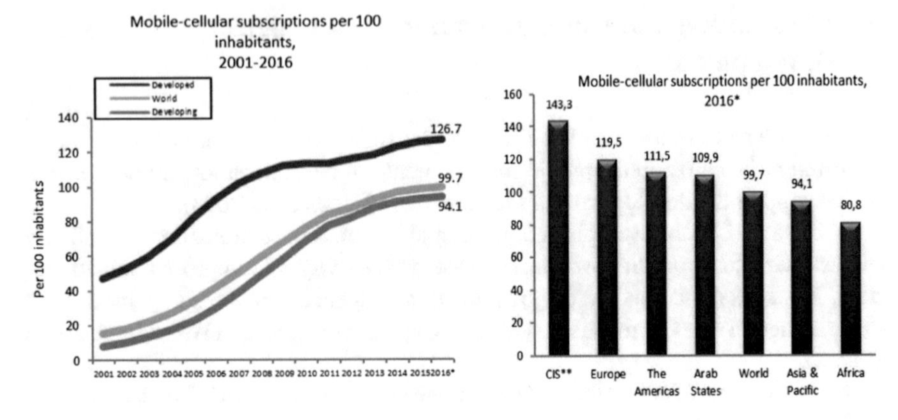

Fig. 3 Mobile-cellular subscriptions per 100 inhabitants, 2001–2016; mobile-cellular subscriptions per 100 inhabitants, 2016*. ITU world telecommunications

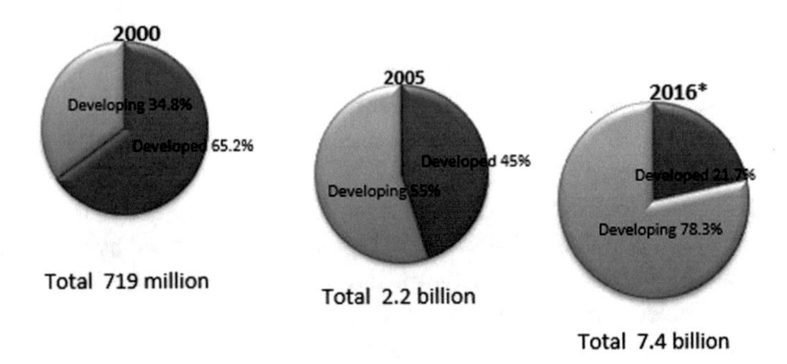

Fig. 4 Mobile-cellular subscriptions, by level of development

the countries in development represent 55% of subscribers and in the same line for 2016 where 78.3% of the subscribers correspond to developing countries.

Figure 5 shows that about half the people in the world are using the Internet (47.1), in developed countries 8 out of 10 people use it compared to developing countries where 4 out of 10 people use Internet.

The amount of individuals using the Internet have grown in the last 15 years. Developing countries have reached 29.4 in 2001 to 81.0 in 2016, mainly in Europe and the Commonwealth of Independent States, countries in development have had growth in the use of Internet 10 times from 2001 to 2016, having the highest growth, the Americas and Africa continues to be the region with the lowest Internet use.

By 2016, 3.9 billion people are not users of Internet resources, although the price of services has decreased considerably. Access to ICT, especially broadband, is part of the Sustainable Development Agenda 2030 in order to reduce the digital divide (International Telecommunication Union, ITU 2016).

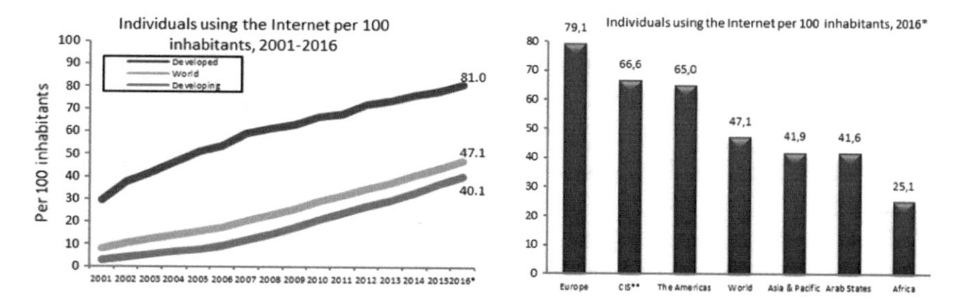

Fig. 5 Individuals using the Internet per 100 inhabitants, 2001–2016/individuals using the Internet per 100 inhabitants, 2016*

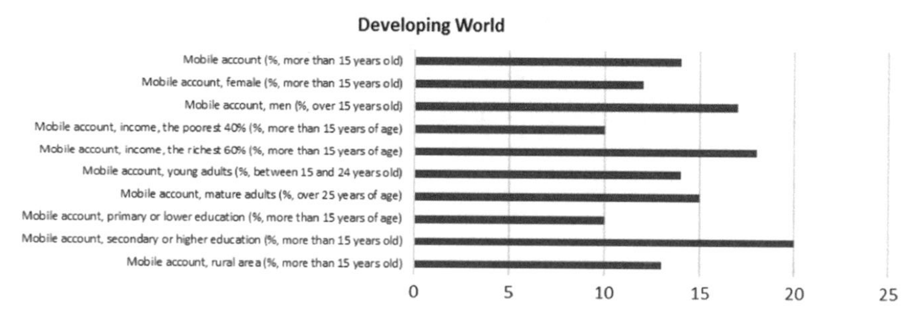

Fig. 6 Use of electronic money in developing world

The use of ICT to promote financial inclusion initiatives has an effect on economic growth in developing countries, establishing a kind of "virtuous circle" among ICTs, the financial system and socioeconomic well-being (Ontiveros et al. 2009).

The development of ICTs has allowed financial inclusion in some countries, as shown in Fig. 6.

Fifteen percent of people over 15 years of age worldwide use a mobile account. Men (17%) outperform mobile accounts compared to women (15%), those with a secondary or higher education have two times more mobile accounts than those with primary or lower education, 18% of the richest people have a mobile account in contrast to 10% of the poorest people who have a mobile account (BBVA Research 2015a, b).

According to the Global Microscope 2016, there were three countries that have stood out with a strategy of financial inclusion in this regard, El Salvador, Honduras, and Nepal. In El Salvador the emergence of mobile money, in Honduras interbank transfers of funds in real time, electronic payment of checks, electronic payments, payment cards, transfers of public funds, cross-border transfers and mobile payments. The Nepal Rastra Bank (central bank), issued new rules on mobile money and banking operations without branches, online banking and business services with electronic cards; all these countries under a rigorous regulation.

By the year 2016, the e-payments indicator Bolivia, Ghana, India, Kenya, Sri Lanka, Tanzania and India is the group of countries with the highest score, among which was measured: “He transacted an account of a financial institution using (% with account, over 15 years)” where 40% of the population of Kenya and 38% of the population of Tanzania do it. In “Using mobile phones to receive money (% over 15 years old)” 67% of the population of Kenya use them, this is because Kenya is taking important steps in the subject of mobile money.

Other Latin American experiences with the use of electronic money are presented below; this case is Mexico through the Bank of National Savings and Services (Bansefi), makes digital payments to the beneficiaries of the program of social inclusion Prospera. In Brazil, the “Bolsa Familia” program, most beneficiaries receive digital payments on a card or bank account, in Uruguay; a law was enacted that reduces the value-added tax (VAT) on transactions with electronic payments through debit cards (Economist Intelligence Unit 2016). In Ecuador, the use of electronic money using a mobile phone, without Internet use, aims at the financial inclusion of approximately 60% of the population that does not have access to financial services (Moncayo and Reis 2015).

3 Banking Innovation

The banking system has suffered various changes throughout the years; one of them is keeping up with technology. By focusing on being more competitive, purposeful, and inclusive in the field.

Group of the World Bank (2016) indicates that more than 40% of the worldwide population has access to the Internet; 74 countries consider that the ICT[1] are necessary and have eliminated unilaterally the tariffs on the goods of capital in the sector of ICT. Other countries follow, while considering computers and smart phones as things of luxury, making possible a greater flexibility of the use of the TIC in the environment. Taking into account, the shortage of time to easily obtain merchandise, products, and financial services, via the Internet. Technological norms in present day is a form of having and communicating information, many financial entities look to use technology as a way to facilitate the way of life for their customers.

The expansion of the Internet, laptops, mobiles, and tablets has popularized the digitalization of a large number of services. Technology has allowed the banking sector to grow, and nowadays going to the bank is becoming less necessary. Through the ICT, we can use ATMs, transfer funds, check products, and check balances and statements, to do online payment, and other kind of transactions that reduce the time of operations and eliminate the physical presence of a bank office.

Banking has been considered the revolution that causes technology, especially in the field of the communications, which is a crucial factor in the banking industry. Banks have included ICT in their operations. Mobile banking nowadays use the cloud computing system which stores information on various devises by accessing ones

[1] Technologies of Information and Communication.

account via the Internet. One is able to accelerate the innovation of processes, reduces costs, improves the experience with the user, monitors and analyses the movements of the banking accounts, supervises transactions, confirms payments, and allows the transfer of money more easily.

Banks' economic activity, innovation, and growth of banks are intrinsically linked to the Big Data; which manipulates enormous quantities of information in order to help the decision makers (BBVA 2016a). Banking industry has many sources of information about their clients, which must be analyzed in order to create new ways to innovate as financial services in communication channels. BBVA (2016a, b) mentions that the digital transformation in banking is easier if it is financially healthy, customers demand, digital products, and the infrastructure will support digital innovation.

John Cohn's assessment reports of tendencies with banking and how in the near future, will be entirely connected. BBVA (2015a, b) illustrates that mobile payment, virtual reality, robots, the Internet; payment systems like bitcoin are among the things that can become the protagonists of the years to come in the world of banking.

Innovation in the banking industry, becomes increasingly necessary, banks should be more creative in order to satisfy to a market that is constantly changing. Where customers are demanding for attention: more diligent with their time, agile, safe, and with a wider range of products and services tailored to their needs.

In order to satisfy these needs, the financial institutions have had to evolve in its form of intermediation. The Internet is a key tool that has allowed evolution to better prepared digital environments and especially adapted to a consumer that increasingly is more familiar with the digital world.

In banking industry, the innovation not only moves to the technological aspect, but to other areas, developing their own initiatives of the environments where they operate and adjusted to customer's needs. By providing a high level of products and services, while having transparent communication channels to satisfy the customer. Banking industry has evolved from a web environment 1.0 to a web of 4.0, with clear examples in the use of disruptive technologies like artificial intelligence.

The change in customers' behavior, as well as the entry of new competitors, are two of the most important risks currently facing the banking industry. It is not only competing among financial institutions but with companies outside their sector; Google, Apple, and Amazon are clear examples of new competitors in the financial industry. While the bank is responding to this threat, only 16% of banks invest in innovation, not being a priority for most financial institutions, according to data from the BBVA Research.

In order to recognize the best innovative practices that were carried out in the banking industry, BAI Global Innovation, every year rewards the best innovative practices around the world. Table 2 shows the winners of the BAI Global Innovation Awards, since 2011–2016.

Since BAI Global Innovation (2017), started to award the best innovation practices in the banking industry, the categories have increased; the organization rewards the best practices in 10 categories. Some categories are innovation in payments and the most innovative community-based banking organization, which were added in the last two years.

Table 2 Winners of BAI global innovation, 2011–2016

Awards	2011	2012	2013	2014	2015	2016
Product and service innovation	Ubank (Australia)	OBC Bank (Singapore)	CaixaBank (Spain) Hana Bank (Korea)	CaixaBank (Spain)	Fidor Bank AG (Germany)	Wells Fargo (United States)
Channel innovation	ASB Bank (New Zeland)	DenizBank (Turkey)	Alior Bank (Poland)	mBank (Poland)	The Bank of East Asia, Limited (Hong Kong, China)	Mizuho Financial Group (Japan)
Innovation in societal and community impact	Punjab National Bank (India) Banco de Credito del Peru		Standard Bank (South Africa)	Fifth Third Bancorp (United States)	Nusenda Credit Union (United States)	Turk Ekonomi Bankasi (Turkey)
Innovation in payments					Idea Bank SA (Poland)	CaixaBank S.A. (Spain)
Innovation in internal process improvement			ZUNO Bank (Czen Republic)	DenizBank (Turkey)	DenizBank (Turkey)	Alior Bank S.A (Poland)
Honorable mention—most innovative non-bank financial services organization				CaixaBank (Spain)	Lending Club (United States)	NerdWallet
Disruptive innovation in banking	Citibank (United States)	Alior Bank (Poland)	Hana Bank (Korea)	Idea Bank S.A (Poland)	Fidor Bank AG (Germany)	CaixaBank S.A. (Spain)
Most innovative community—based banking organization					Nusenda Credit Union (United States)	CBW Bank (United States)
Most innovative bank of the year	CaixaBank (Spain)	First National Bank of South Africa	CaixaBank (Spain)	DenizBank (Turkey)	The Bank of East Asia, Limited (Hong Kong, China)	DenizBank (Turkey)
Honorable mention for disruptive business model			Jibun Bank (Japan)			

Banks, from Europe and United States, stand out amongst their competitors for their capacity for innovation and adaptation to the market, which is evidence by the information contained in Fig. 7.

Figure 7 shows the number of awards by country, United States is notability succeeding in keeping with the times, followed by Spanish banks, promptly by CaixaBank, which has positioned within the banking industry as the most innovative institution. CaixaBank is the financial institution with the largest number of awards received from 2011, with seven awards, in second place is DenizBank with five awards, and Alior Bank which not until 2016 received three awards.

CaixaBank, is an entity that has taken strong steps in the processes of innovation, which has positioned itself as a benchmark in this sector worldwide. By 2014, the entity had a network of more than 9000 terminals, a park of 12.4 million cards issued,

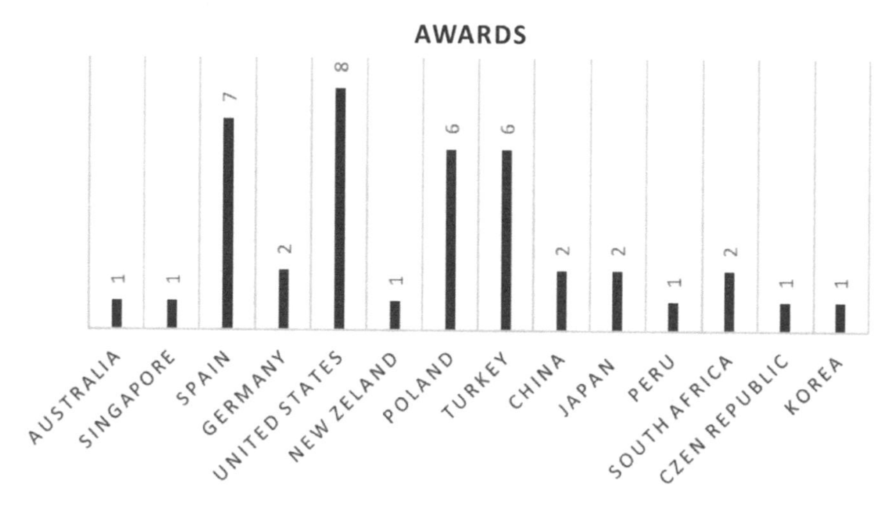

Fig. 7 Numbers of award by country (2011–2016)

and more than 9 million customers banking online, more than 4.2 million of mobile banking customers and more than 10 million downloads in its range of applications for smartphones and tablets. Added to this than CaixaBank is the only financial institution in the world that has its own app store, where they offer more than 40 applications adapted to the different operating systems.

3.1 Challenges in the Banking Industry

The Millennial generation, constitutes a new challenge for the banking industry, the channels of communication and access to information must be adapted to the characteristics of this generation. According to information from the BBVA Research, by the year 2025 the Millennial will compose 75% of the global workforce. These digital natives, on average spend 18 hours a day in electronic media, 46% used paid television, 22% of them spend their time watching traditional TV series, and the 50% of this generation watch at least one online video a day, according to the survey by Koeppel Direct data. Preferences for online media are the following features: 56% for immediate access, 49% multitask, and 44% for the comfort this provided connectivity from anywhere.

About social media and mobile engagement; 71% use hyper-targeted radio like Pandora, iHeartRadio and Spotify, they check their smart phones 43 times per day, and spend 5.4 hours per day on social media. Over 66% of millennial embrace brands on social media and 71% engage in social media daily.

The Millennial Disruption Index (2016), presents that the banking industry is the sector with the highest level of risk, so that 53% of people who belong to this gener-

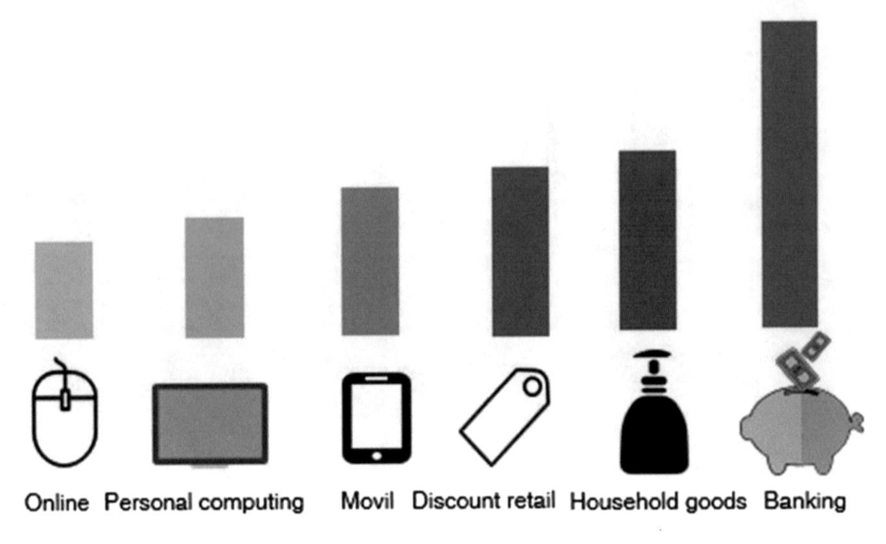

Fig. 8 Risk sectors considered by millennial

ation considered not perceived differentiation between banks and their competition. Topics such as which banks generates lower levels of loyalty amongst their client. The 71% do not care about what the bank does on a regular basis. The changes that will have to carry out the banking industry therefore will be motivated by the trend of consumption of this generation. Since 70% of them considered that the way in which they perform their payments would change completely within five years, 33% considered that in the future they would not need the financial institutions (Fig. 8).

Millennial is a generation that has completely changed the habits and manners of consumption, so it is a decisive factor, it should be considered in terms of innovation. The banking sector should therefore transform their way of doing business, mainly to innovate on its access and communication channels. Banks should need to create fans of their institutions, allowing them to maintain competition in a sector, which increasingly presents major nonbank competitors that have entered with force to break the traditional system of payments.

References

Adxpansion: Curso de Administración de Empresas. http://cursoadministracion1.blogspot.com/2008/11/historia-de-la-banca.html (2016)
BAI Innovation Global Awards: https://www.bai.org/globalinnovations/awards (2017)
BBVA Research: Situación Económica digital. In: https://www.bbvaresearch.com/wpcontent/uploads/2015/08/Situacion_Economia_digital_jul-ago15-Cap5.pdf (2015a)
BBVA Research: http://www.centrodeinnovacionbbva.com/ (2015b)
BBVA Research: Situación económica digital (2016a)
BBVA Research: Centro de Innovación BBVA. http://www.centrodeinnovacionbbva.com/ (2016b)

Blanco, F., Nájera, J., Romero, A.: El efecto de las TIC sobre la productividad en la economía española. In: Berumen, S., Arriaza, K. (eds.) Evolución y desarrollo de las TIC en la economía del conocimiento, p. 261. Ecobook, Madrid (2008)
DiverDocus: https://www.youtube.com/watch?v=vUuEgb0z2YM (2016)
Economist Intelligence Unit.: Microscopio Global 2016. Análisis del entorno para la inclusión financiera. Patrocinado por el FOMIN/BID. Accion y la Fundación MetLife, New York (2016)
Global Finance: http://www.msn.com/es-us/noticias/economia/estos-son-los-mejores-bancos-del-mundo/ss-AAeXowZ?parent-title=el-banco-wells-fargo-retira-bonos-a-exdirectivos-por-apertura-de-cuentas-falsas&parent-ns=ar&parent-content-id=BBzEqRt&fullscreen=true%20-%20image=1 (2016)
Grupo Banco Mundial: http://documents.worldbank.org/curated/en/658821468186546535/pdf/102724-WDR-WDR2016Overview-SPANISH-WebResBox-394840B-OUO-9.pdf (2016)
Guzmán, M.: La banca una historia apasionante. http://eleconomista.com.mx/mercados-estadisticas/2011/04/05/banca-historia-apasionante-I (2011)
Icke, D.: The Biggest secret. https://www.youtube.com/watch?v=crFDjMWE31E (2012)
Ingham, G.: On the underdevelopment of the sociology of money. Acta Sociológica, 41, 1–18. Recuperado de http://journals.sagepub.com/doi/pdf/10.1177/000169939804100101 (1998)
Ingham, G.: The Nature of Money. Polity Press, Cambridge (2004)
International Telecommunication Union, ITU: ICT Facts and Figures 2016. Geneva (2016)
Kemp, S.: Wearesocial.com. https://wearesocial.com/uk/special-reports/digital-in-2017-global-overview (2017)
Lietaer, B.: El futuro del dinero. Errepar, Buenos Aires (2005)
Mendoza, Mendoza: Productos y servicios financieros. Guayaquil, Ecuador, Artes Gráficas Senefelder (2009)
Millennial disruption index: http://www.millennialdisruptionindex.com/ (2016)
Moncayo, J., Reis, M.: Un análisis inicial del Dinero Electrónico en Ecuador y su impacto en la inclusión financiera. Cuestiones Económicas, 31 (2015)
Ontiveros, E., Martín, Á., Fernández, S., Rodríguez, I., López, V.: Telefonía móvil y desarrollo financiero en América Latina. Ariel, Barcelona (2009)
Orzi, R.: Moneda social y mercados solidarios II: la moneda social como lazo social. http://www.campusvirtual.unt.edu.ar/file.php?file=%2F2177%2FMoneda_Social_y_mercados_Solidarios_II_1.pdf (2012)
Vélez Núñez, F., Jaramillo, C.: Análisis de casos bancarios. Ediloja, Loja (2017)
Weber, M.: Historia Económica General, Fondo de Cultura Económica de México (1923)
World Bank: http://www.bancomundial.org/es/results/2013/04/13/ict-results-profile (2014)

Viviana Espinoza-Loayza Professor at the Universidad Técnica Particular de Loja in Ecuador. Ph.D. candidate from the Universidade de a Coruña. Master in financial and banking administration. Engineer in banking and finance. Member of the research group in Finance and Financial Systems of the Department of Business Sciences at the Universidad Técnica Particular de Loja. Her lines of research focus on financial systems and economic development. Loja, Ecuador.

Eulalia-Elizabeth Salas-Tenesaca Professor at the Universidad Técnica Particular de Loja in Ecuador. Ph.D. in Administrative Sciences, Master in Management and Educational Leadership, engineer in Banking and Finance Administration and engineer in Computer Systems and Computer Science. She is also member of the research group in Finance and Financial Systems of the Department of Business Sciences at the Universidad Técnica Particular de Loja. Her line of research focuses on popular finance and solidarity. Loja, Ecuador.

Aurora Samaniego-Namicela Professor at the Universidad Técnica Particular de Loja in Ecuador. Master in financial and banking administration and Engineer in Business Administration. Member of the research group in Finance and Financial Systems of the Department of Business Sciences at the Universidad Técnica Particular de Loja. Her line of research focuses on popular finance and solidarity. Loja, Ecuador.

Part IV
Education

Quality Indicators for the Design and Development of Online Educational Content

María Josefa Rubio Gómez

Abstract This paper analyzes the instructional design for the quality development of this category or area, referred to in two quality assessment models, which have been proposed at international level for online courses. On the one hand, this paper takes into account the quality indicators from the model of*Instituto Latinoamericano y del Caribe de Calidad en Educación Superior a Distancia* (CALED), included in its assessment guide for virtual continuous training courses (*Guía de evaluación para cursos virtuales de formación continua*). On the other hand, the model suggested by the Quality Score Card (*Tarjeta de puntuación, SCCQAP*) of the Online Learning Consortium (OLC–CALED) is also taken into consideration in this paper, emphasizing the differences and similarities between both models.

Keywords Quality · Assessment model · Online courses · Instructional design
Information and communication technologies

1 Introduction

The need for an innovation in the higher education system itself, together with the benefits the application of information and communication technologies can provide us by opening large fields of action for the innovation and renovation of the educational offer through online courses, raise questions about, for example, the quality of those courses. This is an area of concern to us, since the possibilities for tomorrow's education are immense.

It is clear that online education is no longer a novelty, being used nowadays by millions of people and institutions. Nevertheless, we should note that we are still at the beginning, if we consider the more than 800 years of the universities' existence and the recent emergence of the application possibilities for information and communication technologies. It is as if we were living the origins of a new education age. Therefore, we consider necessary to talk about basic parameters for

M. J. Rubio Gómez (✉)
Pontificia Universidad Católica del Ecuador—Sede Ibarra (PUCESI), Ibarra, Ecuador
e-mail: mjrubio@pucesi.edu.ec

M. Túñez-López et al. (eds.), *Communication: Innovation & Quality*, Studies in Systems, Decision and Control 154, https://doi.org/10.1007/978-3-319-91860-0_29

the quality of online education and in order to increase its supply, bearing in mind some essential and internationally recognized standards.

Most of higher education institutions currently offer teaching content through a variety of online programs and tools as an alternative field or support for traditional lessons. Both in this context and in the case of full online courses provided by these institutions, we need to address standards and indicators that guarantee a minimum quality in such courses.

In addition to the above, we must take into account the reality of the globalized world, in which all kinds of virtual courses appear, in many cases offering no quality whatsoever or very few indicators about it. Thus, we set two alternatives out for the assessment of those courses, so that they guarantee a minimum quality that allows users to rely on the numerous offers available to them, both free of charge courses and the ones requiring the payment of enrollment fees.

It should be noted that, in this paper, we will focus only in one of the areas or categories from the two quality assessment models suggested above: on the one hand, the "Quality Score Card" (*Tarjeta de puntuación, SCCQAP*) of the Online Learning Consortium (OLC) (former Sloan Consortium), jointly developed with CALED; on the other hand, the model suggested by the *Instituto Latinoamericano y del Caribe de Calidad en Educación Superior a Distancia* (CALED) itself, included in its assessment guide for virtual continuous training courses (*Guía de evaluación para cursos virtuales de formación continua*). We have selected these two by virtue of our awareness of both models, due to the fact that we have played an active role in the groups that contributed to their initial drafting and continuous improvement. Another reason to take those models into consideration was the fact that they are two valid options for the design and development of online educational content, since both models provide quality certifications at an international level and enjoy the consensus of many institutions.

Similarly, we must specify that we refer in this paper to quality indicators for short-term or continuous training courses. Such indicators may also apply to small units within a degree, as can be the case of a subject. We will not tackle full undergraduate or postgraduate degrees because, even though they resemble regarding some of the requirements, full degrees have to fulfill other requirements that we will not consider here.

The selected models provide several categories or areas that are equally necessary to assess quality, but due to the scope of this paper we will focus this time in only one category, considering that it is among the most important ones, that is, in "Instructional Development and Design" (*Diseño y Desarrollo Instruccional*). We can observe the importance attached to this category or area at a glance, given the score obtained in the global calculation for each model. In the case of the "Quality Score Card" (*Tarjeta de puntuación, SCCQAP*), there are eight categories and it obtains 33 points out of 177. As for the CALED model, it obtains 45 points out of 100 in four areas, that is, very close to 50% out of the global score.

2 Quality Indicators for Instructional Development and Design of Online Courses (CALED Model)

This model was preceded by the one implemented for the "Virtual center for the development of quality standards regarding distance higher education in Latin America and the Caribbean" (*Centro virtual para el desarrollo de estándares de calidad para la educación superior a distancia en América Latina y el Caribe*). Such a model was prepared by a group of experts linked to a project of the same name, funded by the Inter-American Development Bank (IDB) and in which more than 20 institutions got involved, both universities and international bodies. The beginning of this project was based on those institutions' concerns about quality in distance education. In advance of the expansion of this kind of education through new technologies, there was an intention to assist Latin American governments in providing legislation and quality control over such study method, which had already gained substantial acceptance in the 1990s and there were already prospects for its wide development through technology.

The project was led by an organizing committee consisting of the Inter-American Distance Education Consortium (CREAD), the Ibero-American Association of Distance Higher Education (AIESAD), and the Private Technical University of Loja (UTPL), which was also the implementing entity.

This project was completed in 2003 and gave rise to the creation of *Instituto Latinoamericano y del Caribe de Calidad en Educación Superior a Distancia* (CALED) in 2005, which has been working until now on the quality standards and indicators in distance education in its various forms. By doing so, CALED led to complementary models of assessment such as, on the one hand, "Accessible virtual courses" (*Cursos virtuales accesibles*), which was developed with the same basis but complemented with the Methodological guide to implementing accessible virtual curriculum developments (*Guía metodológica para la implantación de desarrollos curriculares virtuales accesibles, Proyecto ESVI-AL 2014*), or, on the other hand, the models for the assessment of distance education programs (Rubio et al. 2010).

CALED's mail objective is to contribute to the improvement of quality in distance higher education regarding the institutions from Latin America and the Caribbean that provide this kind of education. It is composed of the institutions that originally helped with the project "Virtual center for the development of quality standards regarding distance higher education in Latin America and the Caribbean" (*Centro virtual para el desarrollo de estándares de calidad para la educación superior a distancia en América Latina y el Caribe*), as well as other bodies that joined such virtual center. CALED is currently made up of a board of directors, which includes AIESAD, CREAD, Virtual Educa, UDUAL (the Union of Latin American and Caribbean Universities, *Unión de Universidades de América Latina y el Caribe*), OUI (the Inter-American University Organization, *Organización Universitaria Interamericana*), and, lastly, the Private Technical University of Loja (*Universidad Técnica Particular de Loja*, UTPL), which holds the Executive Secretariat. The Advisory

Board is composed of OEA (Organization of American States, *Organización de los Estados Americanos*), OEI (the Organization of Ibero-American States for Education, Science and Culture, *Organización de los Estados Iberoamericanos para la Educación, la Ciencia y la Cultura*), ISTEC (Ibero-American Science & Technology Education Consortium), and ICDE-LAC (International Council for Open and Distance Education—Latin America). Additionally, CALED holds a Latin American Academic Board and a non-Latin American one, both made up of around ten institutions each (CALED 2017).

The interest shown by all the mentioned above has consolidated the different assessment models for quality in distance education in all its various forms, since a wide range of experts have been associated both in order to elaborate, assess and implement such models.

Against this background, we take into account the above mentioned model, as it has been widely agreed and implemented in numerous situations.

2.1 Model Deployment

The model deployment is structured as follows:

- The area selection: a critical factor for the proper functioning of virtual courses.
- The subarea selection: they enable a delimitation of the areas of analysis.
- The standard definition: considered as the level or degree identified as necessary and indispensable so that something can be regarded as acceptable.
- The indicator definition: indicators are seen as the operational scale which helps to identify the standard compliance.

This way, each area is composed of several subareas, these have various standards, and standards cover a number of indicators in order to concretely assess the compliance of quality through the identified standards.

2.2 Structure and Score of the Model

The structure of the model provides four areas (Table 1). As we have already mentioned, in this paper we will focus on the *Instructional Design* area, since we are not able to develop all the components and we consider that this area is the most important one in terms of design and development of online content.

This model, the *Instructional Design* area for virtual courses, includes six subareas that we present here in a synthesized way and without taking into account how they are stated in the original model. In order to facilitate the understanding and the comparison between both models, we will only set out the main ideas about the standards and indicators included in the subareas.

Table 1 Structure and score of the model

Areas	Suggested score
Technology	30 points
Education and training	15 points
Instructional design	45 points
Services and support	10 points
Total	100 points

Source Rubio et al. (2009: 10)

- *Course's Broad Guidelines*. The model points out that the *Course's Broad Guidelines* should be accessible and clearly and accurately drafted. They should also be available to the students from the beginning of the course and they should contain the following necessary background information: a calendar where all the activities and dates of interest are specified, the objectives, the content, the methodology, the materials, and the services to the students. Furthermore, the students must get to know the media available to them from the beginning of the course.
- *Objectives*. They should be presented taking into account the content and the stages to be achieved. The objectives should be explicit and should appear featured in the virtual course from its beginning.
- *Content*. It shall be established according to the objectives previously set and it shall be provided in a dynamic, associative and relational way. The content shall be meaningful, contemporary and appropriate to the students targeted. As short-term courses, the content shall be presented in small units. Besides, the population targeted should be taken into consideration and, if so, the content shall be available on alternative media for its publication, considering the non-availability in terms of permanent access to the internet or a low-speed connection, in the case of some participants.

It must not be forgotten that the learning units should be associated to those resources and activities that help materialize what has been learnt and develop creativity. The activities shall be adapted to the various learning strategies through simulations, case studies, etc.

It is essential to have licenses in order to publish content, whether for creative commons or copyright types.

- *Interaction*. Regarded as a requirement for an effective communication among the different actors, the interaction must be guaranteed by means of the diverse tools and using the context and the objectives previously set. The activities shall be designed in order to encourage communication, collaborative work, exchanges among the actors involved in the process and content interaction. Thus, social relations shall be promoted among the participants involved in the teaching-learning process.

Private communications should also be foreseen, whether through e-mails, chats or other non-communicative media. Likewise, the network rules of "behavior" (netiquette) shall help improve both foreseen and spontaneous interactions.
- *Monitoring and mentoring*. During the course, a monitoring and a follow-up of the activities must be foreseen through the students' browsing statistics, the content visited by those students, the activities carried out and the interactions that have been taken place.
  The mentoring plan, that shall be published and well-known to the students, shall be assessed and, besides, action will be taken on the weak points of its compliance. Continuous guidance to the students shall include monitoring sessions with maximum response times in order to clear up their doubts and, when appropriate, provide a feedback in the assessments. It is also advisable to use different tools for the student guidance (video conferences, e-mail, Skype, debate fora, chats, etc.).
- *Assessment*. Continuous assessment in this kind of courses is considered essential to be able to talk about quality, and therefore there must be a reliable system concerning the record of grades. Such an assessment shall take place in accordance with the objectives or skills to be achieved and it shall take collaborative work into consideration. There shall be self-assessment tests available for the students so that they are aware of their evolution and degree of understanding. It is vital to guarantee the students' identity and the confidentiality of their personal data. Due to various contingencies in the assessment system, some alternative ways of assessment should be considered.

3 "Quality Score Card" Model (SCCQAP)

The "Score Cards" were the fruit of the joint work carried out by OLC and CALED in 2013, taking the models that each institution was using separately. On the part of CALED, the model used was the one already mentioned above and as far as OLC is concerned, the "Quality assessment cards" (Quality Score Cards) were the selected ones. These Quality Score Cards have been designed to measure and quantify elements of quality regarding online courses, through the assessment of each and every quality indicator included in categories that were created to determine the strengths and weaknesses in the implementation of the process. These "Score Cards" or *Tarjetas de puntuación*, as they have been called in Spanish, included aspects from both models but they preserved the original structure from the "Quality Assessment Cards" or *tarjetas de valoración de la calidad*.

"The Online Learning Consortium (OLC) is the main organization from the United States of America devoted to integrating online education in the core practices of higher education, as well as giving advice to institutions and individuals in order to improve quality" (OLC, CALED 2015: 10), and thus its field of action was focused first on the United States. By means of the joint work, the intention was to create a tool covering a wider area and making it possible to use it as well in Spanish-speaking

Table 2 Score card for the assessment of online educational courses

Categories	Suggested score
Institutional support	18 points
Technological support	21 points
Instructional development and design	33 points
Structure of online courses	18 points
Teaching and learning	12
Support to teachers	6
Support to students	33
Assessment	36
Perfect score	177 points

countries regarding online programs. Moreover, an opportunity was given to apply these cards to short-term courses but selecting certain categories and indicators as shown below.

3.1 *Structure and Score of the Model*

The model is structured in 8 categories and 59 quality indicators in the case of short-term virtual courses.

The Score Card for this kind of courses suggests a score of 177 points, disaggregated by the different categories and quality indicators. Each indicator has an assessment of, maximum, three points (Table 2).

We mention here below, in summarized form, the category of study together with its indicators and suggestions, being the latter provided in the model so as to obtain the indicators.

3.1.1 Category: Instructional Development and Design of Online Courses

1. *A series of guidelines are used when it comes to minimum standards for the development, design, teaching and learning of the course (such as some elements from the course academic programs, assessment strategies, remarks and comments from the teachers).*

It is seen as an indication of quality that the institution provides minimum standards for the teachers regarding the course development and its academic program, guiding them as well in its material elaboration and in its basic components. The creation or adaptation of course certifications that assess quality is recommended, in order to incorporate essential standards for the institutional culture as well as other standards for teaching and for good teaching strategies with a view to make the course attractive for students.

In this manner, it is recommended that templates are in place for the academic programs, with easy to follow sections also containing practical aspects.

Clear guidelines and recommendations should be available in order to achieve the minimum standards corresponding to a given course, for example text transcripts of audio and video files.

Furthermore, examples of good practices should be provided.

At the same time, online courses should include the expectations about the teachers' commitments regarding their presence, support and response times.

2. *The instructional materials, the school curricula and the learning results are periodically reviewed with a view to guarantee that they comply with the standards of the online course.*

A basic requirement for regional accreditation authorities of the United States of America is that the institution commits to executing processes that incorporate a systematic review of the results. The main objective of this review is to continuously improve quality and to show that the institution is efficiently fulfilling its mission.

The institution should identify the expected results and assess their degree of achievement. Besides, the institution should present evidence to a continuous improvement in terms of the result analysis made on a regular basis, both of the learning results and the developed improvements. The course materials need to be improved once the courses have been given and assessed, so as to provide a continuous improvement in the achievement of the learning results.

Therefore, in order to talk about quality, the course results should be systematically reviewed to determine if the course is up to date and still relevant. Also, the interested parties in this process should be involved in it.

In addition, it is essential to systematically review and improve the educational material with the purpose of achieving the learning results.

Finally, there is also a need to systematically review the course materials in order to guarantee the relevance of the educational technology and the assessment activities.

3. *The courses are designed in such a way that students develop the necessary knowledge and skills to achieve the learning objectives. These courses may include participation through analysis, synthesis and assessment.*[1]

Courses are considered to be designed in such a way that they strengthen the interactivity so as to master the course objectives and, as a consequence, it is suggested that some aspects regarding the progression are taken into consideration, according to Puzziferro and Shelton which we summarize here below:

[1] Adapted from the report "Quality on the Line: Benchmarks for Success in internet-based Distance Education, 2000. Institute for Higher Education Policy" (OLC, CALED 2015: 18 and 25).

- Having a good command of the information regarding the key concepts and the main terms or ideas;
- Progressively capturing the information by means of practicing learning activities that focus on the memory, as well as building the vocabulary and the information understanding;
- Apply the information to a situation based on problems and allow these to be jointly analyzed and developed;
- Analyzing the information through the recognition of patterns or relations between the information and the problem;
- Creating new knowledge and the ability to reason when it comes to information, applying it to practical situations (OLC, CALED 2015: 54).

Therefore, it shall be verified that the online course contributes to the learning objectives presented, and it can be made through result maps, charts, etc.

A series of educational and regulatory assessment criteria shall be used in order to support the learning results that are being measured and also in order to help students measure their improvement.

The course design should include practical aspects with which the students are able to check and improve their understanding regarding the use of that course by means of comments from colleagues and experts.

4. *The learning objectives describe measurable results.*

The courses should set learning objectives that guide us where we want to go, as otherwise we would end up somewhere else. The objectives focus on what the student has to learn and not on what the teacher is going to teach. Also, the objectives should be clear and deliver measurable results. Many educators use Bloom's taxonomy in order to build the learning objectives, and it has replaced names with verbs: knowledge with to remember; comprehension with to understand; application with to apply; analysis with to analyze; synthesis and assessment with to create and assess, etc. Each of these verbs can be replaced with different ones, as for example to remember with to define, duplicate, list, memorize, repeat, reproduce, etc.

It is recommended that the course developers start with a planning guide or itinerary which may support the learning results.

A further essential point in online education is that the students have the complementary information available in a clear and simple way and stating the course objectives, the concepts, the ideas and the learning results.

Some kind of assistance to teachers is recommended in order to manage the objectives development through workshops, tutorials and documents.

Adequate means should be provided to ensure that the students give regular feedback of the objectives understanding.

Each of the objectives shall describe what needs to be learnt, how it needs to be learnt and how the learning shall be assessed.

5. *The selected assessments measure the learning objectives of the courses and they are suitable for an online learning environment.*

The assessments should be created to measure the level of success achieved by the students, the command of the content and the development of the critical aspect. These assessments should be properly designed and executed, providing teachers with the evidence of learning, the efficiency of the materials used and the information available for the future improvements applied to the course. It is vital for these assessments to match with the course content and to be defined and identified from the beginning of their design. In addition, the assessments may include objective tests and exams, trials, case studies, considerations and group projects.

Moreover, a teacher's training is recommended about how to measure several learning objectives through a single assessment.

Besides, it is advised to use certifications as a guide for the students and as a way of facilitating the rating of the students' papers.

Furthermore, self-assessment tools should be fostered so as to obtain a prompt feedback and also throughout the entire process.

Finally, it is necessary to have a number of mechanisms dedicated to foreseeing that no academic dishonesty occurs, especially during the final reviews and assessments. This can be foreseen through the creativity shown in the papers, since the creation of videos, PowerPoint presentations or websites can be asked instead of doing the standard exam.

6. *The design of online courses promotes the participation of both teachers and students.*

The design of the course should provide a wide range of possibilities for the student interaction with the content, the teachers and their classmates. This shall help the institution maintain a commitment environment. Attractive and collaborative activities need to be designed in order to increase participation, such as blogs, wikis, audio and video files, and collaborative projects.

It is also recommended to provide some kind of support to teachers for the educational design of virtual courses and which underscores the active learning.

Moreover, it is advisable to provide an online course to teachers that could serve as a model in order to suggest interesting activities.

Additionally, the teachers should be trained in the proper use of technology.

The teachers should provide their students with clearly stated expectations regarding the methods and the frequency of their interaction with other students, teachers and content.

7. *The curriculum development is a key responsibility of the teaching staff (in other words, the teaching staff should take part in the development or the decision making processes regarding the online curriculum options).*

Since the curriculum is the most prominent statement that an institution can attest to, regarding what it can contribute to the students' intellectual development and regarding its service to society, the curriculum development should lie in the teachers. They shall be the main responsible for the school curricula, keeping them updated. Thus, they should review and assess those programs, comply with the grading standards, and select the learning resources and the media available.

An alert in this indicator may occur when the institution lacks teaching staff and it has to make use of external suppliers.

It is recommended that the teachers lead the selection and development of online school curricula from the beginning, including the decisions on selecting experts in content and open resources.

The teaching staff should have knowledge of the data concerning the learning results, the students' achievement and graduation rates, the delinquency rates, the cost efficiency, and the actual applicability of the content.

8. *The content is up-to-date and appropriate for the students (the content is adjusted to the target students).*

If all kinds of education require an updating of their curriculum content, it is even more necessary for online education to require a constant review of the content, the learning and assessment activities, the teacher performance and that of the technological platform. There has been a change in the use and application of the information and communication technologies, as well as in the definition of the learning processes, in the teachers' role, in the core resources (materials), in the infrastructure and the access to the networks, in the open use of many resources and in the regular practices of teachers and students. Therefore, the technological resources need to provide flexibility in order to meet the individual and social needs, and to create effective learning environments which enable interaction.

The review of the curriculum content and the execution of the necessary updating and adjustments are recommended at the end of each course, also in order to achieve the quality indicators. Besides, the content needs to be assessed so as to make sure it meets the objectives pursued and it is adequate to the students who have chosen the course. It is also required to identify strategies for several learning styles and include digital resources to provide a better understanding of the content.

9. *A series of alternative means are available to publish content (CDs) for the students who do not have a permanent access to the Internet or have a low-speed connection.*

The availability of an Internet access is a key element in this kind of courses, but we can see that in some countries or places there is no connection at all or the speed is very low. Thus, there is a need to develop alternative ways of material delivery to use in a given moment or circumstance, as for example: to allow downloading materials from the Internet or delivering them in alternative documents, providing CDs, listening material, videos, etc.

It is recommended for support materials to be available in other formats, such as CDs, DVDs, USB flash drives, etc., intended for students with connection problems.

Furthermore, it is necessary to offer the possibility of downloading or printing content and learning tasks directly from the platform.

Lastly, there is a need to remind the students of these alternatives for facilitating access to their content, without having always to connect to the Internet.

10. *A number of activities are set out, adjusted to the different learning strategies (simulations, case studies, etc.).*

Both the teachers' and students' roles are different from the past. The teacher is no longer the only source of knowledge and the student is no longer the passive individual receiving that knowledge. The teacher should be able to turn the course content into a series of meaningful learning activities. Thus, the teacher's tasks include planning a series of activities that are appropriate to different learning styles. The teacher should consider the complexity of the information and the students' prior knowledge, the different learning styles of each pupil and the need for reinforcement or persistence in the activities for some students that may also need a direct support from their teachers.

It is therefore recommended to provide a number of constructivist strategies if we want to talk about quality in these indicators, as for example simulations, case studies, role-playing games, collaborative work, tutorials, etc.

The different learning styles should be identified and the activities should be guided around these styles.

It is also necessary to offer different learning activities from which the students can choose the ones of their interest.

11. *There are alternative assessment systems for those students who do not have a permanent access to the Internet.*

The variable opportunities when it comes to Internet access should not be considered a limiting factor in order to provide an online education for those who wish to have access to it, guaranteeing at least a minimum level of connection. The institutions should analyze the alternatives so as to provide the content or the learning assessment systems when there is no or bad connection.

A further aspect to be considered is the guarantee of the student's identity. Thus, online-on-site tests may be foreseen in approved centers with guarantees of connectivity and authentication or, if it is not possible, monitored written tests.

It is also advised to provide the students with alternative assessments not requiring a permanent access to the Internet. The teachers should allow the students to participate in their assessments in approved centers and with the appropriate guarantees (OLC, CALED 2015: 51–69).

4 Comparative Study of Both Models

Both models are structured so as to be used in the self-assessment of the online courses and they are also offered with the possibility of having external certifications provided by the institutions supported, that is, CALED and OLC.

For a better understanding of the subject, it is necessary to have knowledge of the full models, as the study that we provide here only addresses the subject partially. This does not prevent us from giving credit to the aspects mentioned here, because they can help us with the course design and provide us with standards for the development and assessment of those courses.

In fact, a number of official Mexican bodies have used the Score Cards as an assessment tool used before the approval of the program offer.

Clearly, the Score Cards and the explanations for their understanding aim to provide support so that the courses are well designed. For example, it is recommended that the course developers start with a planning guide or itinerary which may support the learning results. Besides, some kind of assistance to teachers is recommended in order to manage the objectives development through workshops, tutorials and documents. It is also necessary to have a number of mechanisms dedicated to foreseeing that no academic dishonesty occurs, especially during the final reviews and assessments. This can be foreseen through the creativity shown in the papers, since the creation of videos, PowerPoint presentations or websites can be asked instead of doing the standard exam. We also insist on providing some kind of support to teachers for the educational design of virtual courses and which underscores the active learning. It is also advisable to provide an online course to teachers that could serve as a model in order to suggest interesting activities.

As a general assessment and regarding the approach, we could emphasize that the CALED model focuses more on the student, while the OLC-CALED model addresses the institutional forecast in order to tackle quality and the teacher's performance to achieve it. Apart from the recommendations already mentioned, there are some others, as the fact of having already certifications or parameters for the assessment, the recommendation that templates are in place for the academic programs, with easy-to follow sections also containing practical aspects, or including the expectations about the teachers' commitments regarding their presence, support and response times.

Generally speaking, the CALED model makes more explicit the standards and indicators that are to be fulfilled and it focuses more on testing them, while the OLC-CALED model provides indications and guidelines for the interpretation of quality.

The first difference that we encounter regarding the CALED model starts with the area or category name that we have tackled in our study. In the case of CALED, it is just called "Instructional Design", while in the Score Card model it is "Instructional and Development Design".

The structure is also different: in the CALED model there are areas, subareas, standards and indicators, while in the Score Cards we talk about categories and indicators, the latter being more considered as items within those categories.

As we have already mentioned and despite the high score of both models in the area or category of Instructional Development and Design of online courses, it is clear that this area has a higher weighting in the case of the CALED model, with 45% out of the total, while in the Score Card model the area is given a weighting of 19% out of the total. It should be noted that there are eight categories in the Score Card model, while there are only four in the case of CALED.

The Score Card model insists on systematically reviewing the course results to determine if the course is up to date and still relevant, which are two quality signs. Also, the interested parties in this process should be involved in it.

The different indicators in each of the models are differently distributed but are included in the subareas or indicators to a greater or lesser extent. Both models, despite their different approaches, take into account:

- The course's broad guidelines for its design and for the student's knowledge;
- The objectives to be achieved and how they should be attained;
- The content according to the objectives previously set, that is, contemporary and appropriate;
- The materials and their alternative means;
- The interaction activities, adjusted to the different learning strategies;
- The monitoring and mentoring;
- The appropriate assessment to the learning objectives and to the online learning environment, as well as the alternative systems for such an assessment.

5 What Model to Choose?

The approach of each model allows us to choose between one or the other for the design and the assessment of the different online courses, making it difficult to choose one of them. Both models might be used in order to assess the quality of a given online course. The difference lies in what kind of report we want in a given moment, or the assessor's experience or knowledge in one or the other model, as there are some significant differences regarding the coverage, depth and management of each process. What remains clear to us is that we have to use one model fully, since there are a number of aspects included in other areas or indicators than the one we have analyzed and, thus, if we take into account only the analyzed area, we would be biasing our work, we would not have a broad view and we would not be able to establish significant differences or similarities.

It is obvious that the effort made by the Online Learning Consortium (OLC) and the *Instituto Latinoamericano y del Caribe de Calidad en Educación Superior a Distancia* (CALED), when it comes to performing a collaborative and complementary process, has resulted in the Score Cards. Moreover, each of its assessment methodologies was enhanced and achieved an international broader vision. This is a new product that does not cancel the previous ones, and thus both the CALED model (the Score Cards, OLC-CALED) and the initial Quality Score Cards can be used interchangeably.

We have analyzed in this paper the two models that could be more easily used in Latin America, although both have their English version just if an institution from other continent wishes to use one or the other.

In spite of the conditions promised or required by the different state legislations about the secure access to digital environments (2016: code, Art. 39), it is well known that this is not the case, especially in Latin America. Therefore, we consider that the security measures provided by the models presented here are important if we want to talk about quality in online courses.

It is necessary to assess the quality starting from the design and also once the courses developed. The first assessment shall avoid many mistakes and the second one shall allow the updating to be a key principle of the quality and the institutional commitment regarding a continuous improvement.

We cannot forget that the scope of online courses is global. For this reason, the commitment to offer quality goes beyond the specific legislations from the countries providing online courses, although in some cases the country requests the student to be accredited. Our proposal is to begin to use international standards and indicators like the ones we have already presented, as from these ones there can be new local standards if a given country requires them.

Finally, another issue to be pointed out is the fact that the standards and indicators, especially in this online environment, are changeable or complementary according to the possibilities offered by the technologies. But let us not forget that the means is one thing and the objective to be achieved is something else.

References

CALED: http://www.caled-ead.org/ (2017)

OLC, CALED: El proceso de garantía de la calidad para la educación en línea y a distancia. Tarjeta de puntuación (SCCQAP) Evaluación de Programas de Pregrado en Línea. Universidad Técnica Particular de Loja, Loja, Ecuador (2015)

Rubio, M.J., Morochos, M., Torres, J.C., Maldonado, J., Alejandro, J., Ramírez, I.: Guía de evaluación para cursos virtuales de formación continua. Instituto Latinoamericano y del Caribe de Calidad en Educación Superior a Distancia (CALED). Universidad Técnica Particular de Loja, Loja, Ecuador (2009)

Rubio, M.J., Morochos, M., Maldonado, J., Alejandro, J., Ramírez, I.: Guía de autoevaluación para programas de pregrado a Distancia. Instituto Latinoamericano y del Caribe de Calidad en Educación Superior a Distancia (CALED). Universidad Técnica Particular de Loja, Loja, Ecuador (2010)

María Josefa Rubio Gómez Doctor of Philosophy and Sciences of the Education (Universidad Nacional de Educación a Distancia—UNED, Spain). She is Vice-rectorate of the Pontificia Universidad Católica del Ecuador, Campus Ibarra. Deputy Director of the Instituto Latinoamericano y del Caribe de Calidad en Educación Superior a Distancia (CALED). Former DG of the Open and Distance Modality of the Universidad Técnica Particular de Loja (UTPL 1997/2010). Associate Director at Center of the UNED in Balearics and Person in charge of the COIE (1988/1997). Ibarra, Ecuador.

Social Networks as a New University Venue

Diana Rivera-Rogel, Jenny Yaguache Quichimbo, Andrea Victoria Velásquez Benavides and Fanny Paladines Galarza

Abstract The university with the aim of maintaining a bidirectional communication takes advantage of the technological era, as a tool to strengthen not only its teaching-learning process, but also strategic communication at all levels: teacher-student, student-student, and university-student. The 2.0 university walks at the same time as the digital revolution. There are only some of the institutions that are translated in the exemplary adaptation to the diverse possibilities that offers Internet. There are others who are still working on this path; however, the tendency will lead them to improve their digital communicative processes. At present, the most used social networks in the communication of the universities that allow both an informative and communicational approach are Facebook, Twitter and Instagram.

Keywords University · Web 2.0 · Student-pupil · Student-teacher
Communication management · Education

1 Introduction

The university institution like any other organization acts under functional strategic units, each of which is consistent from a higher hierarchical level to an operational line. These functional strategies are linked to educational quality, research and the transformation of the environment. Each university determines according to its pos-

---

D. Rivera-Rogel (✉) · J. Yaguache Quichimbo · A. V. Velásquez Benavides · F. Paladines Galarza
Universidad Técnica Particular de Loja, Loja, Ecuador
e-mail: derivera@utpl.edu.ec

J. Yaguache Quichimbo
e-mail: jjyaguache@utpl.edu.ec

A. V. Velásquez Benavides
e-mail: avvelasquez@utpl.edu.ec

F. Paladines Galarza
e-mail: fypaladines@utpl.edu.ec

M. Túñez-López et al. (eds.), *Communication: Innovation & Quality*, Studies in Systems, Decision and Control 154, https://doi.org/10.1007/978-3-319-91860-0_30

sibilities the intensity that will give these functional areas, and will choose the one that offers more attention to this service.

This service is visualized in the educational quality, students, teachers and administrative satisfaction through internal and external activities, and the impact on revenue and costs. Disseminating the offer of services is the key in any institution and in our university; it leads to attracting students and prestige. Identifying the best channels and tools to achieve the expected reception is the move of the communication cabinets, considering that the audiences are increasingly digitized and especially participatory.

Today, universities around the world use social networks and mobile applications to connect with current and future students, employees and communities of interest; and, they also open spaces for external dialogue to disseminate their academic programs, research and ongoing work.

Undoubtedly, today social networks cross the boundaries of the university campus, from its internal and formal use that is born in the strategic planning of the directions of university communication; the one created by the teachers as the support to the subjects until those external and informal activities are developed by the students. In this mediated world totally dedicated to the active participation of audiences, the university takes advantage of social media to strengthen its brand, improve its reputation and implement actions that generate confidence.

The university communication management explores the diverse ways in which students, teachers and administrative staff can be heard in order to create strong relationships that generate institutional value. This communicational and institutional objective is achieved as long as the university organization has a corporate culture 2.0. Contrary to this, the educational organization would be behind new highly productive communicative processes.

At an educational level, social networks have an effective participation (Donlan 2014), allowing a collaborative learning, bringing together the fascination of the students for the constant use of the various social media tools (Gross et al. 2015). In terms of promotion and dissemination, social networks have been used to attract and retain customer loyalty (Chatterjee 2013); always taking care of the information that is issued (Treem and Leonardi 2013). Both scenarios: the educational and the communicative are those in which the universities put more emphasis in their audience to think over, when choosing a university, or helping the diffusion of their performances through a "Like", a tweet or a retweet.

2 Conceptual Foundations and Context of Social Networks in University Students

At the beginning of the 21st century, the Internet became an important form of communication. E-mail has now become one of the most popular ways of transmitting messages. Since 2010, e-book readers have become increasingly common. "The recent history of communication has had a clear protagonist that is the audiovisual revolution, to which it is now added a second one that is the broadband network" (Majó 2012: 66).

For their advantage, institutions of higher education have Information and Communication Technologies (ICT). The challenge lies in the integration of ICT in teaching/learning processes, but this use of technologies is not new; especially for those universities that extended their education at a distance. For example, communication technologies: satellite television or video conferencing have been in use since the 1950s. Similarly, in some leading universities in the United States such as Stanford, Georgia Tech, or the University of California at Santa Barbara that developed a closed circuit television for their students and teachers; they also recorded their video classes in such a way that are available when needed. However, not all educational institutions have received this impact with the same intensity and at the same time, since there are many social and economic factors that have influenced their adaptation (Sangrà and González 2004).

The major promoter of the term Web 2.0, Tim O'Really highlights two important aspects of this evolutionary process: the ability to take full advantage of collective intelligence and the creation of content produced in a collaborative sense by users (O'Reilly 2012).

If the organizations used to upload their information in the Web, today the users are the ones that do this activity mainly through the social networks. With Web 2.0, information consumers take the position of "prosumers" because they are the ones who produce the information they consume (Maciá and Gosende 2011). This group is the "natives of the digital world" or in English the "digital natives" that are born in the era of the Internet. These Internet users use up-to-date instruments in the market and counteract "digital immigrants", or immigrants, who were born in the pre-digital age (Pisani and Piotet 2009: 39).

Prensky (2010) who coined the term digital natives sees the generation born after 1980 as a generation of people, who grew up surrounded by new mass-consuming technologies have developed another way of thinking, and conceiving the world. Digital natives are characterized by having skills in the use of technology. Most of their life and daily activities are mediated by digital technologies. A digital native knows no other kind of life!

If we have to add in data in the timeline we can say that 2004 is a year of great importance for social networks, Mark Zuckerberg launches Facebook. In February 2004, Chad Hurley, Steve Chen and Jawed Karin create YouTube, and in June of the same year, Google creates a social network called Orkut that ends up being consumed by Brazil and India. In January 2006, the Internet reached 1.1 billion users. Also in July 2006, Twitter was launched, the first microblogging network created by Jack Dorsey. Besides that, in November 2006, Google acquires YouTube for a value of 1.65 billion dollars. And finally, Amazon creates the Kindle e-book reader. In June of 2007 appears the IPhone, first multimedia telephone device with Internet connection. From here much of the Internet traffic is beginning to be established through mobile devices. In January 2011, Facebook reached 600 million users. By 2015, 750 million households worldwide are connected to the Internet; half of these households are from developing countries (Velásquez 2017).

Young people form their own virtual communities, groups of people who share, discuss and get interested in the same subject. Networks work just like a social club among Internet users. "They are Web sites in which an individual space is made available to the user where he can and should configure, design, and fill with content to share it with whoever his interests are" (Dans 2010: 269). In fact, the students' fascination with networks makes it a didactic and strongly communicative possibility.

The forms of communication in the universities have evolved allowing the collaborative learning which implies the creation of spaces of exchange of information. Therefore, "to communicate at the University it is no longer enough only the traditional channels; it is necessary to innovate and create new supports that reach all the agents involved" (Franch and Camacho 2004: 227), besides the need to integrate information technologies (ICT) such as social networks.

Social networks arise based on the interests of potential users. The most used are: Facebook, Tuenti, YouTube, Myspace, Xing, Hi5, Twitter, LinkedIn each one offering their services in different languages. In the educational field their use can be focused on classroom projects, subjects' portfolios, virtual learning communities or communities of teaching practices. The adequate use of social networks allows students to train in different skills, and abilities that contribute to their cognitive, social and personal development in general; also to school learning in particular (Maiz and Tejada 2013). And in the strategic area of communication they allow an agile relationship between the various audiences in the university field.

Table 1 shows the number of users of the main social networks in the world.

Therefore, in order to take advantage of social networks it is necessary to modify the roles of teachers and students, which implies the adaptation of perfectly designed strategies that can influence their decisions, in the sense of teaching–learning and/or communicative approach.

From a study carried out in the University Center, of the Highs of the University of Guadalajara with 414 students of the 14 university careers that are offered in this institution, with the sole purpose of knowing the use of social networks as a learning strategy. It is well known that 71% of students use it to inform peers about school activities; 42% to play; and 90% to communicate with others showing the importance and strength that social networks are taking in the educational process. In addition, the study indicates that some students chose one or more options showing the possibilities of using social networks for other activities, too (Islas and Carranza 2011).

However, another study on the Andalusian Public Universities, which analyzes the use of social networks in institutional communication with university students, indicates that Facebook and Twitter are the most used networks, Estévez et al. (2012: 138) points out:

> In general, universities are not taking advantage of the wide amount of benefits of each of the social networks, and tend to use the same formula of communication for all of them, favoring that the channels should be publicized among themselves, but without providing differentiated content of quality or interest.

Table 1 Users of the main social networks in the world

Name	Creation date	Services offered	Users worldwide
Facebook	2004	Social network that connects people with people around the world. It shares resources like: web pages, photos, videos …	1440 million of users globally
Twitter	2007	Microblogging service. It allows you to send plain text messages of short length, with a maximum of 140 characters called tweets that are displayed in the main page of the user. Customers can subscribe to other consumers' tweets—this is called "follow" and subscriber users are called "followers"	326 millon of users globally
YouTube	2005 owned by Google since October 2006	It is an Internet portal that allows its users to upload and view videos. It has an online player based on Flash. One of its main innovations is the facility to visualize videos in streaming, that is to say, without needing to download the file to the computer. Users, therefore, can select which video they want to watch and instantly play it	1000 million of users globally
Google+	June 28, 2011	Google Plus, Google+ or G+ is Google's social network. Google claims that it is a social platform and not a social network because Google+ lives inside, and throughout the rest of Google products, and services. The mode of operation and use is very different to Facebook or Twitter	500 million of users globally

(continued)

Table 1 (continued)

Name	Creation date	Services offered	Users worldwide
Instagram	October 2010	It is a program or application to share photos with which users can apply photographic effects like filters, retro and vintage colors; Then, share photos on different social networks like Facebook, Twitter and Flickr …	400 million of users globally
WhatsApp	2009 In 2014 it was acquired by Facebook	Basic messaging application. Groups can be created and sent between them an unlimited number of images, videos and audio messages	700 million of users globally
LinkedIn	2002	It is a professional social network, that is oriented more to commercial and professional relationships than to personal relationships. Companies and professionals seek to promote themselves, do networking and business	400 million of users globally (100 million active users)
Line	2011	Send messages, calls and/or video calls through smart phones. Stickers, emoticons and games. Free calls through the network. Mobile social network called Timeline	450 million of users globally
Skype	2003	Free voice communication between users from and to any part of the world. It also uses text, voice (voIP) and video	300 million of users globally

Source Authors' compilation with data from Socialbakers (2015) and Management Training (2015)

On the other hand, when analyzing the different types of communities that can arise within the university-student-university communication process, networks can have different orientations: personal learning networks (allowance to fulfill personal and professional goals); Communities of practice (union of related groups to fulfill concrete tasks); Learning communities (group of students learning together); Subject networks (for consultations, to carry out works, dialogues between teachers and students, between pairs, etc.); Groups of students (to carry out work, use of forums to debate, or walls to leave a message) (Maiz and Tejada 2013); Thematic networks according to the intention of communication (admissions, research, social linkage).

We are in a technological era marked by the use of the Internet, a tool that has been very powerful due to its ease of use, and its accessibility that has allowed the new generations to have at their disposal information of less or greater importance. The young people of the 21st century are part of a generation that has experienced a stage in which technology has made life easier for them in an impressive way; like for instance, the use of tools such as search engines to perform their tasks and obtain new information; and, the existence of social networks to communicate and know. It is thus, that young people have created new ways of expressing themselves: writing, relating … and even they have conceived new ways and rules of behavior; therefore, it is logical to ask what impact this revolution has had in the culture and education of youngsters.

Communication means have been evolving and improving over time, so that people obtain in a simple and easy way the information they require. Currently, there are three main players: The Internet, digital social networks and a generation of young people with particular characteristics.

3 A Generation of University Students with Marked Characteristics

Defining young people seems simple, but complex. It is not an abstract concept, but it must be analyzed from many categories such as age, gender, social components, cultural, economic, etc.

García Canclini (2007) situates the sphere of the juvenile stage between the 14 and 28 years. This phase along with that of childhood and pre-adolescence are fundamental stages in which more than half of their intellectual, behavioral and cultural capacities are developed. It is also the period of human life where one is exposed to the uncritical reception of external influences, that is why “the media must take into account in a special way the possible—positive and negative—effects that they can exert on children and youngsters”.

But this conception has become obsolete; today the youth has been filled with social meanings and practices that give it an identity demanding legitimacy and participation in educational, political, social, economic, cultural and moral decisions. Then the sociological cultural concept of the young person should include a feeling of common belonging to the same age group with certain attitudes and patterns of behavior (the young as a highly valued cultural good). From this point of view the following categorization can be made: pre-adolescence (12–14 years), adolescence (15–17 years), youth (18–24 years) and prolonged youth (25–29 years) (Merino 2010: 37–39).

According to Aparici et al. (2010: 189) “our young people, identified as digital natives differ from their parents and teachers called digital immigrants. Between them, there are generational fractures in the family, the school and daily life". This refers both to the skills young people have in manipulating the various screens and to interests in terms of immediacy, agility and speed with which they access information, entertainment and social networking sites on the Internet. A fact to consider is that this generation of digital natives is self-study, that is, they do not need a teacher to teach them to properly handle the devices that the screens house, but they do it alone; and, when they require a more specific skill, they learn from the Internet.

With respect to this, Professor Prensky (2010: 5) states:

> It is clear that our students think and process information significantly differently from their predecessors. In addition, it is not a temporary habit but is called to prolong in time without interruption, but with increase so that their skill in the handling and use of technology is superior to that of their teachers and educators.

This is how digital natives use—and will always—use screens in a more optimal way with Internet support. They are agile, effective and pragmatic. They look for simple solutions to problems that would represent the end of the world for a digital immigrant. Digital natives quickly learn to use email, take pictures, record videos and broadcast them on social netwosrks like YouTube in seconds; they can write ten text messages in the time an immigrant is barely accessing WhatsApp. But they go further; junior high school students can configure and update smartphone operating systems, efficiently download and use hundreds of applications that greatly simplify the tasks, and activities of everyday life. This generation of young people has also been referred to as Millennials, known in particular as multitask; that is, to perform several activities at the same time.

According to the report of "Tendencias Digitales *Connect your brand with the Millennials*", currently in Latin America a 30% of the population is Millennial, and according to a projection of the consultancy Deloitte in 2025, the 75% will represent the world's workforce. The study *How Millennials See the World?* Made by "Fundación Teléfonica" on millennials applied to young people aged 18–30 from 18 countries in North America, Central America, South America and Europe proposes the keys to understanding this new generation; a generation halfway between classrooms and the world of work. According to this study Millennials from different regions, roughly are very similar. Their greatest influence is the family with 75% in all areas; although, some particularities are shown according to their country of origin and culture; Friends (54% US and Europe); the media hardly influence (20%), and the government only (10%).

In almost all zones, the most used device among the Millennials is the Smartphone (78%); in Central America 54% use it to connect to social networks, send text messages and make calls; 60–80% dedicate their time to the use and consumption of social networks; 60% to text messages and in terms of calls 42% in Europe, 59% in Latin America, and Central America use the telephone to call.

4 Social Networks in the Training Field

Students are the elements of the university structure that benefit most from social networks, not only in leisure, but also in both the educational and communicative field. The Ymedia Agency mentions that of the 2.0 set, the online social networks are the most accepted among university students at an 84% (Agencia Ymedia 2008: 41).

The research "Social networks and higher education: attitudes of university students towards the educational use of social networks, back to review", by González et al. (2016), at the University Rovira i Virgili of Tarragona, Spain, took as a reference six social networks: Facebook, Twitter, Patatabrava, Tuenti, LinkedIn and eLearning Social where it was indicated that Facebook (87.3%) and Twitter (24.1%) are the most used networks in general use, Tuenti–Patatabrava are for university use only, within the professional field LinkedIn is based on labor market and eLearning Social to teaching and training particularly.

Young university students do indeed appreciate that their teachers have contact with them by other means, in addition to the synchronous communication in class, it is considered effective the use of electronic mail or the platform Moodle (virtual environment of teaching learning used in our context); Social networks Facebook and Twitter are the most manipulated networks by students; However, Facebook leads with a big difference to other networks (González et al. 2016).

The incorporation of ICTs in the educational field, and university in particular has caused teachers to feel strange or uncomfortable because there have been discourses and prejudices that tend to see ICT as representatives of dehumanization in which the teacher loses leadership. In this sense, Dussel and Quevedo (2010) point out that in practice it is the opposite since the current demand makes the presence of the teacher more necessary as a guide that accompanies and advises the student in his learning process and in issues of communication and permanent dialogue with the student.

Undoubtedly, educators have the responsibility and the challenge of developing digital skills among young people to use social networks in a critical and constructive way.

4.1 *Student-Teacher Relationship*

Nowadays, there is talk of a new role for the teacher and a new role for the student, precisely because technologies have dramatically broken into university life, the teacher is not the only one who transmits knowledge, so does the student, the classroom is not the only space to learn, social networks are scenarios where the student can communicate with the teacher and his colleagues to solve academic doubts, learn about events, ask for advice on a topic, tutoring, etc.

Social networks are used both inside and outside the classroom. They are technological tools of learning and communication with which you can discover, share ideas, thoughts between teacher and student, obtaining a continuous communication that provides new materials for communication between them without limits of time and space. In this sense Requena (2008: 30) points out:

> Students do not have to wait the next day to communicate with the teacher or to expose some idea to their peers, they simply through social networks have the opportunity to write and share their thoughts whenever they want, and with the recipient they desire, existing a bond of partnership which is not limited by a physical space.

That is, social networks have facilitated student-teacher communication, the student not only communicates with his teacher in the classroom, but outside of it. There is, therefore, a continuous communication enabling a relationship and commitment of the two parties as they are in constant interaction.

One of the first platforms that allowed communication beyond the classroom were blogs, on this subject Requena (2008: 31) states:

> Blogs increase communication between classmates, teachers and even family members. This tool gives people the opportunity to connect, as it happens with social networks to any member of the classroom without any barriers of time or space.

When the student communicates with the teacher through social networks, it has been demonstrated that it is a micro world communication because it is individualized where the student feels more confident, independent, resulting in a student who can communicate effectively. Nowadays, in the university classrooms the digital platforms are used to explain diverse subjects. Previously teachers in geography class to teach a country showed a globe, now teachers through YouTube can show a video of that country, and with Google Earth the exact location; we talk about an interactive communication class where students can talk with teachers because they know the subject in a profound and explicit way.

The teacher-student communication in social networks is a didactic communication that consists of the transmission of knowledge, understanding, dialogue, in the teaching-learning process when they get communicated via online within the interaction there is dialogue, motivation and individualized orientation per se, in a communication mediated by social networks.

The old school has undergone a change with the arrival of Web 2.0, therefore, the interactions between teacher and student have slipped as: student-instructor, student-content, student-student, student-interface. Salinas (2004: 12) points out: "Traditional teaching puts the emphasis on social interaction. Multimedia models strive to provide quality to individual student interaction with materials". Taking into account that within the student-student interaction. There may be an interaction of individual character or between groups of students.

Hernández (2008: 30) calls social networks as "constructivist tools" in direct relation to their possibilities when they are put at the service of interaction in the group, between the group and the teaching staff, within the teaching staff, and everything outside the temporal demands and spatial aspects of a school setting. The virtual environment allows breaking these coordinates to facilitate interaction, share a myriad of files of also varied typology, and communicate in the most similar way to the current, combining sound, video, documents... The teacher must ensure and motivate the diversity of opinions, independence, decentralization, organization, etc.

4.2 Student-Student Relationship

Social networks due to their characteristics can be used by students for various activities within their process of autonomous learning, interaction, motivation and creativity. Even the student creates networks of collaboration and exchange with space-time continuity with their classmates, friends, and close friends or simply with young people of the world with whom they share certain preferences.

Students use social networks to communicate, meet, entertain, have fun, being the space of social interaction preferred by them since they consume digital platforms most of their time, taking into account that social networks are a main tool for their academic learning (Castañeda and Gutiérrez 2010).

All this leads us to think that the Digital Natives enjoy more advantages than the Digital Immigrants; however, digital natives have a lack of extreme importance which is according to Prensky (2010) that the digital age happened too fast affecting their capacity for reflection, the same that is necessary for the construction of mental models from the experience making possible a deep analysis and critical thinking. It is necessary, then, to rethink an educational model according to the needs and demands of the Digital Natives through appropriate languages and strategies which correct the shortcomings and promote the positive aspects of these new generations.

The student-student interaction is a very important dimension in the process of training and communication; it allows you to be constantly updated of the context. This interaction is considered to be of great help and benefit to the student because it can consume and acquire information in terms of an individualized learning, self-study where the student acquires his own knowledge, skills, competences and abilities product of group interaction where the social networks are the fundamental tool.

The students' learning in social networks is collaborative and open where each member contributes with their experience and knowledge with the sole intention of being fed by their peers that are anywhere in the world communicating each other without geographical barriers.

An investigation by Ortiz and Romero (2016) shows that young people relate the concept of social network with communication, spaces, sharing, interaction, experiences and relaxation. The term communication is one of the most mentioned; however, for young people this idea goes beyond sharing a simple message, that is, they recognize that thanks to social networks they can have an asynchronous communication. They even consider that they are spaces where they can carry out multiple activities.

Student must also assume their role within social networks knowing that the use of these technological tools guarantees a better transmission of information and ease of sending and receiving messages in such a way that the individual must know the language, resources and forms with which they can be persuaded to be critical to that situation researching and analyzing what they receive.

The majority of researchers insist on the importance of training the new generations by instilling a critical look at these tools. However, they have failed to find the most appropriate pedagogies, trying to solve the problem of today's young people without understanding the simplest levels of conception where notions and definitions are the ones to be debated. No matter the source of the message or its coding, the important thing is that users have to learn to be critical of what they are receiving.

> In short, youth is a social group which is more receptive to new media and at the same time, potentially more vulnerable; not in vain there is a concern in the legislative area to protect both children and youth from certain content such as violence (Huertas and Elisa 2012: 39).

The educational model must be in constant readjustment for students and people in general and thus to guarantee a technological world that is in continuous innovation, but above all that this world counts on educated people who can use these new and better tools.

5 Social Networks in the University Communicative Field

Several literature proposes some points to consider in terms of the implementation of the University 2.0, all of them have a high involvement of internal audiences. Pedreño (2009) emphasizes the importance of loyalty to teachers in open knowledge in the change of administrative processes to open data, and in collaborative work; no doubt, all this can be done, but it is necessary whenever the institution of higher education has its institutional policies for this effect.

The incorporation of ICT in the educational world has been a latent concern of all the governments in turn, as shown by the programs or initiatives that have been worked in several countries in Latin America, such as Plan Ceibal in Uruguay, Enlaces in Chile, Huascarán Project in Peru, Computer Program to Educate in Colombia, Comprehensive Program Connect in El Salvador, Schools of the Future in Guatemala, Plan for Digital Inclusion in Education, Connect Equality in Argentina, and The Comprehensive System of Technologies for Schools and Community (SíTEC) in Ecuador among others. All of them show the existence of policies that seek to accompany educational transformation and enrich teacher-student communication.

The University as a Higher Education Institution is clear that it must take advantage of the use of social media to offer a more agile and close communication in accordance with the current demands: real time information, aimed at its students, teachers and administrators; but each one is framed within its social reality, therefore the results have not been totally positive, there are still some universities using the Internet with the sole purpose of informing admission and enrollment issues without possibilities of interactivity.

It is impossible to analyze the impact of technologies in the university world without the institutions becoming aware of the implementation of digital culture at all levels, playing a key role communication cabinets in this process. For Paniagua et al. (2012: 695) the main drawback is that the communication cabinets "adopted the internet as an information channel and not as a communication channel". These authors state that 57% of the 78 Spanish universities can be considered cabinets 1.1; 23% have organized a virtual community for the exchange of bi-directional information and only 17 have a 2.0 space. All universities in this country have created profiles on social networks, such as Twitter and Facebook mainly.

The creation of communication cabinets in universities dating back to the 1980s appeared first of all in the largest universities in the United States and Europe. Nowadays, they have grown considerably in the whole society and they have acquired functions of marketing, public relations, graphic design, radio and TV edition, etc. They have changed a lot since their beginnings where their only function was as press cabinets.

Álvarez and Caballero (1997: 85) summarize the activity of communication cabinets in the last decades of the twentieth century as follows:

> The communicative fact is as old as society. What has happened in the twentieth century is that a specialization has arisen in the work related to communication, delimiting the rest of the tasks of the organization. In addition, a new economic assessment of the fact of communicating has emerged. In the past an entity would offer products, services or ideas, now a philosophy and a charitable consultation is mandatory. Organizations no longer relate themselves solely through their material or ideological product.

The universities by their own characteristics—education, research and extension—generate constant information of greater or smaller magnitude. The university communication managers are responsible for communicating and/or disseminating this entire information compendium, analyzing the best channels and communication tools, considering the target audiences. "We cannot forget, also, that we speak of a particularly complex organization which hinders its internal management and its external projection and makes professional intervention in the field of communication more decisive" Losada (2004). And as an educational institution with dispersed publics (students, alumni, teachers, employees, etc.) where each group has different affinities and seeks information that is of interest, each manager establishes his own strategies. For example, the creation of a Fan Page as a means of supporting the campaigns to call for registration and thus avoid confusion with all the information that is generated day by day from within (Paladines 2012).

Therefore, the university, like any other higher education institution does is only manage knowledge communicate their intellectual capital, corporate culture and select the various open devices that facilitate the generation with access to knowledge that the institution generates as a competitive advantage. The lines of communication adopted by universities will depend on the objective of communication and education.

Independently of the organizational design of the universities, the communication cabinets play a very important role, structured according to the organizational possibilities of the institution. There may not be a Community Manager, but there will always be the chief of staff, responsible for building, arranging and managing an online community based on the brand.

5.1 Social Media Interaction with Students

According to the Digital in 2017 report, Global Overview, edited by We Are Social and Hootsuite indicates that 2789 billion users have average social activity, out of a universe of 7476 billion people belonging to 239 countries in the world. The university therefore could not deviate from this trend under the pretension of a Web positioning that is of their interest.

One of the variables of evaluation of the academic ranking Webometrics is the positioning of the universities in the Web, that is to say, the number of links that other web sites have towards which they consider are object of analysis. These types of measurements help higher education institutions to strengthen their strategy of visibility and comparison with the websites of other universities.

According to the academic rankings of Shanghai universities in 2016, they ranked Harvard, Standford and California (Berkeley) as the leading in the field of education; and Webometrics ranked (January 2017), in the impact variable, Massachusetts, Harvard and Santanford. Faced with this reality communication cabinets use social networks to generate traffic, especially in research results that are disseminated among the scientific community.

In a review of the websites of these institutions, Facebook, Twitter and Instagram are the most widely used social networks. Harvard maintains its leadership on all three social media platforms (Table 2).

Quantitative data are nothing more than figures for communication managers, the communicative strategy is to determine according to the profiles, given by metrics, which those potential students, contacts for projects, and financiers are. The idea is to search efficiently and cost-effectively for the best ways to connect with these audiences. The more active the community in networking, the more visibility the university will have with better options for attracting more students.

Involving students in college through those platforms that are an integral part of their daily lives is an essential part of any marketing strategy, recruitment and brand image. If you ignore them, take the risk. Chris Larkin, director of communications, University of Salford, England.

Table 2 Followers on Facebook, Twitter and Instagran at Harvard, Stanford, California and Massachusetts

Universities	Facebook	Twitter	Instagram
Harvard University	5,044,903	733,535	434,083
Stanford University	1,202,864	513,184	238,022
California, Berkeley University	450,790	177,301	–
Massachusetts Institute of Technology	23,899	789,643	–

Source Own elaboration with information of the social networks of each university (May, 2017)

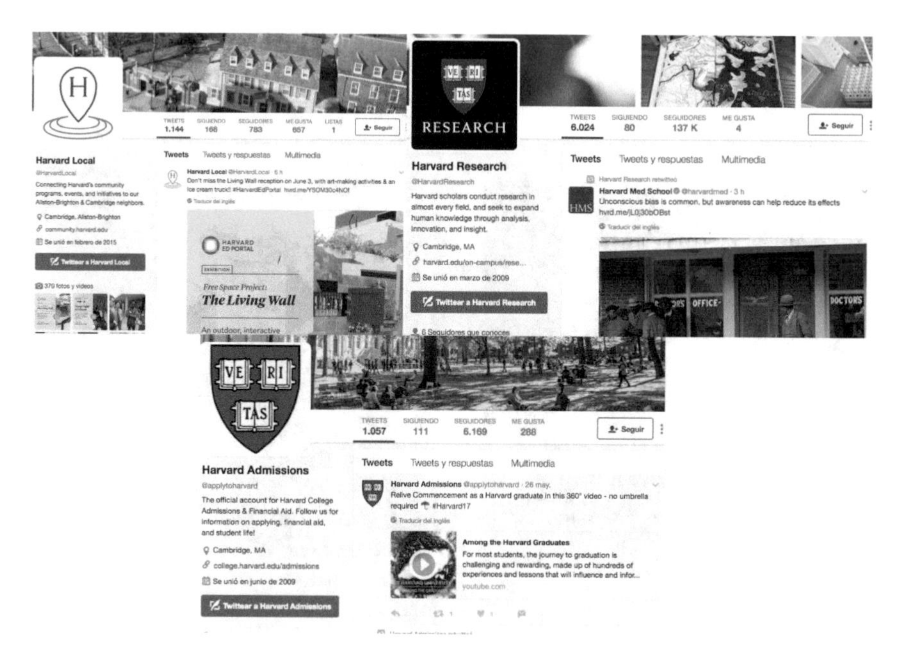

Image 1 Accounts in social networks at Harvard University (http://www.harvard.edu)

According to institutional communication policies, universities create parallel accounts to channel the communicative process. Harvard, for example, has three official accounts, each for functional areas of the university: connection to society, research and student admission (Image 1).

An immediate response to users is the main premise when it comes to managing social networks. They always have questions on administrative and academic issues and all these questions are asked immediately through social networks, hence the importance of efficiently serving to provide customer value. This is a task that requires a lot of attention and time, despite this, it is also a possibility to approach students to know what are the main topics of interest or doubt either.

Besides, it is important that university representatives participate in social communities, forums; blogs that may be popular among students, practicing this will allow to obtain information about what is said about the institution in the network, including providing conflicting information. The orders or dispositions to answer the questions raised should be included in manuals for the use of digital communication so that the same speech or key words can be used for a response.

Some universities take advantage of their flashy locations to create sentimental messages and transmit them through social networks. Today, as a fashion, you can visualize in the institutional accounts photographs of the university campus accompanied by motivational messages. The University of Nottingham of England, for example uploads daily photographs of some of its main facilities. The same happens with the Universidad Técnica Particular de Loja, Ecuador (Image 2).

Image 2 Facebook accounts of the Particular Technical University of Loja and University of Nottingham. *Source* https://www.utpl.edu.ec/ https://www.nottingham.ac.uk/

Universities like other institutions are also exposed to content moderation in some of the formal messages issued or the creation of external ones. Hence the great work of digital communication managers to continuously observe the appearance of their brand with foreign content; To this end, technologies that alert the use of a trademark by other sources should be used instead.

5.2 *Communication 2.0. in Internal Audiences*

The worker's intranets and portals evolve their services to internal social networks, similar to Facebook and Twitter, allowing interactivity between teachers and administrators and streamline collaborative work. Various platforms are used by different organizations. Free or paid applications allow real-time auditory and written communication.

Microsoft, for example, offers companies the 'Microsoft Teams' tool which was designed to help the internal communication processes of organizations using the Office 365 package that is not only linked to the Office suite (Word, Excel, etc.), but also to business Outlook e-mail.

Whatever the tool that the university adopts to communicate with the internal audience, the most important thing is that this instrument works on favor of the relation, the collaborative work and the articulation with the institutional values. The university that manages to communicate with its internal audience will do it effectively with the external ones.

6 The University 2.0 Versus 3.0

The multiple digital tools available on the Internet and the adaptations that new audiences acquire challenge the university to make decisive decisions regarding the communication they generate from their organizations, considering that they are an abundant source of information and that is from the Alma Mater where the knowledge is born.

The emergence of the Internet and its popularization of the Web and the social Web have meant for these institutions the apparition and consolidation of a new paradigm of communication and culture, less vertical and more circular strengthening communication and learning in students.

Communication cabinets in conjunction with academic departments should promote a variety of initiatives that strengthen the participation and interactivity of students and teachers. The benefit of this process will depend on the collaboration of its internal audiences. It is important for the university to recognize that its communication management in social networks has no value if the institution promotes social platforms with unidirectional channels only.

While some universities are just adapted to communication 2.0, the presence of communication 3.0 glimpses and emerges at ease. A communication which requires specialized and interesting information according to the needs of each prosumer since it generates opinion and creates spaces for interacting with other clients who are also prosumers.

Going further, the university will soon adopt all the resources offered by the revolution 4.0, with its products like for instance nanotechnology, neurotechnology, robots, artificial intelligence, biotechnology, among others, that will help in the communication process so that the communication managers, university authorities, teachers and students, must be ready and prepared for it.

References

Álvarez, T., Caballero, M.: Image Sellers. The Challenges of the New Communication Cabinets. Ediciones Paidós Ibérica, S.A., Spain (1997)

Aparici, R., Orozco, G., Férres, J., Osuna, S., Kaplún, M.: Educommunication: beyond 2.0. Gedisa, Spain (2010)

Castañeda, L., Gutiérrez, I.: Social networks and other online fabrics to connect people. Learning with social networks. Educational fabrics for new environments. 17–39 (2010)

Chatterjee, S.: Simple rules for designing business models. Calif. Manag. Rev. **55**(2), 97–124 (2013)

Dans, P.: Internet. Anaya Multimedia, Madrid (2010)

Donlan, L.: Exploring the views of students on the use of Facebook in university teaching and learning. J. Further High Educ. **38**(4), 572–588. https://doi.org/10.1080/0309877x.2012.726973 (2014)

Dussel, I., Quevedo, L.: Education and new technologies: the pedagogical challenges facing the digital world. Faculty J. Curriculum Teach. Training **19**(2) (2010)

Estévez, R., Fernández, I., Noguer, A.: The use of social networks in Andalusian universities: the case of facebook and Twitter. Int. J. Public Relat. **II**(4), 138 (2012)

Franch, B., Camacho, M.: New Communication Stands. Case study: Jaume I University of Castellón. AA.VV. Communication in the University (2004)

García Canclini, N.: Readers, Spectators and Internet Users. Gedisa, Barcelona (2007)

González, J., Lleixà, M., Espuny, C.: Social networks and higher education: the attitudes of university students towards the educational use of social networks, again to exam theory of education. Educ. Cult. Inf. Soc. **17**(2), 21–38 (2016)

Gross, D., Pietri, E., Anderson, G., Moyano-Camihort, K., Graham, M.: Increasing the use of student-centered pedagogies from moderate to higher student improvement and attitudes on biology. CBE Life SciEduc. **14**(4). https://doi.org/10.1187/cbe.15-02-0040 (2015)

Hernández, S.: The constructivist model with the new technologies: applied in the learning process. RUSC. J. Univ. Knowl. Soc. **5**(2), 30 (2008)

Huertas, A., Elisa, F. M.: The adolescent spectator. An approach to how television contributes to the construction of the self. In: Callejo, J., Gutiérrez, J. (eds.) Adolescence Between Screens. Young People in the Communication System. LNCS p. 39. Gedisa S.A. (2012)

Islas, C., Carranza, M.: Use of social networks as learning strategies. Educational Transformation? Opening, University of Guadalajara, Mexico (2011)

Losada, J.: Communication in the construction of university marks. In: Losada, J. (coord.) Management of Communication in Organizations. LNCS, pp. 476. Ariel S.A (2004)

Maciá, F., Gosende, J.: Marketing with Social Networks. Grupo Anaya, S.A., Madrid (2011)

Maiz, I., Tejada, E.: The use of social networks from an educational perspective. In: Barroso, J., Cabero, J. (coords.) New Digital Scenarios. Information and Communication Technologies Applied to Curricular Training and Development. LNCS, pp. 313–315. Ediciones Pirámide (2013)

Majó, J.: Evolution of communication technologies. In: Moragas, M. (ed.) Communication: From the Origins to the Internet. LNCS, vol. 66. Gedisa (2012)

Merino, L.: Digital Natives: An Approach to the Technological Socialization of Young People. Institute of Youth, Madrid (2010)

O'Reilly, T.: Web 2.0 Compact Definition: Retry O'Reilly radar (2012)

Ortiz, G., Romero, K.: Young people of the Autonomous Metropolitan University, Lerma unit and its digital culture. In: Crovi, D. (coord) Digital Social Networks: Meeting Place, Expression and Organization for Young People. LNCS, p. 180. National Autonomous University of Mexico (2016)

Paladines, F.Y.: Management of brand communication in social networks: study of three cases of campaigns with Facebook in Ecuador. Ph.D. dissertation, Universidade de Santiago de Compostela, España (2012)

Paniagua, F.J., Gómez, B.J., Fernández, M.: The incorporation of the communication departments of the Spanish universities to the digital environment. A quantitative analysis. Stud. Journalistic Message **18**, 691–701 (2012)

Pedreño, A.: Universidad 2.0. Vision and strategies of action. Utopias and realities. Online Magazine of published works. http://utopiasrealidades.blogspot.com/2009/08/universidad-20.html (2009)

Pisani, F., Piotet, D.: The Alchemy of Crowds: How the Web is Changing the World. Paidós Ibérica S.A., Barcelona (2009)

Prensky, M.: Digital Natives and Digital Immigrants. SEK International University (2010)

Requena, S.: The constructivist model with the new technologies, applied in the learning process. RUSC. Univ. Knowl. Soc. J. **5**(2), 30–31 (2008)

Salinas, J.: Methodological changes with ICT. Didactic strategies and virtual environments of teaching-learning. Bordón **56**(3–4), 12 (2004)

Sangrà, A., González, M.: Chapter IV. University teachers and ICT: redefine roles and competencies. In: Sangrà, A., González, M. (coords.) The Transformation of Universities Through ICT. LNCS, vol. 75. Editorial UOC, Barcelona (2004)
Treem, J.W., Leonardi, P.M.: Social media use in organizations: exploring the affordances of visibility, editability, persistence, and association. In; Salmon, C.T. (ed.) Communication Yearbook. LNCS, vol. 36, pp. 143–189. Routledge (2013)
Velásquez, A.V.: Consumption and media use of young Ecuadorian university students in times of convergence. Ph.D. dissertation, Universidade de Santiago de Compostela, España (2017)
Ymedia Agency: Users cannot live without the Internet. Interact Commun. Mag. Digit. Mark. **98**, 41 (2008)

Diana Rivera-Rogel Professor at the Universidad Técnica Particular de Loja. She holds a Ph.D. title in Communication and Journalism from the Universidade de Santiago de Compostela (Spain). She was international co-editor of the scientific journal Comunicar. She is the Director of the socio-humanistic department at the Universidad Técnica Particular de Loja. She has also participated in five funded research projects. She was selected by UNESCO and FELAFACS to teach the course for Andina countries. Loja, Ecuador.

Jenny Yaguache Quichimbo Professor at the Universidad Técnica Particular de Loja. She holds a Ph.D. title in Communication and Journalism from the Universidade de Santiago de Compostela (Spain). She was the Director of the Department of Communication Sciences at the Universidad Técnica Particular de Loja. She has also participated in four funded research projects, and his line of research is strategic communication. Loja, Ecuador.

Andrea Victoria Velásquez Benavides Professor of Corporate Image and Audit in Communication of the Universidad Técnica Particular de Loja. Ph.D. in Communication and Creative Industries from the Universidade de Santiago de Compostela (Spain). Her publications include topics related to Consumption and Use of Media, Hearings, Reception, Media Skills, Advertising and Digital Brand. Present member of the Alfamed network and currently works on two research projects: Media Skills in Adolescents, and Media Consumption and Use in Millennials. Loja, Ecuador.

Fanny Paladines Galarza Degree in Advertising at the Universidad Tecnológica Equinoccial (Quito-Ecuador). Executive Master in Management and Strategic Marketing Management at the Business School (EOI) Madrid-Spain. Ph.D. in Communication and Journalism from the Universidade de Santiago de Compostela. Professor of the advertising component in Public Relations and Communication degrees. She currently coordinates the Public Relations Degree. Her lines of research focus on research brand/branding, digital branding, management in social networks, traditional/digital corporate/strategic communication. Loja, Ecuador.

Printed in the United States
By Bookmasters